Encyclopedia of
Television Law Shows

D1230820

ALSO BY HAL ERICKSON
AND FROM MCFARLAND

The Baseball Filmography, 1915 through 2001,
2d ed. (2002; softcover 2009)

"From Beautiful Downtown Burbank": A Critical History of
Rowan and Martin's Laugh-In, *1968–1973* (2000; softcover 2009)

Sid and Marty Krofft: A Critical Study of Saturday Morning
Children's Television, 1969–1993 (1998; softcover 2007)

Television Cartoon Shows: An Illustrated Encyclopedia,
1949 through 2003, 2d ed. (2005)

Syndicated Television: The First Forty Years, 1947–1987 (1989; softcover 2001)

Religious Radio and Television in the United States, 1921–1991:
The Programs and Personalities (1992; softcover 2001)

Television Cartoon Shows: An Illustrated Encyclopedia,
1949 through 1993 (1995)

Baseball in the Movies: A Comprehensive Reference, 1915–1991 (1992)

ENCYCLOPEDIA OF TELEVISION LAW SHOWS

*Factual and Fictional Series
About Judges, Lawyers and
the Courtroom, 1948–2008*

HAL ERICKSON

McFarland & Company, Inc., Publishers
Jefferson, North Carolina, and London

LIBRARY OF CONGRESS CATALOGUING-IN-PUBLICATION DATA

Erickson, Hal, 1950–
Encyclopedia of television law shows : factual and fictional
series about judges, lawyers and the courtroom,
1948–2008 / Hal Erickson.
p.   cm.
Includes bibliographical references and index.

ISBN 978-0-7864-3828-0
softcover : 50# alkaline paper ∞

1. Justice, Administration of, on television — Encyclopedias.
I. Title.
PN1992.8.J87E53   2009
791.45'655403—dc22   2009025181

British Library cataloguing data are available

On the cover: Judge Judy, 1996 (Photofest); gavel ©2009 Shutterstock

Manufactured in the United States of America

*McFarland & Company, Inc., Publishers
Box 611, Jefferson, North Carolina 28640
www.mcfarlandpub.com*

To the memory of my mother,

**Frances "Pudge" Erickson
(1927–2006).**

"Of course I'll join your committee."

Table of Contents

Preface   1

Introduction   3

THE SERIES

General Notes   25
The Entries   28

Bibliography   275

Index   279

Preface

I am not a lawyer, nor do I play one on TV. Before I agreed to tackle this book, my own courtroom experience was limited to serving two years on a federal grand jury (during which there was only one non-unanimous decision, and *that* one was resolved in the next session) and showing up in traffic court to testify against a reckless driver — who never showed up. I am a media historian, and it is from a media historian's perspective that this book has been written.

You won't find a plethora of "legalese" in this book: if you want precise definitions for such terms as "ancillary jurisdiction," "book-entry bond," "collateral estoppel," "factor's lien," "moratory damages," "synallagmatic contract" and "zipper clause," I strongly recommend that you consult a law dictionary. Nor am I prepared to provide translations for such Latin legal terms as *ad colligenda bona, contra proferentium, doli incapax, hostis humani generis, nolle prosequi, praetor peregrinus, res judicata, quantum meruit, suo moto,* and *uti possidetis* (truth to tell, I'm still struggling with *caveat emptor* and *veni vidi vici*). Admittedly, a book of this type will appeal mainly to a "niche" audience of television aficionados and law-show fans — but there's no reason I can't endeavor to broaden its appeal by using words that are readily understandable to everyone.

Before going any farther, I must pay homage to two excellent books in print, both written from the perspective of acknowledged legal experts: Elayne Rapping's *Law and Justice as Seen on TV*, and Paul Bergman and Michael Asimow's *Reel Justice*. In their assessment of the overall accuracy of legal dramas, their descriptions of the "best" and "worst" that the genre has to offer, their citation of real-life litigation that has influenced fictional TV programs and movies, and their recognition of the profound impact that the entertainment industry has had on the American Justice system, both books are outstanding in their field. They are also by their very nature selective, offering carefully chosen examples rather than itemizing every single movie or TV series within the genre. While this approach is perfectly appropriate within the parameters of *Law and Justice as Seen on TV* and *Reel Justice*, it is not the approach that I have taken. In keeping with the tradition of my previous books on baseball films and TV cartoon shows, this current work is all-inclusive: an alphabetical encyclopedia of every regularly scheduled "legal," "lawyer," "courtroom" and "judge" series produced for American television from 1948 through 2008 (with a few exceptions noted in the body of the text).

My primary goal is to provide in-depth background information for the various programs (whenever possible), and to place each program within context of the period in TV history in which it aired (whenever appropriate). In the early stages of the project, I had hoped to include such sideshow attractions as a list of legal-drama clichés, beginning with the ever-popular, "This is a courtroom, not a circus!" (To date, no TV producer has ever had the audacity to tweak this old chestnut by showing a group of red-nosed clowns conducting a slapstick trial, whereupon the leader raps an oversized gavel and exclaims "This is a circus, not a courtroom!") I was also tempted to provide a rundown of the standard lawyer "types" seen on dramatic programs, along with a tally of the most frequently used legal-show plotlines. But in the end, I abandoned such whims and caprices to stick to the job at hand.

Basic research methods included the obligatory burrowing through reams of printed material, exhaustive trolling of the Internet, and viewing sample episodes of TV programs both famous (*The Practice*) and obscure (*They Stand Accused*). My own voluminous collection of vintage TV shows was augmented by rare and precious material available from such facilities as the Museum of Broadcast Communications in Chicago, Illinois, and such on-line services as TV4U and hulu.com. I have also referenced the standard TV history texts, as well a marvelous and very thorough website maintained by the Tarlton Law Library of the University of Texas School of Law. The Tarlton site proved an invaluable launching pad for my own research, as did an entertaining and informative series of articles on the various TV "judge show" cycles of the 1950s, 1980s and 1990s (including such titles as *Divorce Court* and *People's Court*), written between March 27 and October 23, 2003, by Los Angeles media specialist Roger M. Grace for his "Reminiscing" column in *The Metropolitan News-Enterprise*.

For access to otherwise inaccessible material, I offer special thanks to the following individuals: Jim Feeley, David Bobke, Bruce Simon, Robert Ligtermoet and the staff members of Finders Keepers Classics and Shokus Video. Extra-special thanks is herewith extended to my longtime friend Eugene Rubenzer for his incisive "behind-the-scenes" account of the 1980s version of that hardy perennial *Divorce Court*.

And for providing moral support and encouragement throughout this project, there aren't enough words to express my love for my wonderful wife Joanne and my terrific sons Peter and Brian. And how could I possibly forget such loyal and supportive friends as the late Dick Golembiewski, David Seebach, Cari and Carl Bobke, Roger Sorenson, Dale Craven, Lee Matthias, Jane and Mark Martell, Wayne and Rita Hawk, and others too humorous to mention?

So ... let's get started on our sixty-year journey through TV's various and sundry "legal" shows. I hope you like what you see. And if you don't ... sue me.

Hal Erickson
Milwaukee, Wisconsin
Fall, 2009

Introduction

Even the most casual of TV watchers born within the past quarter-century have been inundated by round-the-clock coverage of major courtroom trials and hearings — often to the point of screaming, "Enough already!" That said, it may prove surprising to these fans to learn that cameras and microphones have not always been standard courtroom equipment. In fact, for nearly forty-five years, only a fraction of American court procedures were covered live by the electronic media — simply because, with rare exceptions, it was forbidden to do so.

Up until the mid–1930s, live courtroom broadcasts proliferated on both local and network radio. The first documented example was the Scopes "Monkey Trial" of 1925, in which two legal heavyweights, defense lawyer Clarence Darrow and populist politician William Jennings Bryan, squared off over the issue of teaching Evolution in the classrooms of Tennessee. Most people are familiar with this case by way of the brilliant but heavily fictionalized Jerome Lawrence-Robert E. Lee stage play *Inherit the Wind,* which suggests that schoolteacher John Scopes's arrest on a charge of violating Tennessee's Butler Act (which prohibited the teaching of Evolution in favor of "Divine Creation") was a minor incident that unexpectedly mushroomed into a national *cause celebre.* In truth, the Scopes Trial was the culmination of a carefully mapped-out strategy by the American Civil Liberties Union to publicly challenge the Butler Act. The local authorities of Dayton,

Tennessee, agreed to work hand-in-glove with the ACLU, pre-arranging to place Scopes under arrest in order to force the issue of academic freedom into the courtroom — with the entire nation bearing witness.

Since this was admittedly a "show trial," it made perfect sense to allow the event to be broadcast. Obligingly, Chicago radio station WGN set up four microphones in the Dayton courtroom to provide on-the-spot coverage. At a cost of $1000 per day, WGN rented AT&T cables which stretched from Illinois to Tennessee; thereafter, the broadcast was theoretically available for free to the millions of listeners within the range of the station's 50,000-watt signal.

According to contemporary WGN personality John Williams, "The radio station received the rights to rearrange the way the courtroom was set up. And this was the first time this happened where the media manipulates an event literally the way it's played out.... The relationship of the judge to the prosecution and the defense, all of that changed to accommodate the radio station's microphones."

Providing commentary was legendary Chicago broadcaster Quinn Ryan, who in keeping with the journalistic tradition of the day spoke in fluent hyperbole: "Here comes William Jennings Bryan. He enters now. His bald pate like a sunrise over Key West." For the most part, however, Ryan sat quietly next to a windowsill and allowed the trial to proceed without comment, save for an occasional interpolation to clarify a point of law. Whenever he felt the urge to deliver more in-depth commentary, Ryan moved into a chamber near the courtroom so that the sound of his voice would not intrude upon the testimony. And though he was admittedly on the side on John Scopes (as were most Northern journalists of the time), Ryan did not editorialize on the defendant's behalf, nor did he mock the anti–Evolutionists in the crowd.

As microphones in the courtroom became more commonplace, many commentators drifted away from Quinn Ryan's objectivity and began to exhibit the same sort of bias common to the era's newspaper reporters, who were prone to favor either the defense or the prosecution depending upon which stance would sell more papers. This left-handed form of journalism was mercilessly satirized in such films as *Peach O' Reno* (1931) and *The Trial of Vivienne Ware* (1932), which inferred that radio spielers seldom let facts get in the way of a good story. And what about the defendant's "right to privacy"? Well, since this particular right was not technically guaranteed by the Constitution, no one gave it a second thought.

But media access to the courtroom would be severely curtailed after the 1935 trial of Bruno Richard Hauptmann for the kidnapping and murder of the Lindbergh baby. Seven hundred newsmen and 120 cameramen were among the spectators who crowded into the Flemington, New Jersey, courtroom for what was described as "the Trial of the Century." Also on hand were the representatives of the three major radio networks: NBC, CBS, and Mutual. Of the local radio stations covering the trial, only New York City's WNEW was actually permitted in the courtroom. The logistics of the situation did not allow the authorities to extend to WNEW the same courtesies enjoyed by WGN during the Scopes trial, and the station could not position its equipment to facilitate on-site reporting. Instead, the WNEW reporters issued periodic bulletins to the listeners, staged re-enactments of trial highlights, and interviewed several of the principals involved in the case. A similar procedure was followed by the representatives of the three networks, who were strategically placed in various hallways and anterooms. There were also a number of studio-based radio commentary programs throughout the trial, one of them hosted by famed attorney Samuel J. Leibowitz, a member of Hauptmann's defense team.

Existing newsreel footage of the trial fairly reeks with the "carnival-like atmosphere" so often cited by contemporary viewers. Described by the press as "The Most Hated Man in the World," Bruno Richard Hauptmann had already been found guilty in the court of public opinion, and the newsreels did nothing to discourage this presumption of guilt. The defendant was shown in tight, sweaty closeups, voice quavering and eyes darting about furiously as the evidence piled up against him. Meanwhile, the prosecuting attorney screamed at the top of his lungs and postured for the cameras while the spectators buzzed with approval — a breach of courtroom etiquette that in itself would have likely forced a mistrial under current legal standards. As expected, Hauptmann was sentenced to death, and radio was on hand to cover his execution on April 3, 1936, with up-and-coming Mutual Network announcer Gabriel Heatter achieving overnight stardom as he ad-libbed for nearly an hour, wondering aloud if the unexpected delay in carrying out the sentence was due to a last-minute reprieve or an eleventh-hour confession from the condemned man.

In the aftermath of the execution, many legal experts expressed the opinion that Hauptmann was denied a fair trial. When charges of prosecutorial misconduct could not be proven, these experts redirected their complaints towards the journalists covering the trial, with radio and the newsreels bearing the brunt of the criticism. Particularly damning was anecdotal evidence that the jury members were within earshot of the radio commentators as they shouted into their microphones — a blatant violation of the "limited media access" rule pertaining to sequestered jurors. The alleged journalistic excesses and abuses surrounding the Hauptmann trial were key motivating factors in the 1937 decision by the House of Delegates of the American Bar association to pass Canon 35 of the Canons of Judicial Ethics (later Canon 3A [7] of the Code of Judicial Conduct), prohibiting the use of photography and radio broadcasting in the courtroom except for ceremonial occasions. Though not a binding decision, this ruling was generally honored throughout the United States — and in 1952, it was amended to include banning TV cameras in the courtroom as well.

Though radio fans were now denied the vicarious thrill of eavesdropping on live courtroom trials, the Hauptmann coverage sparked an upsurge of interest in fictional programs wherein trials and hearings were dramatized and lawyers and judges were the principal characters. Prior to 1936, there had been only two major series focusing on courtroom activity. Debuting January 1, 1934, and eventually carried by all three networks (though not at the same time!), *The Court of Human Relations* was described as a "human interest" series by its sponsor, *True Story* magazine. The series offered reenactments of genuine courtroom litigation, presided over by actor Percy Hemus as "The Judge." Just before the end of each broadcast, the home audience was invited to render a verdict — and lest you conclude that this was the very first example of "interactive" programming, we note that the listeners had no way to contact the program, and that the actual verdict had already been reached. On March 31, 1935, another human-interest effort, A. L. Alexander's *Goodwill Court*, premiered over New York station WMCA, moving to NBC on September 20, 1936. This series offered what would later be termed "legal aid" to people

unable to afford proper representation. Not a dramatization, the series was an early example of "reality" programming, as mediator Alexander listened to tales of woe from various real-life defendants (never identified by name, and strongly admonished not to use "inappropriate" language); their cases would be discussed by a panel consisting of genuine sitting judges, who in turn would offer legal advice. While *The Court of Human Relations* managed to survive until 1939, *Goodwill Court* was forced off NBC by the end of 1936 when the New York County Lawyers' Association lodged a protest over the dispensation of free counsel over the air; at the same time, The New York Supreme Court prohibited actual judges and lawyers from appearing on the program, a ban that would extend to virtually all future legal shows.

The first series to truly reflect the public's fascination with the Hauptmann case was Mutual's *Famous Jury Trials*, which began its twelve-year run on January 5, 1936. Though the series trafficked in recreations of actual case histories, the threat of libel or slander suits was sidestepped by focusing almost exclusively on trials of the distant past, with the original participants safely dead. In anticipation of the New York Supreme Court's ruling vis-à-vis *Goodwill Court*, all the judges and lawyers were impersonated by professional actors.

It should be noted that *Famous Jury Trials* was an anthology, with no regular characters. While there were plenty of early radio programs featuring detectives and police officers as heroes, lawyers were barely represented at all. This may have been a response to the then-common assumption, fueled by the motion picture industry, that most lawyers fell into one of two disreputable categories: Shysters and Ambulance Chasers. The apotheosis of the early-1930s movie attorney was the character played by Warren William in 1932's *The Mouthpiece*, based on the once-notorious, recently deceased criminal lawyer William J. Fallon: forever perverting and distorting the law for the purpose of increasing his fame and his bank account, hiring "professional witnesses" who'd say anything for a price, bribing judges and juries, blackmailing the prosecution into silence with compromising photos, and given to such melodramatic gestures as swallowing a vial of poison to prove that it was impossible for his client to have committed murder — then rushing out of the courtroom to have his stomach pumped.

In *We're in the Money*, a landmark overview of Hollywood in the 1930s, author Andrew Bergman suggests that the popularity of the "shyster movie" was a byproduct of Depression-era cynicism: "Their makers delighted in the rogues they were attempting to condemn. And this ambivalence was tied not only to the dismal position occupied by the law in 1931 and 1932 but to the fact that in depicting shysters during the nadir of the Depression they were administering relief, rather than addressing a central problem. In 1932, corruption must have seemed like an old and trusted friend." Virtually the only writer to make any sort of effort to counter the image of the duplicitous shyster was attorney-novelist Erle Stanley Gardner, creator of the fundamentally honest and upright criminal lawyer Perry Mason. One can only imagine Gardner's dismay when, upon adapting the Mason novel *The Case of the Howling Dog* as a motion picture, Warner Bros. engaged none other than Warren William, "Mr. Mouthpiece" himself, to star in the film!

While it was acceptable for the movies to dwell upon dishonest lawyers, the prevailing standards of network radio would not permit exposure to such scoundrels on a weekly basis, certainly not as protagonists; hence the lack of lawyer heroes in radio's formative years. All this changed in the mid–1930s with the rise to prominence of New York's special prosecutor Thomas E. Dewey, who made it his personal mission to expunge his jurisdiction of all corruption and organized crime. Dewey's most infamous target was Mafia kingpin Charles "Lucky" Luciano, who controlled all prostitution activities in New York City. Armed with eyewitness testimony, Dewey was able to convict Luciano on vice, tax-evasion and perjury charges, bundling "Lucky" off to Dannemora Prison and setting the stage for the gangster's ultimate fall from grace. The image of Dewey as a fearless, unstoppable "gang buster" inspired a cycle of popular films in which the hero was invariably a dynamic, thoroughly incorruptible prosecuting attorney, portrayed by such veteran tough guys as Edward G. Robinson, Humphrey Bogart, Chester Morris and Jack Holt: 1936's *Exclusive Story*, 1937's *Marked Woman*, and, in 1938 alone, *Smashing the Rackets, Racket Busters, I Am the Law, Crime Takes a Holiday* and *Gang Bullets*. Radio likewise hopped on the Dewey bandwagon with *Mr. District Attorney*, a "torn from today's headlines" dramatic series which ran variously on NBC and ABC from 1939 through 1952, then enjoyed a two-year revival in off-network syndication. And in 1940, crusading female prosecutor Portia Blake set up shop in the daily, 15-minute serial *Portia Faces Life*, which alternated between NBC and CBS until 1951.

The "purification" of fictional prosecutors extended to defense attorneys as well. On January 3, 1938, NBC's Blue Network (the precursor to ABC) introduced *Attorney at Law*, a daily 15-minute serial starring Jim Ameche as straight-arrow lawyer Terry Regan; when the series was reformatted as a weekly half-hour summer replacement for NBC's *Fibber McGee and Molly*, Henry Hunter assumed the leading role. Later radio defenders included Michael West on the CBS soap opera *Bright Horizon* and the title characters in *The Amazing Mr. Malone, Lawyer Tucker, The Amazing Mr. Tutt, Roger Kilgore: Public Defender* and *Defense Attorney*. The most successful of radio's criminal lawyers was our old friend Perry Mason, who headlined his own series from 1943 through 1955, periodically muscling into *Mr. District Attorney*'s territory by taking on mobsters and racketeers.

Other legal dramas during radio's Golden Age included two sympathetic portrayals of judges, *Her Honor, Nancy James* and *His Honor, the Barber.* There were also such programs as *A Life in Your Hands,* created by *Perry Mason*'s Erle Stanley Gardner and featuring an *amicus curiae* (friend of the court) named Jonathan Kegg, who was permitted to interview witnesses on both sides of a scripted murder trial in order to determine the truth; and the *Dragnet*-like "procedural" series *Indictment,* dramatizing case histories from the files of former New York assistant DA Eleazer Lipsky. Less conventional were two series which used the legal process more or less as a gimmick: *The American Women's Jury*, in which domestic advice was dispensed within the framework of a courtroom hearing presided over by a lady judge and featuring a female defense attorney and male prosecutor (all portrayed by actors), with members of Boston-area women's clubs serving on the mock jury; and *Lawyer Q,* a quiz program built around reenactments of actual court cases, with twelve contestants pulling "jury duty" and sharing a huge cash prize if they entered the correct verdict.

When television became commercially viable in the late 1940s, the restrictive Canon 35 did not as yet apply to live TV coverage of courtroom proceedings. But the bulky equipment required for remote telecasts, coupled with the inadequate acoustics and lighting facilities in most courtrooms, discouraged TV producers from venturing into the real world of litigation. In fact, the first legal program produced for the small screen wasn't live at all, but a filmed series from the enterprising Jerry Fairbanks: *Public Prosecutor,* produced in 1947 and first syndicated to local stations during the 1948–49 season. Like most other Fairbanks productions of the pe-

riod (*Goin' Places with Uncle George, Crusader Rabbit*), *Public Prosecutor* was not designed as a stand-alone series; each 17½-minute episode was intended as a wraparound for a locally produced quiz segment, in which a panel of legal experts was invited to predict the outcome of the story, or the folks at home were encouraged to phone in their guesses.

Other early TV offerings *were* broadcast live, often starting out as local programs before being picked up by NBC, CBS, ABC, or the late, lamented DuMont Network. DuMont in fact was responsible for the first of the nationally telecast legal shows, *The Court of Public Opinion* (later retitled *The Court of Current Issues*), which debuted February 9, 1948. This was also the first of several TV programs which, in the tradition of radio's *American Women's Jury*, presented debates on current issues in the form of a courtroom trial, complete with ersatz "judge" and "opposing counsel." Other examples of this format included *On Trial* (1948–52) and *Politics on Trial* (1952), both from ABC.

Television's first live courtroom drama was the Chicago-based *They Stand Accused*, briefly picked up by CBS under the title *Cross Question* in January of 1949 before settling into a three-year run on DuMont. This was followed in quick succession by *The Black Robe*, aka *Police Night Court* (NBC, 1949–1950), which recreated actual New York City night-court sessions using several of the original participants; another courtroom-reenactment series, *Your Witness* (ABC 1949–50), which like *They Stand Accused* originated from Chicago; and a TV version of *Famous Jury Trials* (DuMont, 1949–1952). With the exception of the ambitiously produced *Famous Jury Trials*, the above-mentioned series never ventured beyond their basic courtroom setting. Networks tended to lavish their biggest budgets on such prestige items as *Studio One* and *The Texaco Star Theater with Milton Berle,* leaving very little money to expend upon the relatively insignificant legal-drama genre; this was also true of the ultra-cheap NBC public affairs series *Four Square Court* (1952), a single-set panel show featuring genuine parole officers and ex-prisoners. Conversely, ABC's live adaptations of the radio dramas *The Amazing Mr. Malone* and *Mr. District Attorney*, both seen in the same time slot on an alternating basis during the 1951–52 season, made extensive use of non-courtroom settings — and it's just possible that the extra added expenditure contributed to the early cancellation of both programs.

Fascinating though these early legal series may have been, they paled in comparison to the real-life drama emanating from New York's Foley Square Courthouse beginning in March of 1950. Senator Estes Kefauver, a Tennessee Democrat with presidential aspirations, had been touring the courtrooms of the nation as chairman of what was officially known as the Senate Special Committee to Investigate Crime in Interstate Commerce, but popularly referred to as the Kefauver Committee. Since the Senator was holding hearings rather than trials, he was able to circumvent the "no cameras in the courtroom" rule, allowing photojournalists full access as the Committee bore down on a variety of underworld figures and corrupt public officials. Kefauver's activities attracted only moderate press notice and public interest until the Committee arrived in New York City, where the Senator arranged for the proceedings to be covered by local TV station WPIX, owned and operated by the *New York Daily News*. Though WPIX was not a network affiliate, the hearings were telecast nationally, with full sponsorship by *Time* magazine.

Public response to the TV coverage was overwhelming, with up to seventeen times the normal daytime audience tuning into the proceedings. This upsurge of interest in the hearings can be attributed to the glittering "all-star" lineup of participants, including not only Estes Kefauver and his second-in-command Charles Tobey but also former New York City mayor William O'Dwyer, Chief Counselor Rudolph Halley, and especially such Mafia figures as Willie Moretti, Joe Adonis, and Frank Costello. Better known as "The Prime Minister of the Underworld," Costello in particular achieved a negative form of superstardom by insisting that his face never be shown on camera: as a result, fascinated viewers were treated to lengthy closeups of Costello's hands, fidgeting with his eyeglasses, crumpled sheets of paper, and a carafe of water. Frank Costello's "hand ballet" was second only in notoriety to the Mob witnesses' repeated invocation of the Fifth Amendment, elevating the phrase "Pleading the Fifth" to household-word status.

The Kefauver hearings helped to make TV's daytime hours attractive to potential sponsors, resulting in a marked increase of network programming between 10 A.M. and 5 P.M. Included in these new offerings were a number of daily soap operas, among them *Miss Susan*, produced in Philadelphia and starring Hollywood actress Susan Peters as a paraplegic attorney. Though *Miss Susan* was cancelled only nine months after its March 1951 premiere, it served as the vanguard for several other legal-themed daytime dramas, including the Chicago-produced *The Bennetts*, a TV adaptation of the radio favorite *Portia Faces Life*, and the most successful of the batch, CBS' *The Edge of Night*, which began in 1956 as a knockoff of radio's *Perry Mason* but quickly took on a life of its own, remaining in production until 1984. The enduring popularity of *The Edge of Night* ultimately spawned two imitations from rival network NBC, but neither *Ben Jerrod* (1963) nor *Hidden Faces* (1968) survived past its first year on the air.

Another spate of legal programs coincided with network television's coverage of the Army-McCarthy hearings, which commenced on April 22, 1954, and ended on June 17 of that year. Audiences were mesmerized by the war of words between demagogic, Red-baiting Senator Joseph Mc-

Carthy and avuncular U.S. Army counsel Joseph N. Welch, climaxing with Welch's eloquent condemnation of McCarthy's bully-boy tactics: "Have you no sense of decency, sir? At long last, have you no sense of decency?" It should not be necessary at this late date to dwell upon the details of these hearings, except to note that they were able to get around Canon 35's "no cameras, no mikes" edict for the simple reason that they *were* hearings and not trials.

Inasmuch as Canon 35 did not technically establish a precedent back in 1937, individual state jurisdictions occasionally permitted TV trial coverage in the 1950s, depending upon the discretion of the judge. In 1953, just one year before Joe Welch sliced Joe McCarthy into ribbons, TV cameramen were allowed for the first time to cover a trial in an Oklahoma courtroom, though home viewers saw only filmed and edited highlights; and in 1955, the year following the Army-McCarthy broadcasts, the first live trial coverage was seen in Waco, Texas. But these were exceptions to the rule, and for the most part the "law junkies" of the era had to be content with dramatized legal shows. Four such series were seen in 1954, when audience interest in Army-McCarthy was at its peak: the live network offerings *The Mask* and *Justice*, the filmed semi-anthology *Public Defender*, and the first in a long line of legal-centric situation comedies, *Willy*.

Nineteen fifty-seven was a watershed year for legal programming, both real and fictional. On February 26, 1957, the McClellan Committee, chaired by Democratic Senator John McClellan (previously a ranking member of the Army/McCarthy hearings) and numbering among its members John F. Kennedy and Barry Goldwater, opened Senate hearings to investigate corruption, illegal activities and Mob influence in America's labor unions — with special scrutiny reserved for the International Brotherhood of Teamsters. Networkcast throughout the eastern United States to an audience of 1.2 million

viewers, the hearings made media stars out of Chief Counsel Robert F. Kennedy and Teamsters president Jimmy Hoffa. The McClellan Committee also directly inspired four dramatic TV series: *The D.A.'s Man, Congressional Investigator, Grand Jury* and *The Witness.*

Nineteen fifty-seven was also the year that Erle Stanley Gardner, at that time the world's best-selling author, came to television with two different series. (Yes, two. Keep reading.) A master blend of fail-safe format and copacetic casting, *Perry Mason* was not only the longest-running legal drama of the 1950s and 1960s, but also elevated the character of the compassionate, dedicated defense attorney to iconic stature, a model for television generations to come. (In 1996, TV critic Gary Deeb would describe Perry Mason as a "one-dimensional legal superman" who evidently had no personal life outside the courtroom; maybe so, but that's what audiences *wanted* back in the Eisenhower era.) Echoes of *Perry Mason* reverberated ever after in such TV lawyer shows as *Lock-Up, Sam Benedict, Judd for the Defense, Hawkins, Rosetti and Ryan, The Mississippi*, and most famously, *Matlock*. Less successful but no less professionally assembled than *Mason* was Erle Stanley Gardner's other TV project of 1957, *The Court of Last Resort*, which unfortunately made the tactical error of denying viewers an appealing protagonist while overstressing the nuts and bolts of legal procedure.

Finally, 1957 witnessed the launching of the first cycle of what would later be designated as the "judge" show. It all began in Los Angeles, where seven commercial stations had been engaged in a cutthroat ratings competition ever since the late 1940s. This ongoing battle for viewer attention resulted in an abundance of popular local programs, with certain LA-based personalities (Liberace, Lawrence Welk, Betty White) achieving national prominence. Among the city's local offerings was *Traffic Court*, introduced as a public-affairs program by KABC-TV in the summer of 1957. Featuring reenactments of actual traffic-court litigation and presided over by real-life LA Municipal Court judge Evelle J. Younger, *Traffic Court* scored an unexpected hit, inspiring CBS to develop a "courtroom simulation" program of its own, *The Verdict is Yours*, which debuted as a Monday-through-Friday daytimer in September of 1957. Intimately involved in both these projects was TV producer and former State Department attorney Selig J. Seligman, who was also a guiding force behind the ABC Prime Time network version of *Traffic Court* (launched in June of 1958 with UCLA law professor Edgar Allan Jones Jr. taking over as judge). The combined popularity of *Traffic Court* and *The Verdict is Yours* led to the ABC daytimer *Day in Court*, also starring "Judge" Jones, in October 1958. *Day in Court* in turn spawned a weekly nighttime version in December 1958, and a daily spinoff, *Morning Court*, in October 1960.

The main drawing card here was "spontaneity." The ABC shows were all scripted to an extent, but the actors were allowed to ad-lib so long as they stuck to the established facts — and if the ad-libs deviated too far from the predetermined continuity, it was not uncommon for the judge to render an entirely different verdict than he had in rehearsal! Conversely, there was no script and no rehearsal on CBS' *Verdict Is Yours* — and no one, not even the series' producer, could predict in advance what the verdict would be, especially since the jury was made up of studio-audience members who'd been instructed to base their decision on the improvisations that they'd just heard for the very first time.

Acknowledging the popularity of the judge-show cycle, *TV Guide* observed in its August 30, 1958, edition: "Not since the quiz craze has there been such excitement over a television format as there is over 'the court show.'" Quoted by UPI on November 11 of that year, *Day in Court* star Edgar Allan Jones Jr. noted: "We're involved in a court cycle, and I'm confident it will last as long as the westerns, maybe longer. They'll go on because they're realistic and don't ham it up." And *Newsweek* had this to say on January 12, 1959: "In the past year and a half, since Station KABC-TV's pioneering *Traffic Court* first appeared, more than half a dozen new, documentary-type courtroom shows have gone on the air in California and several, via the networks and syndication, around the nation." Of the many reasons given for this phenomenon, the three which seem to make the most sense are as follows: (1) The intimacy of the courtroom setting was ideal for the TV medium, and the shows themselves were inexpensive to produce; (2) Denied access to genuine court trials, viewers regarded these TV fictionalizations as the next best thing; and (3) Everyone is a voyeur at heart.

As mentioned in *Newsweek*, the new courtroom shows were just as prevalent in syndication as on the networks — and once again, the Los Angeles TV market was the main breeding ground. First out of the chute was the semi-scripted, wholly fictional *Divorce Court*, telecast live from the studios of KTTV beginning in February 1958, then distributed nationally in the fall of that year — the first series ever to be syndicated on videotape. Then came the thoroughly dramatized *The People's Court of Small Claims* from KCOP, followed by the meticulously scripted *Night Court* from KTLA; the former was issued on videotape by the same distributor handling *Divorce Court*, while the latter was filmed, using a cost-efficient multicamera process. Also syndicated on film were two unscripted offerings featuring "real" people instead of actors: *Divorce Hearing*, an early effort from prolific documentary producer David L. Wolper; and *Parole*, which had evidently lain on the shelf for several years before the sudden popularity of legal programs plucked it from obscurity. Of all these syndicated efforts, the only one to enjoy lasting success was *Divorce Court*, which continued to churn out new episodes until 1969. The series was revived in 1985 in the wake of the second syndicated judge-show cycle instigated by *The People's Court*; and in 1999, yet another edition of *Divorce Court*, this one using real litigants rather than actors, was quickly assembled to capitalize on the *third* cycle of judge shows launched by *Judge Judy*.

While the 1957–60 bumper crop of courtroom programs was entertaining enough, seldom did these shows dwell upon the more controversial legal and moral issues of the era; most TV executives and sponsors were unwilling to risk offending viewers by tackling sensitive subject matter. What changed their minds in the early 1960s was a challenge issued by new Federal Communications Commission chairman Newton R. Minow. In a landmark speech before the National Association of Broadcasters on May 9, 1961, Minow charged that most of what passed for television was nothing more than a "vast wasteland," and that medium was capable of far greater things: "When television is good, nothing — not the theater, not the magazines or the newspapers — nothing is better. But when television is bad, nothing is worse." Though some observers dismissed Minow as a snobbish elitist, others applauded his words as a long-overdue frontal assault on the mindless sitcoms and excessively violent westerns and cop shows that glutted the market. At the same time, the executives of the three major networks perceived that Minow had thrown down a gauntlet, saying in effect "You'd better start upgrading TV in a hurry if you don't want to feel the hot breath of the FCC on your neck."

Beginning with the 1961–62 TV season, the networks regularly offered what were described in the trade as "problem" dramas, dealing with sensitive issues that had previously been avoided or ignored. Examples included the medical series *Ben Casey, Dr. Kildare* and *The Nurses*; the psychiatric dramas *The Eleventh Hour* and *The Breaking Point*; two shows revolving around "concerned and involved" teachers, *Mr. Novak* and *Channing*; a groundbreaking series about inner-city

social workers, *East Side — West Side*; and a smattering of courtroom dramas, beginning with 1961's *The Defenders*. The antithesis of the noncommittal *Perry Mason* format, *The Defenders* tackled such topics as abortion, civil rights, blacklisting, euthanasia and political extremism with unprecedented fervor — and unlike Perry Mason, it was not uncommon for the series' father-son defense team to lose a case once in a while. No less a legal authority than defense attorney Louis Nizer praised *The Defenders* for depicting the lawyer "as an honorable fighter for his client's cause, dedicating his zeal and devotion to serving justice. This is in the great and true tradition."

But in the grand scheme of things, the Problem Drama cycle in general and *The Defenders* in particular were little more than garlic necklaces designed to ward off the vampires of the FCC, so that the networks would be free to continue grinding out the sort of innocuous tripe that viewers claimed to hate but watched religiously all the same. As noted in a 1997 *New York Times* article by Susan Salmans, "Before and after *The Defenders*, television law tended to be more comfortable and less complicated." And not only before and after, but during: two other issue-oriented legal series introduced around the same time as *The Defenders*, *Slattery's People* and *For the People*, were conspicuous failures.

In contrast, the *Defenders*-like *Judd for the Defense* managed to stay afloat for two full seasons (1967–69) with storylines involving such touchy subject matter as mental retardation, criminal insanity, black militancy and the antiwar movement. *Judd*'s comparative success had less to do with content than presentation: protagonist Clinton Judd (Carl Betz) was a crusading firebrand who seemed quite capable of settling cases through sheer physical force rather than verbal dexterity. Though Judd never actually resorted to violence, he was in the grand tradition of what attorney Louis Nizer once mockingly labeled, "The slugging lawyer, who depends on his fists more than on his wits to topple the evildoers." This fictional creation was first cousin to what Nizer described as the "Mr. District Attorney" type, "a roving sleuth of heroic proportions." Judd also bore traces of another TV invention cited by Nizer: the fictional super-lawyers "who are combinations of a tough Marine, Mickey Spillane and a righteous FBI man."

The amalgam of these three stereotypes could be designated "The Maverick Lawyer": brash, opinionated, pugnacious, iconoclastic, playing by the rules only when it suits him, gleefully incurring the wrath of choleric colleagues and sputtering superiors who can think of nothing more original than threatening to revoke his license, and invariably emerging triumphant from the courtroom (whether or not he actually wins the case) by scoring some sort of moral victory. Examples of this type can be found in such series as *The Law and Mr. Jones*, *The Trials of O'Brien*, *Petrocelli*, *Kaz*, *The Eddie Capra Mysteries*, *The Great Defender*, and *Raising the Bar*. Also falling within this category were the hard-riding, fast-shooting "frontier lawyers" who began sprouting up during the TV Western craze of the 1950s, and continued procreating into the 1970s: this subgenre is represented by such shows as *Judge Roy Bean*, *Sugarfoot*, *Black Saddle*, *Temple Houston* and *Dundee and the Culhane*.

If any *real* attorneys ever behaved in the gonzo fashion described above, TV fans of the mid–1960s would never have known it, thanks to a sudden tightening of the prohibitive Canon 35. Ever since Texas had given the okay to the first live trial telecast in 1955, the state had been willing on special occasions to relax the "no electronic media in the courtroom" edict. Though there was resistance to this sort of public access, Texas TV executives could point to the example of the state of Colorado, which had abandoned Canon 35 in 1956. Not having forgotten the tumultuous Hauptmann trial of 1935, certain members of the legal community were worried that that media's alleged infringement of the right to a fair and unbiased trial would result in a multitude of reversed decisions; but as it turned out, no harmful after-effects of any kind resulted from lifting the ban on TV cameras in Colorado's courtrooms.

Such was *not* the case in the trial of high-rolling Texas financier Billie Sol Estes, who in 1962 was indicted by a Federal grand jury on 57 counts of fraud. Because of Estes' close personal ties with Vice President Lyndon B. Johnson, his trial received considerably more than the standard amount of media attention, with the presiding judge giving full reign to media photographers in his courtroom. Found guilty and sentenced to 15 years in prison, Estes argued that the heavy-handed media coverage had deprived him of a fair trial. In their 1965 decision *Estes v. Texas, 381 U.S. 532*, the United States Supreme Court voted 5–4 in favor of the defendant, and his conviction was overturned. As a result, Texas courtrooms would be off-limits to cameras and microphones for the next fourteen years — and with the exceptions of Colorado, Alabama and Washington State, this ban was rigidly enforced throughout America.

"Paradoxically, the Supreme Court overturned the Estes conviction while opening the nation's courtrooms to television access," noted legal historian Mary Kay Switzer in 1997. "The opinion of the court was delivered by Justice Tom Clark. After a review of the situation, Justice Clark acknowledged the fact that TV and radio reporters had the same rights as the general public with regard to access. But he also indicated that the Court's conclusion was then supported by the fact that [most] states considered the use of television in the courtroom improper. Chief Justice Earl Warren in his concurring opinion took the trial judge to task for his political use of television and his lack of control of the courtroom." The four dissenting Supreme Court justices felt that more "experimentation" was needed to fully determine TV's impact; Justice Potter Stewart added, "The suggestion that there are limits upon the public's right to know what goes on in the courts causes me deep concern."

Undoubtedly this judicial debate would have sounded like so much inside baseball to the average TV viewer of the late 1960s. Besides, who needed a dusty diatribe on "right to access" when you could flip on a switch and get Instant Relevance? Having awakened to the fact that much of their audience consisted of people under the age of 30, the three networks decided it was high time to come up with programming that endeavored to speak to the sociopolitically supercharged younger generation. And if the results tended to indicate that the old men in charge of television would forever be a day late and a dollar short when it came to keeping abreast of changing times, at least they were trying their best.

The legal-show genre was of course influenced by the networks' new "relevance" kick, beginning in 1969 with *The Lawyers* and *The Protectors*, two components of the rotating dramatic series *The Bold Ones*. Both these efforts worked overtime to deal with such "meaningful" issues as the Generation Gap, Black Power and the Drug Culture. Even more blatant in their solicitation of the under-30s were the 1970 series *The Young Lawyers* and *Storefront Lawyers*, both populated by actors so young they appeared to have only recently reached puberty. Similarly trafficking in current events and volatile subject matter was *Owen Marshall: Counselor at Law*, the most successful of the new legal series — possibly because the producers were savvy enough to reach out to older viewers as well as "the kids" by casting the patriarchal Arthur Hill in the leading role.

Other new developments of the mid–1970s included a tentative nod towards sexual diversity in the courtroom vis-à-vis *Kate McShane,* the first prime time dramatic series to top-bill a female attorney; unfortu-

nately, *Kate McShane* folded after six episodes, and it would be many, many years before women lawyers truly began to proliferate in Prime Time with such series as *Trials of Rosie O'Neill, Ally McBeal, Judging Amy, Family Law, Any Day Now, Reasonable Doubts* and *Sweet Justice*. The Seventies also witnessed the launching of a mini-movement to fashion weekly series out of popular films about the legal process with the misbegotten 1973 TV adaptation of the 1949 Tracy-Hepburn movie classic *Adam's Rib*. The failure of this project did not prevent future producers from attempting to reshape such theatrical features as *The Client* and *True Believer* for the small screen; but the only movie-to-television transfer to survive past its first season was 1979's *The Paper Chase*, mainly because the TV producers wisely retained the services of the film version's curmudgeonly star John Houseman. Oddly enough, the televised Watergate hearings of 1973, which commanded the attention of millions of viewers — and added such phrases as "To the best of my recollection," "Expletive deleted!" and "Stonewall" to the national lexicon — had very little impact on the fictional legal series of the period, almost as if the network heads were hoping that "Great National Nightmare" would simply go away so they could return to business as usual.

There was, however, one lasting legacy of the Watergate coverage: Their appetites whetted, viewers now demanded that they be given broader access to the American Justice System at work. This was one of the several reasons that the iron grip of Canon 35 — or Canon 3A[7], as it was now known — was showing signs of weakness in the nonfictional courtroom world. After years of flat refusals from authorities to permit TV cameras in Federal and state courtrooms, Miami station WPLG was allowed to cover selected litigation in the Florida courts beginning in 1977. The following year, the Conference of State Chief Justices adopted a resolution recommending that Canon 3A[7] be amended to authorize state supreme courts to permit TV, radio and newsphoto coverage of their proceedings — at the discretion of the judge, with assurance that media journalists be "subject to express conditions, limitations and guidelines" and that they remain unobtrusive and unobstructive. Around the same time, the U.S. Court of Appeals' Judge Alfred Goodwin chaired a committee recommending that broadcasting and photography be allowed.

Those who disagreed with Goodwin felt that media presence would result in a loss of decorum, and that witnesses and jurors would become unduly camera-conscious. Others worried that the coverage would not present a fair picture of the proceedings because the public would seldom see the entire trial. There was also concern expressed over the possibility that certain lawyers and judges would "showboat" for the cameras, as in the Hauptmann trial. While acknowledging this final point, Colorado judge Edward T. Pringle told *TV Guide* that this might actually be a blessing in disguise: if the public sensed that there was a showboater amidst the jurists, they would demand that he behave himself in court, thereby assuring a fairer outcome to the trial. (This turned out to be especially true in Milwaukee, Wisconsin, where the city's most outrageously flamboyant judge was forced to put a lid on his excesses after his constituents were given a good long look at his buffoonish behavior.) At the same time, Pringle and his peers argued that certain trials that should never be given TV exposure, such as those involving rape, pedophilia, child custody battles and sensitive trade secrets.

Most legal historians concur that the floodgates to full TV coverage were opened by three lawsuits that attempted to limit media access to the courtroom. In the matter of 1979's *Gannett Company Inc. vs. DePasqualé*, Judge John Harlan observed that "the right of 'public trial' is not one belonging to the accused, and inhering in the insti-

tutional process by which justice is administered." Chief Justice Warren Burger was prompted in 1980's *Richmond Newspapers Inc. vs. Virginia* to characterize members of the media as "surrogates for the public," with Justice John Paul Stevens adding, "Today, for the first time, the Court unequivocally holds that an arbitrary interference with access to important information is an abridgement of the freedom of speech and of the press protected by the First Amendment." Finally, with 1981's *Chandler et al. vs. Florida*, the Supreme Court affirmed the right of each state to allow "electronic and still photographic" coverage of criminal trials without first getting permission from the accused.

In response to this, the state of Rhode Island decided to give cameras in the courtroom a two-year tryout, beginning with coverage of the first murder trial of Claus Von Bulow — not full live coverage, but lengthy videotaped highlights for exposure on local newscasts. Interviewed by *TV Guide*'s Max Gunther, court/media liaison Andrew Teitz insisted the experiment "worked out fine. My impression was that TV got a lot of compliments from the public and very few complaints. I think the need to compress a long trial into short news excerpts will always cause some distortion, but on the whole, the network's coverage was fair and balanced. The public got as accurate a picture of the trial as it could get, given the time compression." Rhode Island Superior Court's presiding Justice Anthony Giannini not only agreed, but also addressed several of the initial objections to such coverage: "We were afraid that cameras would make judges and others behave differently, but I haven't seen it happening. I haven't seen lawyers grandstanding more than usual. I haven't found witnesses more timid. The cameras haven't been annoying or disruptive."

Similar to the Rhode Island situation, TV courtroom coverage in Florida was handled quite gingerly, with everyone concerned carefully monitoring the results. Ironically, the first Florida trial examined in depth by the media was that of accused murderer Ronnie Zamora, whose attorney Ellis Rubin deployed the now-famous "Kojak Defense," arguing that Zamora had been inspired to kill via overexposure to television violence!

Though most U.S. jurisdictions were allowing media courtroom coverage by 1982, twelve states and Federal Courts remained resistant, holding fast to the argument that journalists had no Constitutional right to televise cases. This prompted the National Association of Broadcasters' First Amendment Counsel Steve Neves to remark, "The fact is, judges and lawyers tend to think of the courts as their own private preserve. They don't really want TV looking over their shoulders." Gradually, however, the objectors backed off, and by 1984 TV cameras (now smaller, quieter and far more flexible than those awkward, noisy piano crates of the 1950s) were regularly covering courtrooms in 44 out of 50 states. Perhaps significantly, 1984 was the same year that lawyers in *every* state were allowed to advertise on TV and radio for the first time, the end result of a 1977 Supreme Court "First Amendment" ruling in the case of *Bates vs. State Bar of Arizona.*

Several other developments of interest were taking shape in the 1980s. For starters, the long-dormant "judge show" cycle was revived in 1981 with the syndicated *The People's Court*, which took advantage of TV's newfound freedom to cover actual courtroom procedure by featuring real litigants and cases, with genuine binding decisions imposed by a bonafide judge, Joseph A. Wapner. The series spawned several imitations, most of them eschewing fact for fiction, with scripted procedures played out by professional actors posing as litigants. Of these, the only one to approach the popularity and longevity of *People's Court* was the 1985 revival of that old warhorse *Divorce Court.*

In a separate development, 1984 saw the premiere of the weekly situation comedy *Night Court,* starring Harry Anderson as a highly unorthodox judge. The streets of Hollywood are paved with the bones of short-lived sitcoms that have attempted to use the legal system as a source of humor: *Willy, His Honor Homer Bell, The Jean Arthur Show, The Paul Lynde Show, Sirota's Court, The Tony Randall Show, The Associates, Park Place, Sara, Foley Square, Eisenhower & Lutz, Trial and Error, The Home Court, Common Law, Sparks, Work with Me, First Years, A. U.S.A., Courting Alex.* As of this writing, only *Night Court* has managed to succeed and flourish, ultimately clocking in at 193 episodes.

Night Court was also emblematic of a 1980s trend in legal shows in which prosecutors rather than defense attorneys were cast in a heroic light. Where once the Perry Masons of the TV world were the knights in shining armor, prosecuting attorneys were finally getting their due by way of such "superstar DAs" as *Night Court's* Dan Fielding (John Larroquette) and Irwin Bernstein (George Wyner) on the NBC cop drama *Hill Street Blues.* In a similar vein, many of the action-show protagonists were more concerned with chasing down and locking up perpetrators than with such trivialities as "presumption of innocence." *Jake and the Fatman* featured a hero who alternated between D.A. and private detective to the extent that there was hardly any difference; and in *Hardcastle and McCormick*, an ex-judge teamed up with a private eye to impose their own special form of justice upon criminals who'd managed to legal-loophole their way to freedom. This peculiar brand of "due process" often bordered on vigilantism: one episode of *Hardcastle and McCormick* found former judge Hardcastle (Brian Keith) allowing his partner McCormick (Daniel Hugh-Kelly) to hector and humiliate a suspected murderer before making a formal arrest, dryly observing, "We can't charge him till Monday anyway." Typically, TV writers made things easy for these self-styled crimebusters by vilifying all criminals as irredeemable scumbags, for whom a life sentence was far too merciful.

In a *TV Guide* article charting this
trend, Sandra Hansen Konte suggested that
the prosecutor-as-hero was a function of the
rising crime rate in America. Whereas a 1967
Harris poll found that 49 percent of the public thought that the courts were "too soft" on
criminals, that number had shot up to 81
percent by 1986; and though a 1972 Field
Institute Survey revealed that 82 percent of
Californians felt crime could be reduced if offenders were "re-educated" instead of incar-
cerated, only 35 percent held that opinion
fourteen years later. Naturally, many real-
life defense attorneys viewed this development with alarm: George T. Davis, whose
clients included serial rapist Caryl Chessman, worried that the crusading-prosecutor
genre might portend a return to "the law of
the rubber hose," adding, "It's so easy to sell
the image of the prosecutor as the protector
of society these days. The public eats it up.
Television is glamorizing that side of the law,
because it has realized that the us-against-
them mentality is the one with which the
average person can most readily identify."

But by 1986 the pendulum had begun
to swing in the opposite direction, thanks in
great part to a new legal drama produced by
Steven Bochco — who, ironically, had been
one of the foremost purveyors of the "D.A.
hero" stereotype in his previous *Hill Street
Blues*. Bochco's ensemble series *L.A Law* not
only restored the defense attorney to pride of
place, but also represented a three-dimensional alternative to the two-dimensionality
of the old Perry Mason image. The lawyers
on *L.A. Law* were multifaceted human beings,
running the gamut from youthful idealism to
weary cynicism. This new breed of defense
attorneys also harbored more than their share
of personal problems, which unavoidably
spilled over into their professional activities.
Additionally, there was far more ethnic and
gender diversity on this series than in previous lawyer shows.

Meanwhile back in the Real World,
courtroom aficionados were enjoying an em-

barrassment of riches thanks to the rise in
prominence of cable television. The pioneering cable-news outlet CNN led the pack with
its up-close-and-personal coverage of the
1984 "gang rape" that which inspired the
1988 feature film *The Accused*, the 1985 retrial of Claus Von Bulow, the 1987 Supreme
Court confirmation hearings for Robert
Bork, and the Congressional investigation
into Iran-Contra. But CNN was in the business of covering *all* late-breaking news, and
could not confine itself merely to the courtroom. What was needed was a "narrowcast"
service specializing in legal coverage on a
round-the-clock basis.

And so on July 1, 1991, Court TV was
born. Representing the combined efforts of
Time-Warner, NBC, Liberty Media and Cablevision, Court TV — a last-minute merger
of two proposed cable channels, American
Courtroom Network and In-Court Television — was originally pitched to prospective
advertisers as "Soap Opera meets C-SPAN."
This was quickly changed to "Real Law instead of *L.A. Law*" by executive-in-charge
Steve Brill, a practicing attorney who had
previously founded *American Lawyer* magazine. Brill expressed the hope that his new
cable channel would alter the disheartening
fact that, while 54 percent of the public could
identify Judge Wapner of *People's Court*, only
9 percent recognized Supreme Court Justice
William Rehnquist. He promised that Court
TV would provide live coverage of as many
as four civil/criminal cases per broadcast day,
with legal experts filling the "dead spots"
between litigation. Foremost among these
experts was Court TV's senior editor Fred
Graham, a veteran law correspondent and
former CBS news anchor.

Another early promise made by Steve
Brill was that Court TV would strive to avoid
such "sensationalism" as the recent trial of
actress Zsa Zsa Gabor, who'd been hauled
into court on a charge of slapping a traffic cop.
Yet sensationalism could not entirely be
avoided during the first of the cable service's

high-profile trials, in which the LAPD cops accused of beating Rodney King were brought before the bench. Court TV's next big attraction was the rape trial of William Kennedy Smith, followed by full coverage of cases involving the Aryan Nation, Jeffrey Dahmer, Christian Brando, Betty Broderick and "The Battling Bobbitts." During its first few seasons on the air, Court TV averaged 125 trials in 75 different municipalities per year.

Not unexpectedly, there were many lawyers who fretted that Court TV would have a long-range adverse effect on the justice system, with the omnipresence of cameras encouraging a "circus-like atmosphere" (here we go again!). Steve Brill countered this objection by citing a survey of 245 judges, 91 percent of whom expressed either a favorable or neutral opinion of Court TV, and 94 percent stating that they were certain that extended TV coverage would have no affect on a trial's outcome.

Court TV was only a moderate success in its formative years, with several major cable companies refusing to carry the service, arguing that there was no room on the existing channel spectrum (this was several years before the 100-plus channel lineups made possible by the digital-cable boom). All this changed after the 1993 trial of Lyle and Erik Menendez, a pair of "Beverly Hills brats" charged with the 1989 murder of their wealthy parents. Initially, the presiding judge wasn't going to allow television coverage of the brothers' trial because he felt the courtroom was too small, but the Court TV technicians won him over with a quick demonstration of their lightweight, versatile equipment.

The Menendez trial was drenched in the sort of sensationalism that the Court TV executives had hoped to avoid, largely because of the controversial defense mounted by Erik Menendez's attorney Leslie Abramson, who insisted that the brothers were driven to kill after being subjected to a lifetime of physi-cal and sexual abuse from their parents. Public response to the TV coverage went far beyond astonishing, resulting in hundreds of thousands of viewers demanding that local cable services start carrying Court TV immediately or risk mass cancellations. *TV Guide*'s Gini Sikes later observed that the Menendez trial "did for Court TV what the Gulf War did for CNN." The mainstream networks, which had studiously ignored Court TV up to this point, suddenly found themselves clamoring for the services of the cable channel's principal commentator Cynthia McFadden, with ABC eventually claiming this particular prize. McFadden was but one of many commentators who achieved celebrity status on Court TV: others included Dan Abrams, Catherine Crier, Nancy Grace, Gregg Jarrett, Terry Moran and Lisa Bloom.

From the outset, Court TV endeavored to add variety to its lineup with such daily and weekly series as *Prime Time Justice* and *Night Court Live*, both of which provided summaries of the day's legal highlights; *Washington Watch*, legal news from the Nation's capitol; and *Lock & Key*, featuring parole and death-penalty hearings. Later original series created by the network included *Legal Café*, "a daily wakeup call to the law in your life" hosted by June Grasso; *Instant Justice*, spotlighting "ordinary people" involved in minor court cases; and such exotically-titled entries as *Hollywood & Crime*, *Saturday Night Solution*, *Smoking Gun TV*, *Psychic Detectives*, *Til Death Do Us Part*, *Texas SWAT*, *World's Wildest Police Videos*, *I Detective* and *Under Fire: Deadliest Police Shootouts*. Court TV's most controversial series was 2000's *Confessions*, in which murderers and rapists described their crimes in intimate and gruesome detail. The series was greeted with universal horror and yanked from the Prime Time lineup almost immediately after its first appearance.

The controversy stirred up by *Confessions* was minimal compared to the hullabaloo surrounding *l'affaire de* O.J. Simpson in June of 1994. To provide a full chronicle of the sequence of events following the discovery of the mutilated bodies of O.J.'s ex-wife Nicole Brown Simpson and her friend Ronald Goldman would blow out this writer's already overtaxed word processor. Suffice to say that the indelible image of ex–NFL star Simpson making a low-speed "end run" on the LA Freeway in his white Ford Bronco, with the police in steady pursuit, was covered by virtually every TV network and cable service in the Known Universe — and in that moment, Court TV's monopoly on 24/7 legal coverage came to an end.

O.J. Simpson's subsequent murder trial not only ripped the crown off the Bruno Richard Hauptmann case as "The Trial of the Century," but also exhumed the old debate over the effect of cameras in the courtroom. Following Simpson's acquittal in 1995, *TV Guide* published the results of a forum to determine if the excessive media coverage had influenced the outcome of the case. Of the celebrity pundits surveyed, ABC's Peter Jennings, NBC's Jack Ford and Court TV's Gregg Jarrett insisted that the coverage had had no effect on the verdict; celebrity interviewer Larry King "didn't think so"; Court TV's Cynthia McFadden "wasn't sure"; political humorist Bill Maher thought "yes," because O.J. was so "camera-friendly"; CBS' Harry Smith and CNN's Geraldo Rivera agreed that the presence of the TV cameras affected the behavior of the lawyers, especially the preening and posturing Johnnie Cochran; *Inside Edition*'s legal analyst Star Jones argued that the excessive TV exposure prompted the quick and surprising verdict; *Good Morning America*'s Bill Ritter felt that TV had permitted viewers to understand the verdict, even if they didn't agree with it; and filmmaker-*provocateur* Michael Moore characteristically sensed a conspiracy to prevent all the facts of the case from coming to light!

The O.J. coverage turned out to have a major effect on fictional courtroom programming. At a 1996 seminar held by Chicago's Museum of Broadcast Communications,

Barney Rosenzweig, producer of *The Trials of Rosie O'Neill*, opined that the Simpson trial had forced TV shows to become more realistic in their depiction of the trial process. "We know that America knows what it really looks like," commented Rosenzweig, predicting that the days of unfettered dramatic license were over and done with. Barely mentioned during this seminar was the fact that the O.J. verdict had split American viewers straight down middle, with one group demanding everyone in every walk of life receive the same "dream team" degree of legal protection that Simpson had enjoyed during his trial, and the other group insisting prosecutors redouble their efforts to prevent high-profile defendants from receiving preferential treatment. In short, the average viewer either came down hard on the side of the Defense or the side of the Prosecution — and wasn't terribly concerned with the subtle shades of gray between these two extremes. Representing the opposite ends of this legal spectrum were two of TV's biggest and most influential producers: Dick Wolf for the prosecution, and David E. Kelley for the defense.

A lifelong fan of the "procedural" cop drama *Dragnet*, Dick Wolf has always championed the efforts of police officials and prosecutors to prevent clever criminals and their cleverer lawyers from evading punishment. Debuting in 1990, Wolf's breakthrough series *Law & Order* returned to the early-1980s concept of the "DA hero" with a vengeance — and by using current events as the series' inspiration, Wolf was able to reconstruct history and make it come out "right" with fictionalized versions of actual court cases. It was not uncommon for the *a clef* version of a particular case to deviate from its source material by having the guilty party receive a swift and sure conviction, principally because Wolf's prosecution team had avoided the mistakes made by their real-life counterparts.

Not everyone was a wholehearted supporter of Wolf's pro-prosecution philosophy. Legal historian Michael Asimow has expressed bemusement over "the message from pop culture [that] the cops, police forensic scientists, and prosecutors are noble and fascinating creatures who have all the right answers, while those on the defense side are faintly repugnant toads who, at best, merit protection under the Endangered Species Act." But Dick Wolf's formula has proven not only successful but prolific. Drawing upon his pre-showbiz background as an advertising copywriter, Wolf has developed a veritable franchise of *Law & Order* spinoffs. "I learned branding from Procter & Gamble," explained the producer in an Internet interview. "There are seven Crests: they're different tastes, but the reason they can do seven is because people know, 'oh, it's Crest. I can buy it.' There are 312 toothpastes out there and none of them are any better than Crest. And the one rule on brand extensions is there is no such thing as a bad brand extension except one that doesn't work.... If you're going to eat soup, you see the can, you go, 'well, I've never had that but I'm sure if I wanted Cream of Mushroom theirs is going to taste as good as any other cream of mushroom.'" Among the most popular of Dick Wolf's TV "soup cans" have been *Law & Order: Special Victims Unit* and *Law & Order: Criminal Intent.*

Like Dick Wolf, David E. Kelley has been fascinated with "legalities" virtually all his life. Kelley's senior thesis at Princeton University was a play based on the Bill of Rights, with each amendment assuming human form: for example, the Second Amendment was characterized as a right-wing "gun nut." Receiving his law degree from Boston University, Kelley practiced real-estate and minor criminal law before turning his hobby of writing into a full-time vocation in 1983. After serving a high-paid apprenticeship on Steven Bochco's *L.A. Law,* Kelley began creating his own TV series, beginning with *Picket Fences,* in which two of the main characters were a curmudgeonly judge and a flamboyant defense attorney.

Unlike Dick Wolf, Kelley's heart has always been with the Defense, as witness his three most famous and successful TV series: *The Practice, Ally McBeal* and *Boston Legal.* Though his lawyer characters tended to be profoundly flawed — the unlucky-in-love Ally McBeal, the conscience-stricken Bobby Donnell, the *very* ethically challenged Alan Shore and Denny Crane — and their clients often appeared to have been dredged up from the compost heaps of humanity, the basic tenets of "presumed innocence" and "equal justice under law" were always at the forefront. And in another deviation from the down-to-earth traditionalism of Dick Wolf, David Kelley is uniquely capable of balancing the harsh realities of the American legal process with the sort of whimsical surrealism and flights of fantasy that had previously been the domain of such animated-cartoon directors as Tex Avery and Chuck Jones.

While Kelley and Wolf dominated Prime Time in the post–O.J. era, the daytime hours were graced with the newest reincarnation of a venerable format. Discussing the behavior of the lawyers during the Simpson trial, Court TV's Tim Sullivan had this to say on the January 20, 1998 edition of *The PBS Newshour*: "The ultimate problem with the conduct of the O.J. trial was not that a camera was present — it was that Judge [Lance] Ito did not exercise authority over those lawyers. The length of the trial, and the public behavior of the lawyers, should have been under his control." *TV Guide* correspondent Jeffrey Toobin said pretty much the same thing in a 1996 article concerning a proposed ban on future courtroom coverage: "[T]he Simpson trial — to understate the case — was not typical. The best way to correct the impressions of our system left by the Simpson case is to show the American people how real courtrooms are run." Toobin added that Ito's lack of control over his courtroom had transformed the proceedings into (that word again!) a circus.

For the sake of argument, let's say you

are a producer of daytime television programs, you're planning a "reality" courtroom
show, and you agree with the sentiments of
Tim Sullivan and Jeffrey Toobin. Now you're
faced with two choices: to simply set up your
cameras in a genuine courtroom and run the
risk of another "attorneys-run-wild" disaster
like the O.J. case; or to build a courtroom set
in a TV soundstage and hire your own judge,
one who will not only exercise full control
over the proceedings, but make a point of that
control as frequently and loudly as possible
so that no one will miss the message. Well,
that choice was easy, wasn't it? You avoid the
risk of things going awry, you build your set,
and you hire feisty former New York family-
court judge Judy Sheindlin, author of the
best-selling autobiography *Don't Pee on My
Leg and Tell Me It's Raining*.

And so, a little less than one year after
O.J. Simpson walked out of Lance Ito's
courtroom a free man, the daily, half-hour
Judge Judy made its syndicated TV debut.
For those still harboring doubts that *Judge
Judy* was intended as a response to the Simpson fiasco, listen to the words of series producer Larry Lyttle. "We broke the proscenium of the courtroom world with the O.J.
trial, and when we pierced that proscenium
we saw stuff we hated," Lyttle explained to
Time magazine. "When Judy showed the audience that she was decisive, that was the
elixir for all the malaise that we'd suffered."

Judge Judy was not only an elixir, but a
very potent one. As observed by Joel Stein in
"Here Come the Judges" in the August 24,
1998, issue of *Time*:

There are few pleasures greater than watching somebody get yelled at. You loved it when
your mom stuck it to your brother, and you
love it now, rubbernecking to see a cop pull a
car over. One of the best spots for catching
good, stern lectures in our authority-free culture is the bench of lower-court judges.
These guys can lay into punks and deadbeats
like *Father Knows Best* on a caffeine jag.
And that may explain the success of Judy

Sheindlin ... the resident scourge on *Judge
Judy*, which as of this month is the eighth
most popular show in syndication. The appeal of TV-judge shows is that they are little
more than highly structured versions of Jerry
Springer, in which the feuding idiots are si-
lenced by a decisive moral authority instead
of a bald bouncer. *Judge Judy* developed this
formula in September 1996, and was followed
a season later by a revival of the '80s show
The People's Court, currently presided over by
former New York City Mayor Ed Koch.

And this was only the tip of the iceberg.
Before long, the daytime-TV airwaves were
swarming with brand-new judge shows fea-
turing superstar jurists of all sizes, shapes,
colors and sexual preferences: *Judge Mills
Lane, Judge Joe Brown, Judge Mathis, Judge
Hatchett, Judge Alex, Judge Maria Lopez,
Judge David Young, Judge Karen, Judge Jea-
nine Pirro, Jones and Jury, Curtis Court,
Cristina's Court* and *Family Court with Judge
Penny*— to say nothing of such format-driven
reality shows as *Jury Duty, Texas Justice, Moral
Court, Style Court, Animal Court* and even *Sex
Court*. And let's not forget the third (and
presumably *not* the last) edition of the ever-
popular *Divorce Court*.

Asked on *PBS Newshour* if the judge-
show genre affected the public's perception
of the legal system, Court TV's Tim Sullivan responded, "I don't think those shows
necessarily do any damage, if viewers understand they're not witnessing an authentic
courtroom situation." But USC law professor Erwin Chemerinsky worried that people
would expect the law to act as "quickly and
superficially" as Judge Judy and her contemporaries, telling *Time* magazine, "They want
to present a case in 30 minutes, and it's
difficult to do that without oversimplification.
The judge in the courtroom is interested in
following the law and creating fair procedures in the court of law. A judge on TV is
only interested in the drama of the proceedings, in good television, and those are obviously different goals." And attorney Bill Hal-

tom, writing for *The Tennessee Law Journal* in December 1999, mocked the public perception that a "celebrity" judge is somehow superior to an "ordinary" one: "Judge Judy is such a big star that she sits right up there on *The Hollywood Squares* Tic-Tac-Toe board right between Whoopi Goldberg and Charo. And the dumb contestants say things like, 'I'll go to Judge Judy to block!' Now you show me a judge who is a star on *The Hollywood Squares,* and I'll show you a truly powerful judge. You never see Sandra Day O'Connor or Ruth Bader Ginsburg on *The Hollywood Squares.*"

Alexander Wohl, adjunct professor at American University's Washington College of Law, offered this analysis of the judge-show craze in "And the Verdict Is...", an article in the November 30, 2002, *The American Prospect*: "The basic reasons for the abundance of these shows are clear: Their costs are low and their viewership high.... But *why* are these programs so popular? In part, no doubt, because the law provides good narratives — stories with beginnings, middles, and ends, not to mention classic dramatic elements like conflict and comeuppance. But this can't fully account for the appeal of reality court shows whose narratives are so much less compelling than those of the more refined and well-scripted prime-time courthouse dramas. Generally the reality shows involve only the parties and a judge, with no opportunity for stories about the machinations, the ethical dilemmas, or even the sex lives of lawyers. Indeed, the focus on accused and accuser, and not on lawyers, is probably a critical ingredient of these shows' success. What they celebrate is 'the law' not as process but as authority — the authority of a judge who settles disputes, no matter how trivial, and puts those who question society's rules in their place....

"Yet as much as these shows distort how the law really works in America ... their success seems to be due in great part to Americans' abiding faith and trust in our legal system. And that particular unreality is distressing. Studies show that Americans continue to highly value their courts and the fact that ordinary people get a hearing in them, but as University of Pennsylvania Law School professor Geoffrey Hazard has pointed out, most believe the adversary system actually interferes with their opportunity to be heard. They don't want lawyers muddling it.... They want Judge Judy."

Professor Wohl's thesis that the American public not only wanted but demanded an uncluttered, authoritative legal process to govern their lives was never truer than in the dark days following September 11, 2001. Millions of angry, frightened and confused Americans who had previously regarded the government, the military and the legal system merely as necessary evils suddenly found themselves turning to those institutions for protection, reassurance and guidance.

In the world of TV lawyer shows, 9/11 and its aftermath had their most profound effect upon a series bearing the acronymic title *JAG*. Debuting in 1995, this weekly series dramatized the activities of the American Armed Forces' Judge Advocate General division, an elite corps of military lawyers who functioned as either defenders or prosecutors, depending on the circumstances. An earlier series focusing on the JAG corps, 1966's *Court-Martial*, had died in infancy, possibly because it wasn't fashionable to be pro-military during the Vietnam era. During its first year on the air it looked as though *JAG* was likewise headed for obscurity; but thanks to major changes in network, time-slot and personnel, the series suddenly began catching fire in its second season, becoming one of the most popular legal programs on the air. Inevitably, the ratings started sagging during Season Six, thanks largely to the heady competition of NBC's *Who Wants to Be a Millionaire*. But in the wake of 9/11, *JAG* reclaimed its audience many times over, and continued riding high for the next four years — a fact that its producer, Donald P. Bellisario, attributed to the public's renewed confidence in, and respect for, America's men and women in uniform.

Once the initial shock of 9/11 receded, viewers became less reactive and more introspective, many finding themselves questioning the opinions and values that they had held dear before the terrorist attacks on New York City and Washington DC shook them out of their complacency. This, too, was reflected in the legal dramas of the early 2000s, specifically those series in which the protagonist had experienced some sort of epiphany. Shows like *The Guardian, Just Legal, Shark, Eli Stone* and *Canterbury's Law* focused on attorneys who had been forced to realign their lives and ambitions because of a devastating professional setback or a shattering personal tragedy

Introspection gradually gave way to heated debate over how the post–9/11 "War on Terror" was being pursued, adding another crack to the political schism in America that had first opened in the previous century during the Clinton impeachment proceedings (derided by industry insiders as "All Monica All the Time"), then grew progressively wider in the aftermath of the controversial 2000 Presidential election. There seemed to be no common ground in the rarefied world of cable television: one was either Liberal or Conservative, and never the twain would meet. The ideological battle that dominated the post–Clinton airwaves was perceived as "good theater," and it wasn't long before even the most fair and balanced of cable outlets had succumbed to the temptation to replace thoughtful debate sessions with adversarial shouting matches. For good or ill, this attitude trickled down to the previously evenhanded Court TV in the late 20th and early 21st century.

Contrary to popular belief, Court TV did *not* reap long-range benefits from its O.J. Simpson coverage of 1994 and 1995. Audiences barraged by O.J. stories from every other TV and radio outlet in the country quickly began suffering from sensory over-

load, and Court TV's ratings suffered accordingly. By the time the verdict was announced, the cable service had been toppled from its throne and was fighting for its life: its advertising rates had fallen to an all-time low of $100 per minute, and much of its broadcast day had been surrendered to infomercials.

Climbing on board to revitalize the network in 1998 was seasoned Viacom executive Henry Schlieff, who pepped up the daily schedule with new nonfiction series along the lines of *Dominick Dunne's Power, Privilege and Justice* and the aforementioned *Confessions*, reruns of such network programs as *Homicide: Life on the Street*, and original made-for-TV movies bearing titles like *Guilt by Association* and *The Interrogation of Michael Crowe*. Schlieff also pumped new oxygen into Court TV's ongoing commentary shows, hiring anchors with opposing points of view and throwing them together in hopes that the fur would fly. The combustible, shoot-from-the-hip Nancy Grace of *Closing Arguments* and Lisa Bloom of *Trial Heat* were a far cry from such comparatively sedate personalities as Fred Graham and Catherine Crier. Many lawyers and legal experts were appalled by the direction which Court TV had taken, their chief complaint being that "presumption of innocence" was no longer a consideration as the new crop of commentators merrily injected their own opinions, intuitions and biases into the various courtroom proceedings. But Schlieff's brash new approach proved successful, providing Court TV with a new lease on life — and during its anything-but-objective coverage of the Michael Jackson child molestation trial in 2004, the service's ratings skyrocketed.

After Jackson's acquittal in 2005, the Court TV executive staff determined that its best hope for continued prosperity lay not in adhering to its original mission of providing straightforward courtroom coverage, but in pursuing the service's new policy of disagreement and diversification. In the years that followed, less and less of the cable service's air space was devoted to live trials and hearings, and more and more time was given over to debate programs and such reality/procedural efforts as *Forensic Files* and *Body of Evidence*. On January 1, 2008, Court TV effectively ceased to exist: promising "revamped" trial coverage and an even broader lineup of reality series, its executives relaunched the service under a new name, TruTV.

This of course did not mean an immediate cessation of TV courtroom coverage on other broadcast outlets. In fact, there had already been a marked increase of such coverage during the protracted probate battle between E. Pierce Marshall and ex-Playboy Playmate Anna Nicole Smith over the assets of Smith's deceased billionaire husband J. Howard Marshall II. Though the Supreme Court ruled in Smith's favor in 2006, the saga was far from over: after Smith's sudden death in 2007, a whole new cycle of litigation commenced over her estate, the paternity of her daughter, the cause of her death, and even the custody of her corpse. Just as the O.J. case had brought worldwide fame to Judge Lance Ito, the Anna Nicole Smith postmortem made an overnight celebrity out of the estimable "crying judge" Larry Seidlin — who immediately began jockeying for his own syndicated courtroom series!

While Seidlin has not fulfilled his dream as of this writing, new judge shows continued to pop up throughout the first decade of the 21st century, as did new fictional legal programs. Though no wide-ranging trends developed during this period, it can be noted that many producers endeavored to enliven the courtroom genre with nuance and gimmickry. In the wake of the U.S. Supreme Court's involvement in the outcome of the 2000 Presidential election, viewers were offered a brace of dramatic series which, for the first time in TV history, focused on Supreme Court justices: *The Court* and *First Monday*. The prolific David E. Kelley issued

forth *Boston Legal*, the first successful mainstream dramatic series built around the antics of a team of delightfully unethical attorneys; this was followed in quick succession by two other programs with morally ambiguous lawyer protagonists, *Shark* and *Canterbury's Law*. The similarly titled *Justice* and *In-Justice* both appropriated the *Rashomon* technique of presenting contradictory versions of the "truth" in the pursuit of acquittal. The popularity of such serialized "need-to-know" series as *24* and *Lost* can be seen as the inspiration for *Damages*, which opened "cold" in the aftermath of a bloody murder, then spent the remainder of its first season building up to this critical and anxious moment. The courtroom genre has also invaded the realm of sci-fi/fantasy with *Century City* and *Eli Stone*; likewise the "reality-competition" genre, with the *Apprentice*-like series *The Law Firm*. And last but not least, the early 2000s witnessed a tantalizingly brief spate of animated lawyer shows, *Gary the Rat* and *Harvey Birdman, Attorney at Law*.

Though there have been several junctures in TV history when it may appear that literally dozens of legal programs have all sprouted up at once, there has never truly been a major "lawyer cycle" in the manner of the Western and quiz-show cycles of the 1950s, the spy-show cycle of the 1960s, the raunchy-sitcom cycle of the 1970s, the Prime Time soap-opera cycle of the 1980s, the adventure-fantasy cycle of the 1990s, or the talent-contest and reality-competition cycles of the 2000s. More than any other TV genre, legal shows have tended to be judged and/or accepted on their own individual merits, rather than as part and parcel of a wider selection of similar shows. Only the "judge" format — a fascinating example of a genre *within* a genre — has resulted in a proliferation of programs. As these words are being written in late 2008, three new judge series have joined the ranks of the nine similar efforts already flourishing in daytime syndication, and undoubtedly more are on the way.

And while Court TV no longer exists, at least not under that name, there remains a thriving market for legal-themed programs within the realm of cable TV. Three excellent examples among the many include Lifetime's *Final Justice with Erin Brockovich*, A&E's *American Justice*, and the Fox News Channel's *The Line-Up*.

Now as in the past, there are plenty of legal experts at the ready to complain about the perceived inaccuracies and distortions inherent in TV's coverage of courtroom proceedings, both fictional and factual. In recent years, the biggest cannonades have been reserved for the treatment of "celebrity" cases; Houston defense attorney Kent Schaffer was far from alone when, in 2007, he chastised such coverage as an "aberration as unrealis-tic as *Perry Mason*." At the same time, many legal pundits have resigned themselves to the fact that TV *must* take occasional liberties to keep viewers tuned in: as Julie Hinden noted in 2004, "The way the law is depicted on television is far from realistic — and rightly so. Television writers and producers are wise to edit out those aspects of criminal and civil cases that are tedious, or simply undramatic."

There is no neat-and-tidy way to wrap up this eight-decade overview of legal programming, except to wonder aloud what the reaction will be from viewers and experts alike when the next "Trial of the Century" comes along. All we can say for certain is, while it will be the first such trial of the 21st century, it definitely *won't* be the last.

THE SERIES

General Notes

This reference book covers virtually all "legal" series nationally broadcast in the United States between January 1948, and December 2008, listed alphabetically. Two or more series bearing the same title (e.g., *Arrest and Trial*, *The D.A.*, *On Trial*), are entered chronologically. In most cases, each entry represents one individual series. Exceptions include the various spin-offs of *Day in Court* and *Law & Order*, which are gathered into single, all-encompassing entries. The same single-entry policy applies to those series which reappeared in "revival" form after their original cancellation, such as *Divorce Court* and *People's Court*.

Each entry lists the "over-the-air" or cable network where the series originally aired. For example, *Ally McBeal* is identified as a Fox program, meaning that it was telecast on the Fox Network, an over-the-air service for anyone with an antenna; while *Final Justice with Erin Brockovich* is identified as a "Lifetime" program, meaning that it was initially seen on the Lifetime Network, a service available only to cable-TV subscribers. Those series that were *not* first-run on network or cable television, but instead were sold directly to local stations, are identified as "Syndicated." Examples of syndicated programs include *Congressional Investigator*, *Eye for an Eye*, and *Judge Judy*.

Each program is designated "network" or "syndicated" and original air dates are listed, from debut to final telecast. If a program is still on a first-run basis as of this writing, only the debut date is given. When precise first and last telecast dates are unavailable, an approximate lifespan is provided (e.g., "2001–2003") or the year that the series first appeared. (For example, the syndicated *His Honor, Homer Bell* is identified as a "1955" series.) Telecast dates are followed by the name of the production company, a list of the significant production personnel (executive producer, creator, etc.), and a cast list.

Most of the names of the major over-the-air and cable networks mentioned in this book will not need to be explained or clarified for readers outside of the United States: Fox is simply "Fox," Discovery Channel is "Discovery Channel," and so forth. However, here is a list of the American networks which are popularly identified only by their initials:

Over-the-Air Networks

ABC (American Broadcasting Company)

CBS (Columbia Broadcasting System)

CW (A combination of "CBS" and "Warner Bros."; came into being with WB and UPN merger [see below] on September 17, 2006.)

ION *see* PAX

NBC (National Broadcasting Company)

PAX (Network created by home-shopping entrepreneur Lowell "Bud" *Pax*son in 1996. Renamed "i.Independent Television" on July 1, 2005, then "ION [Eye-On] Television" on January 29, 2007.)

PBS (Public Broadcasting System)

UPN (United Paramount Network; folded into CW network September 17, 2006.)

WB (Warner Bros. Television Network; folded into CW Network September 17, 2006.)

Cable Networks

A&E (Arts & Entertainment)

CNN (Cable News Network)

E! (E!Entertainment Network)

FX (*Fox Extended* Cable Network; not confused with over-the-air Fox Network.)

MSNBC (Combination of *Microsoft Network* and *NBC*; "sister" cable news service to the over-the-air NBC network)

MTV (Music Television)

TNN (The National Network, formerly The Nashville Network; renamed Spike TV August 11, 2003.)

TNT (Turner Network Television)

USA (The USA Cable Network)

Inclusion Criteria

A few paragraphs ago, it was stated that this book chronicles *virtually* all American legal series. So what *isn't* ncluded?

To begin with, none of the programs which originated on cable's Court TV service are included. The history of Court TV, from its inception in 1991 to its death and "reincarnation" as TruTV in 2008, is detailed in the introduction of this book, which also includes a representative cross-section of the service's daily and weekly series. To list *all* of Court TV's programs would prove redundant, since so many of them cover the same basic territory.

I have also limited this book to American TV productions. With profound regret and reluctance, I had to exclude one of my own personal favorites, *Rumpole of the Bailey*, which though it was seen stateside on PBS was originally produced in England and designed for a predominately English audience. The one exception to this "American only" rule is the 1966 series *Court-Martial*, filmed entirely in England. However, *Court-Martial* qualifies as an American series because it was largely funded by a Hollywood film studio, featured two American stars in the leading roles, and was intended from its inception to be telecast on a "dual-market" basis in both the U.S. and the U.K.

As for my own determination of what qualifies as a "legal" series, my personal rule of thumb is as follows: if a series is primarily legal in nature, or if a significant percentage of in-

dividual episodes deal with legal matters handled by professional lawyers, it is in this book. That said, I realize I have opened myself up to criticism and even condemnation by readers who will take issue over the many series that I have eliminated from the text.

Why, for example, is there no mention of such situation comedies as *Will & Grace, Dharma and Greg, Sex and the City, Girlfriends* and *Almost Perfect*? Did not each of these programs feature at least one lawyer protagonist? Yes, they did, but these series were not *about* The Law. The aforementioned characters were lawyers essentially to explain why they lived in relative luxury, drove fancy cars, were able to eat at the best restaurants, and were always dressed to the nines. Otherwise, they could have been architects, doctors or engineers for all the bearing that their profession had on the series' plotlines. Case in point: *Bachelor Father* (1957–1962) starred John Forsythe as Bentley Gregg, a handsome bachelor who happened to be an affluent Beverly Hills attorney. The money earned from Bentley's practice was used for the upkeep of his posh suburban home, his orphaned niece Kelly (Noreen Corcoran) and his versatile Chinese houseboy Peter (Sammee Tong). It also enabled him to wine, dine and romance an endless parade of beautiful young women, very few of whom were professional associates of the charming Mr. Gregg. Clearly, the hero of *Bachelor Father* was identified as a lawyer *only* so that the average viewer could understand why he was living higher on the hog than most working stiffs: the amount of time that Bentley Gregg was actually seen in a courtroom was infinitesimal.

Other sitcoms that have featured lawyers as principal characters include *Hazel, The Mothers-in-Law, Honey I Shrunk the Kids* and even *My Mother the Car*. One would have to stretch a point on an Inquisition rack to describe these series as legal programs, just as one might risk public ridicule by describing *The Addams Family* as a "lawyer" show on the strength of those random moments when it was explained that the source of Gomez Addams' wealth was his previous life as a defense attorney — all of whose clients were found guilty!

The same exclusionary rule applies to several series which have prominently featured judges in the cast. To be sure, Judge Bradley Stevens, the jurist played by Jim Backus on the vintage sitcom *I Married Joan*, was often seen in his chambers, counseling bickering couples by recalling an amusing incident from his own marital misadventures with his madcap wife (Joan Davis); but the fact that Stevens wore a black robe and wielded a gavel did not make *I Married Joan* a legal program any more than the fact that Andy Taylor wore a sheriff's badge made *The Andy Griffith Show* a cops-and-robbers series. And it would be difficult indeed to argue that the presence of judge characters in such series as *The Virginian, Fresh Prince of Bel-Air, Family, Phyllis* and *Sisters* warrants their inclusion in this book.

I have additionally ruled out those adventure series in which the hero's pursuit of justice was motivated by the fact that he was an ex-lawyer, or at the very least a law-school graduate: *21 Beacon Street, Adventures of Brisco County Jr., Barnaby Jones, The Commish, The Gray Ghost, Ironside, Mancuso, Markham.* Likewise stricken from the record are those series in which the "action" hero happened to be dating a female attorney, notably *Rockford Files, Matt Houston,* and *Walker: Texas Ranger.* Also missing are the scattered efforts bearing such titles as *Acapulco* and *Robin's Hoods,* wherein a crusading protagonist with a legal background dispensed justice outside the usual legal channels by organizing an elite team of crimestoppers. (I have, however, included *Dark Justice,* in which a sitting judge led a double life as leader of a band of nocturnal vigilantes, but only because the hero's covert crimebusting was directly related to his courtroom activities).

Other "missing" programs include *Ed, The Four Seasons* and *Bronx Zoo,* in which the main characters started out as lawyers but opted to go into another line of work before the series proper got under way. In a similar vein, Paul Bryan, the ex-attorney played by Ben Gazzara in *Run for Your Life,* may have occasionally taken on a pro-bono case as a favor to an old friend, but the bulk of the series' episodes focused on the efforts of a terminally ill man to make the most of the time he had remaining on earth. And did it really matter to the plot-lines of *Highway to Heaven* that the itinerant guardian angel played by Michael Landon had been a lawyer in life?

Admittedly, some of the series that I *have* chosen to include in the book are open to debate, notably *The Paul Lynde Show, Sugarfoot* and *Second Chances.* I have endeavored to justify my reasons for including these series in the body of the text, and I hope I have stated my case to the satisfaction of the jury. There is, however, one major exclusion that will most likely inspire a huge percentage of the viewing public, namely the fans of handsome heartthrob David Boreanaz, to organize a kangaroo court and demand vigilante justice: specifically, I have decided not to include the popular fantasy series *Angel.*

Yes, yes, I *know* that the character played by Boreanaz, "vampire with a soul" Angel, spent five full seasons waging war against the forces of evil headquartered in the L.A. law firm of Wolfram & Hart. But in point of fact, Wolfram & Hart was merely a "mortal" front for a band of ancient demons bent upon spreading chaos throughout the world. Even though the otherwordly attorneys who worked for the firm went through the motions of representing a venomous variety of mob bosses, serial killers, crooked politicians and sundry non-human miscreants in the courtroom, their reliance upon Due Process took second place to the frequent application of their own demonic power — a talent that you'll never learn in *any* law school, contrary to popular opinion.

Thus, after extensive self-deliberation, I have concluded that to describe *Angel* as a show about lawyers would be comparable to describing the 1960s spy series *The Man from U.N.C.L.E.* as a show about Della Floria's, the Manhattan dry-cleaning shop which the U.N.C.L.E. agents used as *their* front. Ladies and gentlemen, I rest my case and throw myself on the mercy of the court.

The Entries

A.U.S.A.

NBC: Feb. 4–Apr. 1, 2003. Persons Unknown Productions/20th Century–Fox Television. Executive Producer/Creator: Rich Appel. Cast: Scott Foley (Adam Sullivan); Amanda Detmer (Susan Rakoff); Peter Jacobson (Geoffrey Laurence); John Ross Bowie (Wally Berman); Ana Ortiz (Ana Rivera); Eddie McClintock (Owen Harper).

Its title an acronym for "Assistant United States Attorneys," the half-hour NBC sitcom *A.U.S.A.* was one of a handful of TV legal series with a genuine attorney (or former attorney) at the controls. Producer-creator Rich Appel had graduated from Harvard Law School and clerked for Judge John Walker in New York's Second Circuit Court of Appeals before becoming a Federal prosecutor. In the early 1990s, he kissed the legal profession goodbye to try his hand at writing TV comedy shows, encouraged by the positive response to his comic essays in *The Harvard Lampoon,* and the warm reception afforded his jokes at private law-office testimonials. "It's easier to believe they would let a lawyer write for *The Simpsons* than it is to think they'd let a *Simpsons* writer decide who to indict," he told *The New York Times* in 2003. Appel was eventually promoted to *The Simpsons*' executive-producer staff, and later held down the same job on another animated Fox networker, *King of the Hill,* winning Emmy awards for both assignments.

One of the richest sources for Appel's humor was his own experience as an attorney. "I saw an amazing variety of life come in and out of [the Court of Appeals] building, which is a magnet for every kind of character," he recalled in his *Times* interview. "You'll see a pompous defense lawyer, a brilliant judge, a judge a little off his rocker, DEA agents who are honest, some federal agents who are shady — they all have business being there. As a storyteller, that's a gift." He received even more raw material from his friends in the legal profession, some of whom gave him court transcripts containing far funnier lines and situations than he could ever have dreamed up. "You could literally spend an hour walking around the office and read these hilar-

ious moments…. These were scripts, ready to go." On several occasions, even his former boss Judge Walker provided grist for Appel's mill.

What a shame that *A.U.S.A.* never managed to scale these comic heights. All too often, the series' humor was on the puerile level of the early gag in which a young attorney is shown in the Men's Room, blow-drying his trousers after accidentally urinating on his leg. Compounding the offense, this particular gag was used by NBC to *promote* the show!

Former *Felicity* and future *The Unit* costar Scott Foley headed the cast as Adam Sullivan, one of several newly appointed Assistant District Attorneys in the New York judicial system. According to the series' backstory, Adam had been jockeying for this job for six years — making it all the more unbelievable that he would show up late for his first court case, then force an immediate mistrial by asking one of the jurors for a date. All that saved Adam from being sent back to the minors was his mature and compassionate handling of tricky legal issues, notably the plight of a former Tuskegee Airman who'd failed to inform Social Security that his wife had died in order to use her checks to pay for the funeral. (The series' abrupt mood swings, from broad comedy to poignant drama, were handled with admirable smoothness — one of the few points in the show's favor.)

In the tradition of *Cheers,* Adam harbored an insatiable lust for a woman who by rights should have been his worst enemy: Gorgeous public defender Susan Rakoff (Amanda Detmer), who never missed an opportunity to insult Adam for his "coddling" of criminals. The rest of the cast consisted of one-note sitcom contrivances: Adam's nerdish supervisor Geoffrey (Peter Jacobson); anal-retentive paralegal Wally (John Ross Bowie); sassy, streetwise ADA Ana (Ana Ortiz); and Adam's party-dude roomie Owen (Eddie McClintock).

On the strength of Rich Appel's track record, NBC extended every courtesy to the *A.U.S.A.* cast and crew, even allowing extra production time and money to reshoot the pilot episode in hopes of enhancing the "spontaneity" of the humor. The network also gave viewers ample opportunity to savor the program by playing it in two choice timeslots during its first week on the air. But the show failed to make an impression. Particularly brutal in their assessment of the series were genuine lawyers: writing for the website *Picturing Justice*, Christine Alice Corcos dismissed Adam Sullivan as "another useless stereotyped attorney."

According to industry insiders, NBC had hoped that *A.U.S.A.* would be the legal equivalent to the network's popular medical comedy *Scrubs*. It is therefore no small irony that star Scott Foley joined the cast of *Scrubs* at virtually the same moment that *A.U.S.A.* aired its eighth and final episode.

Accused *see* Day in Court

Adam's Rib

ABC: Sept. 14,–Dec. 28, 1973. MGM Television. Created and produced by Peter H. Hunt; based on the 1949 film of the same name, written by Ruth Gordon and Garson Kanin. Cast: Ken Howard (ADA Adam Bonner); Blythe Danner (Amanda Bonner); Dena Dietrich (Gracie); Edward Winter (Kip Kipple); Ron Rifkin (ADA Roy Mendelsohn); Norman Bartold (D.A. Donahue).

It's always a risky proposition to use a successful motion picture as the basis for a weekly TV series. Only rarely does the TV version rise to the level of the original film, either in quality or popularity: for every *M*A*S*H* or *Buffy the Vampire Slayer*, there are dozens of misfires like *Casablanca*, *The Asphalt Jungle*, *Going My Way*, *Bob&Carol&Ted&Alice*, *Seven Brides for Seven Brothers*, and *A League of Their Own*. This is especially true when the original film's biggest selling card is the presence of a unique, irreplaceable star player.

The 1949 MGM comedy *Adam's Rib* boasted not one but two dynamic personalities in the leading roles. Spencer Tracy and Katharine Hepburn were cast as Adam and Amanda Bonner, a married couple who happened to be lawyers: Adam was a state prosecutor, Amanda a defense attorney. The plot focused on the old "Double Standard," which stipulated that what was acceptable behavior for a man was entirely unacceptable for a woman. A staunch equal-rights advocate, Amanda publicly challenged the Double Standard by defending Doris Attinger (Judy Holliday), who had shot and wounded her philandering husband Warren (Tom Ewell) after finding him in the arms of his paramour Beryl Caighn (Jean Hagen). Amanda argued that if the sexes were reversed, and Mr. Attinger had shot Mrs. Attinger's lover, he would probably have gone free by virtue of the "unwritten law"; it would be only fair and equitable to release Mrs. Attinger on the same grounds. The fun began when Adam Bonner was assigned to prosecute Mrs. Attinger, pitting him against his own wife. Though Adam and Amanda were confident that they could maintain their marital equilibrium even though they were antagonists in court, the "battle of the sexes" judicial battle spilled over into their private lives, with surprising results. The film's witty, incisive screenplay was written by the husband-wife team of Garson Kanin and Ruth Gordon, and directed with style and class by George Cukor.

No one realistically expected the stars of the 1973 TV version of *Adam's Rib* to be Tracy and Hepburn, thus a lot of slack was cut by viewers and critics alike, and the actors were judged on their own merits. Though both were still under the age of 30, Ken Howard (Adam Bonner) and Blythe Danner (Amanda Bonner) had already established themselves as skilled, charming performers; as a bonus, they had demonstrated an appealing rapport in their previous costarring assignment as Thomas and Martha Jefferson in the film version of the Broadway musical *1776*. And of course no one truly expected TV's *Adam's Rib* to scale the lofty heights of the original film: high-priced production talent like Kanin, Gordon and Cukor fell far outside the realm of the average weekly television budget, and besides, there's a world of difference between a 104-minute feature film and a 30-minute sitcom.

Certain key elements from the original *Adam's Rib* were retained for the series. Adam remained a prosecutor, while Amanda continued working for the defense, albeit as a junior rather than full partner of her law firm. As in the film, the Bonners discussed the particulars of the case at hand just before bedtime — the difference being that censorship rules had relaxed since 1949 and they were allowed to retire to a double bed. And true to the source, Adam and Amanda still referred to each other by such pet names as "Pinky" and "Poopsie," and still surreptitiously exchanged affectionate glances in the courtroom.

New to the project were several supporting characters: Dena Dietrich (better known as "Mother Nature" in a series of popular margarine commercials) as Amanda's secretary Gracie; Norman Bartold as Adam's boss, DA Donahue; Ron Rifkin as assistant DA Mendelsohn; and Edward Winter as Amanda's flirtatious law partner Kip Kipple, a character evidently inspired by the original film's Kip Lurie, the implicitly gay songwriter played by David Wayne. In place of the 1949 film's unofficial theme song, Cole Porter's "Farewell Amanda," the 1973 edition of *Adam's Rib* offered a newly minted theme, "Two People," written by Perry Botkin, Jr. and Gil Garfield.

Arguably the best episode was the first one filmed (though not the first shown), a "special" 60-minute remake of the 1949 film. Though lacking the original's casual charm, escalating comic tensions and adroit double-twist climax, this episode, retitled "The Unwritten Law," was not without merit. The supporting cast was particularly appealing, with Madeline Kahn, Allen Garfield and Ms. Timothy Blake competently filling the roles previously played by Judy Holliday, Tom Ewell and Jean Hagen.

The TV version of *Adam's Rib* might have survived past its initial fourteen episodes if, after using the original film's Double Standard plotline as a launching pad, the writers had gone to explore other aspects of the "enemies in the courtroom, lovers in the bedroom" premise. Certainly, future TV legal shows like *L.A. Law* and *The Practice* (both q.v.) were imaginative enough to feature male and female attorneys who were on-the-job opponents but off-duty sweethearts without resorting to the easy-out of having *every* court case focus on gender issues. Unfortunately, the TV *Adam's Rib* started in a rut and stayed there, with every episode telling virtually the same story — Amanda crusading for Feminism, Adam stubbornly defending Male Chauvinism — week after monotonous week. Typical was the episode in which Amanda took legal action against a restaurant that had refused her service because she was wearing a pantsuit; after spending most of the running time advising his wife to get over her outrage, Adam ended up playing the fool when Amanda strongarmed him into wearing a dress in public.

But the biggest strike against the TV version of *Adam's Rib* was simply that the series' time had come ... and gone. In 1949, a film predicated on sexual equality was innovative, cutting-edge entertainment. But by 1973, the Women's Liberation Movement had been firmly entrenched for years, and there was no novelty in a TV series which merely preached to the choir. Brisk and amusing thought it may have been, TV's *Adam's Rib* was an anachronism long before the first episode had been committed to film.

Against the Law

Fox: Sept. 23, 1990–Apr. 12, 1991. Sarabande Productions. Produced by Daniel H. Blatt. Cast: Michael O'Keefe (Simon MacHeath); Suzanne Douglas (Yvette Carruthers); Elizabeth Ruscio (Elizabeth Verhagen); M.C. Gainey (J.T. "Miggsy" Meggs)

The Fox Network's first legal drama, *Against the Law* starred Michael O'Keefe (who in real life came from a family of attorneys) as flamboyant Boston lawyer Simon MacHeath. Originally a member of his former father-in-law's prestigious law firm, Simon went into private practice to become, in the words of *New York Times* critic John O'Connor, "a one-man defense machine." A rebel by nature, Simon had little patience for established legal protocol, and his refusal to abide by the rules and compromise his values frequently got him charged with contempt of court. Like many another maverick TV attorney, Simon was the eternal champion of the underdog — and even if he didn't win every case, he nearly always scored a moral victory. In one memorable episode, a distraught father killed his daughter's rapist after the DA decided not to prosecute the perpetrator. Defending the father, Simon not only established that his client had been arrested only after the police had conducted an illegal search of his home, but also precipitated a hung jury by arguing, "There are times when a case must be resolved not from the law on the book but from the community's standards of good and evil." After viewing this particular episode, one suspects that even if Simon hadn't been so verbally persuasive, he would probably have won over the female jurors with his fabulous wardrobe (not a bad set of threads for a guy who frequently worked pro-bono).

Others in the cast were Suzzanne Douglas as Simon's researcher and assistant Yvette, Elizabeth Ruscio as his secretary Elizabeth, and M.C. Gainey as his hygienically-challenged legman Miggsy. More popular with critics than with viewers (or lawyers, for that matter), *Against the Law* yielded 17 one-hour episodes.

Ally McBeal

Fox: Sept. 8, 1997–May 20, 2002. David E. Kelley Productions/20th Century–Fox Television. Cast: Calista Flockhart (Ally McBeal); Courtney Thorne-Smith (Georgia Thomas); Greg Germann (Richard Fish); Lisa Nicole Carson (Renee Raddick); Jane Krakowski (Elaine Bassell); Gil Bellows (Billy Alan Thomas); Vonda Shepherd (Herself); Peter MacNicol (John Cage); Brooke Burns (Jennifer Higgin); Eric and Steven Cohen (The Dancing Twins); Jesse L. Martin (Dr. Greg Butters); Portia de Rossi (Nell Porter); Lucy Liu (Ling Woo); Albert Hall (Judge Seymore Walsh); Gina Philips (Sandy Hingle); James LeGros (Mark Albert); Tim Dutton (Brian); Robert Downey, Jr. (Larry Paul); Lisa Edelstein (Cindy McAulif); Taye Diggs (Jackson Draper); Regina Hall (Coretta Lipp); Julianne Nicholson (Jenny Shaw); James Marsden (Glenn Foy); Josh Hopkins (Raymond Milbury); Anne Heche (Melanie West); Dame Edna Everage [Barry Humphries] (Claire Otoms); Jon Bon Jovi (Victor Morrison); Hayden Panettiere (Maddie Harrington); Christina Ricci (Liz "Lolita" Bump); Bobby Cannavale (Wilson Jade); Dyan Cannon (Jennifer "Whipper" Cone); Tracey Ullman (Dr. Tracey Clark).

The 60-minute legal dramedy *Ally McBeal* came into being when the Fox Network challenged producer David E. Kelley, fresh from *L.A. Law* and *The Practice* (both q.v.), to create a series that would appeal to the youthful fans of the network's recently departed *Melrose Place*—and one that would provide a viable alternative to ABC's *Monday Night Football* for female viewers between the ages of 18 and 34. Fox made an abstract request for a series about young, ambitious and sexually supercharged business executives; the network also gave Kelley full creative control over the project. Exercising his prerogative and drawing upon his experience as a former attorney, Kelly kept the "young, ambitious, sexually supercharged" part, but served up a series about a group of lawyers, centering around a highly intelligent but incredibly neurotic 28-year-old heroine named Ally McBeal. Originally envisioned as a vehicle for Bridget Fonda, *Ally McBeal* ended up a star-maker for the talented Calista Flockhart, who prior to signing up with Kelley had done most of her best work on stage. Carrying an expensive hour-long series on her skinny little shoulders was quite a daunting responsibility for Flockhart, but she proved more than equal to the task.

Having followed her childhood sweetheart Billy Thomas (Gil Bellows) to Harvard Law School, Ally McBeal ended up one boyfriend short when Billy decided to transfer to Michigan. Nonetheless, she earned her diploma and landed her first job at a male-dominated firm, only to be fired after filing a sexual-harassment complaint against the senior partner. Ally ended up accepting a position at the Boston firm of Cage, Fish & Associates, where as luck would have it, one of her new coworkers was none other than Billy Thomas—now married to attorney Georgia Thomas, a character added to the series at the last minute because Kelley wanted to create a role for former *Melrose Place* regular Courtney Thorne-Smith. Though Ally really, *really* wanted to dislike Georgia, they ended up good friends instead—even when the insecure Billy made it clear that he'd never gotten over Ally and hoped that they could renew their romance.

The "Cage" of Cage, Fish & Associates was John Cage, played by Peter MacNicol, who once described his character as a having a "beehive brain." Known to his colleagues as "The Biscuit" (a childhood nickname of obscure origin), Cage was unquestionably brilliant and boasted an impressive record of legal victories, but his courtroom demeanor was, for lack of a better word, bizarre: he was fond of distracting his opponents with such tactics as using a clicker or noisily pouring water, and he was known to unexpectedly burst into song or shout out non sequiturs like "Poughkeepsie" to compensate for a nervous stutter (Cage also played the bagpipes, which happened to be a peculiar talent of the actor portraying him). The firm's "Fish" was the inimitable Richard Fish (Greg Germann), who adhered to the philosophy, "The law sucks, the law is boring," and who'd become an attorney only to make "piles and piles of money." (Ally and Fish had been fellow students at Harvard, and had never gotten along at all.) Richard was famous in legal circles for his "Fishisms": weird turns of phrase and such verbal shorthand as "Bygones," meaning an emotion that he had discarded because it was of no further value. He was also known far and wide for his curious taste in women, beginning in Season One when he entered into an affair with the much-older Judge Jennifer "Whipper" Cone (Dyan Cannon) because he had a fetish for the "wattles" under her chin. The other principal employee of the firm was Ally's busybody secretary Elaine (Jane Krakowski), who kept her job because she knew where all the bodies were buried. When not on the job, Ally commiserated with her roommate, Deputy District Attorney Renee Raddick (Lisa Nicole Carson), who freely dispensed advice about men to the incredibly naïve heroine.

The episodes were narrated by Ally, who was prone to such homilies as "Law is like Life: the actual practice can give you a yeast infection," and "The best people are always taken: if you don't steal them, you won't have them." Though she exuded confidence in the courtroom, Ally could be childish, petulant, temperamental and downright dunderheaded in her private life — character traits that either increased her likeability or heightened her capacity to irritate, depending upon whether or not the viewer was a Calista Flockhart fan. De-

scribing the character in a *TV Guide* interview, David E. Kelley explained, "Sometimes a character can be professionally very strong and capable, but emotionally very weak. Ally is very, very honest and straightforward. Sometimes it is her own fault."

She also possessed a vividly hyperactive imagination, allowing Kelley to indulge to his heart's content in wild, whimsical flights of fancy — unprecedented in the annals of TV legal shows, but reminiscent of such earlier fantasy-reality hybrids as James Thurber's "The Secret Life of Walter Mitty." Envying her friend Renee's bustline, Ally envisioned herself with rapidly inflating breasts; intimidated by Elaine's legal expertise, Ally suddenly imagined her secretary's head expanding and exploding; after Ally was dumped by a boyfriend, a garbage truck appeared out of nowhere and emptied its contents upon the hapless heroine; and most famously, when the unmarried Ally began thinking about motherhood, she was haunted by the computerized image of a dancing baby. This last-named special effect was in fact fairly well known before *Ally McBeal* even hit the airwaves: created for the Internet by Michael Girard to demonstrate his new animation software product Kinetix Character Studio, the Dancing Baby — aka "Baby Cha Cha" — was chosen by *Ally McBeal*'s co-executive producer Jeffrey Kramer to "explore Ally's biological clock."

The "fantastic" elements of *Ally McBeal* also pervaded the characters' "real" lives. One of the series' setpieces was Cage, Fish & Associates' enormous unisex bathroom, the first of its kind ever seen on television. Of course, many actual workplaces *do* have single, non-gender-specific bathrooms, but the one designed for the series by Peter Politanoff was purely a product of his own imagination, totally bereft of urinals but equipped with a huge central "community sink" so everyone could gather to discuss matters of importance. (For the record, Calista Flockhart insisted that she wouldn't be "caught dead" in a bathroom of this nature!) In another surrealistic touch, the characters regularly gathered at a local bar where singer Vonda Shepherd (playing herself) had an eerily prescient habit of selecting tunes which precisely reflected Ally's current moods; and whenever Ally and Renee were dateless, a pair of nameless twins (Eric and Steven Cohen) were always available as dancing partners, almost as if the two men had taken up permanent residence in the bar.

There were those outside the legal profession who assumed that the cases tackled by Ally and her colleagues were likewise far-fetched. Not so, claimed a number of distinguished legal experts, notably Paul R. Joseph in a 2003 article for the *University of Arkansas at Little Rock Law Review*. Listing several examples from the series, Joseph argued that while they may have seemed improbable or even ridiculous, most were inspired by actual litigation. In the very first episode, Ally complained about a senior male partner grabbing her backside, whereupon he sued under the Federal Disabilities Act, claiming he suffered from a "compulsive disorder"— and won. A later episode in which a terminally ill boy decided to sue God may have come off as a frivolous treatment of a serious issue, but such things were not uncommon in genuine courtrooms; similarly, the episode wherein a child prodigy issued subpoenas to his classmates just so he'd have someone to play with was not so far removed from equally outrageous but nonetheless authentic school-related lawsuits. Another example Joseph cited as perfectly within the realm of possibility was Ally's first case with Cage/Fish, "Rev. Kessler vs. *Man Made* Magazine," which as a takeoff of the Jerry Falwell/Larry Flynt libel suit demonstrated how the law protects the Sleazy as well as the Virtuous. Overall, Joseph applauded *Ally McBeal* for its articulation of a cold, hard legal fact: you can't change the law no matter how much you want to, and sometimes it's better just to forget about "justice" and be satisfied with a huge monetary settlement.

Similar to the examples cited by Paul R. Joseph, another *Ally McBeal* episode used recent real-life litigation as a launching pad for an absurd plot development involving a female news anchor (Kate Jackson), who sued her TV station on the basis of age discrimination— then felt humiliated and rejected because none of her fans wanted to see her nude picture on the Internet. And in an episode touching upon the issue of reverse discrimination, the owners of a French restaurant, sued for exclusively hiring homosexuals, managed to emerge triumphant by pointing out that existing laws did not protect the rights of heterosexuals — and besides, their customers enjoyed being insulted by snobbish gay waiters.

While the legal profession in general appreciated the cogent points of law brought up on *Ally McBeal*, many lawyers tended to side with those media critics who complained that the series set back the cause of Feminism several years: this in fact was the central theme of *Time* magazine's June 1998 cover story on the show. It wasn't just because Ally wore revealing outfits in the courtroom (one surly judge cited her for contempt because he could see her calves); it was also due to the fact that, for all her much-vaunted intelligence and independence, she tended to behave like a scatterbrained schoolgirl whenever trouble came her way. Also, there were times that Ally and her female cohorts seemed to have absolutely nothing but men on their minds, suggesting that they defined themselves by their sexual relationships. It helped not at all when Ally spewed forth such sentiments as, "All I wanted was to be rich and successful with three great kids and a husband waiting at the end of the day to tickle my feet — and I don't even like my hair!"

Speaking for the nay-sayers was attorney Gloria Allred, who told *TV Guide*'s Hilary DeVries, "*Ally McBeal* is a comedy, and they're actors, not role models. Ally doesn't really practice much law, so I can't evaluate her as a lawyer, but most civil cases, including sexual harassment cases, never go to trial.... Ally seems to have a lot more time for her personal life than I do. And all those associates sleeping together? That's very unrealistic and could subject some of them to future sexual harassment suits.... The public loves vulnerability in women, but this is really that old *Playboy* attitude: that a strong professional woman is really just a sex object." And representing those journalists who found fault with the series' message, here's Jacqueline Marino of the *Memphis Flyer*: "It's not that I don't think a man is capable of writing a good female character. It's just that the lovely but neurotic, reproductively obsessed Ally McBeal is hardly the type of woman the thinking man of the '90s would pursue for a lifelong mate. She does, however, have the stuff male fantasies are made of. She is an insecure, pico-skirted, one-dimensional tart who looks for sex in every cappuccino. She needs a big, strong man to come to her rescue." Nor was this opinion confined to American writers: in the British tabloid *The Independent*, Andrew Gumbel felt that the series "half-seriously" suggested "the best way for women to get ahead in the legal profession is to starve themselves and act ditsy around the office."

But these dissenting voices were drowned out by critics who adored the show — at least during its first two seasons on the air. Though he carped that *Ally McBeal* strained to "create a *Seinfeld*-ian lexicon" and took David E. Kelley to task for "never knowing when to say enough," *TV Guide*'s Matt Roush described the property as, "a show and a heroine very much of our time: ironic yet confused, wacky and wistful, self-absorbed and over-stimulated, deserving of a swift kick … but also of a fierce hug." *Entertainment Weekly*'s Ken Tucker praised *Ally* as, "The only freshman show this fall that creates its own genre — its own brave new world." D. G. Cameron of the *Weekly Alibi* website noted that the series "has the fun, zappy feel of one of Fox's former cult hits, *Parker Lewis Can't Lose*, and the more you see of it, the better it works. Perhaps it's because television drama so seldom ventures away from straight-ahead realism or perhaps because the effects add humor to tense situations. Either way, it's a refreshing use of television's narrative flexibility." And *Newsweek* described *Ally McBeal* as "quintessential postfeminist" and "the 90s answer to *Mary Tyler Moore*."

But critical support for the series began to erode during its second and third seasons, when the delicate balance between legal drama and romantic comedy was offset by the added weight of too many characters and too much sexual intrigue. Season Two saw the introduction of Portia de Rossi (an Australian actress who'd actually spent a year in law school) as the icy, combative Nellie Porter, brought in by Richard Fish to improve his firm's business. Also swooping onto the scene was Lucy Liu as super-aggressive businesswoman/attorney Ling Woo, whose entrances were heralded by the "Wicked Witch of the West" theme from *The Wizard of Oz*. Nellie would soon launch a romance with John Cage, while Richard Fish would abandon Judge Cone, wattles and all, for a liaison with Ling. Meanwhile, Ally was smitten by handsome African-American physician Greg Butters (played by a pre–*Law & Order* Jesse L. Martin), who won her over by singing a birthday serenade, only to ultimately break her heart (which by this time had been broken so often that Ally would have been well advised to keep a roll of adhesive tape handy).

The third season marked the departure of Courtney Thorne-Smith, who explained, "The amount of time I spent thinking about food and being upset about my body was insane." (In all fairness, she knew the job was dangerous when she took it: *anyone* would appear chubby standing next to the emaciated Calista Flockhart.) This season was also what one series producer described as "the year of the big splashy shows," highlights of which included a sexual liaison in a car wash. Some observers felt the show had degenerated from quirkiness to salaciousness: setting up their own private practice, Renee Raddick and Judge Cone interviewed potential male employees by ordering them to strip; Elaine "helped" John Cage regain his sexual confidence by groping him in the office bathroom ("You hot little biscuit!"); and Nellie developed a spanking fetish. Even without all this hanky-panky, Season Three managed to alienate most of the fan base with its treatment of longtime regular Billy Thomas, who after years of wishy-washy wimpiness underwent a sudden and total personality change: dyeing his hair, morphing into a repulsive Male Chauvinist Pig, and entering into torrid affairs with his new assistant Sandy (Gina Philips) and dozens of other women. It turned out that Billy's strange behavior was the result of an inoperable brain tumor — a rather artless attempt by the producer to inject "emotional interest" in a series which at this point didn't have any.

The season's climax, in which Billy collapsed and died in the middle of a courtroom, has been referenced by many disgruntled ex-fans as *Ally McBeal*'s "Jump the Shark" moment.

If further proof were needed that the bloom was off the rose, we note that from September to December of 1999, Fox ran a "spin-off" sitcom titled *Ally,* consisting of reruns from *Ally McBeal*'s first two seasons, each episode edited down to 30 minutes with all the courtroom scenes removed. *Ally* was a terrific flop, scotching the producers' plans to syndicate the series in a half-hour format.

There was a glimmer of hope that things would improve during the 2000–2001 season with the addition of film favorite Robert Downey, Jr. as attorney Robert Paul, whose calming presence seemed to curb the series' excesses, at the same time providing Ally with a credible romantic interest. Alas, Robert Paul appeared in only eight episodes due to Downey's ongoing struggle with drug addiction; suffering a relapse in November of 1999, the actor was arrested on a cocaine-valium charge in December. Though producer Kelley and star Flockhart did everything in their power to save Downey from himself, they were forced to accede to Fox's demands that the actor be fired. The ensuing negative publicity completely overshadowed the rest of the season, not to mention the positive contributions of the new regulars, including Taye Diggs as dynamic young attorney (and mutual love interest for Ling and Renee) Jackson Draper, James LeGros as Elaine's steady date Mark Albert, and Anne Heche as John's latest girlfriend Melanie, who suffered from Tourette's Syndrome.

The fourth and final season was blighted by a seemingly endless parade of misfire ideas, beginning with the introduction of child actress Hayden Panettiere as 10-year-old Maddie, the product of one of Ally's eggs that had been donated to a scientific research study and inadvertently planted in another woman. Though not legally bound to do so, Ally allowed her "daughter" Melanie to move in with her, resulting in any number of pointless plot developments. At the same time, Julianne Nicholson joined the cast as the departed Robert Paul's ex-girlfriend Jenny Shaw, a hotshot attorney hired by the firm when she agreed to bring along 72,000 clients from a class-action suit. *That* was funny, but the notion of depicting Jenny as an "Ally clone" was not: once it was established that Jenny faced the same life issues as Ally, the joke was over and done with — not that the writers were aware of the fact. Other creative blunders during this season included the presence of Australian comedian Barry Humphries in his familiar "Dame Edna" female drag as a transvestite client, on whom Richard Fish developed a crush; John Cage curtailing his legal activities to organize a Mariachi band, only to be drawn back into the fold during a tough court case pitting Fish against 21-year-old "Lolita"-like prosecutor Liza Bump (Christina Ricci); the subsequent romance between Fish and Liza/Lolita, resulting in her inviting the thoroughly unscrupulous Wilson Jade (Bobby Cannavale) to join the firm; Ling Woo's appointment to the bench, leading to her own TV "judge show"; and love-starved Ally's predilection for chasing after *anything* in pants, including hunkish electrician Victor Morrison (played by singer Jon Bon Jovi).

By the time Ally decided to quit the firm and move to New York with Maddie (as the ghost of Billy bade her a tearful farewell), most of the viewers had already beaten her to the punch and shifted their attentions elsewhere. Inasmuch as *Ally McBeal* was supposed to have something to do with the Law, it was only fitting that the series was finally killed off by the Law of Diminishing Returns.

The Amazing Mr. Malone

ABC: Sept. 24, 1951–March 10, 1952. Produced by Edgar Peterson. Based on a character created by Craig Rice. Cast: Lee Tracy (John J. Malone).

The earliest prime time network dramatic series to feature a lawyer protagonist, *The Amazing Mr. Malone* boasted a long and impressive pedigree. John J. Malone, "Chicago's noisiest and most noted criminal lawyer," had been introduced in 1939 in *Eight Faces at Three,* the first mystery novel written by the prolific Craig Rice. Despite the manly moniker, Craig

Rice was actually a woman, born Georgiana Ann Randolph ("Craig" was the name of one of her husbands). Specializing in brisk, flippant detective stories, Rice not only wrote under her own pen name, but also ghosted a brace of whodunnits credited to actor George Sanders and striptease artist Gypsy Rose Lee.

Though the "stars" of *Eight Faces at Three* were amateur sleuth Jake Justis and his fiancée Helene, the couple's dour, cynical, hard-drinking attorney friend John J. Malone soon took center stage in his own literary vehicles, usually in the company of his girlfriend Dolly Dove (a professional model) and his long-suffering secretary Maggie Cassidy. Mr. Malone's Perry Mason-like fondness for solving the murders that had been wrongly charged to his clients even spilled over into the works of other writers: in 1963, the character showed up in *People vs. Withers and Malone*, a collaboration between Craig Rice and "Hildegarde Withers" creator Stuart Palmer. In films, Malone was played by Pat O'Brien in *Having Wonderful Crime* (1945), Brian Donlevy in *The Lucky Stiff* (1949), and James Whitmore in *Mrs. O'Malley and Mr. Malone* (1950).

The TV adaptation of the property was spun off from a radio series which ran on both ABC and NBC from 1947 to 1951, variously titled *Murder and Mr. Malone, The Amazing Mr. Malone, John J. Malone for the Defense* and *Attorney John J. Malone*. Frank Lovejoy initially starred in the radio version, followed by George Petrie, who later played supporting parts in the TV adaptation. Both actors smoothed out the rougher edges of the character as conceived by Craig Rice, portraying Malone as a girl-chasing quipster who drank only occasionally. This was also the approach adopted by Lee Tracy for ABC's short-lived TV incarnation of *The Amazing Mr. Malone*, which was telecast live on an alternating-week basis with another ABC legal drama, *Mister District Attorney* (q.v.)

American Justice

A&E: debuted Sept. 15, 1992. Towers Productions. Executive Producer: Jonathan Towers. Host: Bill Kurtis.

Described by its parent network A&E as "the gold standard of criminal justice programming," *American Justice* was a weekly, 60-minute documentary series, tackling important issues from the perspective of the legal system. Each episode zeroed in on a major law-and-order crisis of the 20th (and later 21st) century, with eyewitness testimony from police officers, jurists, victims, perpetrators, and assorted friends, family members and interested parties. The purpose of the series was to clear up the more complex legal ramifications of the case at hand, and to offer an explanation of the outcome that would make sense to the layman. Among the subjects covered were the murder trials of Hurricane Carter and O.J. Simpson (and the subsequent civil trial of the latter), the strange disappearance of atheist leader Madalyn Murray O'Hair, the siege at Waco, the Atlanta Child Murders, the Hillside Stranglers, The TV quiz-show scandals of the 1950s, the executions of Julius and Ethel Rosenberg, the hunt for the Unabomber, and even the messy divorce-court litigation of Donald and Ivana Trump.

Hosted by award-winning journalist and former CBS news anchor Bill Kurtis, *American Justice* featured among its interviewees such celebrity attorneys as Johnnie Cochran, Alan Dershowitz, Gerry Spence, Vincent Bugliosi and William Kunstler.

Animal Court

Animal Planet: Debuted Sept. 28, 1998; final episode taped in 2000. Andrew Solt Productions. Executive Producer: Cindy Frei. Cast: Judge Joseph A. Wapner (Himself); Rusty Burrell (Himself). Also known as *Judge Wapner's Animal Court*.

Five years after stepping down from the syndicated *The People's Court* (q.v.), 79-year-old Judge Joseph A. Wapner emerged from retirement to host this daily animal-centric reality series. *Animal Court* had been in development since early 1996, but only after the formation of the Animal Planet cable network was Wapner able to secure an outlet for his new project. The series delivered just what the title promised: with his customary evenhandedness, the venerable jurist settled disputes between pet owners, determined the actual lineage of so-called "pedigreed" pets, weighed malpractice charges against veterinarians, and so on. The series, explained Wapner, was a reaction to the fact that in an actual courtroom, animal lovers were limited to suing only for the cost of harmed pets, and not for such intangibles as "emotional distress" and "alienation of affection."

Although *Animal Court* quickly became its parent network's most popular offering, some critics wrote it off as a mocking travesty of "genuine" legal programs, citing the series' overuse of jokey episode titles like "Goat Massacre," "Jilted Jockey," "Bye Bye Birdies" and "Where Have All the Worms Gone?" Not so, countered Judge Wapner in a *Time* magazine interview: "These are very serious cases, and people get very emotional about their animals, more emotional than they do about money and people."

Also appearing on *Animal Court* was Wapner's former *People's Court* bailiff Rusty Burrell, establishing a record of appearing on more simulated courtroom programs than anyone else (he'd launched his TV career on the original *Divorce Court* [q.v.] way back in 1958). Conspicuous by his absence was longtime *People's Court* announcer-reporter Doug Llewelyn, who had gone on record dismissing *Animal Court* as "just plain stupid"—adding that Judge Wapner himself owned no pets of any kind.

The Antagonists

CBS: March 26–May 30, 1991. Universal Television Entertainment. Executive Producer: William Sackheim. Cast: David Andrews (Jack Scarlett), Lauren Holly (Kate Ward), Lisa Jane Persky (Joanie Rutledge), Brent Jennings (ADA Marvin Thompson), Matt Roth (Clark Mussinger).

The hour-long CBS drama series *The Antagonists* was a potpourri of timeworn legal-show clichés: the flamboyant defense attorney who bends the rules in pursuit of justice, the by-the-book prosecutor who demands that all "t"s be crossed and "i"s be dotted, and the love-hate relationship between the titular antagonists, who happened to be of opposite genders. Jack Scarlett (David Andrews) was a veteran Los Angeles defender, famous for his unpredictable courtroom tactics and for mouthing such pearls of wisdom as, "Guilt and innocence are entirely relative, in a legal sense," and "There's no such thing as justice." Kate Ward (Lauren Holly) was a prosecuting attorney, who though new to her job had already made her mark in the legal world with her strict adherence to the letter of the law. Although Jack and Kate were invariably opponents, he seldom missed the opportunity to offer her the benefit of his expertise, even surreptitiously dropping hints in mid-trial that would help her make the right judicial moves. In the 90-minute opening episode, Jack went so far as to subtly maneuver Kate into figuring out that his own client, an accused murderess, was guilty as charged — all the while providing his client with the best defense possible, even though he too had only recently been apprised of her guilt.

Outside the courtroom, Jack and Kate engaged in what *Entertainment Weekly* critic Ken Tucker described as "coy romantic banter," with Jack's cool cockiness playing off Kate's strait-laced sobriety in the tradition of the popular detective series *Moonlighting*. Kate of course gave as good as she got, seldom allowing Jack to emerge as the verbal victor; at one point during an argument, Jack made what he thought was a strategic retreat into the men's room — whereupon the undaunted Kate went in right after him. There were hints that the couple's relationship would mature past the talking stage, but the series wasn't on the air long enough to afford this luxury.

The supporting cast of *The Antagonists* included Lisa Jane Persky as Kate's best friend and fellow prosecutor Joanie, Brent Jennings as Kate's politically-charged boss Marvin, and Matt Roth as Clark, a dweebish law student who clerked in Jack's office. Serving as executive producer was William Sackheim, whose legal-drama credentials stretched back to the 1975 two-part, made-for-TV movie *The Law*. All of these talented people were thrown out of work when *The Antagonists* was cancelled after a scant nine episodes.

Any Day Now

Lifetime: Aug. 18, 1998–March 10, 2002. Finnegan-Pinchuk Company/ Paid Our Dues Productions/ Spelling Productions. Executive Producers: Bill Finnegan, Sheldon Pinchuk, Deborah Joy LeVine, Gary A. Randall. Title song written by Bob Hilliard and Burt Bacharach, performed by Chuck Jackson and sung by Lori Petty. Executive producer/creator: Nancy Miller. Cast: Annie Potts (Mary Elizabeth "M.E." O'Brien Sims); Lourraine Toussaint (Rene Jackson), Mae Middleton (M.E. as a child: 1998–2001); Sherri Dyon Perry (M.E. as a child: 2001–2002); Olivia Hack (Rene as a child: 1998–2001); Maya Goodwin (Rene as a child 2001–2002); Donzaleigh Abernathy (Sara Jackson); Chris Mulkey (Colliar "Shug" Sims); Dan Byrd (Colliar as a child); Olivia Friedman (Kelly Sims Williams); Calvin Devault (David Sims); Bronson Picket (Joe Lozano); Taneka Johnson (Lakesha Reynolds); Derrex Brady (Ajoni Williams).

Described as "one of the most honest and provocative series in television history" by the NAACP Image Awards' executive director Ernestine Peters, the Lifetime Channel's *Any Day Now* was originally envisioned as a 30-minute dramedy — a *Wonder Years* for girls. The show was conceived in 1990 by Nancy Miller, a native of Birmingham, Alabama, and at the time a member of the *Law & Order* (q.v.) writing pool. Though Miller had been raised in Oklahoma, she was fascinated by the racial upheavals in 1960s Birmingham, and had initially planned *Any Day Now* as a period piece, seen through the eyes of two little girls growing up in the South during the most tempestuous years of the Civil Rights movement. From the outset, Miller had intended one of the girls to be white, a younger version of herself, while the other girl would be African American — and despite the societal pressures brought to bear upon them, the two heroines were the best of friends. Since as a child Miller had been unable to understand the concept of segregation, she hired two black writers who had lived through those years to bring first-hand credibility to the project.

Gradually *Any Day Now* morphed from thirty to sixty minutes, and from a semi-comic effort to a straight dramatic offering. Also, the bulk of the series was now set in the Present, with the two grown-up protagonists periodically flashing back to their childhood. In this form, the series was optioned by CBS, who commissioned six episodes as a potential summer replacement; but Miller's production company could not afford the cost of these trial episodes, and the project was quietly dropped. Eight years later, two former CBS employees remembered *Any Day Now* and decided to revive it on Lifetime as part of that cable service's first block of original series.

Set in Birmingham (but filmed in Atlanta), the Lifetime version focused on the renewed friendship between two childhood chums, Mary Elizabeth "M.E." Sims (Annie Potts) and Rene Jackson (Lorraine Toussaint). Back in the day, M.E. had been a hoydenish youngster, the product of a rabidly racist white family from the wrong side of the tracks; Rene was the daughter of prominent black civil rights attorney James Jackson (John Lafayette), and a girl blissfully ignorant of Southern racial taboos until her family moved from Detroit to Birmingham. The girls first "met cute" when Rene spotted M.E. stealing cigarettes from a neighborhood store. Though shocked by this behavior, Rene soon became M.E.'s closest friend and vice versa, a relationship that was abruptly shattered when, at age 19, M.E. became pregnant by indolent local boy Colliar Sims (Chris Mulkey). Rene refused to be supportive of M.E. at this crucial moment, and shortly afterward left the South to become a successful and powerful Washington D.C. attorney.

The death of her father brought Rene back to Birmingham, where she found M.E. mired in a dead-end marriage with the amiable but undependable Colliar, aka "Shug," for whom she had borne three children. One of M.E.'s kids had drowned at age five; the surviving youngsters were 12-year-old Davis (Calvin Devault) and teenager Kelly (Olivia Friedman), who was every bit as rebellious and troublesome as her mom had been (she ended up getting pregnant by her boyfriend Ajoni [Derrex Brady], who happened to be black). Shedding her snobbish Eastern-seaboard boyfriend Garret (Harry J. Lennix) — the first of several "Mr. Wrongs" in her life — Rene opted to remain in Birmingham to carry on her father's work and to mend fences with M.E. So inspired was M.E. by Rene's feminist example that she cast off the shackles of dull domesticity to resume her education and become a best-selling author — and ultimately a college professor.

Though series star Annie Potts had grown up in the South, she, like creator Nancy Miller, had had little contact with racial prejudice as a child; the actress had been raised by a black woman, and attended a fully integrated Kentucky school. Also, Potts had not been one of the initial choices to play M.E.; she'd learned about the project when her husband Jim Haryman was approached to direct the *Any Day Now* pilot episode. As it happened, casting the role of M.E. was simpler than finding the "right" Rene, especially since there were quite a few African American actresses who flatly rejected the notion that a black and white girl could ever have been best friends in the segregated South. Lorraine Toussaint had no problem with this, but she initially resisted signing on to *Any Day Now* because she'd been burned by the failure of her previous series *Leaving L.A.* Once on board, however, Toussaint was one of *Any Day Now*'s biggest boosters, telling John Kieswetter of the *Cincinnati Enquirer*, "The show is not just based on entertainment, but on healing."

The toughest bit of casting was finding child actresses who could convincingly pass as the younger versions of Annie Potts and Lorraine Toussaint. In the process of expanding from 30 to 60 minutes, *Any Day Now* padded out its scripts with extended flashback sequences, filmed in sepiatone, in which incidents in the past lives of M.E. and Rene proved to have significant bearing on what was going on in the Present. (Technically speaking, these weren't really flashbacks; each episode paralleled the events of the present with those of the past, the action jumping backward and forward in time.) After extensive auditions, 10-year-old Mae Middleton of Statesboro, Georgia, was selected to play the younger M.E., while 12-year-old Shari Perry of Phenix City, Arkansas, was cast as the younger Rene. When these two girls outgrew their roles, they were respectively replaced by Olivia Hack and Maya Elise Goodwin.

The rest of the casting was along more conventional lines — especially the male characters, who, in the time-honored tradition of feminist TV dramas, tended to be clods or boors or both. Chris Mulkey in particular had the unenviable task of injecting warmth and humanity into the character of M.E.'s habitually unemployed husband Shug, even while the big doofus lounged around the house drinking beer, watching *The Jerry Springer Show*, and humiliating his daughter Kelly. One casting choice was eminently appropriate: Donzaleigh Abernathy, the daughter of Civil Rights activist Ralph Abernathy, was seen in both the "Past" and "Present" sequences as Rene Jackson's mother Sara. That Abernathy was hired on the basis of her talent rather than her name value was proven beyond question in an early episode in which the Northern–born Sara convincingly expressed shock and disbelief at the institutionalized racism in Alabama.

This bit of artistic license was typical of *Any Day Now*, which tended to stretch credibility in its flashback scenes and openly courted a hernia in its weekly efforts to establish a link between what had happened during the protagonists' childhood and what was occurring in their adult lives. Astonishingly, Rene Jackson *never* took on a legal case which didn't have something to do with racial strife or some other burning social issue — and by a mind-numbing coincidence, each one of her cases could somehow be traced back to a specific traumatic incident in her youth. In a typical episode, Rene defended a black councilman accused of corruption, as the action sporadically backtracked to the councilman's days as a fearless 1960s activist, being beaten and hosed down in the streets while demanding the right to vote. After Rene dropped her client when she realized he was guilty as charged, a brick came flying through her window, presumably thrown by a disgruntled black citizen. This Week's Lesson: though the cast of characters may have changed, the issue of race is just as problematic in 1999 as it was in 1963.

Though Lorraine Toussaint insisted that

Any Day Now was about "healing," the habitually self-righteous Rene didn't make things terribly easy for her befuddled pal M.E. (nor did the scriptwriters, who not only saddled M.E. with a lazy husband but also a bigoted uncle who was an active member of the Ku Klux Klan — thus perpetuating the Left Coast perception of a modern South still governed by an "Invisible Empire" that was virtually nonexistent by the end of the 20th century). Though willing to adjust her inherent Southern-fried values to accommodate the winds of change from the North, M.E. often took umbrage with Rene's insistence upon driving sociological points home with a power drill, notably in the episode wherein Rene mercilessly reamed out her white friend for innocently mistaking a black restaurant patron for a waitress. In moments like these, it seemed that Rene and M.E. were not so much human beings as archetypes: Rene personifying the blazing torch of racial enlightenment, M.E. as the epitome of an entrenched white culture in desperate need of re-education. In an otherwise positive review of *Any Day Now*, the *Hollywood Reporter* noted, "Much of it feels sloganlike, with fully loaded symbolism reminiscent of *Forrest Gump*'s best-remembered-moments-of-my-generation approach. Each woman represents her people, as it were, and the friendship looms with hyperimportance — symbolic of what happens, and will happen, to the post-civil-rights nation. This is too much for ordinary women to carry."

Despite its shortcomings, *Any Day Now* deserved credit for its willingness to tackle sensitive subject matter that most over-the-air network programs totally ignored. The series' Season Three finale was inspired by an actual 2001 incident in which a black child who struck a white boy for calling him "nigger" was kicked out of school, while the white child went unpunished. Written by producer Nancy Miller, the two-hour episode found Rene defending a black youth who had killed a white classmate for using the dreaded racial epithet. Her bold defense consisted of putting the N–word itself on trial, arguing that even though it is used freely by black people, the word could be re-

garded as a lethal weapon when spoken by a white person. The strategy worked, and Rene's client got off with time served and a year's probation; but the *real* drama had just gotten under way. Throughout the proceedings, M.E. was upset by the repeated usage of the N-word in the courtroom and the media, but Rene insisted that the word had no real power. In a desperate effort to prove her friend wrong, M.E. declared that Rene's sainted father was a "nigger" — which not only invoked an unpleasant experience in the girls' past (the boycotting of Rene's high-school graduation ceremony when she was named valedictorian), but also forced Rene to finally concede that yes, the word is wrong under *any* circumstances. (Annie Potts and Lorraine Toussaint later told the *Palm Beach Post* that the climax of this episode was the hardest scene they ever played together).

Lasting a respectable four seasons and 88 hour-long episodes, *Any Day Now* came to an end in 2002, not because of viewer drop-off but because costar Annie Potts wanted to spend more time with her family. In addition to several industry awards and nominations for its stars and production staff, *Any Day Now* also earned the YWCA Racial Justice Corporate Award.

Arrest and Trial

ABC: Sept. 15, 1963–Sept. 6, 1964. Revue Productions. Executive Producer: Frank P. Rosenberg. Cast: Ben Gazzara (Det. Sgt. Nick Anderson); Chuck Connors (Attorney John Egan); John Larch (Deputy D.A. Johnny Miller); John Kerr (A.D.A. Barry Pine); Roger Perry (Det. Sgt. Dan Kirby); Noah Keen (Det. Lt. Bone); Don Galloway (Mitchell Harris); Joe Higgins (Jake Shakespeare); Jo Ann Miya (Janet Okuda).

Acknowledged as one of the primary inspirations for producer Dick Wolf's *Law & Order* franchise (see separate entry), the ABC series *Arrest and Trial* was created to cash in on the popularity of NBC's first 90-minute western, *The Virginian*. Adding an extra thirty minutes' running time to the "traditional" hour-long drama series was in 1963 considered a brilliant strategy by the major networks in their struggle to dominate the ratings, attract new sponsors, and compete against the full-length theatrical films which had risen to Prime Time prominence via such weekly network packages as NBC's *Saturday Night at the Movies* and ABC's *Hollywood Special* . Thus, the 1963–64 season not only saw the premiere of the 90-minute *Arrest and Trial* but also an expanded version of the long-running *Wagon Train*, while subsequent seasons yielded the likes of *90 Bristol Court*, *Cimarron Strip* and *Name of the Game*. Unfortunately, few such efforts were able to sustain audience interest for a full hour and a half, and as a result the extended format never really caught on.

Setting *Arrest and Trial* apart from other series of its length was the fact that it was essentially two shows in one. Subtitled "The Arrest," the first 45 minutes of each episode chronicled the efforts by Detective Sergeant Nick Anderson (Ben Gazzara) to gather evidence against a suspected lawbreaker and bring the miscreant to justice. The last 45 minutes, subtitled "The Trial," found defense attorney John Egan (Chuck Connors) endeavoring to secure a "not guilty" verdict for Anderson's prisoner, or at the very least a lighter sentence. Occasionally, the "Arrest" portion would command more screen time than the "Trial" portion, or vice versa, but most often, the two-part equilibrium was sustained.

Since the capture of the (presumed) villain was a foregone conclusion in each episode, injecting even a modicum of suspense or surprise into the "Arrest" half was a tricky proposition; as a result, most of the plot twists occurred during the "Trial" half. One method of keeping viewers tuned in during the first 45 minutes was to hold out the possibility that the suspect wasn't guilty. Since the network's publicity flacks were touting Nick Anderson as an "intellectual" detective rather than the "instinctive" kind seen on most cop dramas, the writers had to work extra hard not to make Anderson come off as impulsive or incompetent as he built his case against the suspect. It was shown that Anderson did the best job he could with the evidence at hand — and if, as in the episode "Isn't It A Lovely View?" (a *Rear Window* derivation), virtually all the clues pointed in the wrong direction, it was hardly the detective's fault if he jumped to the wrong conclusion.

On those occasions in which there was no doubt as to the suspect's guilt, it was carefully established that the suspect had not confessed to attorney John Egan, and that Egan was acting in good faith as he mounted his defense; in these instances, suspense was engendered by the possibility that evil would ultimately triumph. In the episode "Journey Into Darkness" (so blatantly "borrowed" from Dostoyevski's *Crime and Punishment* that the Russian author's name was mentioned several times in the script!), an egotistical murderer blithely tagged along as Anderson investigated the crime, arrogantly confident that even if he was revealed as the culprit, he could sway a jury into believing him innocent. A few moments before the end of his trial, the killer calmly informed Egan that he was indeed guilty, secure in the knowledge that he had won over the jurors. To save Egan from ending up with egg on his face — and to stay within the boundaries of mid–1960s TV censorship — the script contrived to have the killer impulsively blurt out a courtroom confession just before the "Not Guilty" verdict was announced.

In common with its stepchild *Law & Order*, *Arrest and Trial* dealt with a number of issues that were considered controversial in the early 1960s: illegitimacy, prostitution, drug addiction, race prejudice. But unlike the story-driven *Law & Order*, *Arrest and Trial* was essentially character-driven, its episodes carefully tailored to the talents of such powerhouse guest stars as Roddy McDowall, William Shatner and Mickey Rooney, rather than dwelling upon plot intricacies or delicate legal and moral issues. Also, in contrast with *Law & Order*'s evenly balanced ensemble cast, *Arrest and Trial* emphasized its two leading players Ben Gazzara and Chuck Connors, relegating the other recurring characters to the background. (Not that these players were entirely obscure: John Kerr, cast as ADA Barry Pine, had risen to prominence in the stage and film versions of *Tea and Sympathy*, and would ultimately curtail his acting appearances to become a real rather than "reel" attorney; Don Galloway, appearing as Mitchell Harris, went on to enjoy a measure of stardom as Sgt. Ed Brown on *Ironside*; and Joe Higgins, seen as philosophical bartender Jake Shakespeare, would gain widespread recognition as a comical redneck sheriff in a series of late-1960s Dodge car commercials.)

Though he'd turned down several previous series, Ben Gazzara accepted this one primarily because the producers promised to steer clear of stereotypes and clichés: the actor was proud of the fact that at no time did he ever say to Chuck Connors' character, "When I get through with you, you shyster, you'll be back chasing ambulances!" As for Connors, he was so anxious to divest himself of his "Rifleman" image that he cut his hair and shed ten pounds. But Chuck couldn't shed nine inches from his height, and 5'7" Ben Gazzara was all too conscious of the fact that his 6'6" costar towered over him. In his autobiography, Gazzara recalled devising "a survival plan" to de-emphasize the height discrepancies: "I 'helped' the director come to the realization that the most interesting way to stage the scenes between Chuck and me was to keep us as far apart as possible. When we were required to sit near each other, at a table, for instance, I had the prop man place a thick, hard pillow under my derriere."

Like many other series of its era, *Arrest and Trial* stocked its guest casts with young actors on the verge of stardom, thereby boosting its entertainment value for future viewers. Sandy Dennis appeared as an accused murderess in "Somewhat Lower Than the Angels," Robert Duvall was seen as a mentally retarded man suspected of killing a child in "The Quality of Justice," George Segal and Katherine Ross were respectively featured in "He Ran for His Life" and "Signals of an Ancient Flame," and multitalented dwarf actor Michael Dunn (best remembered as the diabolical Dr. Loveless on *The Wild Wild West*) made what was billed as "his TV dramatic debut" in "The Revenge of the Worm."

While *Arrest and Trial* was one of the better offerings of the highly variable 1963–64 season, its 90-minute, two-part structure was regarded by some reviewers as awkward (*TV Guide*'s Cleveland Amory described the series as "more trying than arresting"). More critically, its suicide timeslot opposite NBC's *Bonanza* effectively doomed the series to a brief, 30-episode run.

Arrest and Trial

Syndicated: debuted Oct. 2, 2000. Wolf Films/ MOPO Entertainment/Studios USA Productions/ Universal Television. Executive Producers: Dick Wolf, Hank Capshaw, Robert David Port, Peter Jankowski. Host: Brian Dennehy. Narrator: Steve Zirnkilton.

Although producer Dick Wolf has cited ABC's 1963 dramatic series *Arrest and Trial* (see previous entry) as one of the principal inspirations for his own *Law & Order* (q.v.), Wolf's daily, half-hour *Arrest and Trial* was not a remake of the earlier show. Rather, the series was the producer's first (and thus far only) foray into Monday-through-Friday strip syndication — or as he himself described the project, a "dramatic reality strip."

Produced with the cooperation of several local and national law-enforcement agencies, including the Office of Investigations and the Office of Public Affairs, *Arrest and Trial* concentrated on the pursuit, capture and prosecution of real-life criminals, mostly murderers and sex offenders. Each of the series' 250 half-hour episodes made extensive use of film and video clips culled from genuine courtroom trials, shot in such varied locales as Phoenix, Las Vegas, San Diego, Los Angeles, Miami, Memphis, East St. Louis and Baltimore. This actuality footage was combined with re-enactments of the events leading up to and surrounding the trials, directed by Gerald G. Massimei and featuring many of the original police officials and prosecutors (with the occasional participation of "star" attorneys like F. Lee Bailey). These dramatized portions were filmed with old-fashioned photographic equipment so they would blend seamlessly with the "real" scenes and sustain an overall verisimilitude.

Hosting the series was actor Brian Dennehy, who had been hired at the very last moment as the result of an extensive search for a strong host with proven star appeal. To make Dennehy less an observer and more a participant in the proceedings, his filmed introductory sequences were technically composited into the "reality" footage and dramatizations. *Arrest and Trial's* unseen narrator was Steve Zirnkilton, the familiar opening voice ("In the criminal justice system, etc., etc.") for Dick Wolf's *Law & Order* franchise.

Long after completing its syndicated run, *Arrest and Trial* was picked up for cable-TV play by the Fox Reality Channel.

The Associates

ABC: Sept. 23, 1979–Apr. 17, 1980. John Charles Walter Productions/Paramount Television. Executive Producers: James L. Brooks, Stan Daniels and Ed. Weinberger. Theme song "Wall Street Blues" written by Albert Brooks and performed by Albert King [not B.B. King, as previously reported]. Cast: Wilfred Hyde-White (Emerson Marshall); Joe Regalbuto (Elliot Streeter); Alley Mills (Leslie Dunn); Martin Short (Tucker Kerwin); Shelley Smith (Sara James); Tim Thomerson (Johnny Danko).

You might have needed a magnifying glass and a pair of tweezers to figure it out, but the half-hour ABC comedy series *The Associates* started life as a sequel to the critically acclaimed legal drama *The Paper Chase* (q.v.) John Jay Osborn, Jr., author of the novel on which the film and TV adaptations of *The Paper Chase* were based, had written *The Associates* to chronicle the subsequent professional lives of the law-school students introduced in the earlier book. By the time the property had been giving a going over by the same production team responsible for *Taxi*, what had started as an essentially straight-faced story of a group of young college graduates hired as associates by a prestigious Wall Street law firm had morphed into a standard-issue ensemble sitcom, laughtrack and all.

One of the few remaining vestiges of *The Paper Chase* was the presence of a not-so-benevolent father figure to shepherd the legal novices into the real world. In place of *Paper Chase's* Professor Kingsfield, *The Associates* gave us 81-year-old Emerson Marshall, the haughty but slightly senile senior partner of the firm of Bass & Marshall. As played by British actor Wilfred Hyde-White (most fondly remembered as Col. Pickering in the film version of *My Fair Lady*), Marshall was a walking compendium of eccentricities, with occasional rays of brilliance breaking through the clouds of dotage.

The rest of the cast was a grab-bag of comedy stereotypes, given depth and breadth by a talented young cast of stars in the making. Joe Regalbuto (aka *Murphy Brown's* Frank Fontana) played Bass & Marshall's aggressive Junior Partner Elliot Streeter, whose main ambition in life was to topple Marshall from his throne and take over the firm himself. Alley Mills (latterly *Bold and the Beautiful's* Pam Douglas) was Leslie Dunn, a recent Columbia University graduate and the firm's resident bleeding-heart liberal. Shelley Smith, a popular runway model of the period, was cast as alluring, sharp-witted Boston Brahmin Sara James. And former standup comic Tim Thomerson essayed the role of skirt-chasing mailboy Johnny Danko.

By far the biggest "name" to emerge from the series was Martin Short, here playing Tucker Kerwin, a fish-out-of-water Midwesterner who had trouble fitting in with his Ivy League colleagues, but who nonetheless enjoyed a romance with Leslie Dunn. At this very early stage in his career, Short wasn't given much opportunity to display his comic versatility, though in one episode he was permitted to dash off a wicked Judy Garland imitation!

For all its sitcom trappings, *The Associates* occasionally acknowledged the more serious aspects of the legal profession, notably in the premiere episode wherein the youthful associates suffered pangs of conscience over having to follow the letter of the law by ordering the destruction of a beloved local landmark. Generally, however, the farcical aspects dominated, with such comic setpieces as a judge instructing the plaintiff in a palimony suit to sing an operatic aria in the courtroom to prove that she'd given up a promising career for the sake of her live-in lover. On the one-and-only occasion that the series' kinship with *Paper Chase* was ever acknowledged, Elliot Streeter suffered a severe (and overacted) case of stage fright when he was forced to testify in court against his mentor Professor Kingsfield—played, as ever, by John Houseman.

Aggressively promoted by ABC's publicity department in the weeks prior to its September 1979 debut, *The Associates* was more positively received by critics than by viewers (though a few commentators groused that a lot of capable performers and writers had squandered their talents on some fairly lousy jokes). Of the thirteen completed episodes, only nine were telecast before the series' cancellation — which ironically occurred just before *The Associates* earned two Golden Globe awards.

Ben Jerrod

NBC: Apr. 1–June 28, 1963. Theme music by Trefoni Rizzi. Cast: Michael Ryan (Ben Jerrod); Addison Richards (John Abbott); Jeanne Baird (Agnes Abbott); Lyle Talbot (Lt. Choates); Regina Gleason (Janet Donelli); John Napier (D.A. Dan Joplin); William Phipps (Coroner Engle); Denise Alexander (Emily Sanders); Ken Scott (Jim O'Hara); Peter Hansen (Pete Morrison); Martine Bartlett (Lil Morrison); Gerald Gordon (Sam Richardson).

The first network daytime drama to be telecast in color, the daily, half-hour *Ben Jerrod* was created by soap-opera veteran Roy Winsor (*Search for Tomorrow, Another World*) as NBC's answer to the popular CBS law-and-order serial *The Edge of Night* (q.v.). Titular hero Ben Jerrod (Michael Ryan) was a young, idealistic attorney who gave up a thriving big-city career to search for "a meaningful life" (according to the NBC ad copy) in his hometown of Indian Hill, Rhode Island. Upon his return, he set up a new practice in partnership with his friend and mentor, former judge John P. Abbott (Addison Richards), whose daughter Agnes (Jeanne Baird) was sweet on Ben.

In emulation of *The Edge of Night*, it was planned that *Ben Jerrod* would focus on one specific court case over a period of several weeks. Ben and John's first (and as it turned out, last) client was Janet Donelli (Regina Gleason), who was suspected of murdering her husband. Ensuing complications involved Janet's clandestine romances with several other men, which would seem to indicate that she was as guilty as she looked (one of her paramours, a young druggist, was charged as an accessory — only to die under mysterious circumstances himself). Nonetheless, Ben, whose none too assuring credo was "I'd rather have the truth than an acquittal,"

managed to spring his client after nearly three months' worth of surprises, setbacks and sinister subplots, no thanks to the formidable opposition of D.A. Dan Joplin (John Napier).

Introduced as the first half-hour of a new 90-minute NBC daytime block (the other entries were *The Doctors* and *You Don't Say*), *Ben Jerrod* debuted April 1, 1963 — the same day as a new medical serial called *General Hospital*, which aired in the same timeslot on ABC. Care to guess which of the two competing shows was still running in July? (Hint: it's *still* running in 2009.)

The Bennetts [*aka* The Bennett Story]

NBC: July 6, 1953–Jan. 8, 1954. Written by Ben Barrett. Cast: Don Gibson, Sam Gray (Wayne Bennett); Paula Houston (Nancy Bennett); Jack Lester (Blaney Cobb), Vi Berwick (Speedy Winters); Sam Siegel (George Konois); Beverly Younger (Meg Cobb); and Jerry Harvey, Kay Westfall, Eloise Kummer.

Originating live from the Chicago studios of WNBQ-TV, this daily, 15-minute NBC soap opera was about a small-town lawyer named Sam Bennett, who maintained a modest practice in the fictional community of Kingsport. His wife Nancy and his son Gary rounded out the Bennett family, while most of the subordinate characters were either friends, clients, or Sam's legal opponents.

Several sources list *The Bennetts* as a "comedy," and there is definitely a lighthearted tone in the few surviving kinescopes. From this fragmentary evidence, it would appear that Sam Bennett, while fundamentally honest and ethical, was not above a bit of flim-flam to weed out the truth. One episode found Sam hiring a TV actor to pose as a European art expert in an effort to prove that his client had been defrauded by a shady art dealer — a charade that was promptly exposed by Sam's own son, who had seen the actor on his favorite program the night before!

The Black Robe [*aka* Police Night Court]

NBC: May 18, 1949–March 30, 1950. Created by Phillips H. Lord. Cast: Frank Thomas Sr. (The Judge); John Green (The Clerk).

Though its rather ominous title might suggest an evening with the Spanish Inquisition, NBC's weekly, half-hour *The Black Robe* was actually a contemporary human-interest anthology, dramatizing actual cases heard before the Police Night Court of New York City. Frank Thomas Sr., no stranger to the rigors of live television, was seen as the nameless judge, with John Green as the court clerk. In the interest of authenticity, many of the actual defendants and witnesses of the re-enacted cases appeared as themselves, albeit using assumed names and disguised under heavy makeup. Also appearing were prominent public officials who solicited the viewers' help in solving crimes, righting wrongs and providing humanitarian aid for the city's less fortunate residents.

Produced by Phillips H. Lord, best known for his popular radio series *Gangbusters, The Black Robe* was unable to attract a sponsor despite a sizeable fan following, and folded after less than a year on the air.

Black Saddle

NBC: Jan. 10–Sept. 5, 1959; ABC: Oct. 2, 1959–Sept. 30, 1960. Halmac/Zane Grey Productions/Four Star Productions. Created by executive producers Hal Hudson and John McGreevey. Producer: Antony Ellis. Theme music by Herschel Burke Gilbert. Cast: Peter Breck (Clay Culhane); Russell Johnson (Marshal Gib Scott); Anna-Lisa (Nora Travers).

One of the more noteworthy cross-breedings of the "lawyer drama" genre with the Western format (see also *Dundee and the Culhane, Judge Roy Bean, Sugarfoot* and *Temple Houston*), the weekly, 30-minute *Black Saddle* was the saga of former gunfighter Clay Culhane, who had renounced violence after witnessing the deaths of his brothers in a bloody range war. Seriously wounded himself, Clay was nursed back to health by the daughter of one Judge McKinney. Taking a paternal interest in the young gunman, McKinney spent the better part of a year mentoring Culhane in the particulars of the American Justice System, and as a parting gesture presented Clay with several law books, including his personal copy of Blackstone. Carrying these precious volumes in his black saddlebags, Culhane roamed the West in hopes of persuading others to forsake gunplay and violence, and to place their confidence in proper judicial procedure. Eventually, the novice lawyer settled in the New Mexico Territory town of Latigo, where Marshal Gib Scott represented the Law. Armed with foreknowledge of Culhane's reputation as a fast and deadly shot, Scott had difficulty believing that Clay had reformed, and spent most of the series grimly awaiting the moment the ex-gunman would lapse back into his wicked ways. The animosity between Culhane and Scott was fueled by their rivalry for the affections of Nora Travers, the attractive proprietor of the Madison Hotel — which, along with the Fallen Angel Saloon, frequently doubled as a courtroom whenever Culhane landed himself a client.

Written by series cocreator John McGreevey, the pilot episode of *Black Saddle* (working title: *Lawgun*) was seen on May 23, 1958, as part of the western anthology *Zane Grey Theater*. Titled "Threat to Violence," this seminal episode featured Chris Alcaide as Clay Culhane; for the series proper, Alcaide was replaced by Peter Breck, better known to modern viewers as costar of another, longer-lasting western, *The Big Valley*. Cast as Marshal Scott was Russell Johnson, delivering a far grittier performance than one might expect from the bookish "Professor" on *Gilligan's Island*. Scandanavian actress Anna-Lisa was seen as Nora, while other frontier types appearing from time to time included Hampton Fancher as Scott's easygoing deputy Lon Gillis, J. Pat O'Malley as gimlet-eyed circuit judge Caleb Marsh, Walter Burke as eternally besotted town handyman Tim Potter, and Harry Tyler as grizzled old livery stable owner Steve Rhodes.

It would be refreshing to report that *Black Saddle* emulated its pacifistic protagonist by completely sidestepping violence in favor of legal level-headedness; but knowing what hardcore Western fans expected back in 1959, the producers usually managed to have their cake and eat it too. Example: in "Client: McQueen" (every episode included the word "Client" in the title), Culhane came to the aid of his former employer Senator McQueen (Basil Ruysdael), who had been forced to turn control of his ranch over to crooked foreman Josh Holt (Charles H. Gray). Having been expensively educated in the East, the duplicitous Holt knew just how to twist and pervert the law to suit his own purposes. Only after being defeated in the courtroom did Culhane learn from McQueen's faithful family retainer (Rex Ingram) of a loophole Holt had overlooked; and in the final scene, Clay confronted the wily foreman and publicly exposed him as a cad and bounder. For all intents and purposes, the story was over — but the humiliated Holt had to have his revenge, so he whipped out a hitherto hidden six-gun and prepared to plug Culhane in the back, whereupon our hero calmly shot the villain down in self-defense. Man does not live by Blackstone alone.

Dropped by NBC after a single season, *Black Saddle* was promptly picked up by ABC for additional year's worth of episodes. In 1965, the series resurfaced in rerun form, again as part of an anthology. This time around, *Black Saddle* was a component of *The Westerners*, a syndicated package which also included episodes from such short-lived oaters as *Law of the Plainsman*, *Johnny Ringo* and *The Westerner*.

The Blame Game

MTV: debuted February 8, 1999; in production until 2000 for a total of 130 episodes. Buccieri and Weiss Entertainment/Zoo Productions. Executive Producers: Robert Weiss, Paul Buccieri, Barry Poznick. Cast: Chris Reed (The Judge); Kara McNamara, Jason Winer (Counselors); Adam Zuvich (Bailiff); Richard "Humpty" Vission (Deejay).

The MTV reality/game show *The Blame Game* used the "courtroom" format to determine which member of a formerly romantic couple had been responsible for the breakup. Delivering testimony before zany "Judge" Chris Reed (actually an actor), the couples were represented by counselors Kara McNamara (speaking for the women) and Jason Winer (for the men). Each litigant was given 90 seconds to state his or her case during the opening "Tick Tock Testimony" round. This was followed by the "You Did It, Now Admit It" round wherein embarrassing secrets (not all of them true) were

dredged up for public consumption. Round Three was given over to cross-examination by Kara and Jason, and the fourth and final round had the litigants appear separately in chambers — specifically, the "Karaoke Chamber" — to perform a popular song that encapsulated their fractured relationship (This is the only legal series within living memory to feature an in-house deejay).

Determining the winners and losers of each case were members of the studio audience, rendering their verdict via an interactive keypad system developed by MERIDIA® Audience Response. The winning party would earn a fabulous vacation in Cancun or some other exotic locale; the loser would be forced to apologize on bended knee in a nationally distributed magazine ad, gloatingly photographed by his or her "ex." Of course, the litigants *were* given the option to kiss and make up, but what fun would that have been?

Sort of a *Divorce Court* for the Uncommitted, the daily, 30-minute *Blame Game* was created by Dr. Stuart Atkins (not the diet guru, but a prominent Beverly Hills psychologist), who in a *Washington Post* interview insisted that for all the series' heaping-on of humiliation, the litigants themselves seldom felt humiliated. "They take their relationships so lightly anyway, they just move on to something else." Of course, having one's mug shot plastered all over the pages of *Entertainment Weekly* with the caption "DO NOT DATE THIS BLAME GAME LOSER" might conceivably put a damper on one's future love life.

The Bold Ones: The Lawyers

NBC: Sept. 21, 1969–Aug. 20, 1972. Universal Television. Produced by Roy Huggins. Cast: Burl Ives (Walter Nichols); Joseph Campanella (Brian Darrell); James Farentino (Neil Darrell).

THE BOLD ONES: THE PROTECTORS [AKA THE LAW ENFORCERS]

NBC: Sept. 28, 1969–Sept. 6, 1970. Produced by Jack Laird. Cast: Leslie Nielsen (Sam Danforth); Hari Rhodes (William Washburn).

In its time regarded as an innovation, *The Bold Ones* was the umbrella title for several rotating dramatic series, all produced by Universal Television and telecast by NBC in the same weekly, hour-long timeslot. By scheduling *The Bold Ones*' individual series on an "every-third-week" basis, the studio and network executives hoped to upgrade production values while simultaneously keeping the budgets under control; theoretically, the fewer episodes produced of each component, the higher the overall quality. Each of the three series boasted a separate premise, producer and cast of characters, and each was spun off from a two-hour NBC TV movie — leading some industry wiseacres to dismiss *The Bold Ones* as "three pilots in search of a series."

Components during the inaugural 1969–70 season were *The Doctors*, a medical drama starring E.G. Marshall, David Hartman and John Saxon; and two programs dealing with legal matters, *The Lawyers* and *The Protectors* (also known as *The Law Enforcers*). For the purposes of this book, we will analyze only the latter two offerings.

The Lawyers was preceded on March 6, 1969, by the made-for-TV movie *The Whole World is Watching*. This pilot film dealt with the efforts by the law firm of Nichols, Darrell & Darrell to defend student radical Gil Bennett (Rick Ely) on a charge of murdering a police officer (Steve Ihnat) during a campus demonstration. Burl Ives headed the cast as senior law partner Walter Nichols, a wily old campaigner who had accepted the politically volatile case against his will, but was ultimately persuaded of his client's innocence. (The villain of the piece turned out to be neither the shaggy-haired radical nor the buzz-cut cop, but a glory-grabbing fanatic who was willing to throw Bennett under the bus to create a martyr for "the movement.") Ives' earthy, truculent traditionalism played against the youthful progressivism of his dapper, erudite junior partner Brian Darrell (played by Joseph Campanella) and Darrell's hip, swinging brother Walter, played by James Farentino. (*The Bold Ones* it may have been, but no one was taking any risks with the casting choices!)

To avoid comparisons with *Perry Mason* and *The Defenders* (both q.v.), producer Roy Huggins assured viewers that the subsequent series version of *The Whole World Is Watching* would be neither a weekly whoddunit nor exclusively issue-oriented, though the movie version certainly seemed to lean in both of these directions. As *The Lawyers* developed, the storylines took second priority to characterization, with Burl Ives' Walter Nichols emerging as the predominant personality (which should surprise no one who remembers the actor's scenery-chewing tirades in *Cat on a Hot Tin Roof* and *Desire Under the Elms*). Indeed, the show ended up being so tailored to Ives' talents that the actor was permitted to write much of his own dialogue, including more than one climactic courtroom summation — even though Ives' "legal expertise" stretched back no further than his portrayal of an alcoholic ex-judge in *Let No Man Write My Epitaph* (1960).

The Protectors likewise grew fully formed from a TV movie, *Deadlock*, first telecast on February 22, 1969. Much of the film's running time was given over to a battle of wills between conservative, rule-bound Deputy Police Chief Sam Danforth, played by Leslie Nielsen, and liberal, humanistic District Attorney William Washburn, played by Hari Rhodes (the first African American to appear in a starring role on a legal-oriented series). Faced with the prospect of a full-scale ghetto riot after a black youth was killed by a white cop and an equally destructive white backlash precipitated by the murder of a Caucasian reporter, Danforth and Washburn were forced to lay aside their mutual animosity — the DA regarded the cop as a closet racist, the cop felt that the DA was only interested in political gain — and work together as team to save their (unnamed) city from going up in flames. Though the subsequent *Protectors* series was as much a cop drama as a lawyer show, the show carried over the film's emphasis on the "rock and hard place" position of DA Washburn, who not only faced hostility from the white community, but also charges from the black population that he had sold out to "The Man."

Avoiding the trap of limiting its story material exclusively to the race issue, *The Protectors* focused on several potent legal issues relevant to the late 1960s. One episode, for example, served up a thinly disguised version of the Caryl Chessman case in order to dissect the issue of capital punishment, which was then on the verge of being abolished in the U.S. In contrast with Burl Ives' dominance of *The Lawyers*, neither Leslie Nielsen nor Hari Rhodes was given preference over the other in terms of screen time or dialogue; though nominally the star, Nielsen expressed pride in the fact that he and Rhodes were given equal footing. For his part, Rhodes welcomed the opportunity to play a black power broker who had fought his way up from the ghetto and who wielded his political clout with fairness toward all. He was also pleased that the producers had allowed him to retain his Afro hairstyle: "It'll give the kids on the block something to talk about — the DA with a natural."

Though the *Lawyers* portion of *The Bold Ones* was renewed for two more seasons, *The Protectors* was replaced in the fall of 1970 by *The Senator*, a political drama starring Hal Holbrook as a Bobby Kennedy type. *The Senator* was cancelled the following season, followed by *The Lawyers* in September of 1972. Ultimately, the *Bold Ones* "umbrella" concept was discarded, leaving *The Doctors* to soldier on as a single weekly entry until January 9, 1973.

Boston Legal

ABC: debuted Oct. 3, 2004. David E. Kelley Productions/Dick Clark Productions/20th Century–Fox Television. Executive Producer/Creator: David E. Kelley. Cast: James Spader (Alan Shore); William Shatner (Denny Crane); Candice Bergen (Shirley Schmidt); Rene Auberjonois (Paul Lewiston); Rhona Mitra (Tara Wilson); Lake Bell (Sally Heep); Mark Valley (Brad Chase); Monica Potter (Lori Colson); Ryan Michelle Bathe (Sara Holt); Betty White (Catherine Piper); Julie Bowen (Denise Bauer); Justin Mentell (Garrett Wells); Constance Zimmer (Claire Simms); Craig Bierko (Jeffrey Coho); Gary Anthony Williams (Clarence Bell); John Larroquette (Carl Sack); Christian Clemenson (Jerry Espenson); Tara Summers (Katie Lloyd); Saffron Burrows (Lorraine Weller); Taraji P. Henson (Whitney Rome).

David E. Kelley's heavily hyped sequel to his long-running legal drama *The Practice* (q.v.), ABC's *Boston Legal* (working title: *Fleet Street*) picked up exactly where its predecessor left off.

In the final episodes of *The Practice*, a story arc had been introduced involving Alan Shore (James Spader), the least scrupulous and most ethically challenged member of the Boston law firm Young, Frutt & Berluti. The bottom-feeding Shore had just been dismissed from the firm with a $15,000 severance package, prompting him to sue on the grounds that he had brought at least $6,000,000 in new business to his former colleagues and deserved a much larger piece of the pie. Preparing to act as his own attorney, he temporarily joined the rival firm of Crane, Poole & Schmidt, whose wacked-out senior partner Denny Crane (William Shatner) became fascinated with Shore's pit-bull courtroom tactics. Though he won a huge settlement against Young, Frutt & Berlutti, Shore was blocked from rejoining the firm, whereupon Denny Crane extended him the invitation to sign with Crane, Poole & Schmidt on a permanent basis. Coming along for the ride was paralegal Tara Wilson (Rhona Mitra), who had likewise been booted from Young, Frutt & Berlutti for helping Shore steal valuable documents which he'd planned to use against his ex-partners.

When asked by *The Hollywood Reporter* what he had in mind with *Boston Legal*, David E. Kelley would only reveal that the new show would be "lighter in tone" than *The Practice*, and that he looked forward to being "given the opportunity to evolve and go on." If *Boston Legal* represented Evolution, Charles Darwin would certainly have been surprised.

No sooner had he joined forces with Denny Crane than Alan Shore was up to his old dirty tricks, arranging for an opposing counsel to meet with a prostitute and then blackmailing the poor sap with compromising photos. He also demonstrated his genius for achieving positive results with the most questionable of tactics: hired to save a disgruntled ex-girlfriend who'd tried to murder him from being committed to a mental institution, Shore successfully argued that his very presence as the woman's lawyer was positive proof of her sanity!

Alan Shore may have been a reprehensible human being, but he was in full command of his faculties and always knew exactly what he was doing — and as a committed Liberal, he had genuine empathy for the benighted souls who had been dealt a bad hand by the System. There was, however, some doubt as to the competency, mental or otherwise, of Shore's new partner Denny Crane, the sort of blowhard buffoon who made life tough for other blowhard buffoons. Coasting on the reputation of having never lost a case, the superannuated Crane refused to do any pretrial preparation, often merely giving a cursory glance at piles and piles of research notes and muttering, "Thick file, thick file." His first impulse upon meeting anyone of the opposite sex was to leer lecherously, whisper a few lewd suggestions and shamelessly fondle the unfortunate lady. A reactionary Republican, he carried a gun with him at all times and was prone to such non–PC exclamations as, "You Democrats! Protesting wars, banning guns. If you nancies had your way, nobody would ever shoot anyone, and then where would we be?" Once in the courtroom, Crane's train of thought derailed so often that he appeared to be on the brink of Alzheimer's (which indeed he was). For all that, Crane was given to extemporaneous bursts of brilliance that frequently outshone his partner: in one episode, he won a deposition by convincing his opponent's client that her lawyer wasn't acting in her best interests, and that she would be better off settling immediately.

If versatile movie villain James Spader can be said to have been perfectly typecast as the slimy Alan Shore, it can be argued that the irrepressible William Shatner was born to play Denny Crane. The best way to sum up his Emmy–winning performance is to quote one of his costars, who compared acting opposite Shatner to "working with horses, dogs, and babies all at once. You can't take your eyes off him."

Believe it or not, there were other actors besides Spader and Shatner appearing in *Boston Legal*'s first season. In addition to the aforementioned Rhona Mitra as Tara Wilson, Mark Valley costarred as associate (and later partner) Brad Chase, a former Marine who'd been brought into Crane, Poole & Schmidt by one-

time senior partner Edwin Poole (Larry Miller) to keep the other members in line; Lake Bell was seen as inexperienced associate Sally Heep, whose courtroom demeanor was almost as unkempt as her wardrobe and who apparently kept her job only because she was sleeping with Alan Shore; and Monica Potter joined the cast a few episodes into Season One as junior partner Lori Colson, who eventually became so fed up with Denny Crane's loony, lascivious behavior that she lodged an official complaint against him, just before quitting the firm cold.

Originally John Michael Higgins was cast as senior partner Jerry Austin, but he was quickly replaced by Rene Auberjonois as Paul Lewiston, whose solemn, straightforward approach to his work — not to mention his talent for damage control — was supposed to "ground" a series that was dominated by unpredictable eccentrics. When it became obvious that not even Lewiston could save the firm from descending into chaos, founding partner Shirley Schmidt (Candice Bergen) returned to the fold to restore a semblance of order, blowing off Shore's tentative sexual overtures, terrifying her ex-lover Crane with her razor-sharp intellect, and in general running roughshod over all the principal characters — especially Sally Heep, who was unceremoniously fired shortly after being introduced to the redoubtable Ms. Schmidt. That audiences welcomed the presence of a grownup amidst the overgrown delinquents of *Boston Legal* was proven when ratings shot up 19 percent after Candice Bergen joined the show.

Despite its inherent outrageousness and a handful of plotlines that stretched the limits of believability, the first season of *Boston Legal* was more grounded in reality than one might expect. In a 2004 article for CNN.com, legal expert Julie Hilden had this to say: "In looking at how lawyers are portrayed on television, one of the most interesting new shows is *Boston Legal* in which the lawyer hero has given way to the anti-hero.... Yet the show doesn't so much advocate ethical breaches, as it adulates the magic of courtroom oratory and 'out of the box' thinking. The show recognizes that trial judges have immense discretion, under the law, to do as they see fit. And it reveals the power of those who can convince and persuade in the courtroom by speaking well, and pushing the right buttons for judges and juries — despite the fact that their preparation, their papers, and even the legal basis for their case may be shoddy. In addition, the show quite realistically indicates that the obligation of 'zealousness' on a client's behalf can take a lawyer right up to the ethical brink (or, in the case of Shore, far beyond it).... The underlying message of the show, then, seems to be this: Lawyers' work goes far beyond reading voluminous case files, or hewing to the letter of the law. Indeed, creativity — the ability to go beyond — can be the difference between winning and losing." Hilden regarded this as a "good message," concluding, "If there does end up being a *Boston Legal* effect on its audience, let's hope that the effect is to prod real-life lawyers to be more creative."

Fielding complaints from viewers who were angry that the series' first season was cut short so ABC could launch the medical drama *Grey's Anatomy* in the *Boston Legal* timeslot, star James Spader explained that production had shut down prematurely because the cast was "exhausted," promising to make up for the shortage of Season One episodes with five additional hour-long installments during Season Two. Spader did not mention that producer Kelley, secure in the knowledge that *Boston Legal* boasted a solid fan base, was formulating plans to transport his series to a sort of judicial Twilight Zone, where anything could happen and frequently did. To quote Rob Owen of the *Pittsburgh Post-Gazette*, "After a first year that felt tentative and bloodless, Kelley and Co. have found the right tone for *Boston Legal*, abandoning all sense of reality for what is often flat-out comedy, with an added pinch of political irreverence."

With both Lake Bell and Monica Potter out of the picture as *Boston Legal* entered its second season, several new actors were added to the cast. The most "normal" of the newcomers was Julie Bowen as Denise Bauer, a senior associate with partnership aspirations who was going through a painful divorce. Denise would later have an affair with colleague Brad Chase

and bear his child, going on an extended maternity leave from which, as of this writing, she has yet to return (though Brown was still officially listed as a regular as late as 2008). Outside of Denise Bauer, it was strictly Silly Season so far as the other new regulars were concerned. Ryan Michelle Bathe and Justin Mentell were respectively cast as Denise's young "go-fers" Sara Holt and Garett Wells, who in a combined effort to prove their loyalty to their boss attempted to trap the attorney of Denise's ex-husband in a compromising position — and never mind that the attorney was also a priest! Even as Sara was seducing the prelate-prosecutor, Garrett was finding an outlet for his own sexual impulses by sleeping with a rival law firm's paralegal in a vacant filing room. Though both Sara and Garrett were gone by the end of Season Two, rest assured that they weren't forgotten.

The zaniest of the new characters was former *Golden Girls* regular Betty White as elderly Catherine Piper, who applied for the position of Alan Shore's secretary even though she'd despised Alan ever since he was an obnoxious nine-year-old neighbor kid. Catherine left her mark on the firm first by killing a particularly troublesome client, then by pulling off a series of supermarket robberies — but all was forgiven when her coworkers realized that her antisocial behavior was merely a bid for attention. Ultimately Catherine was kept on the payroll to deliver sandwiches and cakes to the lawyers, though personally this writer would hesitate to eat anything from Catherine's kitchen without a lab analysis. While Catherine Piper might have been a bit light in the belfry, she was Sanity Personified compared to the trigger-happy Denny Crane, who by the end of the second season had managed to shoot four people — all justifiably, with the possible exception of his therapist, whom he plugged twice. (Only on *Boston Legal* did the lawyers spend as much time in jail as their clients.)

Joining the cast for Season Three were a brace of token "normals": prickly defense attorney Jeffrey Coho (Craig Bierko) and enigmatic associate Claire Simms (Constance Zimmer), whose favorite phrase "ick and double ick" was most frequently applied to the Roman Fingers and Russian Hands of her boss Denny Crane. Then there was Clarice's secretary Clarence Bell (Gary Anthony Williams), alias Clarice Bell, alias Cleavant Bell, alias Oprah Bell — a multiple-personality transvestite who unexpectedly ended up as his lovely employer's romantic interest. And let's not forget attorney Jerry "Hands" Espenson (Christian Clemenson), introduced as a one-shot character during the third season but added to the regulars the following year. Suffering from Asperger Syndrome, Espenson found it impossible to relate to other people on a normal level, or to control his bizarre hand gestures and unexpected shouts of "Whoop! Whoop!" in the courtroom; but Jerry's genius for interpreting the finer points of the law placed him in the unique position of being network television's first Lawyer Savant.

Future generations stumbling upon random episodes from *Boston Legal*'s second and third seasons will likely cluck their tongues, shake their heads condescendingly and ask, "Did people really *believe* this stuff?" The answer, of course, is that they didn't — nor did David E. Kelley expect them to. In *Boston Legal*, the producer totally gave in to the sort of surrealism that he'd only experimented with in his earlier *Ally McBeal*, deliberately creating a world which so closely resembled a Tex Avery cartoon that one half-expected Denny Crane's jaw to drop to the ground and his eyes to bulge out of their sockets whenever he met a pretty girl. Likewise in the cartoon tradition, Kelley loved to break down the Fourth Wall with self-referential dialogue indicating that the characters knew full well that they were on a weekly TV show. One character might ask another, "Where have you been during this episode?," while another would describe his client as "next week's guest star." And at one crucial moment, Denny Crane turned to his colleagues and griped, "I'm tired of my Alzheimer's being a story point."

Not only did *Boston Legal* make constant references to itself, but in one classic instance the series went so far as to conjure up memories of William Shatner's professional past. In the third-season episode "Son of the Defender," the son of a man who had been murdered back in 1957 burst into the offices of Crane, Poole & Schmidt wired with explosives, demanding a re-enactment of the trial in which Denny Crane had won an acquittal for the murderer. At this point, we flashed back to a kinescope of the 1957 *Studio One* drama "The Defender," in which the father-and-son attorneys were played by Ralph Bellamy and William Shatner (for more on this vintage drama, see the entry for the TV series *The Defenders*). As the action alternated between the full-color "present" and the black-and-white "past," we were informed that the 1957 case was the first ever won by Denny, who used a somewhat underhanded "reasonable doubt" strategy that earned him the harsh disapproval of his father — resulting in Denny being disowned by dear old Dad! Of course, "Son of the Defender" took considerable liberties with its *Studio One* source material in order to accommodate the contemporary framework; during the "flashbacks," James Keane replaced Steve McQueen, the actor originally cast as the defendant — who in the rewriting process was transformed into a homosexual (the defendant, not McQueen). But what could have been a mere showoff stunt paid off in episode's climax, wherein the normally conscienceless Denny Crane was forced to own up to the misdeeds of his past.

For all its cheeky irreverence, *Boston Legal* seldom kidded around when it touched upon hot-button issues; as witness his previous series *Picket Fences* (q.v.), Kelley was skilled in the art of luring viewers into his parlor with his off-balance sense of humor, then hitting them squarely between the eyes with controversy before they had a chance to escape. Cases presented on *Boston Legal* involved a teenage AIDs victim suing her school for teaching an abstinence-only policy; an image-conscious private school refusing to enroll a child prodigy because he had no "smile" facial muscles; a church-controlled hospital that wouldn't administer emergency birth control to a rape victim on religious grounds; a teacher sued for censoring a controversial cable news channel on the school's TV; a sex researcher subjected to repeated arrests

52 Boston Legal

on charges of perversion; a Jewish man taking his Christian coworkers to task for holding Bible meetings at the workplace; a Sudanese native charging the U.S. government with negligence for failing to come to his country's aid; a black girl denied the right to audition for the lead in the musical *Annie*; a Democratic congressman sued by his Republican constituents for failing to keep his campaign promises; Christians vs. Wiccans over the celebration of Halloween; a doctor accused of euthanizing patients in the wake of Hurricane Katrina; and Denny Crane facing a crisis of political conscience when he sued Homeland Security for placing him on the "No Fly" list.

Unlike other TV producers and personalities who espouse a "fair and balanced" policy when dramatizing contemporary issues, David E. Kelley made no secret of his Liberal bias. Any character on *Boston Legal* with a right-of-center political outlook was made to look like a boob or a bigot, while the scripts took arbitrary potshots at such targets as George W. Bush, Trent Lott, Tom DeLay, Glenn Beck and the Fox News Channel with a reckless abandon not seen since the days of *The Smothers Brothers Comedy Hour*. Not surprisingly, Kelley's unapologetic pro–Liberal leanings won him as many enemies as friends, but even those who disagreed with him found his honesty to be a refreshing contrast to TV's overabundance of ideological fence-sitters who seemed too timid to take *any* sort of political stand for fear of alienating this or that demographic group.

And speaking of demographics, who but David E. Kelley would be bold enough to virtually ignore the all-powerful 18-to-49 year old viewer base in favor of mature TV fans who would never see 50 again? As noted in 2006 by the *Pittsburgh Post-Gazette*'s Rob Owen, "Most prime-time series prefer to focus on youthful characters in hopes of attracting viewers from the advertiser-coveted 18–49 age group. But, instead, Mr. Kelley gives the most screen time to older characters, played by TV veterans.... [T]he focus is on stories involving the show's more mature performers, including Candice Bergen, 59, and Rene Auberjonois, 65. That

may hurt the show's demographic ratings — the No. 39-ranked prime-time series season-to-date in households, *Boston Legal* ranks No. 56 among viewers 18–49 — but it has preserved the integrity of Mr. Kelley's vision, which, *Ally McBeal* notwithstanding, has never been dependent on pretty young things."

Owen went on to quote Robert Thompson, director of the Center for the Study of Popular Television at Syracuse University, who regarded *Boston Legal* as a better show than in seasons past because Kelley had abandoned both political balance and "youth worship." "He's said, 'Forget it, I'm going to go straight for the audience that's obviously watching and don't bother with 18-to-29-year-olds,'" observed Thompson. "He's done the same thing in regards to politics. *Boston Legal* has really got some juice and energy to it. It's got a point of view. It doesn't look like it's trying to follow the network rules of diversifying the portfolio."

The cast changes made in the series' fourth season reflected Kelley's loyalty to his older fans. It would have been quite easy to replace departing regulars Rene Auberjonois, Mark Valley, Julie Bowen, Constance Zimmer and Craig Bierko with a batch of faceless twentysomethings merely to woo younger viewers. But the only newcomer who could lay claim to being under 30 was British actress Tara Summers, cast as recent law school grad Katie Lloyd. Otherwise, the new regulars were seasoned veterans: 36-year-old Saffron Burrows as Lorraine Wilder, worldly defense attorney and ex-lover of Alan Shore; and 37-year-old Taraji P. Henson as Whitney Rome, an unorthodox New York City litigator. And in anticipation of Season Four, Kelley reserved his biggest publicity blitz for the actor cast as Chase, Poole & Schmitt's new senior partner Carl Sack: 60-year-old John Larroquette, who already had two popular legal series (*Night Court* and *McBride*) under his belt before he ever set foot on a *Boston Legal* soundstage.

Defiantly marching to the beat of its own drummer, *Boston Legal* not only survived innumerable premature predictions of its imminent demise, but also accumulated dozens of industry awards, including the Emmy, the Peabody and the Golden Globe.

Burden of Proof

CNN: Oct. 1995–Dec. 2001. Cast: Greta van Susteren (Herself); Roger Cossack (Himself).

An outgrowth of CNN's wall-to-wall coverage of the O.J. Simpson trial, the network's first daily legal program *Burden of Proof* enjoyed what *Variety* described as a "modest success"—and several industry awards—during its six-year run. Per its title, the half-hour series objectively examined various contemporary legal issues, deftly translating "lawyer talk" into simple layman's terms. Representative episode titles included "Murder in Greenwich," "Bidding for Ballots: Democracy on the Block" and "Constitutional Rights for Victims." CNN legal analyst Greta Van Susteren, whose near-nonstop commentary throughout the Simpson proceedings earned her the celebrity status that ultimately led to stardom on the rival Fox News Channel, cohosted *Burden of Proof* with Roger Cossack, himself a distinguished legal scholar who subsequently brought his own special brand of expertise to ESPN's coverage of the courtroom entanglements of pro athletes Michael Vick and Barry Bonds.

Cain's Hundred

NBC: Sept. 19, 1961–Sept. 11, 1962. Vandas Productions/MGM Television. Executive Producer: Paul Monash. Produced by Charles Russell. Music by Jerry Goldsmith. Cast: Mark Richman (Nicholas "Nick" Cain).

The 60-minute NBC drama series *Cain's Hundred* is included in this book because its protagonist was an attorney and its episodes dealt with the dispensation of justice. That said, it must be noted that *Cain's Hundred* had more in common with *The Untouchables* than, say, *Perry Mason*.

Mark Richman (latterly billed as Peter Mark Richman) starred as Nicholas "Nick" Cain, a criminal lawyer who had spent many years manipulating the justice system to keep his gangland clients out of prison. The constant exposure to the seamier aspects of organized crime (are there any that *aren't* seamy?) and an overall disgust with the sort of people he was forced to associate with caused Cain to have an epiphany, whereupon he chucked his private practice to become a Federal agent. It was now Cain's sworn duty to track down and prosecute the 100 men whom he (and the Government) regarded as the most powerful movers and shakers of America's Underworld — herein obliquely referred to as "The Organization." Cain and his team of special agents knew they had their work cut out for them: there wasn't much that a subpoena or arrest warrant could do to counteract the physical and psychological tactics used by The Organization to terrorize potential witnesses into silence. Almost without fail, however, each episode ended with at least one courageous crime victim or disgruntled mob functionary putting his or her life on the line to check another of "Cain's Hundred" off the list. (The series was usually realistic enough not to show the targeted criminal actually behind bars; Cain was satisfied just to get an indictment from the Federal Grand Jury.)

The individual episodes unfolded in a semi-documentary fashion, with many of the criminals and crimes based on actual people and events, albeit with names and places changed to protect the NBC legal department. The guest-star roster included such dependables as Martin Gabel, Ed Begley Sr., Sam Jaffe, Edward Andrews, Herschel Bernardi, Robert Culp, Telly Savalas, Fritz Weaver, Walter Slezak, David Janssen, Cloris Leachman and Ricardo Montalban, along with the usual quota of stars-to-be like Robert Duvall, Barbara Eden, Robert Vaughn, James Coburn and Robert Blake. And in a fascinating break from the early-1960s tendency to stock all TV series with predominately white faces, the episode "Blues for a Junkman" provided meaty roles for Dorothy Dandridge and Ivan Dixon.

Cain's Hundred was similar in style and content to another 1961 "gangbuster" series, ABC's *Target: The Corruptors*. Thanks to stiff competition on rival networks, neither of these worthwhile series survived past a single season. *Cain's Hundred* ended its run with only 29 episodes in its manifest, leaving 71 known criminals still roaming about at will.

Canterbury's Law

Fox: debuted March 10, 2008. Apostle Productions/Canterbury Productions/Sony Pictures Television. Executive Producers: Denis Leary, Jim Serpico, Walon Green and Mike Figgis. Created by Dave Erickson. Cast: Julianna Margulies (Elizabeth Canterbury); Ben Shenkman (Russell Cross); Keith Robertson (Chester Grant); Trieste Dunn (Molly McConnell); Aidan Quinn (Matt Feury).

A cursory glance at the finished product might suggest that the Fox Network legal drama *Canterbury's Law* had been conceived from the outset as a vehicle for former *E.R.* costar Julianna Margulies. Instead, the first drafts of the pilot script featured a *male* protagonist named Oscar; the gender switch occurred only after it was decided that network television had far too many male lawyers already. Accordingly, "Oscar" was transformed into Elizabeth Canterbury, a "sexy and savage" defense lawyer operating out of Providence, Rhode Island, who specialized in defending controversial and unpopular clients, usually those wrongly accused of pedophilia or adolescent murders.

Fox insiders described *Canterbury's Law* as a courtroom variation on the cable-TV cop series *The Closer,* in that the leading character was a tough, demanding woman whose single-minded dedication to her career left her precious little time to deal with her own personal problems and character flaws. It could not be denied that Elizabeth Canterbury had more private demons than you could shake a subpoena at. She bullied and browbeat everyone, from distinguished judges to underpaid security guards. She frequently worked out her own troubles through her clients, who suffered in the process. She was unable to expunge her guilt over the mysterious disappearance of her own son during one of the few moments she wasn't paying attention to him. She cheated shamelessly on her feckless husband Matt, a law professor and former Assistant District Attorney. And she was forever on the brink of relapsing into the alcoholism that had nearly destroyed her career on previous occasions.

Julianna Margulies first got wind of *Canterbury's Law* when she ran into the series' executive producer, actor Denis Leary, at their dentist's office. Taking one look at the script cover, the actress agreed to star as Elizabeth Canterbury — thanks to a typographical error which suggested that the series would be telecast not by the over-the-air Fox Network, but by its sister cable service FX, which had built its reputation on such bold, edgy projects as *The Shield, Nip/Tuck,* and Denis Leary's own *Rescue Me.* Even after overcoming the disappointment of being relegated to the comparatively less innovative Fox hookup, Margulies was drawn to the fact that Elizabeth was unapologetically unlikable, a departure from her more sympathetic characterizations of *ER*'s long-suffering Nurse Carol Hathaway and *The Sopranos*' self-destructive real estate agent Juliana Skiff. To prepare for the role, the actress sat in on actual juries and followed a Federal prosecutor on his daily rounds.

Though the pilot of *Canterbury's Law* (directed by Mike Figgis, who promptly joined the production staff) was completed in early 2007, the series didn't make it to the Fox nighttime schedule until March of 2008, thanks to delays brought about by a Hollywood writer's strike and Julianna Margulies' pregnancy (she refused to have her condition written into the script and preferred to wait until after delivering to resume shooting). During this hiatus, much more than the leading character's sex was changed. Linus Roache had originally been cast as Elizabeth's cuckolded husband Matthew Canterbury, but by the time the series resumed production in August of 2007, Roache had accepted a bigger role as ADA Michael Cutter on NBC's *Law & Order* (conversely, one of that series' producers, Walon Green, was now on the *Canterbury's Law* payroll). At the personal request of Margulies, Aidan Quinn was brought in to replace Roache; the character's name was altered to Matt Feury, presumably to point up the schism that divided Elizabeth and her husband. Also, Jocko Sims had initially been cast as Chester Fields, a congressman's son who had renounced his dad's politics to join Elizabeth's legal team. But when *Canterbury's Law* finally reached the series stage, the character name had been altered to Chester Grant, and Sims had been succeeded by Keith Robertson.

Managing to remain with *Canterbury's Law* from its inception were Ben Shenkman as Elizabeth's partner Russell Cross, who'd been passed over for the office of DA after crossing swords with his corrupt boss, and Trieste Dunn as lawyer Molly McConnell, one of the few people who'd ever stood up to Elizabeth's tirades and lived to tell the tale.

Case Closed

USA: debuted Sept. 17, 1993. In production until 1994, for a total of 26 episodes. Four Point Entertainment/Wolfson-Arbus-Ross. Executive Producers: Fred Wolfson, Loreen Arbus, Helaine Swredloff-Ross, Ron Ziskin, Shukri Ghalayini. Cast: Stacy Keach (Host).

Not to be confused with the 2005 *anime* series of the same name, the USA Network's *Case Closed* was a "legal-centric" reality series hosted by actor Stacy Keach. Viewers were invited to submit examples of justice delayed and justice denied, whereupon teams of private investigators, accompanied by camera crews, were dispatched to resolve these problems (assuming, of course, that there was genuinely a problem to be resolved). In the course of the series' 26 hour-long episodes, a man accused of murder was exonerated with new evidence, a youngster abducted by his father was located, child-support deadbeats were tracked down and forced to pay up, a stalker was prevented from harassing a woman, etc. There were also human-interest stories that had little to do with legal issues, notably the episodes in which a grateful man was able to relocate the elusive Good Samaritan who had once saved his life, and a teenager in need of a kidney transplant was reunited with his long-absent biological father. Most of the footage of these incidents was authentic, though juiced up a bit with the occasional reenactment.

In addition to the above-mentioned acts of closure, the series provided valuable tips to viewers as to how to protect themselves against becoming crime victims. Some of the sagest advice was passed along by former criminals, who definitely knew whereof they spoke.

Celebrity Justice

Syndicated: Sept. 2, 2002–Sept. 9, 2005. Harvey Levin Productions/Time Telepictures Television. Executive Producers/Creators: Harvey Levin, Lisa Gregorisch-Dempsey. Hosts: Holly Herbert, Carlos Diaz. Correspondents: Pat LaLama, Jane Velez-Mitchell, Dany Simon, Ross McLaughlin, Dan Simon, Rob Rosen, Omar Lugones, Caroline Bula, Andre Hempkins, Chrissy Albice.

This daily, syndicated reality series allowed Us Regular Folks a glimpse into the legal entanglements of wealthy and glamorous celebrities — or, to quote the advertising blurb, "Famous Faces, Famous Cases." Befitting the series' magazine format, each half-hour episode included several compartmentalized segments. "You Be the Judge" was staged in the form of a mock trial, with viewers invited to render judgment on a current celebrity court case via on-line voting (a lot of air time during the series' third season was given over to the Michael Jackson child-molestation trial). "CJ Spin" detailed the strategies employed by "dream team" law firms. "Equal Justice?" could easily have been renamed "Do the Rich and Famous Receive Preferential Treatment from the Justice System?" (Not always, as witness some of the draconian verdicts rendered by hostile judges.) "Real Estate Wars" offered examples of celebrities doing battle against their neighbors — and vice versa — over such issues as property lines, litter and noise pollution. "Celebrity Wills" disseminated the dispensation of money and property to famous families (one such segment focused on Bob Hope's will — and no, he didn't take it all with him). "Court Appearances" actually referred to the wardrobe worn by celebs while appearing before the bench. And "Power Players" offered thumbnail resumés of prominent attorneys.

Hosting *Celebrity Justice* were Los Angeles newscaster Holly Herbert and Las Vegas sports-entertainment reporter Carlos Diaz. The series' co-creator was Harvey Levin, previously a legal analyst for *The People's Court* (q.v.) and a legal reporter for the Los Angeles and New York NBC affiliates, a job which allowed him to wear outlandish disguises to avoid recognition in the courtroom (or perhaps to attract it!). Levin described *Celebrity Justice* as "a labor of love": it took him seven years to get the show on the air, and then only after calling in favors from contacts he'd made as a reporter.

In active distribution for three years, *Celebrity Justice* earned a 2005 Genesis Award from the U.S. Humane Society for its coverage of the Hollywood pet scene. Also, excerpts from the series were given generous exposure on CNN, whose correspondent Daryn Kagan interviewed Levin on several occasions. Of more long-range value, it was *Celebrity Justice* that led to the establishment of another, even more successful pipeline for celebrity news and gossip: Harvey Levin's TMZ (Thirty Mile Zone) interactive website, the forerunner of the equally well-received daily syndicated magazine *TMZ on TV* (2007–).

Century City

CBS: March 16–30, 2004; Universal-HD: Dec. 23, 2004–Dec. 31, 2005. Heel & Toe Films/Universal Television. Executive Producer/Creators: Ed Zuckerman, Katie Jacobs, Paul Attanasio. Cast: Nestor Carbonell (Tom Montero); Viola Davis (Hannah Crane); Ioan Gruffudd (Lukas Gold); Kristin Lehman (Lee May Bristol); Eric Schaeffer (Darwin McNeil); Hector Elizondo (Martin Constable).

"*The Practice* meets *The Twilight Zone*." One suspects this was the selling angle used to sway CBS in favor of the futuristic legal drama *Century City*. Set in 2030 AD, the series took place in a gleaming, high-tech version of old Los Angeles (which had been destroyed decades earlier in a 7.1 earthquake), where the law firm of Crane, Constable, McNeil & Montero handled difficult cases reflecting the era in which they occurred. Viola Davis starred as the firm's founder and senior partner Hannah Crane, with Hector Elizondo as Hannah's venerable colleague Martin Constable, Eric Schaeffer as arrogant, libidinous attorney Darwin McNeil, Nestor Carbonelli as junior partner and former California congressman Tony Montero, Kristin Lehman as first-year lawyer Lee May Bristol, and Ioan Gruffudd as Lukas Gold, the "conscience" of the firm.

In the World of Tomorrow, America boasted 52 states, Oprah Winfrey was the Na-

tion's president, there was at long last universal healthcare, and everyone lived to a *very* ripe old age. Of more relevance to the lawyer protagonists, court cases were now heard before hologram judges, while virtual-reality secretaries handled the paperwork. But for all these sci-fi trappings, *Century City* avoided the clichéd "Wow, it sure is weird living in the Future!" approach; each legal case was approached in a businesslike, "So what else is new?" fashion, as though the series was actually being filmed some three decades hence.

"We're not doing science fiction," explained series co-creator Paul Attanasio in a *New York Times* interview. "We're just peeking around the corner into the future. Call it 'science fact.'" Elaborating on this, Attanasio's partner Ed Zuckerman (who wrote the pilot episode), told the *Science Fiction Weekly* website that the plots were merely extensions of, and elaborations on, what was going on in the world of 2004 (or 2003, when the series actually went into production as a potential midseason replacement). Each episode featured two oddball law cases, one serious, one lighthearted. In the opener, the lawyers defended a man who was being prosecuted by the Feds for the illegal importation of cloned human embryos to cure his ailing son — who turned out to be a clone himself, thereby technically making the boy his father's twin brother! Other episodes involved a case of "virtual rape"; an aging member of a boy band sued by his fellow singers for not taking a drug that would make him appear forever young; the dilemma of a mentally challenged man who faced death if the implant designed to raise his I.Q. wasn't removed (his family opposed this procedure); a woman with a penis sued for non-disclosure by her flustered male lover; and, in a variation of the ongoing "steroids in professional sports" issue, a baseball star fired for being fitted with a bionic eye after an accident.

Not surprisingly, *Century City* developed a rabid cult following, not all of whose members lived in their parents' basements. Critics were less enchanted, noting that despite its mid-21st-century milieu, the main characters were cut from the usual contemporary TV-stereotype cloth: the sage father figure, the driven career woman, the young idealist, the sleazy womanizer, etc. It was also pointed out that, for all the miraculous technological advances of 2030, scientists still hadn't figured out how to clear up the smog that surrounded Los Angeles, nor how to grow hair on Hector Elizondo's head.

Though nine hour-long episodes were filmed, only four were telecast before *Century City* was cancelled by CBS. The remaining five finally saw the light of day when the Universal-HD digital service rebroadcast the series beginning in late 2004.

Christine Cromwell

ABC: Nov. 11, 1989–Feb. 17, 1990. Wolf Films/Universal Television. Executive Producer: Dick Wolf. Cast: Jaclyn Smith (Christine Cromwell); Celeste Holm (Samantha Cromwell); Ralph Bellamy (Cyrus Blain); Rebecca Cross (Sarah).

One of three rotating components of the *ABC Mystery Movie* omnibus series (the other two during the 1989–90 season were the Burt Reynolds private-eye opus *B.L. Stryker* and a revival of the Telly Savalas vehicle *Kojak*), *Christine Cromwell* was firmly grounded in the "lawyer as sleuth" format established decades earlier by the likes of *The Amazing Mr. Malone* and *Perry Mason* (both q.v.). Former "Charlie's Angel" Jaclyn Smith was suitably dazzling in the role of Christine Cromwell, who though born into a wealthy family had charted her own professional course in life, earning both a law and a business degree. While working as a public defender, Christine landed a job as a financial counselor at the San Francisco investment firm run by Cyrus Blain (Ralph Bellamy in his final recurring TV role). Catering to a rich and privileged clientele, our heroine frequently had to deal with high-profile crimes like murder and embezzlement. With loyal contacts on both sides of the law, not to mention a number of valuable connections within the legal system, she was able to solve whatever mysteries arose with cool efficiency and with nary a hair out of place on her lovely head. Also in the cast were Celeste Holm as Christine's mother Samantha,

whose many marriages had left her even wealthier than when she started, and Rebecca Cross as Ms. Cromwell's wide-eyed secretary Sara.

The five 90-minute episodes of *Christine Cromwell* were seen on a semi-monthly basis until February of 1990. The series proved to be the latest in a long line of failures for producer Dick Wolf, who fortunately was able to redeem himself a few months later with a little item called *Law & Order* (q.v.).

Civil Wars

ABC: Nov. 20, 1991–March 2, 1993. Steven Bochco Productions/20th Century–Fox Television. Executive Producers: Steven Bochco, William Finkelstein. Cast: Mariel Hemingway (Sydney Guilford); Peter Onorati (Charlie Howell); Debi Mazar (Denise Ianello); David Marciano (Jeffrey Lassick); Alan Rosenberg (Eli Levinson).

Ever on the prowl for new and groundbreaking TV concepts, producer Steven Bochco came up with the ABC entry *Civil Wars*, the first weekly, 60-minute dramatic series about divorce lawyers.

As the series began its two-season run, Mariel Hemingway and Alan Rosenberg were seen as Sydney Guilford and Eli Levinson, successful Manhattan divorce attorneys. After suffering a nervous breakdown, Eli opted to work on less stressful cases (e.g. those not involving divorce) and his duties were taken over by a new member of the firm, Charlie Howell (Peter Onorati). Having previously clashed with Mr. Howell as a courtroom opponent, the uptight, work-obsessed Sydney was nonetheless attracted to the impulsive, laid-back Charlie, and he to her. In true theatrical "backstairs romance" fashion, even as Sydney and Charlie grew closer, a love affair developed between their street-smart secretary Denise Ianello (Debi Mazar) and messenger boy Jeffrey Lassick (David Marciano)—who, wouldn't ya know it, turned out to be a wealthy eccentric in disguise. Alas, by the time the series moved into a second season, the marriage between Denise and Jeffrey was on the rocks, and it looked like Sydney and Charlie were about to get a couple of clients from within their own ranks. An additional crisis loomed on the horizon when Charlie found out that his father was suffering from prostate cancer.

Civil Wars' bittersweet approach to its subject matter was summed up by the opening credits, in which formal portraits of blissful married couples were contrasted with grainy footage of the same couples acrimoniously dissolving their unions in court. Similarly, the more serious and traumatic divorce cases were offset by moments of high comedy: in one episode, a woman wanted to divest herself of her spouse because he was convinced that he was Elvis; in another, a husband-wife wrestling team showed up before the judge in full costume and makeup.

Created by William Finkelstein, a former divorce lawyer who had previously worked on Steven Bochco's other legal series *L.A. Law* (q.v.), *Civil Wars* adhered during its initial season to a pattern that was becoming increasingly familiar for Bochco's less successful efforts: the critics adored it (*TV Guide*'s Myles Callum was enraptured by the "grabby, colorful writing"), but the audience didn't. The series ranked 95 out of 127 network shows, slaughtered in the ratings by NBC's *48 Hours* and CBS' *Quantum Leap*. An effort to boost viewership by hammocking the series between *Home Improvement* and *Roseanne* on ABC's well-attended Tuesday night schedule helped a little, but not enough.

Characteristically, Bochco decided to jolt viewers out of their apathy with one of the programming stunts for which he'd become famous. Telecast September 30, 1992, the episode "Grin and Bare It" contrived to have Sydney Guilford pose in the nude for a photographer. By 21st Century standards, the all-too-brief glimpse of Mariel Hemingway's *derriere* was a model of decorum, but in 1992 it brewed up plenty of controversy. Hemingway herself told *Entertainment Weekly* that the nude scene was "pretty nerve-wracking" and that she was "pretty shy," even though she'd previously appeared *au naturel* in both *Playboy* magazine and the movie *Star 80*. Notwithstanding the actress' trepidations and the expected outcry from the Clean-Up-TV zealots, the episode accomplished its purpose, increasing *Civil Wars'* viewership by 1.1 million.

Still, the series, now seen on Wednesdays, was unable to make a dent in the ratings of NBC's *Law & Order* (q.v.), and in March of 1993 its 36-episode run came to an end. Happily, it wasn't a total loss for creator William Finkelstein, who not only returned to *L.A. Law*, but also brought the characters of Eli Levinson and Denise Iannello — and the actors who played them — along with him. In particular, Alan Rosenberg's witty, wise and gloriously idiosyncratic portrayal of Eli Levinson, previously singled out for critical praise during his *Civil Wars* tenure, was credited as one of the new ingredients which "saved" the flagging *L.A. Law* from a premature demise.

In a 2001 TV interview with Larry King, Mariel Hemingway opined that *Civil Wars* hadn't clicked with the public because "it was just, like, one year before its time" and "kind of controversial." That the actress, who'd won a Golden Globe for her portrayal of Sydney Guilford, was in real life not quite as glib or articulate as her TV character was not too surprising. It was, however, a bit of a surprise that Larry King did not latch onto the divorce angle of *Civil Wars* and invite his favorite guest Mickey Rooney to join the discussion.

The Client [*aka* John Grisham's The Client]

TNT: Sept. 17, 1995–Apr. 16, 1996. Michael Fimerman Productions/Judith Paige Mitchell Productions/New Regency/Warner Bros. Television. Executive Producers: Michael Fimerman, Judith Paige Mitchell, Arnon Milchan. Cast: JoBeth Williams (Reggie Love); John Heard (D.A. Roy Foltrigg); Polly Holliday (Mama Love); David Barry Gray (Clint McGuire); Ossie Davis (Judge Harry Roosevelt).

Undaunted by such earlier failed efforts as *Adam's Rib* and *Eddie Dodd* (both q.v.) to transform popular courtroom movies into weekly television series, the TNT cable network unveiled this weekly, hour-long adaptation of the 1994 feature film *The Client*. When the series ended up faring no better than its predecessors, chances are that someone at the network was at last daunted.

Based on the John Grisham novel of the same name, the original film version of *The Client* starred Susan Sarandon as Reggie Love, a Memphis–based family lawyer specializing in helping those unfortunates who had fallen between the cracks of the legal system, especially children. In the movie, Reggie's client was 11-year-old street kid Mark Sway (Brad Renfro), whom the FBI was pressuring to reveal what he knew about the Mafia–ordered murder of a senator. As publicity-hungry Federal prosecutor Roy Foltrigg (Tommy Lee Jones), known as "The Reverend" for his habit of spouting Scripture (often incorrectly) at every opportunity, devised a strategy to loosen Mark's lips in open court — which of course ran the risk of signing the boy's death warrant — Reggie took a warmer, more maternal approach to win Mark's trust, even inviting the youngster to stay in the home she shared with her loving, supporting mother (Micole Mercurio). Other principal players in the film were Reggie's versatile legman Clint Von Hooser (Anthony Edwards) and sympathetic judge Harry Roosevelt (Ossie Davis). Like many another John Grisham movie adaptation, *The Client* was riddled with plot holes and ultimately descended into improbable melodrama; but the essential humanity of the lawyer protagonist remained in the audience's collective memory long after the final credits rolled.

The TV series retained Reggie Love's compassion for the Underdog, her endearing habit of extending the courtesy of her home and hearth to her young clients, her ongoing battle against alcoholism, and her bitter child custody battle with her ex-husband, a prominent doctor. Also, the series' two-hour pilot covered much of the same territory as the film (a disenfranchised kid witnessing a murder), albeit with a entirely different plot and outcome. Changes wrought upon the property began with its locale, as Reggie's practice moved from Memphis to Atlanta, presumably because it was cheaper to film on location in that city. Replacing Susan Sarandon was JoBeth Williams as Reggie, with John Heard taking over from Tommy Lee Jones as Reggie's perennial adversary Roy Foltrigg (less of a comic caricature

than in the movie), David Barry Gray spelling Anthony Edwards as Reggie's assistant Clint (now younger, handsomer, and given the surname McGuire), and Polly Holliday in place of Micole Mercurio as Momma Love. Only Ossie Davis repeated his screen characterization of the stern but goodhearted Judge Roosevelt.

Unfortunately, the TV adaptation of *The Client* shared many of the same flaws as its theatrical-film predecessor, compromising its good intentions and three-dimensional characterizations with ludicrous plot twists and B-movie action finales. Such shortcomings could arguably be overlooked in a one-shot movie, but it was harder to do so with a twenty-episode series. *Variety* noted that *The Client* "translates uneasily to the small screen," and viewers tended to agree, failing to show up in sufficient numbers to prevent the series' cancellation after a single season. (Reruns of *The Client* were later shown on another cable outlet, Court TV).

Close to Home

CBS: Oct. 4, 2005–May 11, 2007. Jerry Bruckheimer Television/Warner Bros. TV. Executive Producers: Jerry Bruckheimer, Jim Leonard, Jonathan Lipman. Cast: Jennifer Finnigan (Annabeth Chase); Kimberly Elise (Maureen Scofield); David James Elliott (Chief Prosecutor James Conlon); John Carroll Lynch (Steve Sharpe); Christian Kane (Jack Chase), Conor Dubin (Danny Robel); Barry Shabaka Henley (Det. Lou Drummer); Dwayne Adway (Off. Keith Macklin).

Though the CBS legal drama *Close to Home* was created by Jim Leonard, the weekly, hour-long series bore the unmistakable escutcheon of executive producer Jerry Bruckheimer, the king of Prime Time procedural shows (*CSI, Cold Case*, et al.). It was also a thick and steaming slice of Mid-American Gothic, peeling off the veneer of Heartland respectability to reveal that the "folks next door" could be just as evil and homicidal as any big-city mobster or gangbanger.

Jennifer Finnigan starred as Annabeth Chase, a brilliant county prosecutor living in Indianapolis. Returning from a 12-week sick leave, Annabeth learned that she had been passed over for promotion and that her new boss was Maureen Scofield (Kimberly Elise).

Expressing doubts that new mom Annabeth would be able to handle tough legal issues without letting her hormone-driven emotions get in the way, Maureen indicated that she was thinking of limiting Annabeth's case load. Naturally, this made our heroine work all the harder in pursuit of justice, even if it meant putting a strain on her relationship with her construction-worker husband Jack (Christian Kane).

If we are to take *Close to Home* at face value, Indianapolis was a hotbed of criminal mischief, brimming over with gang violence, serial killings, sex maniacs, prostitution, kidnapping, spousal abuse, vigilantism, political corruption and real estate fraud. "The creeps don't scare me," Annabeth insisted at one point. "It's the so-called decent guys. You know … the ones who go to church and join the Rotary club while they hide in plain sight. They're the ones who scare me." With that in mind, it is no surprise that her most formidable foes were those Pillars of Society with skeletons rotting in their closets. Of course, Annabeth didn't work in a vacuum: assisting the heroine in her battle against the behemoths of crime were fellow prosecutor Steve Sharpe (John Carroll Lynch), paralegal Danny Robel (Conor Dubin), detective Lou Drummer (Barry Shabaka Henley), and cop Keith Macklin (Dwayne Adway), who doubled as Maureen Scofield's love interest.

The unavoidable family resemblances among the various Jerry Bruckheimer TV projects might have inspired certain critics to dismiss *Close to Home* as *CSI: Indianapolis*, prompting star Finnigan to head the nay-sayers off at the pass: "It sort of scares me that it's being talked about as a legal procedural, because everybody's going to go, 'Oh, another one!' But it's so much more emotional." Whatever critics might have thought, the audience embraced *Close to Home* from the outset, the series opening the 2005–2006 season as CBS' top-rated Friday night show. But by the early spring of 2006 the colors had faded, prompting Bruckheimer to tinker with the format. The season finale not only found the hitherto undefeated Annabeth Chase losing an important case, but also contrived to kill off her husband Jack in a

drunk-driving accident. (Somewhere along the line it had been determined that Annabeth's home life was the series' least interesting aspect—and besides, a beautiful young widow was an alluring magnet for handsome, charismatic male guest stars.)

Season Two began four months after Jack's death, with Annabeth returning from another leave of absence to find out she and Maureen were now under the command of Chief Prosecutor James Conlon (David James Elliott), a brash and bullying New Yorker. As the season progressed, Annabeth's criminal-prosecution load was shunted to the background in favor of the political intrigue surrounding Conlon's bid for the governor's chair, a campaign doomed to collapse in the wake of a police corruption scandal. Though *Variety* had already noted that *Close to Home* was "near death" as the season wore down, the producers valiantly attempted to jump-start the series with a cliffhanger finale, as Annabeth grimly pursued the higher-ups responsible for the assassination of her friend and colleague Maureen Scofield.

We'll never know if Maureen was avenged, nor even if Annabeth herself survived the post-assassination deluge. *Close to Home* was cancelled in May of 2007, one of two Bruckheimer-produced legal series to get the axe within the same year (the other was Fox's *Justice* [q.v.]).

Common Law

ABC: Sept. 28–Oct. 19, 1996. Witt/Thomas Productions. Executive Producers: Rob LaZebnik, Paul Junger Witt, Tony Thomas, Paul Junger Witt, Tony Thomas, Gary S. Levine. Theme music performed by Los Lobos. Cast: Greg Giraldo (John Alvarez); Megyn Price (Nancy Slayton); Carlos Jacott (Pete Gutenhimmel); David Pasquesi (Henry Beckett); Diana-Maria Riva (Maria Marquez); Gregory Sierra (Luis Alvarez); John DiMaggio (Francis).

Latching onto the fact that Greg Giraldo, the standup-comic star of the ABC legal sitcom *Common Law*, was in real life a graduate of Harvard Law School prompted several reviewers to pigeonhole the series as "semi-autobiographical." Possibly there's some truth in this, even though the series was actually created by veteran TV comedy writer-producer Rob La-

Zebnik (*Empty Nest, Blossom, The Simpsons* etc.).

Giraldo was cast as attorney John Alvarez, who (like the star) was an Hispanic raised in Queens who'd worked his way through Harvard to earn a law degree, and who (unlike the star) was a member of an upscale, WASPish Manhattan law firm. Having fallen for his attractive workaholic associate Nancy Slayton (Megyn Price), John shared an apartment with Nancy, a fact that the couple was forced to keep secret from their professional colleagues, who discouraged interoffice affairs out of fear of potential sexual harassment charges, and from John's barber father Luis (Gregory Sierra), an old-country traditionalist who frowned upon "mixed couples."

The rest of the cast included Carlos Jacott as the obligatory second-generation yuppie lawyer Pete Gutenhimmel, David Pasquesi as stereotypical power-hungry associate Henry Beckett, and John DiMaggio as standard-issue stupid messenger boy Francis, given to uttering such howlers as describing a trip to the bathroom as "shaking hands with the President." Virtually the only character who didn't appear to have been stamped out on the sitcom assembly line was sassy secretary Maria Marquez (Diana-Maria Riva), who claimed to have gleaned her organizational skills from talk show host Ricki Lake (it doesn't always have to make sense to be funny).

More might have been made of the culture clash between the Latin–flavored John Alvarez and his uptight, all-white legal associates, or the "opposites attract" romance of the laid-back John and the hard-driving Nancy. Instead, the series dwelt at great and irritating length over John's "Notice me, dammit!" eccentricities: His beard, his earring, his habit of strumming on a guitar when mulling over a sticky problem, and his office decorated in psychedelic-sixties fashion. One not only wondered why his ultraconservative law firm would ever have hired him — especially after the opening episode, in which he nearly blew an important case in order to bail out an old friend — but why *anyone* would give this looney-tune any job more taxing than working the counter of a comics store.

Each of the series' ten episodes bore a title beginning with "In the Matter Of…." Of those ten, only four half-hour installments of *Common Law* were telecast before ABC fulfilled the prophecy of TV critic Tom Shales, who described the series as "another of the new season's Eminently Expendables."

Confession

ABC: June 19, 1958–Jan. 13, 1959. Director: Patrick Fay. Host: Jack Wyatt.

Confession was one of the many locally produced half-hour "fillers" used by ABC to plug the gaps in its sparsely populated Prime Time schedule of the late 1950s. Taped at the studios of Dallas ABC affiliate WFAA (now historically famous as the first TV station to break the news of the JFK assassination), the series attempted to determine the root cause of criminal activity and to prescribe a possible cure. In each episode, host Jack Wyatt interrogated a convicted criminal, whose presence before the cameras had been vetted and approved by local police officials. Once the interview portion was over, a panel consisting of a lawyer, a clergyman, a psychiatrist or psychologist, and a penologist or sociologist would discuss and analyze the interview, then offer suggestions that might empower the lawbreaker to redeem himself. "This may sound corny," Wyatt noted at the time, "but the authorities tell me we've actually helped criminals change their ways."

Perhaps because ABC realized that practically no one was watching (certainly not with such alternative choices as NBC's *You Bet Your Life* and CBS's *The Garry Moore Show*), *Confession* was permitted to traffic in subject matter that was normally taboo on the censor- and sponsor-dominated TV scene of the era. An article in the December 30, 1957, issue of *Time* magazine devoted ample print space to an interview between the unflappable Jack Wyatt and a 22-year-old transvestite, noting that this episode and others which had spotlighted junkies, prostitutes and sex murderers were the sort of fare that "Manhattan might have smothered with a grey-flannel gag." In the Dallas TV market, *Confession* was telecast live, assuring that the WFAA switchboard would light up like a Christmas tree at the conclusion of each episode.

Debuting locally in early 1957, *Confession* was seen variously on ABC's Tuesday and Thursday lineup from June of 1958 through January of 1959; its replacement, in case you're curious, was the paranormal anthology *One Step Beyond*.

Congressional Investigator

Syndicated: 1959. Sandy Howard Productions. Cast: William Masters (Investigator August); Edson Stroll (Investigator Thompson); Stephen Roberts (Senator Endicott); Marian Collier (Sue the Secretary).

The 1957 television coverage of the racket-busting McClellan Committee hearings (see **Introduction**) not only proved fascinating to the public but also profitable for TV producers. Two syndicated series in particular were created to capitalize on the Committee's assault upon labor racketeering: 1958's *Grand Jury* (q.v.), and 1959's *Congressional Investigator*.

Set in Washington, D.C., *Congressional Investigator* proposed to tell "the story behind the Fifth Amendment," as a team of government-appointed investigators dug up sufficient evidence to drag those pesky "Plead the Fifth" lawbreakers into Federal court. Working on behalf of Kefauver–like Senator Endicott (Stephen Roberts), chief investigators August (William Masters) and Thompson (Edson Stroll) probed an exhausting array of crimes both major and minor.

The synopses of the series' 39 half-hour episodes would indicate that nothing escaped the eagle eyes of the two protagonists. Their targets not only included corrupt labor officials, but also illegal gambling, usury, extortion, industrial espionage, bootlegging, "fixed" sports events, dishonest car repairmen, crooked politicians, disc jockeys who accepted "payola," teenage gangs, murder-for-hire, and even something as comparatively benign as the Old Badger Game. Two of the more intriguing episodes found the Investigators bearing down on a phony psychiatrist (described as a "Psycho-

Quack," which sounds more like a cartoon character) and a particularly slimy sharpster who bamboozled people into purchasing over-priced bomb shelters! But lest the variety of subject matter suggest that *Congressional Investigator* was working on an unlimited Federal budget, be assured that producer Sandy Howard was as tight with a dollar here as in his other TV efforts of the period (see the entry for the 1958 version of *Night Court*). The series spent the bulk of its time in offices, interrogation cells, and courtrooms, with most scenes shot in single, multicamera takes; footage of the investigators actually tracking down the villains was at a premium.

Though the series received national distribution, mainly to such major markets as Chicago (where the NBC affiliate ran it on a nightly basis at 12:30 A.M.), *Congressional Investigator* is today so obscure that a number of reference works have listed the title as an "unsold pilot."

Conviction *see* Law & Order (The Franchise)

The Court

ABC: March 26–Apr. 9, 2002. John Wells Productions/Warner Bros. Executive producers: Carol Flint, Charles Haid. Cast: Sally Field (Justice Kate Nolan); Pat Hingle (Chief Justice Amos Townsend); Chris Sarandon (Justice Lucas Voorhees); Diahann Carroll (Justice Angela DeSett); Miguel Sandoval (Justice Roberto Martinez); Christina Hendricks (Betsy Tyler); Nicole DeHuff (Alexis Cameron); Hill Harper (Christopher Bell); Craig Bierko (Harlan Brandt); Josh Radnor (Dylan Hirsch).

The ABC legal drama starring Oscar–winning actress Sally Field and bearing the title *The Court* bore very little resemblance to the property as originally conceived. According to the earliest press releases sent out in May of 2001, the series, slated to air on Mondays, was to have been executive-produced by Oliver Goldstick and Michael Dinner for Touchstone Television, with Field heading the cast as iron-willed Liberal Supreme Court Justice Audrey Karlin. Despite the actress' top-billed status, the proposed series was not specifically about Field's character but instead offered a behind-the-scenes look at the workings of the Court as seen through the eyes of the nine Justices' "vibrant, young" legal clerks. Alicia Witt had been tapped to play Justice Karlin's "earnest and ambitious" clerk Claudia Padgett, while Billy Burke was cast as "clever and charming" Top Kappis, who clerked for Audrey Karlin's staunchest rival, Conservative Justice Martin Van Horn, played by second-billed Brian Cox. Presumably, a romance was to have developed between Claudia and Tom despite their different backgrounds (she was born into wealth, he was strictly working-class) and the diametrically opposite politics of their respective bosses.

By the time *The Court* finally made the ABC Prime Time schedule ten months later—not on Monday, but Tuesday night—only the title, the star, and the Supreme Court setting had been salvaged from the original concept. The package had been rerouted to Warner Bros. Television, with John Wells Productions of *West Wing* and *ER* fame now in the driver's seat, and Carol Flint and actor Charles Haid handling the executive-producer reins. Instead of focusing on the Court's clerical staff, the spotlight was now firmly fixed upon Sally Field, as newly installed Justice Kate Nolan (this name change must have occurred at the very last minute: several of the publicity bulletins issued to local ABC affiliates listed Field under *both* character names).

Not only was Kate no longer a Liberal, she didn't even have a track record as a judge. The "independent" governor of an unnamed midwestern state, Kate had so impressed the President with her sincere idealism, her refusal to grind ideological axes, and her even-handed legislation of hot-button legal issues, that he personally selected her to replace a popular Supreme Court justice who had died suddenly in a car accident. The Court, you see, was cleft right down the middle between Liberal and Conservative, and it was felt that Kate, who presumably wouldn't waste time currying favor with either side, would be the ideal "swing" voter. (The *USA Today* reviewer was bemused by this premise: "No one explains why any

president, given the chance to tip the court, would prefer imbalance.")

Despite her gubernatorial track record, which of course was the reason she got the job, Kate had to be carefully coached by her staff to give noncommittal statements to the media, and it was understood that she would do nothing to stir up controversy. Is it any surprise, then, that Kate's fellow "Supremes" initially regarded the novice jurist as a naïve babe in the woods? And given the fact that the star was Sally "Norma Rae" Field, is it any surprise that her *very first* case would be such a political hot potato—a challenge to Ohio's "three strikes and you're out" life-sentencing procedure—that she couldn't help but cast a vote that automatically labeled her a feisty, opinionated maverick in the eyes of the public? Never mind that in real life Kate wouldn't have been allowed to participate in this decision since she hadn't had time to hear the arguments; the script contrived to have her force a rehearing of the case, which though it accomplished its dramatic goal of making the other Justices grudgingly respectful of her tenacity and common-sense values, also would not have actually occurred.

And while we're on the subject of authenticity, why is it that most fictional Supreme Courts are populated by Justices who in addition to holding contrary political views are also bitter enemies outside chambers? The notion that people can disagree ideologically while still remaining friends (as proven by the famously harmonious out-of-court relationship between Anthony Scalia and Ruth Bader Ginsburg, to cite but one example) seems absolutely foreign to most TV writers: thus, the Conservatives on Kate Nolan's court, notably Chief Justice Townsend (Pat Hingle) and Justice Martinez (Miguel Sandoval), were forever staring daggers and muttering dark oaths at such Liberals as Justices Voorhees (Chris Sarandon) and Desett (Diahann Carroll), and vice versa. This party-line animosity trickled down to the clerical staff, which included left-wing Alexis (Nicole De Huff), right-wing Christopher (Hill Harper), and irritating devil's-advocate Dylan (Josh Radnor)—who, in the words of *USA Today*, endlessly engaged in "crashingly dull debates masquerading as dialogue."

In an effort to include a character who would act as "our" eyes, explaining and interpreting the legal machinations for the benefit of us ignorant laymen, *The Court* featured Craig Bierko as Harlan Brandt, a former lawyer (and former student of Kate Nolan) turned investigative reporter. Forever lurking on the sidelines as he and his assistant Betsy Tyler (Christina Hendricks) filmed a series of TV specials about the clients in Supreme Court cases, Brandt had an intrusive habit of gumming up the judicial procedure in his relentless pursuit of story material. He also devoted far too much of his time dredging up dirty little secrets in the Justices' private lives, self-righteously insisting that he was merely trying to put "a human face" on the highest court of the land. Brandt's ongoing narrative commentary was evidently intended to imbue him with an "Everyman" quality—which it did, if you regard Every Man as a gossip-mongering buttinsky.

The Court debuted around the same time as the thematically similar *First Monday* (q.v.), and was regarded as marginally the better of the two efforts. Unfortunately, the series, each episode of which was to have been rerun on ABC's cable-TV sister Disney Channel within eight days of its network telecast, could not even muster up enough viewers for one weekly showing. Of the six episodes filmed, only three were shown before *The Court* was permanently adjourned by ABC, allowing Sally Field to return to more secure ground as a semi-regular on NBC's still-popular medical drama *E.R.*

Court of Current Issues

DuMont: Feb. 9, 1948–June 26, 1951. Created and produced by Irvin Paul Sulds.

Debuting under the title *Court of Public Opinion*, this early DuMont network series adopted its more familiar title *Court of Current Issues* several months into its run. The weekly, half-hour series would seem to be the first of several public-affairs programs to use the framework of a courtroom trial in order to discuss the important issues of the day. With a genuine judge or attorney on the bench, two people representing the opposite sides of an issue would act as "defense" and "prosecution." Once the hearing portion of the show was over, a jury of 12 audience members delivered a verdict as to which "consul" presented the best argument.

From June of 1949 until its network cancellation two years later, *Court of Current Issues* was telecast on Tuesday evening opposite NBC's Milton Berle, meaning that the only people watching were the guys in the DuMont control booth. After it left the network, the series continued for some time as a local New York offering; during these final weeks, producer Irvin Paul Sulds invited members of the Armed Forces to serve on the studio jury.

Court of Last Resort

NBC: Oct. 4, 1957–Apr. 11, 1958; ABC: Aug. 26, 1959–Feb. 17, 1960. Paisano Productions/ Walden Productions. Executive Producer: Jules C. Goldstone. Producer: Elliot Lewis. Cast: Lyle Bettger (Sam Larsen); Paul Birch (Erle Stanley Gardner); Carleton Young (Harry Steeger); Charles Meredith (Dr. Lemoyne Snyder); Robert H. Harris (Raymond Schindler); S. John Launer (Marshall Houts); John Maxwell (Alex Gregory); Robert Anderson (Park Street Jr.)

NBC's eagerness to option the weekly, half-hour legal anthology *Court of Last Resort* in 1957 had less to do with CBS' recent acquisition of *Perry Mason* (q.v.) than the fact that the creator of both properties, Erle Stanley Gardner, was at that time the world's biggest-selling author. *Court of Last Resort* was inspired by a series of articles written by Gardner for *Argosy* magazine, beginning in 1948 and originally bearing the blanket title "Is Clarence Boggie Innocent?" Having learned that a man suffering from mental illness had been sentenced to life imprisonment for murder in 1935 on the flimsiest of evidence, Gardner organized an ad-hoc panel of criminal law experts to determine if justice had truly been done. To make a long story short, it hadn't—and in 1948 the hapless defendant was pardoned, largely due to Gardner's efforts. Ten years' worth of similar articles followed, with Gardner and his colleagues re-opening several "closed" cases, reassessing the evidence and courtroom testimony, determining if confessions had been freely given or coerced, grilling key witnesses to find out whether they had lied or made honest mistakes, scouring the hinterlands for other witnesses who had never come forth at all during the original trial, and ultimately providing Due Process for those to whom it had been long denied.

In addition to Erle Stanley Gardner, the seven-person "Court of Last Resort" consisted of *Argosy* publisher and "administrator of the Court" Harry Steeger; lawyer, medical doctor and "medicolegal expert" LeMoyne Snyder; high-profile private detective Raymond Schindler; law professor and onetime FBI employee Marshall Houts; lie-detector expert Alex Gregory; and trial lawyer and chairman of the Texas Law Enforcement Foundation, Park Street Jr. These worthies appeared as themselves at the end of each episode of *Court of Last Resort*, and were impersonated by professional actors in the dramatized segments (see the cast list above), albeit merely on a recurring basis. The only character who appeared regularly was fictional special investigator Sam Larsen (Lyle Bettger), who doubled as the Court's legman and the series' narrator.

"According to the case for the prosecution…" intoned Larsen at the beginning of each fact-based episode, whereupon the crime of the week was re-enacted, followed by the arrest and conviction of the main suspect (his name changed a la *Dragnet* to protect the innocent). The rest of the episode usually found Larsen, assisted by one or more of the Court's "experts," piecing together a new case for the defense, the end result being either total exoneration of the accused or the establishment of enough contradictory evidence to provide "reasonable doubt" of his guilt. The first episode filmed (but the tenth one shown) was a dramatization of the Clarence Boggie case, which had inspired Gardner to establish the Court back in 1948. In this one, a mental defective named "Clarence Redding" had the misfortune to be in the vicinity of a vicious rape-murder, which occurred in a small town that couldn't afford a regular

police force but prided itself on "solving its own crimes." Incensed that Redding had been virtually railroaded into prison, magazine publisher Harry Steeger (Carleton Young) dispatched Sam Larsen to get the wheels of justice rolling. After determining that the local constable and the victim's father had surrendered to public opinion and personal prejudices in their haste to convict Redding (the characters were shown to be confused and misguided rather than malicious), Larsen decided to follow up the defendant's infantile ramblings about a man in a "red shirt"—whereupon the authorities instantly closed in on a hitherto unsuspected local serial killer, who at first glance appeared to be so deranged that it was a wonder he hadn't been carted away years earlier on general principles! Once the truth was known, Redding's original persecutors sheepishly apologized, and all was right with the world again.

The other episodes more or less followed this formula, combining step-by-step procedural detective work with histrionic acting and sudden bursts of melodrama, just to make sure that everyone was awake. Almost invariably, the Judicial System itself was shown to be infallible; blame for any miscarriage of justice was placed squarely on individual witnesses and law enforcement officials, whose shortcomings ranged from stupidity to outright villainy. For an occasional change of pace, the Court of Last Resort sought not to prove the innocence of a convicted felon, but to plead for clemency if the felon had managed to rehabilitate himself behind bars. In "The Conrad Murray Case," the title character was irrefutably guilty of a cold-blooded double murder. But after serving 34 years of a life sentence, Murray, now a model prisoner and self-educated legal expert, petitioned for a pardon, persuading even his warden that the State should be lenient. Acting on behalf of the Court of Last Resort, Sam Larsen did everything in his power to help Murray gain his freedom—on condition that he truly deserved it. (Incidentally, although all character names had again been changed, it was patently obvious that "Conrad Murray" was supposed to be Robert Stroud, the infamous Birdman of Alcatraz, who despite innumerable appeals was never released from "The Rock" and died behind bars in 1963.)

Existing episodes of *Court of Last Resort* reveal that the series maintained the same high production standards as Erle Stanley Gardner's more celebrated *Perry Mason*; All it lacked was a charismatic Raymond Burr type in the lead. In his efforts to remain objective throughout each episode, protagonist Sam Larsen came across as an empty suit, and that is precisely how he was played by Lyle Bettger — who, though firmly established as one of Hollywood's best and most convincing "heavies" (catch his bravura nastiness in 1950's *Union Station* sometime), was never quite as effective when playing a character on the right side of the law.

After completing its single-season NBC run in April of 1958, *Court of Last Resort* vanished from sight, reemerging on rival network ABC some sixteen months later, with rebroadcasts of the original 26 episodes.

Court TV — Inside America's Courts

Syndicated: Sept. 18, 1993–Feb. 8, 1997. Court TV/New Line Television. Hosts: Gregg Jarrett, Jami Floyd, Chris Gordon. Reporters: Cynthia McFadden, June Grasso, Kristin Jeannette-Myers.

A syndicated offshoot of the Court TV cable service (see **Introduction**), *Court TV — Inside America's Courts* started out as a weekly, hour-long review of the past week's most significant courtroom cases. Each such case was given approximately eight minutes' airtime per episode, using taped highlights provided by the series' parent cable network. In 1995 the series became a daily, half-hour "strip," with the addition of a weekend wrap-up show, *Court TV: Justice This Week*. By that time, the show's original eight-person staff had expanded to fifty, and production had moved from the Court TV facilities to a separate studio.

Inside America's Courts boasted its highest ratings during the O.J. Simpson civil trial. Perhaps appropriately, the series ceased production the day after that trial ended on February 4, 1997, whereupon most of the staffers returned to their former Court TV stamping grounds.

Courthouse

CBS: Sept. 13–Nov. 15, 1995. Kedzie Productions/Columbia Pictures Television. Executive Producer/Creator: Deborah Joy LeVine. Cast: Patricia Wettig (Judge Justine Parkes); Annabeth Gish (Lenore Laderman); Robin Givens (Suzanne Graham); Nia Peeples (Veronica Gilbert); Bob Gunton (Judge Homer Conklin); Brad Johnson (Judge Wyatt E. Jackson); Michael Lerner (Judge Myron Winkelman); Jennifer Lewis (Judge Rosetta Reide); Jeffrey D. Sams (Edison Moore); Nia Peeples (Veronica Gilbert); Dan Gauthier (Jonathan Mitchell); Jacqueline Kim (Amy Chen); Shelley Morrison (Nell); Cotter Smith (Andrew Rawson); George Newbern (Sam); John Mese (Gabe).

"*ER* in the Courtroom, but with the sophisticated adult sexuality of *NYPD Blue*." That's how series creator Deborah Joy LeVine described the weekly, 60-minute legal drama *Courthouse* to *New York Times* correspondent John O'Connor in October of 1995, one month after the new CBS series' heavily publicized premiere … and one month before its abrupt demise.

ER in the courtroom? Well, *Courthouse* certainly kept apace of that popular medical drama with a wealth of intersecting characters, the ongoing clutter and confusion of an overstocked caseload, and nonstop camera movement. The adult sexuality of *NYPD Blue*? Alas, there was none of the gratuitous nudity which distinguished that celebrated cop series, but no one could fail to pick up on the smoldering sexual tension between Justine Parkes (Patricia Wettig), the gorgeous, no-nonsense presiding judge of the fictional Clark County Courthouse, and rebellious, handsome-hunk Montana transplant Judge Wyatt Earp Jackson (Brad Johnson) — to say nothing of the blazing clandestine romance between DA's Office investigator Suzanne Graham (Robin Givens) and up-and-coming prosecutor Edison Moore (Jeffrey D. Sams). And elsewhere, sex was never far from the thoughts of Public Defender Lenore Lederman (Annabeth Gish), a pampered congressman's daughter who had been forced to grow up in a hurry (professionally speaking) when reassigned to the Sex Crimes unit, nor of Juvenile Court judge Rosetta Reide (Jennifer Lewis), a lesbian who lived in mortal terror of being outed at any moment.

But let's back up a bit and dispense a few more details about *Courtroom*. As mentioned, it took place in "Clark County" (though it was filmed to large extent at LA City Hall), and the main character was Judge Justine Parkes. The two Criminal Court judges on Parkes' staff, the aforementioned Wyatt Jackson and the pompous, rules-are-rules Sterling Conklin (Bob Gunton), were eternally at each other's throats. In addition to those listed above, the other staffers included Family Court judge Myron Winkelman (Michael Lerner), a neurotic bug who cared more about getting a good parking space for his Porsche than the proper administration of justice; public defender Veronica Gilbert (Nia Peeples), forever challenging the System and ending up in big trouble for her efforts; and Veronica's self-absorbed prosecutor boyfriend Jonathan Mitchell (Dan Gauthier).

The series exploded onto the scene with one of those big, splashy, "water cooler conversation" openers that proliferated on TV in mid–1990s: before the horrified eyes of a packed courtroom (including all the principal characters), a philandering judge and a convicted felon were both savagely murdered. Episode Two did its best to sustain the excitement of Episode One with a thinly disguised dramatization of the O.J. case, in which a famous athlete was charged with murder (oh, yes, there was also a racist cop acting as witness — but he didn't put in an appearance until a later, unrelated episode). Other episodes dwelt upon such torn-from-the-headlines issues as child molestation, crack dealing and prison overcrowding. Through it all, Judge Parkes injected her own sentiments and sense of justice into the proceedings, at one point goading a rapist into confessing just as his case was about to be thrown out of court — and just *after* the scriptwriter had contrived to justify the Defense's initial efforts to set the rapist free in two pithy sentences: "If this guy can get a fair trial, everybody can. And that's good to know." To quote the *New York Times*' review of *Courtroom*, "Complicated cases are ... reduced to a quick-fix solution ... and the generally appalling system is often reduced to comforting aphorisms."

Perhaps the producers opted for "quick fixes" because of a gut feeling that the show wouldn't be around very long. *Courtroom* was plagued with backstage problems from the outset, not least of which was a reported clash of personalities which led the writers to shunt Robin Givens' character to the background while building up the role played by Nia Peeples. The breaking point came when, disappointed over the direction her character was taking (or not taking), series star Patricia Wettig asked to be released from her contract. Since Wettig had been the main selling point when *Courthouse* was initially pitched to CBS, the network concluded that without its star, there was no point in extending the series beyond its first nine weeks on the air (as it was, only eleven episodes had been filmed before Wettig's exit). Besides, it was hard to justify clinging to a project that was universally panned by viewers and critics alike, with *Variety* expressing the majority opinion by labelling *Courthouse* "strictly a rehash of ground covered by other legal-themed series."

Courting Alex

CBS: Jan. 23–March 29, 2006. Touchstone Television/Paramount Network Television. Executive Producers: Rob Hanning, Seth Kurland, Eileen Conn. Cast: Jenna Elfman (Alex Rose); Dabney Coleman (Bill Rose); Josh Randall (Scott Larson); Jillian Bach (Molly); Josh Stanberg (Stephen); Hugh Bonneville (Julian).

The fact that this CBS sitcom went into production under such working titles as *The Jenna Elfman Show* and *Everything I Know About Men* should have tipped everyone off that its treatment of legal matters would be of rather low priority. Conceived as a vehicle for *Dharma and Greg* star Jenna Elfman, *Courting Alex* cast the pixyish leading lady as Alex Rose, a high-powered attorney with little time in her schedule for such abstracts as warmth and spontaneity — or for that matter, any sort of love life. All this changed when Alex met handsome young lawyer-entrepreneur Scott Larson (Josh Randall) when they were on opposite sides of a case. The humor arose from the clash of temperaments between the uptight, all-business Alex and the suave, laid-back Scott, which inevitably led to romance (or would have if the series had been around long enough). Inasmuch as *Dharma and Greg* had been predicated on the relationship between an impulsive, free-spirited heroine and a starchy, conservative hero (also an attorney), it was clear that *Courting Alex* deliberately reversed the roles to show off Jenna Elfman's versatility. This didn't quite happen, however, thanks to inconsistent scriptwriting, with the character of Alex Rose going from tough-and-pragmatic to "Dharma redux" and back again without rhyme, reason or warning. (Evidently, the producers were unwilling to alienate Elfman's fans by having her stray *too* far from her established TV persona.)

"Bland on bland" was *The San Francisco Chronicle*'s terse assessment of *Courting Alex*, though the reviewer heaped a great deal of praise on costar Dabney Coleman, cast as Alex's thrice-married father and boss Bill Rose. Also given credit for enlivening some rather dull proceedings were Jillian Bach as Alex's diminutive, down-to-earth assistant Molly, Josh Stanberg as Alex's stodgy law partner and erstwhile boyfriend Stephen, and Hugh Bonneville as the requisite eccentric neighbor Julian. (Lauren Tom, Brady Smith and Rhee Seehorn had appeared in the series' original, unsold pilot.) The show's eight televised episodes (out of the twelve filmed) came and went so quickly that there are those who still believe Jenna Elfman went into retirement after *Dharma and Greg*.

Court-Martial

ABC: April 8–September 2, 1966. Roncom/MGM/ITC/ABC. Executive producers: Roy Huggins, Robert Douglas, Bill Hill. Cast: Bradford Dillman (Capt. David Young); Peter Graves (Maj. Frank Whittaker); Kenneth J. Warren (M/Sgt. John MacCaskey); Angela Browne (Sgt. Yolanda Perkins); Diane Clare (Sgt. Wendy).

Although filmed at Pinewood Studios in Buckinghamshire, England, *Court-Martial* qualifies as an American TV series because it was largely produced with American funding, designed as one of the many "dual market" series (*The Vise, Dick and the Duchess, The Third Man, Shirley's World* et al.) seen on both British and U.S. television in the 1950s, 60s and 70s.

In fact, *Court-Martial*'s two-part pilot was lensed exclusively at Hollywood's Universal City as part of the NBC anthology *Kraft Suspense Theatre*, where it originally aired as the series' two-part premiere on October 10 and 17, 1963. Written by Lester and William Gordon and directed by Buzz Kulik, "The Case Against Paul Ryker" and its followup "Torn Between Two Values" provided a showcase for Lee Marvin as Sgt. Paul Ryker, an American accused of treason during the Korean war. In Part One, Bradford Dillman was seen as U.S. Army Captain David Young of the Judge Advocate General's (JAG) Office, who prosecuted Ryker during his court-martial and secured a guilty verdict. Part Two found Captain Young insisting that Ryker be given a second hearing on grounds that the sergeant hadn't received a proper defense; this request was granted, and in true switched-gears JAG fashion Ryker's former prosecutor was now his defense attorney, determined to save his client from hanging. With second, fourth and fifth billing going to Vera Miles (as Ryker's wife), Lloyd Nolan and Murray Hamilton, sixth billing was reserved for Peter Graves as Young's superior officer, Major Frank Whitaker. Filmed in color, this two-parter was later combined into a single feature and released theatrically by Universal as *Sergeant Ryker*—which, according to film critic Andrew Sarris, "created considerable ill will by charging first-run movie prices for an attraction so obviously designed for television that the audience can almost see the test patterns."

Notwithstanding, *Sergeant Ryker* garnered enough critical and audience praise to warrant a weekly TV series, with Bradford Dillman and Peter Graves (now nestled in a more desirable second-billed spot) reprising their roles as Captain Young and Major Whitaker. Beyond moving production from Hollywood to Buckinghamshire, there were other alterations made in the property before its weekly debut: Universal bowed out of the project, with MGM and ITC taking over; the series dropped its color photography in favor of more cost-efficient black and white; and the action was moved both geographically and chronologically from Korea in the 1950s to the battlefields of World War II. New cast members included Kenneth J. Warren as Young's and Whitaker's chief aide, Master Sergeant MacCaskey, and Angela Browne as their secretary Sergeant Yolanda, later replaced by Diane Clare as Sergeant Wendy. Remaining the same was the basic premise, with the JAG officers alternating between prosecuting and defending military clients in the courtroom.

Most of the hour-long episodes adopted a procedural approach, with the first half hour devoted to a painstaking investigation of the facts at hand, and the final thirty minutes comprised of the trial itself. To satisfy TV fans on both sides of the Atlantic, the guest-star roster was drawn from both American and British talent pools (Cameron Mitchell, Anthony Quayle, Sal Mineo, Diane Cilento, Darren McGavin, Martine Beswick), with a few stray Canadians like Donald Sutherland tossed in for balance.

First seen on London's ATV from September 12, 1965, through April 15, 1966, *Court-Martial* debuted in the US on April 8, 1966, as a midseason replacement for ABC's *The Jimmy Dean Show*. Of the series' 26 episodes, only 20 were seen by American audiences before *Court-Martial's* Friday-night timeslot was handed over to *Twelve O'Clock High* on September 9, 1966. Though written off as a mere "filler" by ABC, the series was more highly regarded in England, where it earned a BAFTA award (that country's equivalent of an Emmy) for Best Dramatic Series. Even so, it would be nearly three decades before weekly television returned to the secular world of military justice vis-à-vis the infinitely more successful *JAG* (q.v.).

Crazy Like a Fox

CBS: Dec. 30, 1984–Sept. 4, 1986. Wooly Mammoth Productions/Columbia Television. Created and executive-produced by Frank Cardea, George Schenck, John Baskin and Roger Shulman. Cast: Jack Warden (Harrison "Harry" Fox Sr.); John Rubinstein (Harrison Fox Jr.); Penny Peyser (Cindy Fox); Robby Kiger (Josh Fox); Lydia Lei (Allison Ling, Season One); Patricia Ayame Thomson (Allison Ling, Season Two); Robert Hanley (Lt. Walker); Theodore Wilson (Benny).

Essentially a detective series, CBS' *Crazy Like a Fox* warrants an entry in this book by virtue of having an attorney as one of the protagonists — and, in a perverse sense, because of the series' breezy disregard for the letter of the law in favor of the spirit.

The attorney was Harrison Fox Jr. (John Rubinstein), a serious-minded young chap possessed of a rock-ribbed Conservatism beyond his years, who hoped to make his mark in the legal world with a small San Francisco–based practice, if only to provide a livable income for wife Cindy (Penny Peyser) and son Josh (Robby Kiger). But try though he might, Harrison found it impossible to navigate the "proper" legal channels thanks to his wack-job of a father, veteran private eye and con artist Harry Fox Sr. (Jack Warden). Beyond the cosmetic contrasts between father and son — Fox Jr. drove a gleaming new BMW while Fox Sr. tooled around in a battered 1975 Coupe de Ville, Fox Jr. wore tailored tweeds while Fox Sr. preferred baggy, food-stained old suits, Fox Jr. strove to honor proper business protocol while Fox Sr. delighted in barging into rooms without any prior warning, and so on — the two men held diametrically opposite views of lawful procedure. Whereas Harrison followed all the rules to the tiniest detail, Harry was a born rule-bender, the ends always justifying the means so long as he could bring criminals to justice.

A typical *Crazy Like a Fox* episode would begin with loose-cannon Harry showing up uninvited at Harrison's office, ostensibly to solicit legal advice but actually for the purpose of dragooning the uptight Fox Jr. into Fox Sr.'s latest murder investigation, usually by strong-arming his son into accepting the main suspect (almost invariably an old pal of Harry's) as a client. By episode's end, both Foxes had been threatened, chased, shot at and otherwise inconvenienced — but by golly, the Guilty had been punished and the Innocent spared, and Harry could once again take paternal pride in the fact that he had financed his son's law-school education ... and incidentally, pick up a fat retainer. And once again, Harrison was given the opportunity to moan, "I don't like people with guns. They scare me. I'm an attorney"—as if that would stop his Old Man from starting up the cycle all over again the following week.

Discussing the genesis of *Crazy Like a Fox*, executive producer Frank Cardea told *TV Guide* interviewer Jack Hicks, "We wrote the pilot around Jack [Warden]. Harry looks weathered, knows all kinds of people on both sides of the law. He's a street-educated P.I. who's withstood the tests of time…. Beneath that crusty exterior, Jack gives Harry Fox warmth, substance, balance. People *care* about him…." Fine. But where did Harrison Fox come from? "We *knew* Harrison," Cardea went on, "a 'Yuppie lawyer,' if you will, bright, up-and-coming, beautiful family…. We [Cardera and co-producer George Schenck] *were* Harrison, collectively. But Harry we've come to discover as we've come to know Jack. Harry Fox *is* Jack Warden." Incredibly, as originally conceived Harry Fox was even *more* brash and flamboyant, replete with obscenely colored sports jackets and other such trappings. But Warden would have none of that: "Just give me a cigar and a nice gray suit and I'll give you what you want."

Such was the chemistry between Warden and costar John Rubinstein that it was easy to forget the series' other regulars. For the record, in addition to the aforementioned Penny Peyser and Robby Kiger, Leila Lei and later Patricia Ayama Thomson were seen as Harrison's unflappable secretary Allison, while the San Francisco Police Department was represented by Robert Hanley as Lt. Walker.

Inasmuch as the series was initially scheduled on Sunday evenings right after *Murder She Wrote*, it should surprise no one that *Crazy Like a Fox*'s biggest fan base consisted of people over the age of 40, mostly ladies who yearned to mother sweet young Harrison Fox and reprimand unsweet old Harry for being so rough on his poor son. Cancelled after two seasons and 37 hour-long episodes, *Crazy Like a Fox* was briefly revived on April 5, 1987, with a two-hour "reunion" movie filmed in England, *Still Crazy Like a Fox*.

Crime & Punishment

NBC: June 16, 2002–July 17, 2004. Wolf Films/ Shape Pictures/Anonymous Content/Universal Television. Executive Producer: Dick Wolf.

Produced by *Law & Order*'s Dick Wolf, *Crime & Punishment* (working title: *Trial & Error*) should not be confused with Wolf's 1993 series of the same name. Beyond the fact that the earlier series was not a legal program but a cop drama, the 2002 edition of *Crime & Punishment* was not technically a drama at all, but instead an unscripted docudrama focusing on genuine criminal cases and featuring actual prosecutors from the San Diego District Attorney's office. Come to think of it, it wasn't a docudrama either—at least not according to Dick Wolf, who preferred the designation "drama-mentary."

That said, *Crime & Punishment* bore the unmistakable stamp of the fictional entries in Wolf's *Law & Order* franchise (see separate entry). Each hour-long episode opened with the requisite "In the criminal justice system…" off-screen narration; the "cast" was introduced via group photos, with the main prosecutor afforded star billing; and the show's intertitles (rendered in the familiar *L&O* white letter/ black background format) established the time and location of each successive scene, underlined by that ubiquitous *ching-ching!* background music.

For all its documentary underpinnings, *Crime & Punishment* was edited and paced exactly like a Dick Wolf dramatic series, with tight closeups of the principals and rapid cross-cutting between highlights (the producer used these same post-production techniques for his syndicated reality series *Arrest & Trial* [q.v.]). The action was artfully reassembled for maximum dramatic effect, usually by interrupting the courtroom proceedings with flashbacks to the pretrial negotiation-and-discussion sessions; and so as not to confuse the viewers with too much "legalese," the lawyers' arguments were whittled down to the bare bones. However, there was no restaging of the basic events: the crime at hand was never shown, except through official police photos and when the jury reached its verdict, the reaction of the defendant was the "real deal."

To enhance the authenticity and immediacy of the proceedings, *Crime & Punishment* was shot on HD video by documentary filmmaker Bill Guttentag, with the naturally unsteady camerawork (as opposed to the deliberate bobbing and weaving in the *Law & Order* efforts) one might see in a nightly newscast. "The way the show has been shot is a tribute to both technology and the art that Bill brings to presenting reality in a totally unique way," proclaimed Dick Wolf. "The footage he has captured is extremely compelling and sometimes disturbing — but it is also very human." Guttentag was equally generous in his praise of Wolf: "This is truly the real life *Law & Order*— everything is real, there are no interviews, no narration, and no reenactments — everything is real as it unfolds on the screen."

Everything including the "actors": as mentioned, the series featured actual members of the San Diego DA's office. The show's basic cast included Supervising Deputy DA Eugenia Eyherabide; Dan Goldstein, Jill DeCarlo, Chris Lindberg and Garry Haenhle from the department's Family Protection Unit; Lisa Weinreb, Mark Amador, and Michael Runyon from the Gang Unit; Michael Groch, from the High-Tech Crimes Division; and Blaine Bowman, from the Superior Court Division. Likewise, all of the series' defendants were the genuine article. In the debut episode, an inordinately ill-tempered fellow was charged with killing his wife, even though no body or evidence had been found; it seems that the gentleman had indiscreetly told everyone exactly what he was planning. Subsequent weeks offered a vast and disgusting array of baby killers, pedophiles, rapists and other such wretches. Some fifty trials were filmed by Bill Guttentag for the series' first season, from which thirteen — undoubtedly the most lurid — were chosen for telecast.

In keeping with Standard *Law & Order* Operating Procedure, the prosecutors on *Crime & Punishment* were invariably the heroes, the defendants nearly always the villains — even before their guilt had been firmly established. Not surprisingly, Dick Wolf received 100 percent support from the deputy DAs who appeared on the show, and was given hosannahs even by those attorneys who represented the "other side": quoted in *TV Guide*, San Diego public defender Steve Carroll congratulated the producer for capturing "the essence" of each trial and doing a "decent job" with the project. Other observers had reservations about Wolf's black-and-white treatment of the criminal justice system. "As reality TV, it's riveting, addictive and well told," commented *Time* magazine's James Poniewozik in his critique of *Crime & Punishment*. "As a civics lesson, it's manipulative and tendentious. We have access only to the DAs, so the presumption of innocence, unpopular with crime-show viewers anyway, gets 86ed, and every emotional cue prods us to root for 'guilty'— even the show's title. You don't get ratings making *Accusation & Verdict*."

Completing a healthy two-year run in 2004, *Crime & Punishment* was rebroadcast by cable's MSNBC beginning in 2006.

Cristina's Court

Syndicated: Debuted Sept. 11, 2006. Schordinger's Cat Productions/20th Television. Executive Producer: Peter Brennan. Cast: Judge Cristina Perez (Herself).

It isn't often that you come across a TV judge who has achieved stardom in two different languages — but then, Cristina Perez isn't your run-of-the-mill TV judge. Born in New York to a family of Colombian immigrants, the beauteous, blonde-haired Perez didn't speak English until she was ten years old, then more than made up for lost time by earning degrees at UCLA and Whittier College. Establishing herself as a top-rank jurist and best-selling author, Perez made it her mission to "raise the consciousness" of the Latino community, emphasizing the necessity of a strong work ethic and unswerving pride in the community's cultural roots. Her earliest foray into television was as presiding judge on *Corte de Pueblo*, the local Los Angeles Spanish–language version of the long-running *People's Court* (q.v.). (Unlike the original series, the cases on *Corte de Pueblo* were reenacted.) She went on to become the first female judge to star in a nationally telecast

Spanish–language courtroom series, Telemundo's *La Corte de Familia*, and at the same time cohosted a weekly legal program on Radio Univision. Although she remained largely monolingual on TV, Perez developed enough of a non–Hispanic fan following to warrant a daily syndicated English–language courtroom show — and thus *Cristina's Court* came into being in the fall of 2006.

Like its template series *La Corte de Familia*, the half-hour *Cristina's Court* zeroed in on legal disputes between married couples and family members: typical episodes included "Spoiled Son's Unpaid Loan" and "My Tire-Slashing Ex Stole My Class Ring." Though she brooked no nonsense from the litigants and was known to display a fiery temper when pushed too far, Perez was careful not to emulate the more vituperative style of Judge Judy (q.v.), with whom she shared the same executive producer (Peter Brennan). "There are a lot of 'TV Court Show' styles and the ones I like the least are the ones that mock the litigants," explained the Judge in a 2007 *New York Post* interview. "We've seen some great family tragedies on *Cristina's Court* and I'm not going to feel comfortable belittling someone's story." Indeed, Perez frequently asked the plaintiffs and defendants to retire to her chambers where she could calmly and without rancor help resolve their problems in private (or as private as it was possible to get with the cameras running). "I want to make sure whoever comes to my courtroom is a little bit stronger and united … and I don't think TV ratings are enhanced by the decibel level." She did, however, apparently think that the occasional celebrity would hype the viewership: during the series' 2007–2008 season, the litigants included ex–*Munsters* star Butch Patrick and popular street entertainer Robert Burck, better known as The Naked Cowboy.

Taped at the Houston studios of Fox–owned KRIV-TV, *Cristina's Court* was syndicated by 20th Television, often showing up on two different stations in the same market — one of the many fringe benefits of "multiple ownership."

Curtis Court

Syndicated: Sept. 2000–Sept. 2001. King World Productions. Produced by Karen Bosnak and Lee Navlen. Cast: The Honorable James E. Curtis (Himself, judge); Anthony Pasquin (Himself, bailiff).

A short-lived and barely-remembered entry in the daily-courtroom-series genre, the syndicated, half-hour *Curtis Court* was presided over by former Riverside County (California) Deputy DA James E. Curtis. Admitted to the California State Bar in 1989, the soft-spoken Curtis had been involved in a number of high-profile cases before joining Riverside's Juvenile Crimes division. Just prior to retiring from office in 1999, Curtis branched out as a motivational speaker and founded the Justice Project Incorporated, a consulting firm dealing in law-enforcement relations. He also enjoyed some media exposure as host of a radio call-in show.

According to its star, the purpose of *Curtis Court* was to explain and clarify the intricacies of the legal system so the Average Joe or Josephine could understand them without consulting a law dictionary. In a further effort to simplify matters, Curtis eschewed the usual practice of featuring several cases within a single episode, generally focusing on a single trial, frequently punctuated with visits to the crime scene and films of the evidence. In other respects, however, *Curtis Court* was the same mixture as before. Adhering to the rules established by such court-show brethren as *People's Court* and *Judge Judy* (both q.v.), litigants in small-claim cases were strictly limited as to how much they could collect in damages: the "ceiling" was $3000. Also, Judge Curtis had his own "Rusty Burrell" in the form of bailiff Anthony Pasquin, a New York City police officer.

Though his series was discontinued in 2001 after a single season, James E. Curtis remained in the public eye by joining the ever-growing roster of legal commentators on cable's Court TV, where he co-emceed the series *Closing Arguments* with Lisa Bloom. On a lighter note, Curtis also hosted the semi-satirical reenactments of the Michael Jackson trial on the E! Entertainment channel, apparently on the assumption that a distinguished African American jurist would be less prone to adverse criticism for participating in such a frivolous project.

The D.A.

NBC: Sept. 17, 1971–Jan. 7, 1972. Mark VII Ltd. Productions/Universal. Executive Producer: Jack Webb. Cast: Robert Conrad (Deputy D.A. Paul Ryan); Harry Morgan (Chief Deputy District Attorney H. M. "Staff" Stafford); Ned Romero (D.A. Investigator Bob Ramirez); Julie Cobb (Deputy Public Defender Katherine Benson).

Produced by Jack Webb, *The D.A.* was a game attempt to utilize the quasi-documentary "procedural" method popularized by the producer's long-running cop drama *Dragnet* within the confines of the legal-show genre. As noted by Webb biographer Michael Hayde, the series was conceived as "the other half of *Dragnet:* a courtroom drama presented from the viewpoint of the Los Angeles District Attorney's Office (which, naturally, provided technical assistance)."

Before the series proper got under way, Webb and NBC tested the waters with a brace of TV–movie pilots, *D.A.: Murder One* (originally telecast December 8, 1969) and *D.A.: Conspiracy to Kill* (January 11, 1971). Both films starred Robert Conrad, fresh from his successful run on the tongue-in-cheek adventure series *The Wild Wild West*, as dynamic young deputy DA Paul Ryan, who regularly pulled double duty as both investigator and prosecutor. A man who tended to follow "crazy" hunches while carefully remaining within the limits of his authority, Ryan invariably nabbed the perpetrator in record time (four days was the usual span), then saw to it that justice was served in the courtroom. Acting as Ryan's legman was DA Investigator Bob Ramirez, a role played by Carlos Romero in *Murder One* and Armando Silvestre in *Conspiracy to Kill*. When *The D.A.* finally achieved weekly series status, Conrad retained the role of Paul Ryan, while Ned Romero filled the well-worn shoes of Bob Ramirez. For the weekly version, two additional regulars were added: former *Dragnet* costar Harry Morgan as Ryan's gruff but supportive boss, Chief Deputy "Staff" Stafford, and Julie Cobb (daughter of Lee J. Cobb) as Ryan's friendly adversary, Public Defender Katherine Benson.

Each 30-minute episode was more or less divided into two halves, the first half focusing on the investigation of the crime and the second on the trial and conviction. Beyond this dramatic device, *The D.A.* adhered slavishly to the *Dragnet* formula, with Robert Conrad providing deadpan offscreen narration while his character Paul Ryan laboriously gathered evidence and rounded up witnesses, and with a plethora of technical jargon to satisfy the "legal junkies" who comprised most of Jack Webb's fan following. (Sample dialogue: "Well, you do anything worthwhile for the taxpayers today?" "Could be. I got me a '459' going on in Halder's court.")

When *The D.A.* premiered in September, 1971, the series was bracketed on the Friday night schedule with another Jack Webb effort, *O'Hara: U.S. Treasury*— though thanks to the vagaries of the TV marketplace, *D.A.* was on NBC while *O'Hara* was on rival network CBS. Despite its proven audience appeal, *The D.A.* fared poorly for a variety of reasons: with rare exceptions, half-hour dramatic series had become passé by 1971 (one of the few 30-minuters left on the air was Webb's own *Adam-12*, which was represented on *The D.A.* with the crossover episode "The Sniper"); younger viewers of the period were turned off by programs with a pro–Establishment stance; star Robert Conrad was better suited for action roles than his current coat-and-tie assignment; and most damaging of all, the series was scheduled smack-dab opposite ABC's *The Brady Bunch*.

Cancelled after 15 episodes, *The D.A.* was replaced by a little item called *Sanford and Son*. It would take another eighteen years before TV producer (and lifelong Jack Webb fan) Dick Wolf was able to seize and expand upon the *D.A.* format for his own *Law & Order* (q.v.)

The D.A.

ABC: March 19–Apr. 9, 2004. Shephard/Robin Company-Warner Bros. Television. Executive Producers: Greer Shephard, Michael M. Robin, James

Duff. Cast: Steven Weber (DA David Franks); Bruno Campos (Deputy DA Mark Camacho); Sarah Paulson (Chief Deputy DA Lisa Patterson); J.K. Simmons (Deputy DA Joe Carter); Michaela Colin (Jinette McMahon).

Not to be confused with the failed Jack Webb series of the same name (see previous entry), *The D.A.* was an entirely different failed series, assembled by the same production team responsible for the cutting-edge medical drama *Nip/Tuck*.

Set in the Los Angeles District Attorney's office, the weekly, 60-minute series bore a passing resemblance to the "dirty laundry" *Nip/Tuck* formula by spending less time in the courtroom than in probing the character flaws of the principal players (this despite the fact that the series' technical consultant was former LA district attorney Gil Garcetti). Steven Weber was top-billed as youthful DA David Franks, the office's head man, whose pursuit of justice was often hindered by his lust for publicity and his political ambitions. The office's chief investigator — and resident hothead idealist — was Deputy DA Mark Camacho (Bruno Campos), who was not above twisting a few influential arms to serve the Cause — and who frequently halted mid-investigation to make long-winded speeches about his incorruptibilty. The love-hate relationship between Franks and Campos quickly came to a head during the opening episode, in which a leak in the DA's office at the height of an investigation of the Russian Mafia resulted in three embarrassing murders, placing Franks' re-election in jeopardy and intensifying Camacho's disdain for his colleagues. Others in the cast were Sarah Paulson as Franks' feisty second in command, Chief Deputy DA Lisa Patterson; J.K. Simmons as Patterson's verbal sparring partner, Deputy DA Joe Carter; and Michaela Colin as Franks' ruthless campaign manager Jinette McMahon. Though billed as a guest star, Felicity Huffman (*Desperate Housewives*) was prominently featured in the bulk of the episodes (each of which bore a variation of the title "The People vs...") as Charlotte Ellis.

Existing evidence indicates *The D.A.* spent a long time on the shelf before ABC gave the series a four-week run in the Friday evening timeslot usually reserved for *20/20*. The network's publicity people indicated that the show had been slated for a limited run from the outset, proclaiming that it would initially be seen for "just four murders, four cases and four weeks," and suggesting production would immediately resume if the audience so demanded. Most TV critics smelled a rat, opining that ABC had taken one look at the first four *D.A.* episodes and decided to throw in the towel before the series even opened: "*DOA* would be more like it," began the review provided by *USA Today*'s Robert Bianco, who summed up the finished product as "terribly ordinary."

Damages

FX: Debuted July 24, 2007. KZK Productions/FX Productions/Sony Pictures Television. Executive Producers: Todd A. Kessler, Daniel Zelman, Glenn Kessler. Cast: Glenn Close (Patty Hewes); Rose Byrne (Ellen Parsons); Ted Danson (Arthur Frobisher); Tate Donovan (Tom Shayes); Zelijko Ivanek (Ray Fiske); Anastasia Griffith (Katie Connor); Noah Bean (David Connor).

No one has ever summed up the credo of cable's FX Network better than its president, John Landgraf, in an interview for *Entertainment Weekly*: "For many years, network television had a very simplistic formula. Good guys win and bad guys lose. That just isn't the way the world works. If crime did not pay, no one would do it." Indeed, the FX output schedule was rife with iconoclastic series featuring "heroes" who, if not blatantly corrupt (detective Vic Mackey in *The Shield*), irredeemably larcenous (the family of grifters in *The Riches*) or scrupulously unscrupulous (gossip-rag editor Lucy Spiller in *Dirt*) were at the very least deeply flawed and profoundly troubled individuals (the protagonists of *Nip/Tuck* and *Rescue Me*). With this in mind, the barracuda-like lawyer heroine of FX's *Damages* must have taken one glance around at her network neighbors and sighed, "There's no place like home."

Packaged by the same people responsible for HBO's crime-family saga *The Sopranos*, *Damages* was initially planned as a spinoff of FX's *The Shield*, with Glenn Close repeating her role from the earlier series as gimlet-eyed LA police captain Monica Rawling. As the project took shape, the backdrop of a corruption-ridden California police force was abandoned in favor of an ethically challenged Manhattan law firm. Now Close was cast as Patty Heyes, a highly respected and greatly feared plaintiff's attorney who thrived on going directly for the jugular and tying the legal system in knots of Gordian complexity. Every bit as ruthless and conniving as the crooked defendants she attacked in the courtroom, Heyes spent most of the series' first season pursuing a nine-figure class action suit against billionaire Arthur Frobisher (Ted Danson), a corporate CEO who made Ken Lay look like Mahatma Gandhi. Heyes' legal team included her idealistic protégée Ellen Parsons (Rose Byrne) and her favorite whipping boy Tom Shayes (Tate Donovan). Acting in Frobisher's defense was human rattlesnake Ray Fiske (Zelijko Ivanek), who evidently was not above commissioning a few murders to win acquittal for his client.

To whip up audience interest from the outset, *Damages* opened with a shocker, as Ellen Parsons, wild-eyed, half-naked and spattered with blood, ran through the streets of New York, raced into her office, and in a voice drenched with terror declared, "I need a lawyer!"—just before she was arrested for the murder of her fiance. The rest of Season One was a buildup to this critical moment, with fragmentary clues and morsels of exposition revealed on a "need to know" basis. This story structure was described by producer Kessler as having the grim inevitablity of a Greek tragedy: "You're able to play up the drama of knowing where you're heading." And this being an FX effort, the dividing line between "Good" and "Bad," gossamer-thin to begin with, became all but invisible as the story progressed.

It was planned that each successive season of *Damages* would open with a similarly tantalizing "teaser," and that Patty Heyes would pursue a single prosecution per year, just as she'd previously devoted thirteen hour-long episodes to the destruction of Arthur Frobisher. As of this writing, Season Two of *Damages* has yet to materialize, but industry insiders are cer-

tain that FX will revive the property as soon as it is economically feasible—and assuming that the Hollywood production machine is not slowed down by another writer's strike, as happened in 2008. (Ironically, when Glenn Close won a Golden Globe for her performance as Patty Heyes, she refused to cross picket lines and was obliged to watch the televised awards ceremony from a neighborhood bar.)

Dark Justice

CBS: Apr. 3, 1991–Apr. 4, 1994. David Salzman Entertainment/Televisio de Cataluna/Magnum Productions/Lorimar Television. Executive Producers: David Salzman, Jeff Freilich. Cast: Ramy Zada (Justice Nicholas "Nick" Marshall, Season One); Bruce Abbott (Judge Nicholas "Nick" Marshall, Seasons Two and Three); Dick O'Neill (Arnold "Moon" Willis); Clayton Prince (Jericho "Gibs" Gibson); Begonya Plaza (Catalina "Cat" Duran); Vivivan Vives (Maria Marti); Carrie-Ann Moss (Tara McDonald); Janet Gunn (Kelly Cochrane); Kit Kincannon (D.A. Ken Horton); Elisa Heinsohn (Samantha "Sam" Collins).

There are undoubtedly a number of real-life judges who, frustrated by a legal system that seems to be a revolving door through which criminals can pass unmolested, would love to indulge in the vigilante antics of Judge Nicholas "Nick" Marshall on the TV series *Dark Justice.* However, most of these real-life judges would probably not be able to squeeze into Marshall's tight black leather jacket—nor even fit into the saddle of his hopped-up motorcycle.

One of the youngest and handsomest jurists on the bench, Nick Marshall had worked his way up from police officer to district attorney, only to become bitterly disillusioned with traditional Law and Order procedure after the scumbags who set the car bomb that killed Nick's wife and daughter were allowed to walk out of the courtroom with nary a slap on the wrist, thanks to one of those pesky "technicalities." Determined to bring the full weight of the law down on those well-connected crooks who'd managed to loophole their way out of prison, Marshall began to lead a double life: black-robed judge by day, leather-clad vigilante by night. Rarely resorting to violence,

Marshall preferred to set up elaborate sting operations, wherein the bad guys would be trapped, incriminated and defeated by their own greed. And in the spirit of fair play, the long-haired Judge would even warn his victims-to-be that he was on their trail; after a particularly reprehensible scofflaw managed to slip through his fingers in the courtroom, Marshall would say in a calm, measured tone, "Justice may be blind … but it can see in the dark."

No "lone wolf" he, Judge Marshall worked with an elite team of crimefighters known as The Night Watchmen (even though we never saw any of them winding a clock or eating a sub while watching TV). Among Nick's cohorts during the series' three-year run were reformed con artist/counterfeiter and health-club owner Moon Willis (Dick O'Neill), ex-street punk turned special-effects whiz Gibs Gibson (Clayton Prince), and a variety of female assistants. During the first season, sexy day-care owner Cat Duran (Begonya Plaza) was part of the team; when Cat was killed in a gun battle, her replacement was former Interpol agent Maria Marti (Viviane Vives), who after moving to Boston was herself replaced by Sam Collins (Elisa Heinsohn). Other occasional distaff members of the Night Watchmen (technically Night Watchwomen, one supposes) were Marshall's faithful secretary Tara McDonald (Carrie-Anne Moss); onetime private eye Kelly Cochran (Janet Gunn); and Kari-Linn (Joanne Haas), a waitress in the bar purchased by Moon Willis during Season Three. Rounding out the regulars was Kit Kincannon as DA Ken Horton, who, unaware of Marshall's dual identity, was understandably perplexed over the huge number of "clever" lawbreakers who ended up being stupidly hoist on their own petard.

A casual viewer might assume that *Dark Justice* took place in a large, unnamed American city. Actually, the first season was lensed in Barcelona, Spain, with Ramy Zada — who though born in America qualified as a Spanish actor for the purposes of the producers' filming permit — starring as Judge Marshall. When the 1992 Summer Olympics caused a sharp increase in labor costs, the producers left Barcelona and relocated to Los Angeles. Reportedly, Zada was unable to make the move due to professional commitments (production of the series did not resume for nearly a year), thus he was replaced during the final two seasons by Bruce Abbott.

In the United States, *Dark Justice* aired as a component of CBS' late-night *Crimetime After Primetime* package, along with such other adventure series as *Sweating Bullets* and *Silk Stalkings*. Though production wrapped in early 1993 after 65 hour-long episodes, reruns of *Dark Justice* remained on the *Crimetime After Primetime* schedule until April of 1994.

The D.A.'s Man

NBC: Jan. 3–Aug. 29, 1959. Mark VII Productions Ltd. Executive Producer: Jack Webb. Produced by Frank LaTourette. Based on the book by James B. Horan and Harold Danforth. Cast: John Compton (Shannon); Ralph Manza (Assistant DA Al Bonacorsi); Herb Ellis (Frank LaValle).

With his long-running *Dragnet* entering the home stretch in 1958, producer Jack Webb was interested in developing newer, fresher TV properties. Ever since he'd played erstwhile private detective Pat Novak on radio, Webb had wanted to produce a TV series centering around a tough, maverick private eye. The popularity of such series as *Peter Gunn* and *The Thin Man* made the P.I. genre more saleable than ever in 1958, but by this time Webb was firmly committed to *non*-private, "official" police procedure as typified by *Dragnet*. It was Frank LaTourette, in charge of program development for Webb's Mark VII Productions, who located a property which combined the best of two genres by featuring a P.I. hero who, despite a streak of fierce independence, worked exclusively for the New York City District Attorney's Office. Written by James B. Horan and Harold "Dan" Danforth, *The D.A.'s Man* was a nonfiction novel based on Danforth's real-life experiences as a special investigator for the NYC Special Rackets Prosecutor and the Manhattan DA's office.

In the TV version of the Horan-Danforth book (for which Danforth functioned as technical adviser) the hero, renamed Shannon, was an ex-private eye turned undercover agent who specialized in gathering evidence necessary to bring big-time criminals before a Federal Grand Jury for indictment. In the first episode, "Sammy's Friend," it was established that Shannon would be given a free hand to use whatever means necessary to entrap scofflaws on behalf of his boss, Assistant DA Al Bonacorsi (played by Ralph Manza). There was only one condition: Shannon was not allowed to carry a gun or any sort of identification. While this made it easier for him to impersonate a variety of petty crooks and rootless drifters as he infiltrated such operations as a mob-connected gun rental store, a phony union, a crooked insurance firm and a prostitution ring, the lack of weaponry and credentials also meant Shannon could be killed at any moment — not only by the Mob, but also by the Police. (After all, how could the cops possibly know that he was working on their side?)

John Compton, star of *The D.A.'s Man*, was a virtual unknown who most likely was hired *because* he was unknown: since Shannon's survival depended upon his anonymity, it was better to cast an unfamiliar face than a high-profile actor. Far more recognizable was the series' guest-star lineup, which included such detective-show perennials as Jack Kruschen, George Macready, John Beradino, Harry Dean Stanton, Paul Burke, Robert Cornthwaite and Johnny Seven. Among the scriptwriters were future *Twilight Zone* stalwarts Richard Matheson and Charles Beaumont, while the directors included producer Jack Webb and his *Dragnet* costar Ben Alexander.

Sponsored by Chesterfield cigarettes (no, there weren't any episodes in which Shannon exposed an illicit tobacco operation), the weekly, half-hour *D.A.'s Man* premiered January 3, 1959, as a replacement for the NBC quiz show *Brains & Brawn*. According to Jack Webb biographer Michael Hayde, neither viewers nor critics were impressed: *Variety* dismissed the pilot episode (written by James E. Moser of *Medic* fame) as "[not] polished enough to make the viewer forget that he was seeing something he'd seen on the TV and motion picture screen time and time again." But though *The D.A.'s Man* folded after 26 episodes, the "undercover man" premise was viable enough to warrant

another, more popular series: *Tightrope!*, which debuted in the fall 1959 with future *Mannix* star Mike Connors in the lead — and which, unlike *The D.A.'s Man*, downplayed the hero's links with the "Feds" to focus instead on action and suspense.

Day in Court

ABC: Oct. 13, 1958–June 24, 1965. Produced by Selig J. Seligman and Gene Banks. Cast: Edgar Allan Jones Jr. (Presiding Judge); William Gwinn (Presiding Judge — occasional). The cast of the 1964–65 "continuing drama" version included Keith Andes, Betsy Jones-Moreland, Jan Shepard, Tom Palmer, Ann Doran, Maggie Mahoney, Gloria Talbott, Russel Arms, Coleen Gray and Gerald Mohr.

ACCUSED
ABC: Dec. 10, 1958–Sept. 30, 1959. Produced by Selig J. Seligman. Cast: Edgar Allan Jones Jr. (Presiding Judge); William Gwinn (Presiding Judge — occasional); Tim Farrell (Bailiff); Jim Hodson (Clerk); Violet Gilmore (Reporter).

MORNING COURT
ABC: Oct. 10, 1960–May 12, 1961. Produced by Selig J. Seligman. Cast: William Gwinn (Presiding Judge).

When ABC expanded its daytime schedule with "Operation Daybreak" in the fall of 1958, the network's executives were well aware they didn't have the budget nor production wherewithal to create the sort of news/talk and soap-opera lineups flourishing on NBC and CBS. Thus, in addition to its established late-afternoon favorites *Who Do You Trust?*, *American Bandstand* and *The Mickey Mouse Club*, the network trafficked in inexpensive game shows and musical programs. "Operation Daybreak" offered one additional modestly-priced series: a spinoff of a popular local Los Angeles attraction, produced by onetime State Department attorney Selig J. Seligman, which ABC hoped would match the popularity of Seligman's earlier CBS courtroom daytimer *The Verdict Is Yours* (q.v.).

Though it was familiar only to viewers within range of Los Angeles' KABC-TV, *Day in Court* was in a sense a proven commodity, being an outgrowth of another KABC "local" which went national in June of 1958, *Traffic Court* (q.v.). Like *Traffic Court, Day in Court* specialized in reenactments of actual court cases, with actors as the defendants, plaintiffs and witnesses and featuring a roster of genuine lawyers (including rookie attorney David Glickman, who was also frequently seen on the syndicated *Divorce Court* [q.v.]). Carried over to *Day in Court* was *Traffic Court*'s presiding "judge" Edgar Allan Jones Jr., a boyishly handsome 39-year-old UCLA law professor who taped his appearances on a tight three-day schedule so as not to interfere with his teaching duties. (Since he was not a member of the State Bar of California, Jones was able to avoid the "judicial ethics" quandary that had compelled his *Traffic Court* predecessor, Judge Evelle J. Younger, to leave the series when it got a sponsor.)

Backed up by the rhythmic pounding of a bass drum, each half-hour episode of *Day in Court* opened with the booth announcer promising the series would offer "actual trials and hearings, simulating the workings of justice." A team of third-year UCLA law students handled the research end of the series, poring over dozens of published appellate decisions or applications of principles in restatement to come up with the weekly quota of single-page story synopses. Though the series used professional writers to maintain the continuity and dramatic flow of each episode, much of the dialogue was improvised during rehearsal. The actor-litigants were supplied with the principal facts of the actual case at hand (invariably a civil or criminal complaint), while "Judge" Jones made his ruling based on the summary provided him by the writers. Occasionally, Jones would reverse the decision he'd made in rehearsal if the actors forgot to repeat the vital points of the case; and in more than one instance, an ambitious member of the writing staff, unhappy with Jones' decision, would ask permission to alter those facts so that the Judge would be forced to come up with a different verdict.

For a man who had never presided over a genuine court case — and in fact had never set foot in a TV studio before *Traffic Court*— Edgar Allan Jones was one of show business' best and quickest adlibbers, especially whenever he re-

alized the actors had inadvertently left out several pages of testimony. In addition, he came in handy whenever the writers were confused or mistaken over a specific aspect of the law: one network executive referred to Jones as "our built-in technical adviser." He was also a built-in editor, going through each script to eliminate unnecessary or inappropriate verbiage: "Writers are not lawyers, and they inevitably set up things that are not admissible as evidence or that don't fit in with the substance of the law." So crucial was Edgar Allan Jones to the success of *Day in Court* that the series' ratings (quite impressive for third-string ABC) would drop noticeably on the two days per week Jones was spelled by "alternate judge" William Gwinn, a former law professor turned TV announcer who handled the series' domestic-relation cases.

The popularity of *Day in Court* inspired ABC to serve up a weekly Prime Time version, which premiered under the same title on December 10, 1958, then was rechristened a few weeks later as *Accused*. Edgar Allan Jones and William Gwinn continued alternating as judge, with a regular cast of professional actors appearing as bailiff, court clerk and reporter. Though *Accused* was cancelled in the fall of 1959, the daytime original continued to flourish, prompting ABC to create *another* spinoff, this one titled *Morning Court*, which aired Monday through Friday from 11 to 11:30 A.M. during the 1960–61 season. By now too busy to tackle a second daily series, Edgar Allan Jones took a pass on *Morning Court*, thereby giving William Gwinn his own solo starring vehicle.

Of the hundreds of *Day in Court* telecasts seen during its seven-season run, the one that most people remember, and the one that was restaged most often "by popular demand," was its Christmas Day show, which was not based on any court record but instead was lifted from a different and much older literary source. On this one, an immigrant couple named José and Maria came into Judge Jones' courtroom to lodge a complaint against a cruel motel owner who had refused to give them a room, claiming there were no vacancies. The couple was forced to spend the night in the motel garage, where the pregnant Maria gave birth to a child named ... shucks, you're way ahead of us, aren't you? (Alas, this little parable no longer exists, nor does any other episode of *Day in Court:* every one of the series' videotapes was "wiped" thirty days after telecast.)

In 1962 Gene Banks became producer of *Day in Court*, immediately instituting a number of changes. Where previously most of the actors playing the litigants were nondescript Hollywood day players, Banks, a lifelong movie buff, assembled a new talent pool of veteran performers with "name" value. Appearing in a variety of one-shot roles were silent-film stars Neil Hamilton, Viola Dana and Betty Bronson, B-picture stalwarts Heather Angel, Gertrude Michael, Isabel Jewell, Ted DeCorsia, Tom Brown, Madge Meredith, Iris Adrian and Claire Carleton, radio personalities Ransom Sherman and Barbara Luddy, and even dialect comedian Benny Rubin. These actors, noted Banks affectionately if somewhat inaccurately in a *TV Guide* interview, "constitute a lot of talent that is not being used, and we are lucky to get them for scale. We try to give them star treatment — which some of them demand — because they were once at the top of the nation's marquees. Most of them are so delighted about seeing old friends that it's a ball working with them."

Banks' most radical change was dictated by the popularity of ABC's first serialized drama, *General Hospital*, which premiered in April of 1963. It was decided in the fall of 1964 to add *Day in Court* to a 90-minute soap opera block along with *General Hospital* and a new arrival, *The Young Marrieds*. Inasmuch as *Day in Court*'s self-contained, fact-based episodes did not conform with the "continuing story" format of its two companion series, the program was completely overhauled. On September 30, 1964, the series launched its first three-part serialized storyline, a work of total fiction about a woman charged with child desertion. Later story arcs ran anywhere from five to ten episodes, featuring a cast of regulars headed by Jan Shepard and Tom Palmer as the opposing attorneys.

There may have been a few hardcore soap fans who applauded this change, but UPI columnist Rick Du Brow was not one of them: "It is irritating to watch how ABC-TV's respectable afternoon show, *Day in Court*, has been turned into a soap opera in the current network trend toward serials." Likewise displeased by this turn of events was Edgar Allan Jones, who quit the show cold when it became "the continuing story of *Day in Court*," relinquishing his robe and gavel to faithful old William Gwinn. "*Day in Court* and television have lost a remarkable performer in Jones," concluded Du Brow, and audiences agreed. The "new and improved" series hobbled along to rapidly declining ratings until it was finally put out of its misery in the summer of 1965. (Edgar Allan Jones retired in 1991, after 40 years at UCLA — then promptly launched a new career as a novelist.)

The Defenders

CBS: Sept. 16, 1961–Sept. 9, 1965. Herbert Brodkin/Plautus Productions. Created by Reginald Rose. Executive Producer: Herbert Brodkin. Cast: E.G. Marshall (Lawrence Preston); Robert Reed (Kenneth Preston); Polly Rowles (Helen Donaldson); Joan Hackett (Joan Miller).

Outside of *Perry Mason* (q.v.), *The Defenders* is the best-remembered and most highly regarded legal series of TV's black-and-white era. That *Perry Mason* is still flourishing in reruns while *The Defenders* is generally available only to archivists may say a lot about the American Viewer's overall preference for a series in which the Good Guys are always right and always win, over a series in which there are seldom any easy answers or pat solutions.

The genesis of *The Defenders* was a real-life manslaughter trial held in New York in 1954. Among the jurors deliberating within the confines of the Foley Square Courthouse was a young TV writer named Reginald Rose. Fascinated by the "terrific, furious" eight hours spent in the jury room, Rose used the experience as the basis for a TV play, *Twelve Angry Men*, which served as the seventh-season opener for the CBS anthology *Studio One* on September 20, 1954, and subsequently yielded a widely-performed stage piece and two feature films (as well as an "alternate" theatrical version, *Twelve Angry Women*).

A little over two years later, Rose returned to the courtroom-drama genre with another *Studio One* contribution, a two-part drama originally telecast February 25 and March 4, 1957, under the title *The Defender*. Ralph Bellamy and William Shatner headed the cast as Walter and Kenneth Pearson, a father-son attorney team who took on the defense of a belligerent and supremely uncooperative murder suspect, played with blazing intensity by a pre-stardom Steve McQueen. Though Walter Pearson was convinced that his client was guilty, son Kenneth argued that this wasn't as important as establishing "reasonable doubt" during the trial, which he did with an audacious grandstand play. While the conservative Pearson Sr. expressed harsh disapproval of his son's flamboyance, Pearson Jr. persisted in the argument that providing their client with the best possible defense superceded any other consideration. To underline this point, *The Defender* ended with neither the lawyers nor the audience truly knowing whether the acquitted McQueen was guilty or not (This open-ended denouement made it possible for the producers of the much-later TV series *Boston Legal* [q.v.] — which also starred William Shatner — to use excerpts from the original kinescope of *The Defender* as flashbacks in the 2007 episode "Son of the Defender").

Not long afterward, Reginald Rose decided to shape an entire weekly series around *The Defender*, a bid to provide television with a courtroom series that would deal with legal issues in subtle shades of gray rather than the dogmatic black-and-whites of *Perry Mason*, and would also enable Rose to indulge in the sort of penetrating social commentary he did so well. Even though network and sponsorial interference had previously diluted such powerful Rose pieces as "Thunder in Sycamore Street" (*Studio One*, March 16, 1954), which started life as a slashing attack upon race prejudice but ultimately emerged as a non-specific indictment of prejudice in general (the African-American character who'd originally been victimized by bigots was transformed into a white ex-convict), the writer was confident that Herbert Brodkin, executive producer of *The De-*

fenders (its title had been pluralized during the development stage), would avoid artistic compromises as much as possible. Besides, both Rose and Brodkin had been assured that CBS *wanted* a series that would provoke thought and stir up controversy, if only to disprove the recent assertion by FCC chairman Newton Minow that television was a "Vast Wasteland."

Not that CBS was being entirely altruistic. As noted by Reginald Rose to TV historian Jeff Kisseloff, the series made the network heads "nervous as hell, but whenever they were criticized for doing terrible shows they would say, 'Yeah, but we do *The Defenders*. We don't always do crap,' and that was the only real reason they kept it on." Michael Dann, the CBS programming head who would always take credit for demanding that the network accept *The Defenders* in the first place — and for battling with his fellow executives to keep it on — admitted to Kisseloff that CBS president William Paley had ultimately given the green light to the series principally because the previous offering in the Saturday night 8:30–9:30 slot, the detective opus *Checkmate*, was doing so poorly that anything would have been an improvement.

According to Dann, *Defenders* might have debuted earlier than the fall of 1961, "but had been kept off the air because [network head] Jim Aubrey said, 'We don't want that junk on the air. It's just pleading causes.' Every episode of *The Defenders* dealt with a social issue.... It became a hallmark around the world." The series also proved to Dann that "you should never rule against the producer. Pat [Weaver] once said to me, 'The function of bureaucrats is to pick the producer and support the producer. Once in a while he'll be wrong and you will be right, but about ninety-nine times in a hundred he'll be right, because his life is on the line.'"

Ironically it was Jim Aubrey, the man so opposed to *The Defenders*' very existence, who frequently cast himself in the role of White Knight crusading to uphold the series' integrity. In Robert Metz' book *CBS: Reflections in a Bloodshot Eye*, an unnamed CBS executive, citing the brouhaha surrounding a particularly volatile *Defenders* episode, recalled, "There wasn't an advertiser who would touch it, and we wound up at a meeting with a rep from every agency in town in the CBS boardroom, and the head of the network [Aubrey] asking why they didn't have more guts. An hour later he was bawling out programming for ever having authorized it."

By the time *The Defenders* went to series, Ralph Bellamy and William Shatner were too busy to recreate their roles, though Shatner would frequently turn up as a *Defenders* guest star, notably in the episodes "The Invisible Badge" and "The Cruel Hook" in which the actor played prototypes of his starring role as Assistant DA David Koster in another Brodkin-produced series, *For the People* (q.v.). (Another *Defenders* episode, "The Traitor," would be developed into Brodkin's short-lived but much-admired 1967 series *Coronet Blue*.) Chosen to play the father-son law firm's senior partner, now renamed Lawrence Preston, was E.G. Marshall, a consummate professional previously seen as "Juror #4" in the 1957 film version of *Twelve Angry Men* and defense attorney William Thompson in another two-part Reginald Rose TV drama, "The Sacco–Vanzetti Story" (*NBC Sunday Showcase*, June 3 and 10, 1960). Lawrence Preston's son, still named Kenneth, was played by comparative newcomer Robert Reed, with whom Marshall had previously worked in a Chicago stage production. Though the basic character conflict of the original *The Defender* remained, with the hidebound Preston Sr. usually insisting upon following the letter of the law no matter what the circumstances while the more idealistic Preston Jr. was willing to circumvent the rules in pursuit of a "greater" justice, the warm rapport and mutual respect exhibited by E.G. Marshall and Robert Reed played much better than the uneasy détente between Ralph Bellamy and William Shatner. Also unlike their predecessors, the two Prestons were able to leaven the seriousness of their surroundings with a healthy sense of humor—a byproduct, one suspects, of E.G. Marshall's fondness for breaking up his fellow actors with smutty one-liners and non sequiturs during rehearsal (and sometimes even

during filming, if a well-circulated *Defenders* blooper reel is any indication!).

Critics enthused over *Defenders* as an "antidote" to the artistic license exhibited on *Perry Mason*: whereas Perry invariably won his cases and his clients were uniformly innocent, it was not unusual at all for the Prestons to lose in court and for a client to be irrefutably guilty. In a 1964 interview, Reginald Rose addressed the fundamental diffrerence between his series and *Perry Mason* by stating, "the law is the subject of our programs: not crime, not mystery, not the courtroom for its own sake." But to blithely compare *Perry Mason* unfavorably to *The Defenders* is to misunderstand the purpose of both series. *Perry Mason* never pretended to be anything more than entertainment; *Defenders* strove to make a weekly "statement," and often as not such statements do not come wrapped in tidy little happy-ending packages. Also, as noted by media historian Ida Jeter in 1977, Perry Mason took the detective-story approach of uncovering new clues, while Lawrence and Kenneth Preston worked along more orthodox legal lines with the existing evidence; it was hard to clear a client at the last minute when all you had was what was placed before you on the exhibit table. Finally, while Mason was obsessed with proving his clients' complete innocence, the Prestons usually strove more realistically to secure their clients a lighter sentence or a lesser charge. Both series' approaches were viable, and both succeeded admirably: the viewer has the right to prefer one over the other, but it is the height of arrogance to automatically suggest that *Defenders* was "better" *only* because it was different.

For all its adherence to judicial authenticity, *Defenders* was capable of being every bit as artificial as *Perry Mason*, especially in its "solutions" to weighty sociological issues. In the words of Ida Jeter, "*The Defenders'* overriding message seems to have been of accommodation to the realities of life ... [the series'] ideology is based primarily on a belief that men and women, once they are fully aware, once they replace blind passion with enlightened compassion, will do the right thing." In Real Life, ideologues seldom alter their viewpoint even when circumstances alter cases; we could name any number of American politicians or foreign despots as examples. And it goes without saying that not every controversy or mystery is instantly resolved by a single courtroom decision, as witness *Roe vs. Wade* or the O.J. case. Conversely, the characters in *The Defenders*— even the murderers and the misguided bigots — invariably learned an Important Lesson at the end of each episode, benefiting and sometimes growing from the experience. Unlike the smugly self-confident Perry Mason, the Prestons seldom had all the answers; even so, few episodes of *The Defenders* ended on a question mark.

The Defenders not only stirred up controversy but actively courted it, dealing with subject matter that was generally *verboten* on network television in the early 1960s. In the very first episode, "Quality of Mercy," the Prestons defended a doctor accused of euthanizing a Down's Syndrome baby (tactlessly labeled a "Mongoloid Idiot" in the original *TV Guide* listings). Subsequent episodes dealt with a cop who became an urban hero by killing the man accused of molesting the cop's daughter, only to find that the victim was innocent; an elderly Jew who murdered the ex–Nazi responsible for the extermination of the old man's family; the "wall of silence" surrounding doctors accused of malpractice; the selfish motives of certain liberal whites who became involved in black causes; the sanctity of the church confessional; the dangers of legally prescribed narcotics; admission of evidence gathered by wiretapping; the draconian "three strikes and you're out" law for repeat offenders; and various stories involving illegitimacy, homosexuality, separation of Church and State, union corruption, neo-fascism, pornography and even cannibalism. A favorite plot device found the Prestons attempting to save the life of a murderer by pleading temporary insanity: this was used most effectively in the two-part "The Madman" and in "The 100 Lives of Harry Sims," the latter featuring Frank Gorshin as a schizophrenic nightclub impressionist who bumped off his girlfriend, then insisted that "James Cagney" actually committed the foul

deed — transforming into Cagney before the jury's eyes.

The single most controversial episode of *The Defenders*' first season was "The Benefactor," in which the Prestons came to the defense of an illegal abortionist. Written by Peter Stone (better known for such lightweight films as 1963's *Charade*), the episode was seen by only a fraction of its intended audience on March 28, 1962, with several CBS affiliates either flatly refusing to air the program or shifting it to a later timeslot. (This writer recalls the excitement stirred up in Milwaukee when the city's Hearst-owned CBS station vetoed the episode, whereupon it was broadcast on a two-week delay over a low-rated independent UHF outlet.) Wisely, the episode took no sides in the abortion issue, though the script came under fire for depicting the abortionist as sympathetic and the patient's husband as an insensitive jerk. For many observers, this was the one major failing of *The Defenders*: stacking the deck in favor of the Prestons' client by depicting the lawyers' opponents as narrow-minded fools, which some interpreted as a standard ploy of show-business Liberals — even though producer Herbert Brodkin was an avowed conservative Republican.

While there was some validity in this criticism, a good number of *Defenders* episodes were scrupulously fair to both Defense and Prosecution. Also, the writers were careful not to characterize the Prestons as pompously self-righteous, but instead fully aware that they had to consider all sides of an issue before charging into the courtroom: in one episode, Lawrence Preston commented, "Every time I go in with a sweeping Constitutional issue, I like to have something to hedge my bets with." (It should be noted that the series had an on-site legal expert in the form of production manager Hal Schaffel, a former lawyer.) Finally, the stars themselves were permitted to register comments or complaints if a particular episode went "over the edge": when he read the script for 1963's "The Hour Before Doomsday," in which the villain was a demented evangelist, Robert Reed asked to be written out of the episode, explaining, "Some of the things this wild man spouted were the same words the minister might say in my own church."

In one respect, however, everyone on the *Defenders* staff was on the same page: the Hollywood Blacklist of the 1950s and early 1960s, which denied radio, TV and film work to actors, writers and directors accused of "subversive" political beliefs (usually on the flimsiest of evidence), was an abomination that needed to be permanently extinguished. In addition to regularly hiring such "radical" performers as Lee Grant, Howard da Silva, Sam Wanamaker and Jack Gilford — all of whom had suffered mightily at the height of the McCarthy Era — the series took its most defiant stance in the Emmy–winning 1964 episode "Blacklist." Jack Klugman guest starred as an unemployed actor who, after ten years' exile because of his appearance as a hostile witness before the House Un-American Activities Committee, came to the Prestons for legal assistance when a right-wing activist group attempted to block the revival of his career.

To be sure, some compromises were made with the material: Jack Klugman himself had never been blacklisted, and in the first draft of the script his character was making his comeback not with a TV or movie job, but in a Broadway play — a plot point that everyone found laughable, since the legitimate theater was one of the few branches of show business that was *not* affected by the Blacklist (this was rewritten so that Klugman's character was more credibly in danger of losing a film role). Nonetheless, there was no question where the episode's creators stood on the issue, just as there'd been no question when the producers of *The Defenders* hired John Randolph, an actual blacklistee who hadn't worked in movies, TV or radio since 1951, for three different episodes within a two-year period, beginning with 1964's "All the Loud Voices." Randolph recalled the circumstances surrounding his own comeback in the 1977 compendium *TV Book*: "When I arrived on the set, I said to [the actors] in a very loud voice, 'I'm glad I have a chance to work again after being blacklisted for ten years.' Nobody said anything. I never knew why they decided to use me until many

years later. The author of [*Blacklist*], Ernest Kinoy, told me that they were sitting in conference, casting, when my name came up. One of the vice presidents immediately said, 'No, no, John Randolph can't work; he's on the list.' Kinoy then said, 'How can we do a show on blacklisting with our left hand and with our right hand, the very next day, blacklist?' That was the reason I was hired."

Like John Randolph, the bulk of the *Defenders* supporting players were recruited from the ranks of New York actors — logically enough, since the series was lensed in the Manhattan-based Filmways Studios. Joan Hackett enjoyed one of her earliest leading roles in the recurring character of Ken Preston's girlfriend Joan Miller, while Polly Rowles, whose Broadway credits stretched all the way back to Orson Welles' Mercury Theater, appeared during the first season as Lawrence Preston's secretary Helen Donaldson. Guest stars included such Broadway and off–Broadway fixtures as John Cullum, Brenda Vaccaro, Richard Kiley, Eileen Heckart, Tom Bosley, and Doris Roberts, as well as such "bi-coastal" favorites as Kevin McCarthy, Fritz Weaver, Cloris Leachman, Mercedes McCambridge, Viveca Lindfors and James Coburn. And unlike most other series of its time, *The Defenders* provided meaty, respectable roles for African-American performers, notably Ivan Dixon, Ossie Davis, and James Earl Jones, the latter cast as a Martin Luther King–style activist in the episode "The Non-Violent."

Others making appearances in the series' 132 hour-long episodes were playwrights Emlyn Williams and Marc Connelly, future sportscaster Heywood Hale Broun, silent screen icon Lillian Gish (delightful in the episode "Grandma TNT" as a little old lady who threatened to blow up a bank unless she received exactly $1342.76!), British luminaries Edward Woodward and Glynis Johns, and a host of stars-to-be, among them Robert Redford, Dustin Hoffman, Martin Sheen, Martin Landau and Daniel J. Travanti. Some of the casting choices that remain most vividly in the memory include Peter Fonda and Jon Voight as college fraternity members accused of orchestrating a fatal hazing in "The Brother Killers"; Milton Berle as a suicidal comedian in "Die Laughing"; and, in an unforgettable scene from "The Hidden Jungle," *All in the Family's* Carroll O'Connor as a former mental patient accused of murder on evidence provided by a female witness — played by Jean Stapleton!

Though the series earned several Emmy Awards and a 1963 Golden Globe, *The Defenders* was only a modest ratings success, never even ranking in the Top 30 programs. Still, the series was popular enough to enter the pop-culture mainstream: its impressive opening-credits overhead view of the New York County Courthouse, and its stirring theme music by Leonard Rosenman, were endlessly imitated and spoofed in everything from Alka-Seltzer commercials to the 1963 James Garner film *The Wheeler-Dealers*; and, in that most coveted of honors, the series warranted no fewer than two parodies within the same year in the pages of *Mad* magazine.

But when *The Defenders* was cancelled by CBS in 1965, few local stations picked up the series' syndicated rerun package, perhaps theorizing that controversy was a dish better served on a weekly rather than daily basis, and that hardened network executives were better suited to "taking the heat" than regional station managers. And when the *Perry Mason* reruns entered syndication in 1966, the *Defenders* manifest was effectively squeezed out of the market and drifted into obscurity — at least until 1997. This was the year that *The Defenders* was revived as a two-hour TV movie titled *The Defenders: Payback*, which aired October 12 on the Showtime cable service.

Producer Stan Rogow and writer-director Andy Wolk had encountered a lot of resistance selling this property to over-the-air networks, mainly because defense attorneys in general had gotten a "bad rap" after the O.J. Simpson trial — and besides, the real heroes on TV courtroom series of the late 1990s, as typified by *Law & Order* (q.v.), were the prosecutors. Thus, whereas the Prestons were perceived as heroes for taking on unpopular clients and articulating unpopular issues back in the 1960s, they might be regarded as the heavies in

the new prosecutorial climate. Fortunately, Showtime's president Jerry Offsay was a big fan of the original *Defenders*, so he green-lighted the TV movie as the pilot for a projected series. Signing onto the project with the breathless enthusiasm of a college drama major was 86-year-old E.G. Marshall, who had earlier discussed a reunion movie with Robert Reed which came to nothing, and in any event was abandoned after Reed's death in 1992 (the two actors had been briefly reunited — though not as father and son — in a 1969 episode of Reed's starring sitcom *The Brady Bunch*). It was explained in *Payback* that Lawrence Preston's son Ken had passed away, and now the elder Preston was partnered with his *other* son Don, played by Beau Bridges. The new film emphasized the solidarity of a multigenerational family by casting Martha Plimpton as Don's daughter (and Lawrence's granddaughter) M.J. Preston, the firm's new junior partner.

The plot of *Payback* was apparently inspired by one of the original series' earliest episodes, "The Attack." John Larroquette, midway between his own legal-TV starring assignments *Night Court* and *McBride* (both q.v.), was cast as the Preston's client, who had murdered his daughter's rapist. In the grand tradition, the film used its basic storyline as a forum to disseminate a larger social issue, in this case the court-ordered placement of known sex offenders in suburban neighborhoods. Interviewed by Warren Berger of the *New York Times*, scripter Andy Wolk promised that, if *Payback* succeeded in launching a series, the new *Defenders* would challenge the viewer's sensibilties by refusing to offer "easy resolutions" to dicey legal dilemmas; as for Showtime head Jerry Offsay, he publicly expressed the hope that a "prestige" item like *The Defenders* would attract more big-name film actors to cable television. As it happened, *Payback* was indeed a success, warranting a followup TV movie, *Defenders: Choice of Evil*, first seen over Showtime on January 18, 1998. Making his last-ever appearance before the cameras, E.G. Marshall was in fine fettle in this story of a murder that may have been caused by a bureaucratic foul-up.

A third *Defenders* movie, *The Defenders: Taking the First*, aired on October 25, 1998, two months after E.G. Marshall's death; this time, Don and M.J. Preston soldiered on alone to defend four teenage boys accused of committing a hate crime fomented by a rabid white supremacist. Alas, without E.G. Marshall, Showtime determined that *The Defenders* had no future, thus *Taking the First* was the final entry in this short but sweet revival.

Divorce Court

1. Syndicated: 1958–1961. NTA/Guild Films. Produced by Jackson Hill. Cast: Voltaire Perkins (Judge); Bill Welsh (Commentator).
2. Syndicated: 1965–1966; 1969–1970. Storer Television. Cast: Voltaire Perkins (Judge); Colin Male (Commentator).
3. Syndicated: 1985–1991. Storer Television/Blair Television. Executive producers: Donald Kushner, Peter Locke, Bruce McKay. Cast: William B. Keene (Judge); Jim Peck (Commentator).
4. Syndicated: debuted 1999. Monet Lane Productions/20th Television. Executive producers: Mark Koberg, Jill Blackstone. Cast: Mablean Ephraim (Judge: 1999–2006); Lynn Toler (Judge: 2006–); Sgt. Joseph Catalano (Bailiff).

There is a prevalent belief that because *Divorce Court* made its first appearance in 1958, the series is the longest-running legal program in existence. In truth, the series has not been in *continuous* production since 1958 — and beyond that, there have been *three* different syndicated series with the title *Divorce Court*, the first two bearing only a surface resemblance to the third.

The original weekly, hour-long *Divorce Court* debuted as a local Los Angeles program on February 26, 1958, telecast live from the studios of KTTV. Its avowed purpose was to stem "the rising tide of divorce," though suspicions persist that the series was also intended to make a heap of cash for its creators. Producer Jackson Hill, who attended several actual divorce cases in preparation for the show, had wanted to hire a real judge to preside over the KTTV courtroom, but was unable to find any genuine jurists willing to risk conflict of interest issues by appearing on a sponsored series. So Hill settled for 64-year-old Voltaire Perkins,

a practicing lawyer and part-time actor who specialized in playing judges in such films *A Man Called Peter* (1955), *Over-Exposed* (1956) and *The Solid Gold Cadillac* (1957). Perkins was described by the producers as "a man of 40 years' experience in the law"; but once he became established on *Divorce Court*, he never worked as an attorney again (his final TV appearance was in 1970's *The Young Lawyers*—as a judge). Cast as *Divorce Court*'s on-the-spot commentator was veteran L.A. radio and TV announcer Bill Welsh, who had joined KTTV in 1951 as director of sports and special events. Welsh tackled his new assignment with the same enthusiasm he'd exhibited while providing blow-by-blow descriptions of KTTV's weekly wrestling matches and the station's coverage of the annual Rose Bowl parade.

Though the cases dramatized on *Divorce Court* were fictional, the lawyers were genuine (though not all of them were genuine *divorce* lawyers). At the outset, the series drew upon a rotating pool of eight lawyers, a number that expanded to 300 as the show progressed; each man was paid between $165 and $200 per appearance. Many of the attorneys were rookies just starting in the business, or had retired from practice: among those appearing on a semi-regular basis were Paul Caruso, who went on to become a famous trial attorney; David R. Glickman, later the president of the L.A. Chapter of American Board of Trial Advocates; and Joseph Max Wapner, the father of future *People's Court* judge Joseph M. Wapner. Ironically, the man in charge of recruiting these lawyers was none other than Rusty M. Burrell, a real-life bailiff who played himself on the original *Divorce Court*, and who from 1981 to 2000 enjoyed a profitable association with the younger Joseph Wapner as bailiff on both *People's Court* and *Animal Court* (see separate entries for these two programs).

All the plaintiffs on *Divorce Court* were professional actors, mostly unknowns recruited from various theater workshops in the L.A. area (though there were a few "names" such as former silent-screen luminary Francis X. Bushman). These performers were largely chosen for their ability to think on their feet: there was no rehearsal or run-through on the series, and no formal script. Each actor was given three separate typewritten sheets, the first consisting of facts about the person he or she was playing, the second containing a list of lies the actor would try to pass off as truth, and the third a tally of facts the actor would attempt to conceal but would admit during cross-examination. Similarly, each attorney was given a booklet with three important facts to be brought out in the course of the trial; it was understood that the lawyers would keep their facts secret from each other and from the actors playing the opposing spouses.

Otherwise, everything was completely ad-libbed, including the judge's final decision, which was based solely on evidence presented in the course of the episode and was not revealed until the very last moment. In an interview with *TV Guide*, producer Jackson Hill explained, "Thus we achieve a kind of superreality and spontaneity not found anywhere else in television. The contest between lawyers is a genuine one, as is the decision of the judge. We're not so much putting on a show as a real situation." Occasionally, it became a bit *too* real: in one episode, an actor playing an aggrieved husband became so swept up in The Moment that he accused his "wife" of infidelity — a charge that didn't appear on *any* of the fact sheets. And in an excerpt used by KTTV to promote the series for syndication, the "son" of a divorcing couple suddenly jumped off the witness stand and began pummelling the actor playing his stepfather!

With so much excitement occurring on the set, it's amazing that the actors remembered to break for commercials. Actually, it wasn't up to them at all; as "break time" approached, the floor director flashed an upside-down cue card reading "OBJECTION" just beyond camera range — whereupon one of the attorneys would rush up to the bench with an objection, and Bill Welsh would unctuously whisper into his microphone "While the attorney is conferring with the judge, we have time for this commercial message." (In later years, the director would hold up a sign reading "CONFERENCE," whereupon the judge himself would insist that both lawyers approach the bench —*just* in time for that commercial message, of course).

It should surprise no one that the original local version of *Divorce Court* met with harsh disapproval from the Los Angeles County Bar Association, which at one point attempted to prohibit lawyers within their jurisdiction from appearing on the series. The Association's biggest gripe was that the show was far too preoccupied with sex and sensationalism to pass muster as acceptable entertainment. In a 2003 article by TV historian Roger M. Grace, frequent *Divorce Court* attorney David R. Glickman cheerfully admitted the charges were true: "What sells? Sex and violence. It's the same today." There was also the perception that the series lowered the dignity of its genuine attorneys, who in turn may or may not have been in violation of professional ethics by using the show for "self-advertising and self-laudation." As it turned out, this charge didn't hold water because the lawyers appeared anonymously and did not technically advertise themselves.

Telecast in L.A. on Wednesday evening, *Divorce Court* performed so admirably opposite such stiff network competition as *Kraft Theater* and *U.S. Steel Hour* that the decision was made to nationally syndicate the series. In the fall of 1958, it was put on the market by Guild Films as the first-ever series to be syndicated on videotape, with kinescopes (films taken directly from the TV monitor) provided to local stations without tape facilities. *Divorce Court's* biggest and best customer was the fledgling NTA Film Network, created in 1956 as an alternative to the combined output of NBC, ABC and CBS. *Divorce Court* was offered to major-market independent stations as part of a weekly NTA block with several other first-run syndicated efforts, including *How to Marry a Millionaire*, *This is Alice* and *Man Without a Gun*. Those network affiliates who purchased the series tended to run it in Prime Time, often pre-empting the "weaker" offerings from their parent network. (In Milwaukee, *Divorce Court* was seen over the CBS affiliate on Fridays at 9 P.M. during the 1959–60 season, replacing a low-rated network anthology which went by the name of *The Twilight Zone*.)

The 130 hour-long episodes of *Divorce Court* taped between 1958 and 1961 were rebroadcast on a daily basis through the facilities of Storer Television until the mid–1960s. Storer also handled the syndication of the weekly, half-hour *Divorce Court* revival of 1965, and the daily 30-minute "strip" version taped in 1969. Photographed in color, the 1965 and 1969 editions both followed the same production procedure as the 1958 original, and both still featured Voltaire Perkins as the judge. With Bill Welsh having moved on to other things (including three terms as president of the Hollywood Chamber of Commerce), the commentator duties were taken over by ex-private detective and former Cincinnati TV personality Colin Male, best known to trivia buffs as the offscreen voice heard at the beginning of such CBS sitcoms as *The Dick Van Dyke Show* and *The Andy Griffith Show*.

The popularity of the syndicated *People's Court* in the early 1980s inspired producers Donald Kushner, Peter Locke and Bruce McKay to take *Divorce Court* out of mothballs for a brand-new version, packaged by Storer TV and Blair Television and first made available to local markets in 1985. Like the 1969 version, this one was seen on a daily half-hour basis, with five episodes produced during each marathon taping session. With Rusty Burrell preoccupied with *People's Court*, the real lawyers seen on the new *Divorce Court* were largely recruited via classified ads placed in legal publications: they, like the actors playing the plaintiffs, were paid SAG union scale, plus residuals.

Replacing the late Voltaire Perkins as the presiding jurist was 59-year-old William B. Keene, a former private practitioner, L.A. deputy district attorney and Superior Court judge. Familiar to the general public via his participation in the high-profile trials of Charles Manson and "Freeway Killer" William Bonin, Keene was no stranger to television; he had appeared as one of the attorneys on the old daily CBS courtroomer *The Verdict Is Yours* (q.v.), and had also been among the first judges to allow TV coverage of his trials. Having re-

tired from the bench just before production of *Divorce Court* began, Keene was hired for the series over 50 possible applicants; he hadn't even been required to audition, but instead had been recommended for the assignment by a fellow member of the California Judges Association. (Many of the lawyers appearing on *Divorce Court*—once again, not exclusively divorce lawyers—were members of the firm of Morgan, Wenzel & McNicholas, for which Keene functioned as counsel.) Another fresh face on this version of *Divorce Court* was the series' latest host-commentator: Milwaukee's own Jim Peck, previously the emcee of such daytime game shows as *You Don't Say* and *Three's a Crowd*.

When plans for the new *Divorce Court* were announced, quite a few representatives of the legal community were surprised, regarding the warhorse property as an anachronism. Chicago divorce attorney James T. Friedman told *TV Guide:* "Marital misconduct is hardly ever an issue any more. It isn't even an issue in child custody cases, unless the child was harmed by it. Mostly, we deal with complex property valuations. It's more like probate or estate law…. It's not juicy. There's no under-the-bed or over-the-transom testimony, no parade of mother-in-law and girl-friend witnesses. I do divorce law exclusively, and it's been at least two years since I've tried a case on grounds." But the producers of the new version had obviously taken this into account: the 1985 *Divorce Court* dealt less with individual cases and connubial intrigue than with broader social issues. Each episode focused on a contemporary theme, inspired by current events and obviously chosen on the basis of controversy: child molestation, mental and physical disability, homosexuality, surrogate parenthood, the rights of the elderly, impotence, political extremism, pornography, and so forth. Occasionally, such old standbys as adultery and mental cruelty still figured into the action; but in the new era of "no-fault" divorce, such mundane grounds for separation could not possibly hold the audience's attention as effectively as such hot issues as transvestism and S&M.

As in years past, Judge Keene made his

final decision based on the evidence at hand, with none of the actors knowing ahead of time how the story would come out. Also as before, the series was completely unrehearsed. But unlike the earlier editions of *Divorce Court*, this one was fully scripted, and the actors were expected to have their lines totally memorized prior to taping. Eugene Rubenzer, an actor who appeared as a plaintiff in a 1988 episode, recently recalled the experience to this author: "I received a call about two weeks before the shoot and was asked if I was available for the shoot date and time. They told me about the 'no cue cards, no teleprompter, nonstop five-camera shoot' and then I was told the name of my character. Also, that we had to provide our own wardrobe.

"The next day, I got my script via Federal Express and this was the first time that I knew the size of my role. There was no rehearsal. On the day of the shoot, the cast was walked through the set and told how to approach the stand. Since most of the time we were seated until called up, there really was no blocking rehearsal needed. At that time I also found out when my show (of five being taped that day) was scheduled (it was the last in the day). I then went to my own dressing room and some time during the day the director showed up. We talked briefly and he got my lean on the character. In total, that was it."

Also, the original series' three "fact sheets" — truth, lies, and facts revealed in cross-examination — were dispensed with this time around. "The only information we got was the script itself," notes Rubenzer. "Anything that came out as a lie or whatever was up to the interpretation of the actor. We had no idea of the outcome, and part of this was because the judge himself didn't decide the outcome till the end of the episode. A lot of it relied on the abilities of the actors to create their roles and stay as letter perfect to the script as possible." If any mistakes were made, they were corrected on the spot: "Otherwise, it was 'real time'."

Some observers felt that the new, socially-conscious *Divorce Court* was neither an improvement on the original nor a true reflection of the work accomplished by genuine domes-

tic courts. To this, William B. Keene responded that the show had created its own reality, its own "jurisdiction," citing "the law we have developed here on the show." The public didn't seem to care one iota if the show was an accurate representation of actual divorce litigation: seen in 90 percent of the country's TV markets, the series proved to be a solid hit, especially with female viewers age 25–54. A somewhat surprising booster of the series was eminent psychoanalyst Walter E. Brackelmans, though he expressed a few reservations to *TV Guide*: "It's entertaining, full of good guys and bad guys, and psychologically not invalid. My objection is that the public isn't adequately informed that is is *just* entertainment.... I think a disclaimer is needed, something like 'Despite what you have seen, we can now have divorce with dignity.'"

But dignity was not the series' strong suit, as evidenced by later episodes featuring such guest stars as comedian Charles Nelson Reilly (as himself!) and multipart stories wherein the attorneys (now played by actors) were romantically involved with their clients. However, viewers lapped up *Divorce Court* until production finally wrapped after nearly 700 episodes in 1991.

Five years later, the judge-show format enjoyed a renaissance thanks to the success of Judge Judy Sheindlin's daily, syndicated shout-fest. In addition to encouraging several other authentic jurists to headline their own shows, this latest cycle also spawned a brace of revivals: brand-spanking-new versions of *The People's Court* and, in 1999, *Divorce Court*. Distributed by 20th Century–Fox Television, the latter series emulated the 1985 version in that it was a half-hour daily, with an honest-to-goodness legal professional serving as judge: Mississippi-born Mablean Ephraim, who before entering private practice as an attorney specializing in personal-injury and family-law cases had been a correctional officer in the Women's Division of Prisons. "My partner was at a meeting of the lawyers," Ephraim recalled to interviewer Larry King in 2000. "One of the entertainment lawyers there told her that Fox was looking for a judge to revive *Divorce Court*. She thought that I would be perfect for the job [and] passed the information on to me. I contacted the 20th Television executives, went for an interview on Thursday, called them on Wednesday, and the following Wednesday I had the job."

In a major stylistic break from all previous versions of *Divorce Court*, the 1999 revival was neither scripted nor performed by actors, but instead featured genuine divorced couples who had already dissolved their relationship and were appearing before "Judge" Ephraim to argue over such post-marital issues as division of property. Of course, the series' producers wouldn't think of letting these warring couples leave the studio without first recalling the events leading up to the split, especially if extramarital hanky-panky was involved. (The series had obviously long since forsaken its mission to "stem the rising tide of divorce.") Most often, Judge Ephraim ended up favoring one litigant over the other in her binding-arbitration decision, though sometimes she granted a joint decree. Very occasionally, she encouraged the couple to seek reconciliation, inviting them to return to the series at some later date to see if they'd managed to iron out their differences. (And if they had, would *you* tune in? Get real.)

When 20th Television and Mablean Ephraim were unable to come to contractual terms in 2006, she was replaced by Lynn Toler, a graduate of Harvard and the University of Pennsylvania who had served a six-year term as municipal judge in Cleveland Heights, Ohio, and had done a previous "TV judge" turn on the short-lived *Power of Attorney* (q.v.). Though Judge Toler tended to take a feminist slant when discussing domestic problems with the litigants, she was scrupulously fair to both sides. On certain occasions, Toler sensed that a couple wasn't "ready" to appear on *Divorce Court* and advised them to leave the set, go home and try to patch up their differences; in some episodes, she refused to adjudicate at all if the plaintiffs were too loud and out of control. Clearly, Toler was interested in taste and decorum so far as the series' format would allow: still, the show boasted its share of embarrassing moments, such as when former child star Gary

Coleman showed up to end his marriage to Sharon Price.

At the same time she was appearing on *Divorce Court*, Lynn Toler also headlined her own reality series, *Decision House*, which aired on MyNetwork TV beginning in the fall of 2007. This effort could be described as a "Pre–Divorce Court," with Toler endeavoring to nip selected couples' marital problems in the bud with the help of a team of domestic experts. Referencing the litigants on her other series, Toler explained that *Decision House* was "what I would like to do with those people if I had more time and we'd gotten there earlier." (Tread softly, Judge: You may be biting the hand that's feeding you.)

Divorce Hearing

Syndicated: debuted in 1958. Andrews-Wolpers-Spears Productions/Interstate Television. Executive Producers: David L. Wolper, Ralph Andrews, Harry Spears. Directed by Harry Spears. Research: Al Blake. Cast: Dr. Paul Popenoe (Himself).

Like the more famous *Divorce Court* (see previous entry), *Divorce Hearing* was a local Los Angeles program that entered national syndication in the fall of 1958. Unlike *Divorce Court*, which used professional actors as litigants, *Divorce Hearing* featured actual couples whose marriages had disintegrated. The avowed purpose of the series was elucidated by an off-screen announcer during the opening credits: "DIVORCE HEARING ... presented in the belief that divorce is America's greatest danger to the home and the community ... and that understanding is the greatest weapon against divorce."

There were few who questioned the sincerity of series creator Dr. Paul Popenoe, who since 1930 had headed the American Institute of Marital Relations, the first marriage clinic in the United States. Though not a real doctor — his highest academic degree was an honorary one from Occidental College — Popenoe was nonetheless a recognized and respected family counselor, as well as the author of the widely syndicated newspaper column "Your Family and You" and the popular "Can This Marriage Be Saved?" feature for *Ladies Home Journal*.

Dedicated to preserving the institute of marriage, Popenoe had successfully steered thousands of couples away from divorce court and toward reconciliation (of course, there were a few failures along the way: one of his less successful clients was actress Lana Turner, who was rumored to own a wash-and-wear wedding gown). In addition to patching up broken relationships, Popenoe liked to cast himself in the role of matchmaker: in the late 1950s, his frequent appearances on Art Linkletter's various TV series earned him national face recognition as the first advocate of computer dating, linking up couples via UNIVAC. (That Dr. Popenoe was also a leading proponent of the questionable science of Eugenics is a matter best ignored in this entry, else I'll start railing on Margaret Sanger and other such sacred cows.)

Divorce Hearing was one of the earliest TV projects of producer David L. Wolper, best known today for his award-winning *National Geographic* specials and the groundbreaking dramatic miniseries *Roots*. In his autobiography, Wolper recalled that Paul Popenoe "took this show seriously. He really wanted to help people." Each episode was staged in the manner of a standard court hearing, with Popenoe seated at a judge's bench and the battling couples — identified only by such initials as "Mr. and Mrs. M" — standing before him, separated by a railing. It was Popenoe's mission in life to let the couples air their differences, then find some way to repair the cracks, offer practical solutions and ultimately rescue the marriage. Lest any viewer question the authenticity of the proceedings, the announcer assured us, "This unrehearsed divorce hearing is not a re-enactment…. The couples appearing are the real people."

True enough, as far as it goes. But whereas Dr. Popenoe was interested only in promoting marital harmony, Wolper and his partners Ralph Andrews and Harry Spears were interested only in entertainment — and at the risk of traumatizing the reader, it cannot be denied there is nothing more entertaining to some people than watching other people air their dirty laundry in public, usually at the top of their lungs. Thus, the producers did everything they could to ensure that the couples on *Divorce Hearing*—even those who had separated amicably—were at each other's throats from start to finish.

Wolper and his partners brought the husbands and wives into the show's Emperor Studios facilities via separate entrances and kept them apart until filming began. Andrews and Spears then "briefed" the litigants by goading them into remembering why they were so angry with each other: "We wanted confrontation, and we got it." By the time poor Dr. Popenoe was able to meet with the warring couple, both parties had been whipped into a frenzy and were loaded for bear — and *Divorce Hearing* often as not degenerated into a 1950s edition of *The Jerry Springer Show*.

Existing episodes do not include the "money scene" described by David Wolper in which a fed-up wife took a sock at her stinking-drunk husband, but there are still fireworks aplenty. The ceaseless bickering between the couples is amusing in its own sick, twisted way, suggesting there is no one more naturally witty than a hateful person: after her husband complains that she never serves him a decent breakfast, one flustered woman shoots back, "With my Irish temper and red hair you're liable to get your breakfast over your head!" In the face of such vitriol, the nonplussed Dr. Popenoe appears to be on the verge of booting the couple out his chambers and going into another line of work: instead, he valiantly continues to advise the combatants that they still have time to iron out their differences by paying a visit to the American Institute of Marital Relations (plug, plug) — and failing that, their fractured marriage can serve as an example to others.

Two couples were featured per half-hour episode of *Divorce Hearing*, allowing local stations to split each entry in half and offer the series in a 15-minute format if necessary.

Dundee and the Culhane

CBS: Sept. 6–Dec. 13, 1967. Filmways Television Productions. Created by Sam Rolfe. Executive Producers: Sam Rolfe and David Victor. Cast: John Mills (Dundee); Sean Garrison (The Culhane).

One of a handful of series combining the Legal and Western formats (see also *Black Saddle, Judge Roy Bean, Sugarfoot* and *Temple Houston*), *Dundee and the Culhane* was created by Sam Rolfe, the man behind the classic Richard Boone vehicle *Have Gun, Will Travel*. In the tradition of *Have Gun's* erudite gunslinger Paladin, *Dundee* revolved around an unorthodox western hero who apparently had no first name. And just like Rolfe's and coproducer David Victor's previous endeavor, the espionage series *The Man From U.N.C.L.E.*, *Dundee and the Culhane* tried to pep up a well-worn TV genre with a tongue-in-cheek approach.

Dundee was a well-spoken, peace-loving Britsh attorney whose law offices were located in Sausalito, California, just across the Bay from San Francisco (there's that *Have Gun, Will Travel* connection again). His young legal apprentice Culhane tended to use his guns and his fists rather than his intellect, making him an amusing contrast to his pacifistic boss. Together, Dundee and The Culhane roamed the American Southwest, hoping to establish proper legal procedure in a land where justice was usually dispensed by the barrel of a six-shooter or at the end of a rope. Each episode underlined the law angle by including the word "Brief" in the title: "The Vasquez Brief," "The Murderer Stallion Brief," and so on.

According to Stephen Lodge, who worked on the series as a costumer, *Dundee and the Culhane* had been specifically tailored to the talents of distinguished British actor John Mills, and securing his participation was considered a major coup by CBS. Mills agreed to star in the series so that he and his wife could live in California while his daughters Hayley and Juliet were pursuing their Hollywood careers. In his autobiography, Mills recalled that he was so anxious to do the show he was willing to sign a five-year contract with CBS. The British press chided Mills for "selling out" to Hollywood, but the actor ignored these slings and arrows, secure in the knowledge that no British TV company could offer him more than a fraction of what CBS was willing to pay.

With their "Dundee" firmly in place, the producers began casting about for a suitable "Culhane." Someone named Rick Falk played the character in the pilot film, but in the series proper the role went to Sean Garrison, a ruggedly handsome actor who'd been on the fringes of stardom since the early 1960s.

When not filming at General Service Studios in Hollywood, the *Dundee* production unit could be found on location in Apacheland, Grand Canyon, Scottsdale and Old Tucson in Arizona, and closer to home in Chatsworth and Thousand Oaks, California. Though the series required John Mills to emote "in the middle of the Arizona desert in a temperature at 108 degrees," the grand old trouper never wilted. During the first day of shooting, Mills, faced with the prospect of eating the standard movie-company box lunch under a blistering sun, politely let it be known he preferred the British tradition of a daily four o'clock teatime. Much to the actor's surprise, an impeccably dressed waiter arrived at 4 P.M. sharp with a full tea service and a selection of cucumber sandwiches, scones and pots of honey. The hardened production crew was enchanted by this quaint ceremony, and thereafter the 4:00 teatime was faithfully honored during every shooting day, even if the actors were in mid-scene.

Five of the thirteen hour-long episodes of *Dundee and the Culhane* were filmed when Sam Rolfe, tired of quarrelling with the network heads over story content, quit the show. John Mills was concerned about the series' future, but was assured that production would continue. Little did he know what CBS programming chief Michael Dann later admitted to a *TV Guide* interviewer: the network was so displeased with the direction the series was taking that they'd decided to cancel *Dundee and the Culhane* even before its premiere on September 6, 1967. Apparently this decision took everyone but Dann and his fellow executives by complete surprise: John Mills didn't even learn of his series' demise until he read about it in the *London Daily Mail*.

The few available episodes of *Dundee and the Culhane* indicate the show was not quite as hopeless as Mike Dann believed it to be. Its major failure was that it didn't take full advantage of its premise, emerging as just another TV

Western—of which there were far too many during the 1967–68 season.

The Eddie Capra Mysteries

NBC: Sept. 8, 1978–Jan. 12, 1979; CBS: July 26–Aug. 30, 1990. Universal Television. Executive Producer/Creator: Peter S. Fischer. Produced by James McAdams. Cast: Vincent Baggetta (Eddie Capra); Wendy Phillips (Lacey Brown); Ken Swofford (J.J. Devlin); Michael Horton (Harvey Winchell); Seven Ann McDonald (Jennie Brown).

A late-1970s entry in the "lawyer/sleuth" genre popularized years earlier by *Perry Mason* (q.v.) and its ilk, *The Eddie Capra Mysteries* starred Vincent Baggetta as the title character, a recent NYU School of Law graduate working for Devlin, Linkman & O'Brien, a firm specializing in criminal cases. Much to the dismay of his conservative boss J.J. Devlin (Ken Swofford), the headstrong and rebellious Eddie Capra (we knew he was rebellious because he had the longest hair in the office) thought nothing of cutting traditional legal and police procedure into little paper doilies as he attempted to solve the murders for which his clients had been indicted. In time-honored fashion, each episode opened with a graphic depiction of the homicide-of-the-week, whereupon Eddie Capra proceeded to unearth clues that the authorities had overlooked, and to interview witnesses and other concerned parties who had a myriad of reasons to avoid the truth. The climax generally found Mr. Capra arranging to have all the suspects in the same room—usually the courtroom—and by deductive logic zeroing in on the actual culprit. Featured in the cast were Wendy Phillips as Eddie's secretary, assistant and sometimes sweetheart Lacey, Seven Ann McDonald as Lacey's perky young daughter Jennie, and Michael Horton as Eddie's law-student legman Harvey.

Though some viewers felt Eddie Capra was merely "Columbo as a lawyer," the series had a closer affinity to another Richard Levinson-William Link creation, the 1975 TV version of *Ellery Queen*. Several of the scripts for *The Eddie Capra Mysteries* had actually been written for *Ellery Queen*, but had gone unfilmed when that series was cancelled after a single season. Eddie Capra himself was gone after 13 hour-long episodes on NBC, though a handful of reruns showed up on CBS in the summer of 1990 as a temporary replacement for *Wiseguy.*

Eddie Dodd

ABC: March 12–June 6, 1991. Columbia Pictures Television. Executive Producers: Bob Goodwin, Peter Rosten, Clyde Phillips, Walter Parkes, Lawrence Lasker. Cast: Treat Williams (Eddie Dodd); Corey Parker (Roger Baron); Sydney Walsh (Kitty Greer); Annabelle Gurwitch (Billie).

The 60-minute ABC legal drama *Eddie Dodd* was based on the 1989 theatrical feature *True Believer*, which starred James Woods in the title role, a character inspired by famed San Francisco civil-rights attorney J. Tony Serra. A flaming radical lawyer of the 1960s, Eddie Dodd had grown dumpy and disillusioned over the years, beaten down by a system he'd hoped to change but never could. In the film, Eddie returned to the land of the living when his hero-worshipping yuppie protege Roger Baron, played by Robert Downey Jr., urged him to take on the "lost cause" case of a Korean man, in prison for a gang killing he might not have committed and facing the death penalty for the murder of a fellow convict (this was loosely based on J. Tony Serra's real-life defense of Chinatown resident Chol Sol Lee). Eddie's efforts to recapture his youthful idealism were repeatedly compromised by the realities of the legal world: in one scene, his angst soared into the stratosphere when he was forced to cut a deal to spring a client. Also cast in *True Believer* was Margaret Colin as Eddie's girlfriend, private detective Kitty Greer.

Taking over from James Woods in the TV series *Eddie Dodd* was Treat Williams, who was not quite as alienated and self-loathing as the film's Eddie (nor did he sport Woods' endearingly anachronistic ponytail). Also, the original Eddie's far-left political views were totally expunged, presumably so as not to alienate a sizeable chunk of potential viewers—which may explain why his practice had moved from the Liberal cloud-cuckoo-land of San Francisco to New York City. Though Eddie Dodd still felt like a used-up relic of a forgotten era, it was less due to the current political climate than the fact that he was rapidly approaching the age of forty. (If we follow this logic, Mr. Dodd was practicing law in his mid-teens back in the 1960s!)

Still, the new Eddie Dodd specialized in taking on cases other attorneys had written off as unwinnable or too controversial. In the first episode, he defended a former lover (Susan Blakely) who'd confessed to assisting the suicide of her terminally ill husband. Subsequent cases involved an alleged police coverup after the killing of a unarmed black youth, an elderly judge whose dementia may have been coloring his rulings, and a volatile senator accused of corruption. Still seeking out Eddie's clients was his youthful aide Roger Baron, now played by Corey Parker along even preppier lines than Robert Downey, Jr. Also, the TV version of Roger was more cautious and conservative than his movie counterpart. In the assisted-suicide episode, Roger confronted Eddie on the matter of mercy killing: "The laws of New York have no exception to this. She's guilty. Case closed." Eddie hotly countered with "We can't argue the law here. We have to argue right and wrong." Variations of this scene played out between Eddie Dodd and Roger Baron week after week, suggesting that it was Roger and not Eddie who needed the occasional booster-shot of Righteous Idealism. Of the other regular cast members, Sydney Walsh's interpretation of Kitty Greer more or less followed the pattern established by the film's Margaret Colin. New to the property was the character of Eddie's secretary Billie, a Madonna wannabe played by Annabelle Gurwitch.

Eddie Dodd debuted in March of 1991 as a temporary replacement for ABC's on-hiatus *thirtysomething*. "[A]n overblown show full of hot air" was the verdict of *Entertainment Weekly*'s Ken Tucker. All the viewers said was, "When is *thirtysomething* coming back?." The answer: "Two and a half months," which is the length of time the six-episode *Eddie Dodd* remained on the air.

The Edge of Night

CBS: Apr. 2, 1956–Nov. 28, 1975; ABC: Dec. 1, 1975–Dec. 28, 1984. Procter & Gamble. Created by Irving Vendig. Produced by Werner Michel, Don Wallace, Charles Fisher, Erwin Nicholson. Head writers: Irving Vendig, Henry Slesar, Lee Sheldon. Cast: John Larkin, Larry Hugo, Forrest Compton (Mike Karr); Teal Ames (Sarah Lane Karr); Ann Flood (Nancy Pollock Karr); Victoria Larkin, Kathleen Bracken, Kathy Code, Emily Prager, Jeannie Ruskin, Linda Cook (Laurie Ann Karr Lamont Dallas); Donald May (Adam Drake); Maeve McGuire, Jayne Bentzen, Lisa Sloan (Nicole Travis Drake Cavanaugh); Juanin Clay, Sharon Gabet (Charlotte "Raven" Alexander Jamison Swift Whitney Devereaux Whitney); Dixie Carter (ADA Olivia Brandeis "Brandy" Henderson); Louis Turrene (Tony Saxon); Frances Fisher (Deborah Saxon); Denny Albee (Steve Guthrie); Don Hastings (Jack Lane); Mary Alice Moore (Betty Jean Lane); Betty Garde, Peggy Allenby, Katherine Meskill (Mattie Lane); Walter Greaza (Winston Grimsley); Carl Frank, Mandel Kramer (Police Chief Bill Marceau); Joan Harvey (Judy Marceau); Teri Keane (Martha Marceau); Heidi Vaughn, Renne Jarrett, Laurie Kennedy, Johanna Leister (Phoebe Smith Marceau); Ronnie Welch, Sam Groom, Tony Roberts (Lee Pollock); June Carter, Fran Sharon (Cookie Pollock Thomas Christopher); Ed Kemmer (Malcolm Thomas); Burt Douglas (Ron Christopher); John Gibson, Allen Nourse (Joe Pollock); Ruth Matteson, Frances Reid, Kay Campbell, Virginia Kaye (Rose Pollock); Millette Alexander (Gail Armstrong/Laura Hillyer/Julie Jamison); Wesley Addy (Hugh Campbell); Ed Holmes (Det. Willie Bryan); Larry Hagman (Ed Gibson); Karen Thorsell (Margie Gibson); Ray MacDonnell (Phil Capice); Lisa Howard, Mary K. Wells (Louise Grimsley Capice); Conrad Fowkes (Steve Prentiss); Liz Hubbard (Carol Kramer); Val Dufour (Andre Lazar); Lauren Gilbert (Harry Lane); Lester Rawlins (Orin Hillyer); Albert Grant (Liz Hillyer Fields); Penny Fuller, Joanna Miles, Millee Taggart (Gerry McGrath); Keith Charles (Rick Oliver); Barry Newman (John Barnes); Bibi Besch (Susan Forbes); William Prince, Cec Linder (Ben Travis); Alice Hirson (Stephanie Martin); Irene Dailey (Pamela Stewart); Richard Clarke (Duane Stewart); Alan Feinstein (Dr. Jim Fields); Ted Tinling (Vic Lamont); John LaGioia (Johnny Dallas); Pat Conwell (Tracey Dallas Micelli); Alan Gifford (Sen. Gordon Whitney); Lois Kibbee (Geraldine Whitney); Anthony Call (Collin Whitney); Bruce Martin (Keith Whitney, aka Jonah Lockwood); George Hall (John); Mary Hayden (Trudy); Hugh Reilly, Simon Gregory (Morlock); Elizabeth Farley (Kate Reynolds); Nick Pryor, Paul Henry Itkin (Joel Gantry); Ward Costello (Jake Berman); Dorothy Lyman (Elly Jo Jamison); Dick Shoberg, John Driver

(Kevin Jamison); Michael Stroka (Dr. Quentin Henderson); Lou Criscuolo (Danny Micelli); Louise Shaffer (Serena Faraday, aka Josie); Doug McKeon (Timmy Faraday); Dick Latessa (Noel Douglas); Niles McMaster (Dr. Clay Jordan); Brooks Rogers (Dr. Hugh Lacey); Helena Carroll, Jane Hoffman, Laurinda Barrett (Molly O'Connor); Herb Davis (Lt. Luke Chandler) Tony Craig (Draper Scott); Joel Crothers (Dr. Miles Cavanaugh); Holland Taylor (Denise Cavanaugh); Terry Davis (April Cavanaugh Scott); Robin Groves (Maggie); Joe Lambie, Tom Tammi (Logan Stoner); Irving Lee (Calvin Stoner); Yahee (Star Stoner); Kiel Martin (Raney Cooper); Marilyn Randall (Theresa); Dick Callinan (Ray Harper)); Micki Grant (Ada Chandler); Dennis Marino (Packy Dietrich); Gwynn Press (Inez Johnson); Dorothy Stinette (Naidne Scott); Eileen Finley (Joannie Collier); Lee Godart (Elliot Dorn); Dan Hamilton (Wade Meecham); Michael Longfield (Tank Jarvis); Ann Williams (Margo Huntington); Susan Yusen (Diana Selkirk); Lori Cardille, Stephanie Braxton (Winter Austen); Mel Cobb (Ben Everett); Wyman Pendleton (Dr. Norwood); Lori Loughlin, Karrie Emerson (Jody Travis); Sonia Petrovna (Martine Duval); Joey Alan Phipps, Allen Fawcett (Kelly McGrath); Bruce Gray (Owen Madison); Dennis Parker (Police Chief Derek Mallory); Shirley Stoler (Frankie); Mark Arnold (Gavin Wylie); Kim Hunter (Nola Madison); Ernie Townsend (Cliff Nelson); Larkin Malloy (Schuyler Whitney/Jeff Brown); Leli Ivey (Mitzi Martin); Mariann Aalda (Di Di Bannister); George D. Wallace (Dr. Leo Gault); Patricia Andrews (Chad Sutherland); Ray Serra (Eddie Lormer); Karen Needle (Poppy Johnson); Chris Jarrett (Damian Tyler); Victor Arnold (Joe Bulmer); Leah Ayres (Valerie Bryson); David Brooks (Jim Dedrickson); Richard Borg (Spencer Varney); David Froman (Gunther and Bruno Wanger); Catherine Bruno (Nora Fulton); Meg Myles (Sid Brennan); Alan Coates (Ian Devereaux); Charles Flohe (John "Preacher" Emerson); Mary Layne (Camilia Devereaux); Norman Parker (David Cameron); Pat Stanley (Mrs. Goodman); Willie Aames (Robbie Hamlin); Michael Stark (Barry Gillette); Ronn Carroll (Stan Hathaway); Derek Evans (Mary Stillwater); Steven Flynn (Davey Oakes); Pamela Shoemaker (Shelley Franklyn); Chris Weatherfield (Alicia Van Dine); Sandy Faison (Dr. Beth Correll); Jerry Zacks (Louis Van Dine); Cyd Quilling (Claire Daye); Ralph Byers (Donald Hext); Jason Zimbler (Jamey Swift); Ken Campbell (Russ Powell); A.C. Weary (Gary Shaw); Kerry Armstrong (Tess McAdams); Kelly Patterson (Hollace Dineen); John Allen Nelson (Jack Boyd); Julianne Moore (Carmen Engler); Bob Gerringer (Del Emerson); Christopher Holder (Mark Hamilton); Jennifer Taylor (Chris Egan); Amanda Blake (Dr. Juliana Stanhower); Hal Simms (Announcer).

But for a few creative differences between the Procter & Gamble Manufacting Company and mystery writer Erle Stanley Gardner, the long-running daytime drama *The Edge of Night* might have been titled *Perry Mason*— and Raymond Burr might never have happened.

In 1955, radio's top soap-opera writer Irna Phillips collaborated with future *All My Children* creator Agnes Nixon to develop television's first half-hour serial, *As the World Turns* (previously, no such TV series had ever run any longer than 15 minutes). Sponsor Procter & Gamble, who'd pretty much controlled the entire broadcast-serial genre since 1932, went whole-hog on the idea, not only securing an afternoon CBS timeslot for *As the World Turns* beginning April 2, 1956, but also ordering one of their advertising agencies to prepare another half-hour "soap" to premiere the same day on the same network. In stark contrast with the standard family-oriented *As the World Turns*, P&G wanted the companion program to be a mystery serial — and a daytime version of *Perry Mason* was their first choice, since they'd enjoyed considerable success sponsoring a five-per-week *Perry Mason* radio serial for the past dozen years. But *Mason* creator Erle Stanley Gardner wasn't happy with the way his property had been handled by P&G, so he set up his own production company to pitch a weekly, Prime Time *Perry Mason*, over which he could exercise full creative control (see separate entry for this series).

Only momentarily stymied, P&G contracted Irving Vendig from the radio *Mason*'s writing staff to create a similar (if not lookalike) property for daytime television. Thus, the character of crusading defense attorney Perry Mason was reconfigured as crusading Assistant District Attorney Mike Karr, with John Larkin, who'd played Mason on radio since 1947, cast in the role. Similarly, Perry's faithful secretary Della Street served as the model for Mike Karr's fiancée Sara Lane, while Mike's law partner Adam Drake was essentially Mason's private-eye associate Paul Drake, right down to the last name. Since the new series was scheduled for a 4:30–5 P.M. (EST) timeslot, less than three hours before CBS switched to its Prime Time manifest, it was titled *The Edge of Night*.

Though the series ostensibly took place in the fictional city of Monticello, an "average-sized" midwestern community with an inordinately high level of criminal activity and political corruption, it was obvious to most observers that *The Edge of Night* was actually set in Procter & Gamble's home base of Cincinnati, Ohio. Indeed, for many years the series' opening credits featured a daytime panorama of the Cincinnati skyline, upon which the night literally fell with the aid of a wipe-dissolve. (Ironically, for several years the Cincinnati CBS affiliate WKRC-TV obstinately scheduled children's programming in the 4:30 P.M. slot, televising *Edge of Night* in kinescope form on a one-day-delay basis in a less desirable 10 A.M. berth.)

During its first seven years on the air, *Edge of Night* concentrated on crime stories. At least one major murder trial was featured per year, usually involving the principal characters. (Poor Martha Marceau, the wife of Monticello's police chief, found herself facing the electric chair on two separate occasions!) During Irving Vendig's tenure as head writer, the mystery angle was downplayed. The innocence of the accused killer was never in doubt, since the viewer knew the identity of the guilty party from the outset; and in keeping with the series' *Perry Mason* roots, the actual culprit was invariably tricked or cajoled into a public confession by the intrepid DA hero. Beyond such core characters as Mike Karr and Sara Lane, new characters were regularly brought in to suit the needs of each new story arc, most of these lasting anywhere from 12 to 24 months; the characters who registered best with the viewers were carried over into future plotlines. To offset the more lurid aspects of the stories, Mike and Sara Karr were depicted as an intensely devoted couple, paragons of virtue and decency both. So that audiences didn't get tired of this sweetness-and-light, a marriage with more spark and tension was arranged between Karr's partner Adam Drake and mercurial heiress Nicole Travis.

On February 22, 1961, the series' writers sent many loyal viewers into a state of near-catatonia when they killed off Mike Karr's beloved wife Sara in a car crash. CBS received so many anxious and hysterical calls after this episode that actress Teal Ames had to go on the air the following day to assure her fans that she was still very much alive, and had merely been written out of the series so she could return to private life. Within a year, however, viewers were mollified by the romance and subsequent marriage between DA Karr and reporter Nancy Pollock, a union that endured from 1963 until the series' cancellation 21 years later (though by that time the role of Mike Karr had been assayed by three different actors, with Forrest Compton holding down the job for the longest period).

Thanks to its late-afternoon timeslot, *Edge of Night* was CBS' most popular serial among male viewers. This demographic began to chip away when the series was moved back to 3:30 P.M. on July 1, 1963, and virtually disappeared when P&G demanded an even earlier 1:30 P.M. slot in order to create a 90-minute soap-opera block with two other CBS serials, *Love is a Many-Splendored Thing* and *Guiding Light*, beginning September 11, 1972. By this time, *Edge of Night* had undergone several interior changes. In 1966, Erwin Nicholson joined the series as producer, bringing along veteran mystery writer Henry Slesar. Two years later, Slesar succeeded Irving Vendig as head writer, whereupon the crime-story continuities evolved into whodunnits, withholding the identity of the real villain until the very last minute. Slesar also introduced some of the series' quirkier and more flamboyant characters, notably the much-married Raven Alexander and headstrong ADA Brandy Henderson, with whom Adam Drake had an affair while his wife Nicole was missing and presumed dead. And though there were still the occasional showcase murder trials, these became less frequent as the series veered more towards the standard sexual and romantic soap-opera themes, with such tried-and-true plot devices as infidelity, schizophrenia, amnesia and "evil twins."

Though fans of the series, notably *The Edge of Night Homepage*'s creator Mark Faulkner, regard the Nicholson-Slesar years as "the Golden Age," the series' viewership began diminishing as the series entered the 1970s. Even so, CBS would probably have kept the show on its afternoon schedule had not Procter & Gamble opted to expand *As the World Turns* to 60 minutes beginning December 1, 1975, forcing the network to relinquish a half-hour of airtime. That same day, P&G moved *Edge of Night* to ABC, where it remained for the next nine years — remained, but did not flourish. Already suffering from sagging ratings, the series was further hampered by the unwillingness of several ABC affiliates to clear its timeslot and give up their own, more profitable local and syndicated shows. By the early 1980s, less than 62 percent of ABC outlets were running the show.

In an effort to rejuvenate the ailing property, the writers moved the series' location — or at least made it appear to have moved. The "average-sized" town of Monticello had expanded considerably since 1956, thus the standard opening skyline shot of Cincinnati (which had been refilmed a number of times over the years) was no longer appropriate. Beginning June 16, 1980, each episode opened with a spectacular view of the Los Angeles skyline, leading more impressionable viewers to conclude that the action had shifted to LA as well.

The skyline shot was dispensed with entirely in 1983, the same year Lee Sheldon took over as head writer. Despite an encouraging bump in the ratings from 1980 to 1982, the series was showing its age and was regularly being out–Nielsened by the hipper youth-centric offerings from ABC and NBC. Determined to attract new viewers, Sheldon accelerated the pace of the action, with added emphasis on the younger and prettier characters. He also hearkened back to the crime-ridden continuities of the earlier days with an abundance of 1980s-style violence and a staggering increase in the number of on-camera death scenes — not to mention a sudden resurgence of boisterous courtroom confrontations. But nothing could stem the rate of audience attrition nor justify the ever-increasing production costs, and on December 28, 1984, *Edge of Night* came to an abrupt halt.

Producer Erwin Nicholson had ended the series on a cliffhanger, in hopes of repackaging the property for first-run syndication. While this never happened, videotaped episodes from *Edge of Night*'s final four seasons were rebroadcast on cable's USA Network from August 5, 1985, through January 19, 1989: and beginning in 2006, the AOL Video Service offered downloads of several *Edge of Night* installments produced between 1979 and 1984.

Eisenhower & Lutz

CBS: March 14–June 20, 1988. MTM Productions. Executive Producers: Allan Burns, Dan Wilcox. Cast: Scott Bakula (Barnett M. "Bud" Lutz Jr.); De-Lane Matthews (Megan O'Malley); Patricia Richardson (Kay "K.K." Dunne); Rose Portillo (Milly Zamora); Leo Getter (Dwayne Spitler); Henderson Forsythe (Barnett M. "Big Bud" Lutz Sr.).

While skimming through the pages of this book, the average reader might come to the conclusion that the streets of Tinseltown are clogged with the corpses of unsuccessful legal sitcoms. The failure of *Eisenhower & Lutz* is particularly painful to report inasmuch as the series was created by the usually reliable Allan Burns, whose other credits include *The Mary Tyler Moore Show* and *Lou Grant*— to say nothing of the imperishable *The Bullwinkle Show*.

First and foremost, it must be noted that there was no Eisenhower in *Eisenhower & Lutz*. The name of the legal firm where most of the action took place had been dreamed up by zany professional sign painter "Big Bud" Lutz (Henderson Forsythe), who hoped to improve business for his disreputable lawyer son Bud Lutz Jr. (Scott Bakula of *Quantum Leap* fame) by adding the name of a non-existent partner — and a dead president at that — to the operation. Bud Jr. needed all the help he could get: having graduated from the Southeast School of Law and Acupuncture with the lowest possible honors, he'd already come a cropper practicing his trade in Las Vegas, and was now reduced to hanging up his shingle in an abandoned hot-tub showroom, located in a rundown strip mall just outside Palm Springs, California. As a further blow to his ego, most of Bud Jr.'s prospective customers wanted nothing to do with him, insisting upon talking to "Mr. Eisenhower."

Those clients the younger Lutz *was* able to scare up were gathered from the dregs of society, and sometimes even lower than that. His coworkers and friends included his sweetheart Megan (DeLane Matthews), who worked as a waitress at the nearby Kon Tiki cocktail lounge; his never-paid but loyal Latina secretary Milly (Rose Portillo); eager-beaver law student, part-time sushi delivery boy and volunteer "gofer" Dwayne (Leo Getter), who took on variety of office responsibilities to help keep Eisenhower & Lutz afloat; and Bud Jr.'s former high-school flame Kay (Patricia Richardson), who never tired of trying to steal our boy away from Megan and persuading him to sign up at Griffin, McKendrick & Dunne, her own law firm. By the time all these supporting characters were introduced, the series was half over.

Though John J. O'Connor of *The New York Times* found the show "very funny" and favorably compared Scott Bakula's "dopably affable" style to comedian Steve Martin, few others shared O'Connor's enthusiasm. Patched into CBS' Monday night schedule as a replacement for *Newhart* (which had moved to an earlier timeslot), *Eisenhower & Lutz* survived only a scant eleven 30-minute episodes. Perhaps the show might have lasted longer if Dabney Coleman, Allan Burns' first choice for the role of sleazy ambulance-chaser Bud Lutz Jr., hadn't turned down the part because he felt the character wasn't sleazy *enough*.

Eli Stone

ABC: debuted Jan. 31, 2008. Berlanti Television/ Touchstone Television. Executive Producers: Marc Guggenheim, Greg Berlanti, Chris Misiano. Cast: Jonny Lee Miller (Eli Stone); Laura Benanti (Beth Keller); Natasha Henstridge (Taylor Wethersby); Victor Garber (Jordan Wethersby); Sam Jaeger (Matt Dowd); Loretta Devine (Patti); Julie Gonzago (Maggie Dekker); James Saito (Dr. Chen).

Eli Stone was a legal-series variation on the "spiritual awakening" theme that had already done yeoman duty on such series as *Joan of Arcadia* and *Dead Again*. Jonny Lee Miller starred as Eli Stone, a cold-blooded young corporate attorney who worked for a presitigious San Francisco firm that thrived on screwing the "little guy" in favor of rich fat cats; (in case you need reminding: there are no lovable oil executives or real-estate developers on network TV), and who himself believed only in "the Holy Trinity of Armani, Accessories, and Ambition." All this changed when Eli began experiencing weird, MTV–style visions of his deceased ne'er-do-well father (Tom Cavanagh) and of singer George Michael performing a plaintive rendition of "Faith." Even more disturbing for Eli were such bizarre hallucinations as being buzzed by a huge fighter plane in the middle of downtown Manhattan. Convinced that he'd had an epiphany, Eli instantly changed his mercenary ways, telling off his pompous boss Jordan Weathersby (Victor Garber) and largely forsaking his lucrative practice in order to start representing underdog clients against such predators as the U.S. Military Establishment, Big Tobacco and the Agribusiness Empire. In the words of *Milwaukee Journal-Sentinel* TV critic Joanne Weintraub, Eli Stone became "television's latest daydream believer."

Weathersby and his daughter Taylor (Natasha Henstridge) were certain the poor boy's hallucinations were evidence of insanity. But Eli's neurologist had a grimmer prognosis: Mr. Stone was suffering from an inoperative brain aneuryism, the same kind that had killed his father ("You have conjoined butts in your head, which makes you a double butt-head." Who says that neurologists have no bedside manner?) Seeking a second opinion, our hero consulted funky acupuncturist Dr. Chen (James Saito), who bought into Eli's belief that his visions were evidence of a "higher calling" and offered to help interpret those visions. Also aiding and abetting Eli as he carried out his new mission in life were his idealistic assistant Patti (Loretta Devine) and his hero-worshipping protegee Maggie (Julie Gonzago).

Because of its casual crossovers from the Real to the Surreal and back again, *Eli Stone* was likened to the earlier quirky legal comedy *Ally McBeal* (q.v.)—undoubtedly delighting series creator Marc Guggenheim, a former Boston lawyer who admitted *Ally* had been his primary inspiration. Curiously, for all the series' evangelical overtones, Guggenheim insisted it was possible to practice spirituality without being overtly religious. This assertion would seem to be in line with the series' trendy Liberalism, exemplified by the writers' predilection for dwelling upon the *Cause du Jour*—which in at least one instance brought down the full 60,000-member wrath of the American Academy of Pediatrics. In the opening episode, Eli represented a mother who claimed the preservatives in a vaccination serum had made her son autistic. At first demanding that *Eli Stone* be cancelled, the A.A.P. calmed down and insisted upon a televised disclaimer so that viewers would not be frightened out of immunizing their children. ABC agreed to the extent of adding a reference at the end of the episode to the Centers of Disease Control and Prevention website—a weak and insufficient gesture according to one immunologist, who in a *USA Today* article commented, "I only hope that people see [*Eli Stone*] as the fantasy that is is."

Equal Justice

ABC: March 27, 1990–July 24, 1991. Orion Television. Creators/Executive Producers: Thomas Carter, Christopher Knopf, David A. Simons. Cast: George DiCenzo (DA Arnold Bach); Cotter Smith (Dep.DA Eugene "Gene" Rogan); Kathleen Lloyd (Jesse Rogan); Jane Kaczmarek (Linda Bauer); Joe Morton (Michael James); Sarah Jessica Parker (Jo Ann Harris); Barry Miller (Pete "Briggs" Brigman); Debrah Farentino (Julie Janovitch); James Wilder (Christopher Searls); Jon Tenney (Peter Bauer); Lynn Whitfield (Maggie Mayfield).

Introduced with two limited tryout runs in the Spring and Summer of 1990 before its official unveiling in January 1991, the ensemble legal drama *Equal Justice* was conceived as the flip side of the popular *L.A. Law* (q.v), focusing on a group of underpaid public prosecutors rather than a team of expensive defense lawyers. In both cases, the producers were careful to broaden the series' appeal with a cast of good-looking and charismatic actors.

ABC greenlighted the series based on the track record of its creator, former actor Thomas Carter (best remembered as teenage basketball whiz James Hayward in *The White Shadow*). Of the few African-American directors working in television in the early 1990s, Carter was

unquestionably the most successful, helming a string of TV pilot films that invariably matriculated to weekly series: *Miami Vice, St. Elsewhere, A Year in the Life, Call to Glory, Midnight Caller*. *Equal Justice* was packaged by Orion Productions, a moviemaking concern which despite several successful films (soon to include the back-to-back Oscar winners *Dances with Wolves* and *Silence of the Lambs*) had spent themselves to the brink of poverty, and was counting on Thomas Carter to help pull them out of the doldrums. Carter responded with a series based on the experiences of one of his best friends, who'd worked in the Harris County (Texas) District Attorney's office, where scores of eager young associates marked time before graduating to private practice. It was a tough, unglamorous workplace, and Carter hoped to replicate this gritty ambience with a realistic, utterly deglamorized legal series. He also strove to avoid the usual TV-law cliches: "I didn't want to do a show where the lawyer won all the time. I didn't want to feel forced to go into the trial phase every time, because most cases are plea-bargained. I wanted to show how most cases are disposed of in the hallways, in the bathrooms."

Set in the Pittsburgh DA's office, *Equal Justice* boasted a remarkable roster of stars in the making. Heading the cast was George Di Cenzo as DA Arnold Bach, whose basic honesty and integrity was continually put to the test by the harsh realities of the legal and political system. Cotter Smith costarred as Bach's protege Gene Rogan, chief Deputy DA and head of the office's Felony Bureau; though Rogan admired Bach, he frequently questioned his mentor's ethics, and ultimately ran against him to win the DA position himself. Jane Kaczmarek (see also *The Paper Chase* and *Raising the Bar*) portrayed Linda Bauer, beleaguered head of the Sex Crimes Unit, while Jon Tenney was seen as Linda's younger brother and occasional courtroom adversary, Public Defender Peter Bauer. Joe Morton was cast as chief prosecutor Michael James, an African American whose close friendship with the Caucasian Gene Rogan was often strained by the racial tensions that permeated their profession; though James tended to come off as too decent and objective to be true, the fact that he was a diehard opera buff provided a fascinating extra dimension to what could have been a one-note character. Representing the office's "young blood" attorneys were Sarah Jessica Parker as countrified *naif* Jo Ann, Peter Bauer's girlfriend; Barry Miller as the obligatory clueless chauvinist Briggs; Debrah Farentino as the idealistic Julie; and James Wilder as the ambitious Christopher. Also seen were Kathleen Lloyd as Gene Rogan's supportive wife Jesse and Lynne Whitfield as black investigative reporter Maggie Mayfield, whose romance with Michael James was cut short when she perceived him as a "sellout" to the white power structure.

Recent viewings of the first few *Equal Justice* episodes would seem to bear out the critics' early complaints about the series. In its efforts to convey the genuine confusion and cacaphony of a big-city DA's office, the series worked too hard to emulate the multiple-storyline, intersecting-character hustle and bustle of such other ensemble series as *L.A. Law* and *Hill Street Blues*, ending up biting off more than it could chew. Also, the important sociopolitical issues raised in each episode (usually along the lines of "money talks and B.S. walks") were muddily handled, generally playing second fiddle to the protagonists' personal demons — which in turn tended to blunt the sharp edges of their characterizations. Happily, the series got better as it went along, ultimately finding its bearings during the early months of 1991, with credible character development and a clearer approach to the weightier and more complicated issues facing those characters (even though these issues were seldom resolved by episode's end). The fact that order was finally growing out of chaos was recognized by *New York Times* TV critic John J. O'Connor, who noted that the characters moved "nimbly through a criminal justice system that can most charitably be described as overburdened…. The *Equal Justice* mix is tricky, at times even impossible. But Mr. Carter and his feisty repertory company are scoring more and more frequently and impressively."

It's too bad that the series finally hits stride

at the same time its parent company Orion Pictures was beginning its swift and irreversible descent into bankruptcy and dissolution. Despite garnering such industry honors as two consecutive Emmy Awards for Thomas Carter's direction, *Equal Justice* expired after only 26 hour-long episodes.

Eye for an Eye

Syndicated: 2003–2007. Atlas Worldwide Syndications/National Lampoon Inc. Cast: Judge Akim Anastopoulo (Himself); Kato Kaelin (Host); Sugar Ray Phillips (Bailiff).

"Sometimes Justice is a baseball bat!" proclaimed the ads for the syndicated *Eye for an Eye*. Well, what else could one expect from a TV courtroom series coproduced by *The National Lampoon* and costarring America's most famous couch potato?

Debuting on a weekly basis in the fall of 2003, *Eye for an Eye* was a vehicle for former South Carolina state prosecutor "Extreme" Akim Anastopoulo, so named for the outrageous "payback" verdicts he levied in his capacity as judge. "Let the punishment fit the crime," sang Gilbert & Sullivan's Mikado, and "Extreme" Akim certainly shared that philosophy. If a male spousal abuser was brought before Judge Anastopoulo, he stood a very good chance of being ordered to act as a "test dummy" in a women's martial arts class. If a bird fancier was found guilty of disturbing the peace, it wasn't unusual for the Judge to order the man to eat worms. And if some dimbulb thought it was funny to ridicule obese persons, he would be sentenced to spend the next several weeks wearing a cumbersome fat-suit in public. Should any defendant complain about these penalties, he might have to answer to Judge Anastopoulo's beefed-up bailiff, former world middleweight champion Sugar Ray Phillips. And yes, "Extreme" Akim *did* wield a baseball bat instead of a gavel, perhaps to prove that he meant business — or perhaps to clue us in that he regarded the whole thing as a farce.

After two seasons as a weekly, *Eye for an Eye* became a daily half-hour strip in the fall of 2005, clearing 80 percent of the country's local markets (in Los Angeles, the series served as a replacement for the cancelled *Tony Danza Show*, if that's your idea of an improvement). By this time, the series' hosting duties had been assumed by onetime O.J. Simpson houseguest Kato Kaelin, again demonstrating the talent and versatility that has made him the idol of dozens. Though *Eye for an Eye* ceased production in 2007, there are still as of this writing a few independent stations showing the reruns in the wee hours of the morning, sandwiched in between the arguably more edifying half-hour pitches for Bosley Hair Products and StairMaster.

Family Court with Judge Penny

Syndicated: Debuted Sept. 8, 2008. 44 Blue Productions/Program Partners. Executive Producers: Rasha Drachkovitch, Stephanie Drachkovitch. Cast: Judge Penny Brown Reynolds (Herself).

Of the three new daily "judge" shows premiering in the fall of 2008, the syndicated *Family Court with Judge Penny* was the only one to focus exclusively on domestic arbitration. Executive producer Stephanie Drachkovitch explained pragmatically that the previous courtroom series that had fared best in the marketplace were those in which there was some family relationship between plaintiff and defendant. Even so, Drachkovitch also noted that the format of *Family Court* had not been decided upon until the production company had hired their "star" judge.

The jurist in question was Penny Brown Reynolds, one of four children born into an impoverished Southern household. While Reynolds' hard-knock childhood had imbued in her a compassion for the Underdog, the fact that she was able to rise above the deprivations of youth and attain a law degree also gave her a strong respect for the American work ethic — and an equally strong disdain for people who used the miseries of childhood as an all-purpose excuse for antisocial behavior. Like her television colleagues Judge Glenda Hatchett (q.v.) and Nancy Grace, Reynolds spent many years in the Fulton County (Georgia) court system; prior to being appointed to the bench in 2000, she was the first African American to serve as Executive Counsel to the Governor of Georgia. Her initial taste of TV fame occurred during a memorable appearance on Dr. Phil McGraw's daily chatfest in 2007; thereafter, she was a much sought-after legal commentator on cable's Fox News Channel. When approached by 44 Blue Productions to host her own daily, half-hour series, Judge Penny agreed on condition that the series concentrate on family problems arising from bad parenting, domestic abuse, custody battles, monetary and property judgements, and lesser "intramural" disagreements: "In my court, families come first." As a bonus, since Reynolds had become an ordained minister just before taping started, it could be argued that *Family Court with Judge Penny* was America's first "faith-based" courtroom show (though to her credit, she never evangelized on the air).

The series started with a bang when, during production of the pilot episode, the studio was rocked by 5.4 California earthquake, a terrifying moment captured on tape and quickly shipped out to every and entertainment-oriented TV show in the country. None the worse for her experience, Judge Penny quipped that she intended to "shake up" the courtroom-TV genre — and she certainly did her best to live up to that promise. The series opener did *not* include the earthquake footage, but was instead a heart-tugger in which two teenagers sued their "sports crazed mom" for forcing them to pay $1220 for a sports camp which neither wanted to attend. Only occasionally interrupting the litigants, Judge Penny allowed them to air their differences to one another, thereby setting the tone of the series as a glorified therapy session — albeit with an applauding studio audience. In this and subsequent episodes, the Judge evinced an inclination to be fair to both sides and appreciate both arguments: even after making her decision, she advised the winning parties to be more tolerant of the "losers" in the future, and try to settle their differences in private. Each episode included a "Penny for Your Thoughts" segment, wherein viewers were invited to call or e-mail in their opinions of the Judge's verdict; and at the end of the half hour, an off-screen announcer would fill in the viewers as to what had happened to the litigants after they left the courtroom.

"Provocative, emotional, inspiring" were the superlatives used by distributor Program Partners when *Family Court with Judge Penny* entered syndication in September of 2008, with several major TV–station groups (Fox Television, Sinclair Broadcasting, CBS, Hearst-Argyle, Tribune Broadcasting, Capital Broadcasting) eagerly snatching up the series despite the ever-growing number of similar programs that were glutting the daytime airspace.

Family Law

CBS: Sept. 20, 1999–Apr. 15, 2002, with a special finale on May 27, 2002. Paul Haggis Productions/ Columbia TriStar Television. Executive Producers: Paul Haggis, Stephen Nathan, David Shore. Cast: Kathleen Quinlan (Lynn Holt); Julie Warner (Danni Lipton); Christopher McDonald (Rex Weller); Dixie Carter (Randi King); Christian de la Fuente (Andres Dioz); Salli Richardson (Viveca Foster); Merrilee McCommas (Patricia Dumar); Michelle Horn [replacing the pilot episode's Rosemarie Martin] (Cassie Holt); Blake Rossi [replacing the pilot's Elliott and Jordan Dolling] (Rupie Holt, Season One); David Dorfman (Rupie Holt, Seasons Two and Three); Tony Danza (Joe Celano); Orla Brady (Naoise O'Neill); Meredith Eaton (Emily Resnick).

There was no middle ground with CBS' *Family Law.* Viewers either adored or detested the series, and the sides were drawn strictly along gender lines. Created by Paul Haggis, whose other legal-drama credits include *L.A. Law* and *Michael Hayes* (both q.v.), the series came into being not long after Haggis went through a divorce: coincidentally or otherwise, the eons-old Battle of the Sexes was at the core of the project. And though Haggis was a certified male, the women on *Family Law* were the series' most powerful and compelling characters. But the series was not so much ardently pro-feminist as it was stridently anti-male — perhaps the quintessence of what *Entertainment Weekly*'s Ken Tucker has labeled the "All Men Are Weasels" school of television.

As you read the following synopsis, we invite you to try a little experiment. Imagine that the genders are switched, so that females are males and vice versa. Then ask yourself if,

under those circumstances, *Family Law* would ever have gotten any farther than the proposal stage.

Kathleen Quinlan starred as marital attorney Lynn Holt, who was forced to rebuild her career and her life from the ground up when her husband and law partner Michael left her for another woman and stole their joint practice, claiming virtually all the staff and clients and gutting their Century City office —*and,* as if to rub more salt in the wound, Michael moved into new headquarters right next door. In true Hell Hath No Fury fashion, Lynn set up her own practice, dedicated to helping other women who'd been betrayed, battered or bamboozled by the men in their lives. As a symbol of her new-found misandry, Lynn defiantly ripped the "Men's Room" sign from the office lavatory.

The firm's junior associate was Danni Lipton (Julie Warner), the only member of Lynn's former practice who hadn't defected to Michael, while the rest of her staff included loyal secretary Patti Dumar (Merrilee McCommas) and — believe it or not — a man, handsome young Chilean–born "gofer" Andres (Christian de la Fuente). Three episodes into the first season, Lynn hired a new paralegal, Viveca Foster (Salli Richardson). Explained producer Stephen Nathan: "We wanted a character who didn't know her place — somebody far too pushy and arrogant for her position, but someone you can love at the same time."

In order to make ends meet, Lynn sublet office space to Randi King (Dixie Carter), a virago of a divorce attorney who'd served prison time for murdering her husband (the rat had it coming, of course). "I hate men, and I play very dirty!" declared Randi, as if to remove all doubt. Another of Lynn's tenants was Rex Weller (Christopher McDonald), an arrogant criminal attorney who'd been reduced to advertising on an access-cable station after his partners skipped the country with all of his firm's money. When Viveca passed her bar exam, Rex showed he couldn't be trusted by hiring her away from Lynn. Meanwhile, Randi carried on a torrid affair with the much-younger Andres — who returned to Chile to avoid being arrested by the immigration authorities, but not before proposing marriage to Randi. Alas, upon his return to America, Andres was exposed as a chauvinist swine who already had a wife and child tucked away in his home village.

At the outset of the second season, Tony Danza joined the cast as Joe Celano, a brash civil-rights attorney whom Lynn had met when he was representing the Guatemalan woman who claimed that Lynn's adopted child was "stolen." So impressed was Lynn by Joe's tireless championing of the Underdog that she made him her partner, blithely shutting out her faithful coworkers (a strangely callous move for a woman who'd gone into business for herself because she'd been stabbed in the back). But for all his Liberal compassion, Joe was utterly unscrupulous and self-serving: a typical man. As for Danni, she was busily trying to adopt a young boy whom she'd rescued from his abysmal brute of a father; ultimately forced to choose between job and family, Danni exited the firm. (According to insiders, actress Julie Warner's reason for quitting the show was that Tony Danza had been billed above her, even though she had seniority: the guys get all the breaks, don't they?)

Forced to downscale her operation at the beginning of Season Three, Lynn moved her staff to the seedy tourist community of Venice, California, setting up shop in the offices of a failed dot-com company (which probably went under because there weren't any women running things). Having evidently concluded that the road to success was to keep hiring obstreperous females, Lynn took on two new employees, abrasive Irishwoman Naoise (Orla Brady)—later murdered in a gangland ambush—and combative 4'3" dwarf Emily (Meredith Eaton). By comparison, the redoubtable Randi, who had mellowed considerably after adopting her granddaughter, was a candidate for sainthood. Meanwhile, Viveca fell in love with Joe and resigned from Rex' firm after bearing Joe's child—which of course he wanted no part of, the jerk. And in the final episode, which aired as a "special," Lynn and Rex were about to get an annulment after undergoing a hasty marriage for the purpose of making a legal point to some fathead of a judge who undoubtedly preferred his women pregnant, barefoot and in the kitchen.

A number of critics (guess which ones) did nip-ups over *Family Law*, but an equal number grew awfully tired of its male-bashing, awfully fast. When *TV Guide* slammed the series as "tiresomely formulaic," it's a safe bet that the magazine's chief complaint had to do with the writers' habit of *always* casting a negative light on the Brotherhood of Man. Even when the series offered a female villain, it was seldom really her fault. In a typical episode, Danni was disgusted by her client, a woman whose heavy drinking caused her to miscarry and was subsequently charged with murder. Still, Danni stifled her own feelings in order to pursue the LARGER ISSUE: a woman's right to do what she wants with her own body.

Oddly, the usual-suspect pressure groups never lodged a complaint about the series' slanted view of all things sexual; the one episode that sparked the most controversy dealt with the entirely different but no less inflammatory issue of gun control. In the March 9, 2001, installment "Safe at Home," a young boy figured out the combination of his mother's safe, removed the gun hidden therein, and killed a playmate. When time came to choose the summer reruns for the 2000–2001 season, sponsor Procter & Gamble deemed "Safe at Home" too controversial for replay and pressured CBS to remove it from the manifest. Series creator Paul Haggis, who wrote the episode, went directly to *The Wall Street Journal* and griped, "The advertisers are sending the creative community a very clear message that they want nothing that challenges the viewer, nothing that is thought-provoking." Also chiming in to chastise Procter & Gamble was Writers Guild of America president John Wells, calling the sponsor's demand "a serious threat to the creative rights of all artists in our industry." CBS ultimately chose to run the episode, whereupon P&G dropped its sponsorship.

But it was neither controversy nor lack of sponsor support that drove *Family Law* off the air. Never exactly a champion in the ratings, the series played to fewer and fewer viewers with each passing month, leaving only a straggling few to mourn when the show was cancelled at the end of its third season. (Yeah, right. It's always the ratings. More likely it was one of those rotten, scumbag misogynists at CBS. All men are weasels.)

Famous Jury Trials

DuMont: Oct. 12, 1949–March 12, 1952. Cast: Jim Bender (Prosecuting Attorney); Truman Smith (Defense Attorney). Some sources list Donald Woods as the series' narrator.

Famous Jury Trials orginated as a stark, no-frills radio anthology, debuting over the Mutual network on January 5, 1936. Promoted as "the dramatic story of our courts, where rich and poor alike, guilty and innocent, stand before the bar of justice," each half-hour episode (45 minutes during the 1936–37 season) dramatized an actual trial, "disinterred from judicial archives." Predating CBS' historical drama *You Are There* by over a decade, the series took the listener back in time, sometimes as far as 200 years, with a contemporary radio reporter (played by such actors as Roger DeKoven and DeWitt Bride) covering a celebrated trial as if it were unfolding before his eyes. This format was later streamlined, with the reporter appearing in brief cut-ins to bridge the time gaps in the story, and the courtroom scenes themselves carried by the dialogue of the lawyers, litigants, witnesses and judges — words frequently lifted directly from original court transcripts. After receiving instructions from the judge ("Be just and fear not, for the true administration of justice is the foundation of good government"), the jury retired for the verdict, which was withheld until the very end of the program to heighten suspense. The series ended its radio run on June 25, 1949, by which time it had moved from Mutual to ABC.

Four months later, *Famous Jury Trials* was revived for television as a weekly, live half-hour, courtesy of the DuMont network. Though it lagged behind NBC and CBS in terms of budget and viewership, DuMont boasted the most advanced and sophisticated technical operation in the business, its slick and innovative camerawork and elaborate staging effects putting many another live program of the era to shame. From all existing evidence, *Famous Jury Trials* was well up to DuMont's lofty standard — and as the first production to emanate from the network's massive new downtown Manhattan facilities, the series benefited from state-of-the-art sound and picture quality, as well as the combined talents of some of New York's most versatile (if not best-known) character actors.

There were, however, a few logistical problems along the way. In order to make the TV series both visually and verbally stimulating, the trial scenes were dramatically enhanced through the use of extensive flashbacks — each of these related from the viewpoint of the person testifying, meaning that the "facts" presented were often biased and contradictory. At episode's end, the audience was shown what *really* happened, and the verdict was revealed. Actor Frankie Thomas, who made several appearances on the series, recalled in an online interview from the early 2000s that the format of *Famous Jury Trials* "created frenzy" for himself and his fellow performers. "The show opened in a courtroom with someone testifying, and faded out to a flashback of the events covered in the testimony. But of course the flashback involved the same actor or actress seen in the initial courtroom scene, and the problem was that the different sets were quite far apart in a large studio. The actors quickly became breathless running set to set."

TV's *Famous Jury Trials* remained on the DuMont docket until May of 1952. For several months in 1950, the series was seen directly opposite a similar property on ABC, *Your Witness* (q.v.)

Feds

CBS: March 5–Apr. 9, 1997. Wolf Films/Universal TV. Executive Producers: Dick Wolf, Arthur W. Forney, Michael S. Chernuchin. Cast: Blair Brown (U.S. Atty. Erica Stanton); Adrian Pasdar (Asst. U.S. Atty. C. Oliver Resor); John Slattery (Chief Asst. U.S. Atty. Michael Mancini); Dylan Baker (Special Agent Jack Gaffney); Regina Taylor (Asst. U.S. Atty. Sandra Broome); Gracie Phillips (Asst. U.S. Atty. Jessica Graham); John Rothman (Agent Katz); George

DiCenzo (Tony Garufi); George Martin (Charles Resor); John Vengimiglia (Alfonse Bucci); Frank Senger (Tommy Iradesco); Scott Cohen (Rod Nesbitt).

One of the lesser-known courtroom series from *Law & Order* paterfamilias Dick Wolf, *Feds* was set in the Manhattan Federal Prosecutor's office. Blair Brown starred as hard-driving U.S. Attorney Erica Stanton, the person in charge of the office and the second most powerful law enforcement officer in the country (the Attorney General is, of course, the first). Stanton's staff included prosecutor Michael Mancini (John Slattery), whose principal target was Mob chieftan Tony Garufi (George DiCenzo, previously one of the "good guys" on *Equal Justice* [q.v.]), who may or may not have murdered Mancini's family; Civil Rights specialist Sandra Broome (Regina Taylor, previously a costar on *I'll Fly Away* [q.v.]); "white-collar crime" expert C. Oliver Resor (Adrian Pasdar); all-purpose prosecutor Jessica Graham (Gracie Phillips); and FBI agent Jack Gaffney (Dylan Baker).

The first CBS series broadcast in the widescreen letterbox format, *Feds* came into being because, according to producer Wolf, "The FBI has been done, but the federal judicial system has never been done before. So it's a very, very rich area of the law, especially since it's major leagues of prosecution. We're dealing with crimes that local jurisdictions just don't get to deal with, everything from kidnapping to terrorism to the Mob to taking on a major tobacco company for criminal violations." For some observers, however, the show's main selling card was not its story material but the atypical casting of star Blair Brown, herein far removed from her whimsical *Days and Nights of Molly Dodd* characterization. Described by *Entertainment Weekly* as "about as cuddly as Attorney General Janet Reno," Erica Stanton was "kind of hard-ass" to Brown's way of thinking, "But what I like about her is that she's not a neurotic, driven careerist like most women on TV or in movies." Unfortunately, neither Blair Brown, nor the series' format, nor even the wide-screen photography were enough to keep *Feds* on the job for any longer than six weeks.

Final Appeal: From the Files of Unsolved Mysteries

NBC: Sept. 18–Oct. 16, 1992. Cosgrove/Meurer Productions. Executive Producer: John Cosgrove. Host: Robert Stack.

A spinoff of the spectacularly successful reality program *Unsolved Mysteries*, which ran eleven years on two different networks, the considerably shorter-lived legal series *Final Appeal: From the Files of Unsolved Mysteries* was like its parent program hosted by Robert Stack, exhibiting the same stentorian gravitas he'd virtually patented as Eliot Ness on *The Untouchables*. Among the more memorable segments on *Unsolved Mysteries* were those that reopened criminal cases in which the possibility existed that the "guilty" party might actually have been wrongly convicted. *Final Appeal* took this premise to the next level, scrutinizing the case at hand from the viewpoint of both the prosecutors and the defense attorneys. The final verdict was levied by the viewers at home, who were invited to vote by mail, phone, or fax as to whether the defendant was innocent or guilty as charged.

Final Appeal didn't last long enough for viewers to find out if they actually had any influence on the outcome of the appeals — though in retrospect, we now know that the series' most famous reopened case, the 1944 court-martial of fifty African-American sailors charged with mutiny for refusing to handle live ammunition after a disastrous storage-facility explosion in Chicago, resulted in a Presidential pardon for 37 of the defendants in 2000.

Final Justice with Erin Brockovich

Lifetime: Debuted Jan. 17, 2002. Executive Producers: Kathy Williamson, Eric Schotz, Bill Paolantonio. Theme song cowritten and performed by Wynonna Judd. Cast: Erin Brockovich (Herself).

Erin Brockovich was just another unemployed divorcee when, in 1991, she talked her way into a job as file clerk for Masry & Vititoe, the California law firm that had recently handled her auto-accident case. Though she had no formal legal training, the 32-year-old

Brockovich ended up the firm's star employee when she came across the medical records that ultimately led to a $333 million class-action suit against Pacific Gas and Electric for its alleged contamination of a small town's water supply with the toxic carcinogen Chromium VI. This inspirational "David and Goliath" story culminated in the 2000 theatrical feature *Erin Brockovich*, with Julia Roberts winning an Oscar for her virtuoso portrayal of the title role.

As for the real Erin Brockovich, she was approached by dozens of TV producers who wanted her to star in her own consumer-advocate series, but she turned them all down. Then along came the representatives of the Lifetime cable network, who proposed a documentary series with a strong "empowerment" angle, featuring true stories of ordinary women who fought back against the System and won justice for themselves and others. In explaining why she agreed to host *Final Justice*, Brockovich told *TV Guide*'s Robert Abele, "These women don't just sit there and say 'Poor me. I can't go on with life.' They come forward to make a change for themselves, and in the process, they make a change for all of us." On a more personal level, she hoped the series would convince certain skeptics that Erin Brockovich actually existed and was not merely a Hollywood fabrication.

Each hour-long episode of *Final Justice with Erin Brockovich* featured three different stories of women taking legal matters into their own hands and slaying various courtroom dragons with what the series' host described as "dogged persistence." And if those viewers who tuned in to see if the very attractive Ms. Brockovich really looked like Julia Roberts were in any way disappointed, the lady amply compensated for that disappointment by exhibiting the same brash brassiness — and wearing virtually the same revealing wardrobe — as her big-screen counterpart. If there was any complaint to be leveled against *Final Justice*, it was for its lack of suspense: you *knew* that the Good Gals would win every time.

First Monday

CBS: Jan. 15–June 7, 2002. Belisarius/Paramount Network Television. Creator/Executive Producer: Donald P. Bellisario. Cast: Joe Mantegna (Justice Joseph Novelli); James Garner (Chief Justice Thomas Brankin); Charles Durning (Justice Henry Hoskins); Camille Saviola (Justice Esther Weisenberg); James McEachin (Justice Jerome Morris); James Karen (Justice Michael Bancroft); Gail Strickland (Justice Deborah Szwark); Stephen Markle (Justice Theodore Snow); Lyman Ward (Justice Brian Chandler); Hedy Burgess (Elle Pearson); Randy Vasquez (Miguel Mora); Christopher Wiehl (Jerry Klein); Joe Flanigan (Julian Lodge); Linda Purl (Sarah Novelli); Brandon Davies (Andrew Novelli); Rachel Grate (Beth Novelli); Charles Bierbauer (Himself); Sandra Prosper (Kayla Turner); Mark Costello (Court Crier); Liz Torres (Janet Crowley).

One of two "Supreme Court" dramas to premiere on network television in 2002 (see also *The Court*), CBS' weekly, hour-long *First Monday* took its title from the Court's traditional yearly opening session on the first Monday in October. Joe Mantegna headed the huge ensemble cast as Joseph Novelli, a "moderate liberal" attorney newly appointed as an associate justice on a U.S. Supreme Court evenly divided along political lines: four Conservatives, four Liberals. Because of his track record for nonpartisanship, Novelli was meant to be the "swing" vote in difficult decisions. Our hero soon realized he had his work cut out for him as he tackled his very first assignment: a tricky capital-punishment case involving a retarded man who, facing execution for a murder committed when he was sixteen, had been struck by lightning and now argued that it was "cruel and unusual punishment" to be electrocuted twice!

Here as elsewhere, Novelli butted heads with ultraconservative Chief Justice Thomas Brankin (James Garner), a wily Oklahoman described by legal columnist Chris Corcos as an amalgam of Byron White and William Rehnquist, who apparently regarded the highest court in the land as a glorified athletic field (indeed, the Justices began each session by joining hands and in unison vowing to "make history," coming off like a superannuated high school football team). Of the other Justices, the only ones who ever had more than few seconds' airtime per episode were Brankin's chief ally, the elderly and infirm Henry Hoskins (Charles Durning); Estelle Weisenberg (Camille Saviola), a Liberal along the lines of Ruth Bader Ginsberg; and Weisenberg's African-American colleague Jerome Morris (James McEachin), a Thurgood Marshall clone. The supporting cast also included an assortment of young and attractive legal clerks, notably feisty Liberal Elle Pearson (Hedy Burgess) and smooth-talking Conservative Miguel (Randy Vasquez). (The various romantic intrigues amongst the clerks were trivial at best and not worthy of discussion here.) Evidently in the interests of credibility, the series featured a number of real-life "power" attorneys in cameo roles, including Gerry Spence, Vincent Bugliosi, Barry Scheck, and those two ringmasters from the O.J. circus, Marcia Clark and Johnnie Cochran; also appearing as himself was CNN's senior Washington correspondent Charles Bierbauer.

Many TV critics and legal experts complained that *First Monday* was a stacked deck, with the Liberal characters coming off as paragons while the Conservatives were muddle-headed reactionaries, often lacking a basic understanding of the laws at hand and habitually making politics rather than justice their first priority. In a typically loaded sequence — eloquently skewered by courtoom-drama historian Michael Asimow on the *Picturing Justice* website for its totally inaccurate representation of court procedure — a transsexual brings a case before the Court, whereupon the Conservative Justices mercilessly browbeat the poor guy face-to-face rather than speaking through his attorney, asking in accusatory tones if he wanted to be "castrated like a steer." In fairness to series creator Donald P. Bellisario, whose pro-military *JAG* (q.v.) brought a refreshing sense of balance to the predominantly left-leaning world of Prime Time television, it must be noted that the Liberals on *First Monday* were prone to ponderous speechifying and irritating self-righteousness, which was every bit as obstreperous as the Conservatives' more blatant biases.

Julie Salamon of *The New York Times* described *First Monday* as "cheesy," adding, "The machinations of the court are commented on, to be sure, but the characters feel like types pulled out a drawer…. The show serves its constitutional issues three ways: cute, vulgar, sentimental…." The media-research staff of the Tarlton Law Library wasn't quite so gracious, grousing that the series suffered from "terrible writing" and "awkward acting": "It was a miracle that it survived more than a month." Virtually the only good word for *First Monday* was registered by the National Italian American Foundation, for the series' "positive portrayal" of its protagonist Joseph Novelli — and this despite a story arc in which Novelli came under fire for his uncle's alleged Mob connections.

First Years

NBC: March 19–Apr. 2, 2001. NBC Studios/Studios USA Television. Executive Producers: Jill Gordon, Mark B. Perry, Ken Topolsky. Cast: Samantha Mathis (Anna Weller); Mackenzie Astin (Warren Harrison); Sydney Tamiia Poitier (Riley Kessler); James Roday (Edgar "Egg" Ross); Ken Marino (Miles Lawson); Eric Schaeffer (Sam O'Donnell); Bruce Winant (Bruce); Kevin Connolly (Joe).

The sales pitch for the NBC sitcom *First Years* might well have been "*Friends* with lawyers." It wasn't quite that simple: though it admittedly resembled the long-running *Friends*, the series was actually a clone of a British sitcom … which admittedly also resembled the long-running *Friends*.

Making its BBC2 debut on March 18, 1996, the half-hour comedy series *This Life* was about a quintet of first-year law school graduates, male and female, who lived together in what the network described as a "posh South London house." Because of their youth and inexperience, the five fledgling lawyers were stuck with all the cases no one else wanted, such as evicting sweet little old ladies and closing down day-care centers. The series stirred up a storm in a teacup with its references to casual sex and substance abuse, and its frank and nonjudgmental treatment of gay relationships; also, critics took aim at the migraine-inducing "innovational" camerawork, with the correspondent for *The London Daily Mirror* complaining, "The show looks as if the cameraman suffers from the shakes or has watched too many episodes of *Hill Street Blues* and *ER*." But the series was extremely popular with its 18-to-34 year old target audience, especially for its cynicism about traditional British values. *This Life* also made overnight stars of its attractive ensemble players: Daniela Nardini as the cheerfully promiscuous Anna, Jason Hughes as the closeted-gay Warren, Amita Dhiri as erudite Indian lass Milly, Andrew Lincoln as Milly's long-haired beau Edgar "Egg" Ross, and Jack Davenport as the wealthy Miles.

The American version, *First Years*, switched the series' locale from South London to San Francisco's Haight-Ashbury district, where the five principals lived in one of those humongous old houses that no one but TV characters can really afford. Outside of changing the Indian Milly to the African-American Riley (played by Sydney Tamiia Poitier, the daughter of guess which Oscar–winning black superstar), the main characters were fundamentally the same, with Samantha Mathis as Anna, Mackenzie Astin as Warren (still gay, still not quite "out" yet), James Roday as Egg and Ken Marino as Miles. Also featured on *First Years* was Eric Schaeffer as veteran attorney Sam O'Donnell, the five young lawyers' bullying mentor at Hoberman, Spain, McPherson & O'Donnell.

Lasting 32 episodes (not including a reunion special in 2007), the British *This Life* was cancelled after a single season, not because of its controversial content but because the five leading actors had become too popular and too expensive for BBC2 to afford. The nine-episode *First Years* was cancelled after three weeks because everyone was watching its Fox Network competition *Ally McBeal* (q.v.)

Foley Square

CBS: Dec. 11, 1985–Apr. 7, 1986. Shukovsky English Entertainment. Creator/Executive Producer, Diane English. Executive Producers: Saul Turtletaub and Bernie Orenstein. Cast: Margaret Colin (Asst. DA Alex Harrigan); Hector Elizondo (DA Jesse Steinberg); Vernee Watson-Johnson (Denise Willums); Michael Lembeck (Peter Newman); Cathy Silvers (Asst. DA Molly Dobbs); Sanford Jensen (Asst. DA Carter DeVries); Israel Juarbe (Angel Gomez); Jon

Lovitz (Mole); Richard C. Serafian (Spiro Papadopolis).

The CBS legal sitcom *Foley Square* is historically significant as the first major TV credit for Diane English, who later created and produced the popular Candice Bergen vehicle *Murphy Brown*. While developing her saga of the men and women of the Manhattan District Attorney's office (located at the titular square), English was intent upon creating a comedy series with strong feminist leanings. "Every script is written, produced, and overseen by women," she informed Bill O'Halloran of *TV Guide*. "We are getting the women's viewpoint." True, the series' executive producers, veteran TV comedy writers Saul Turtletaub and Bernie Orenstein, were both male-type persons: but according to English, Saul and Bernie "admit they are learning a lot about what modern young women think."

Something of a prototype for Murphy Brown, *Foley Square*'s main character was a single, headstrong and highly intelligent young woman with a masculine-sounding first name: Alex Harrigan, played by Margaret Colin. As one of the office's three assistant district attorneys, Alex was answerable only to her DA boss, lifelong public servant Jesse Steinberg—a curiously Semitic appellation for *any* character played by Hector Elizondo. Others on Jesse's staff were wide-eyed rookie ADA Molly Dobbs (Cathy Silvers) and the aggressively ambitious and terminally annoying ADA Carter DeVries (Sanford Jensen), characters who came off as preliminary sketches for *Murphy Brown*'s Corky Sherwood and Jim Dial.

Appearing respectively as Alex's sarcastic secretary Denise and wacky neighbor Peter were Vernee Watson-Johnson and Michael Lembeck. (The presence of Cathy Silvers and Michael Lembeck on the same sitcom had a peculiar piquancy: the actors' comedian fathers Phil and Harvey had previously costarred as Sgt. Ernie Bilko and Cpl. Rocco Barbella in the classic military sitcom *The Phil Silvers Show*.) Also seen were Israel Juarbe as Angel, an ex-con on the road to redemption as the office messenger; Richard C. Serafian as Spiro, owner of the neighborhood coffee shop where all the regulars congregated (the day will come, perhaps in our lifetime, that all the regulars in a sitcom will bypass the neighborhood coffee shop and congregate in a Burger King or a Walgreen's); and, making furtive appearances as a private investigator known as Mole, *SNL*'s Jon Lovitz.

Despite a surfeit of talent on both sides of the camera, *Foley Square* never caught on with viewers, a problem the cast members attributed to the writers' obsession with Alex Harrigan's romantic travails at the expense of the humor inherent in the series' law-office setting. Diane English agreed, feeling the show should have concentrated more on Alex as the ADA and less as "the girl who still hasn't found a man." But this became a moot point when *Foley Square* was cancelled after fourteen episodes.

For the People

CBS: Jan. 31–May 9, 1965. Herbert Brodkin/Plautus Productions Inc. Executive Producer: Herbert Brodkin. Produced by Joel Katz. Cast: William Shatner (Asst. DA David Koster); Jessica Walter (Phyllis Koster); Howard da Silva (DA Anthony Celese); Lonny Chapman (Det. Frank Malloy).

Filmed in New York City, the 60-minute *For the People* was a spinoff of the award-winning legal drama *The Defenders* (q.v.)—or more specifically, a spinoff of two *Defenders* episodes, "Invisible Badge" (first aired November 24, 1962) and "The Cruel Hook" (November 2, 1963). Both starred William Shatner (who had appeared in the 1957 two-part *Defenders* pilot on the CBS anthology *Studio One*) as a fiercely dedicated young assistant district attorney, albeit with a different character name in each episode. In "Invisible Badge," ADA Charles Terranova faced possible disbarment when accused of accepting a bribe; and in "The Cruel Hook," ADA Cliff Sellers underwent a crisis of conscience upon discovering that his older brother (played by Ed Asner) was a murderer.

In *For the People*, these prototypical prosecutors were melded into Manhattan ADA David Koster, whose passion for justice was often hazardous to his health and career. In the opening episode "To Prosecute All Crimes," Koster was so obsessed with bringing down powerful racketeer Johnny Peck (Patrick Mc-

Vey) that he was willing to cut deals with vicious convicted criminals in order to achieve his goal. Everyone around Koster worried that his single-minded vendetta against Peck might end up destroying him: his boss, DA Anthony Celese (Howard da Silva), warned, "You're a prosecutor. Don't try to be the judge and jury as well"; a frustrated judge gave Koster hell for trying to make Pack "a vagrant without visible means of support"; and even David's sympathetic wife, professional musician Phyllis Koster (Jessica Walter), sarcastically lamented that her husband was "in love" with Pack. About the only person who was on the same page as Koster in his tireless battle against the forces of evil was his chief investigator, detective Frank Malloy (Lonny Chapman).

Like *The Defenders*, *For the People* was topheavy with penetrating social issues, for which no pat answers were ever provided. The series was also a busy way-station for actors on their way up, including Martin Sheen, Philip Bosco, Jaime Sanchez, Lesley Ann Warren, Tony Bill, and Al Freeman Jr., as well as a safe harbor for veteran performers who'd previously been frozen out of film and TV work by the infamous Blacklist, including Lee Grant, Ned Glass, Lloyd Gough and John Henry Faulk. Of the series' 13 episodes, the most fascinating— at least to 1960s TV buffs—was "Act of Violence," the conclusion of a two-part crossover with another Brodkin–produced series, *The Doctors and the Nurses*.

Hastily inserted into CBS' Sunday–night schedule as a midseason replacement for the faltering sitcoms *My Living Doll* and *The Joey Bishop Show*, *For the People* has sometimes been characterized as another of the network's "prestige" sacrificial lambs, offered up against the indomitable NBC western *Bonanza* with full realization that it would quickly be cancelled but would still garner critical praise as a "quality" effort. Herbert Brodkin himself fed this assumption, telling *TV Guide*'s Edith Efron that CBS asked him to put *For the People* together in a fast six weeks as a potential replacement for either *Mr. Broadway*, which was seen on Saturday, or *The Reporter*, which aired on Friday. "And to our horror we discovered that our time period was opposite *Bonanza*. We never expected to be put into a time period where we didn't have a chance. It never got a rating at any time. It was dropped at the end of 13 weeks. It was unintelligent programming — and a waste of $2 million." But CBS programming executive Mike Dann, who'd frequently stood up for Brodkin against the monolithic resistance of his network colleagues, disagreed: "Herb's trouble is that he's got kind of a monopoly on this particular kind of show. They [including *Defenders* and *Doctors and the Nurses*] all started declining at one time. Despite what he says, *For the People* did not fail because of the time slot. It had *Ed Sullivan* as a lead-in — and *Candid Camera* following it. People who usually stay right through that sequence of programming went to the trouble to switch it off."

It was up to another CBS executive, Bruce Lansbury, to bring some objectivity to the issue: "*For the People* at least had the intent of trying to do a show about things that needed talking about…. At its worst, it was head and shoulders above most TV programming. It made people angry. It made people discuss. It raised issues that *needed* to be raised."

As for star William Shatner, he was disappointed but not devastated by *For the People*'s abrupt cancellation in the spring of 1965. Now he was free to mull over a few other TV projects that had been submitted to him … such as that weird new science-fiction series being developed over at Desilu….

For the People

Lifetime: July 21, 2002–Feb. 16, 2003. Cumulus Productions. Creator/Executive Producer: Catherine LePard. Executive Producers: Kim Moses, Ian Sanders, Sheldon Pinchuk. Cast: Debbi Morgan (DA Lora Gibson); Lea Thompson (Chief Deputy DA Camille Paris); A Martinez (Michael Olivas); Derek Morgan (Thomas Gibson); Matthew Richards (Zach Paris); Frank Grillo (J.C. Hunter); Julian Bailey (Scott Wilson); Cecilia Suarez (Anita Lopez).

Created by *Seventh Heaven*'s Catherine LePard, the second TV legal drama to bear the title *For the People* appeared on cable's Lifetime Network in the summer of 2002. "In keeping with Lifetime's strong tradition of female programming, *For the People* will do for women's

legal issues what [the medical drama] *Strong Medicine* does for women's health," insisted the network's senior vice-president of programming Kelly Abugov. "The new drama will tackle subject matters such as adoption, custody battles and gender discrimination, all of which entertain and resonate with the viewers.…" But series costar Lea Thompson took a less lofty view of things. With the breezy insouciance she'd previously exhibited as star of the sitcom *Caroline in the City*, Thompson told *Entertainment Weekly* that *For the People* was "a fantasy": "[Costar] Debbi Morgan is this conservative black woman who is the district attorney of Los Angeles, and I'm this liberal-minded white deputy DA. There's no glass ceiling."

And that about summed it up. Debbie Morgan did indeed play Los Angeles DA Lora Gibson, the newly appointed boss of chief deputy DA Camille Paris. Convinced that she was on brink of being fired, the left-leaning Camille began cleaning out her desk; but right-leaning Lora liked Camille's style and kept her on. Despite their diametrically opposite ideologies — Lora had been swept into office on a "get tough on crime" platform, while Camille believed in tempering justice with compassion — the two women often as not found themselves in agreement when dealing with such cases as the attempted murder of a homophobic talk show host by an angry woman who held the man responsible for the killing of a gay activist.

Of course, there was an abundance of heated arguments between the two protagonists, usually falling neatly and thematically into the "point-counterpoint" category, with both ladies getting equal time (though Camille invariably made the final decision). And there was a wealth of inter-office intrigue, usually fomented by Lora's enthusiastic chief ally, prosecutor Anita Lopez (Cecilia Suarez), who dearly coveted Camille's job; and by public defender Michael Olivas (A Martinez), who as the Fates and the Scriptwriters would have it was Camille's ex-husband. When they actually had time for private lives, Lora spent hers with her bookish professor husband Tom (Derek Morgan) and her two daughters, while Camille took care of her nephew Zach (Matthew Richards), the 15-year-old son of her recovering-addict sister. At the end of each hour-long *For the People* episode, Debbi Morgan and Lea Thompson stepped out of character to advise women who were interested in legal careers to contact the proper schools and professional organizations.

The series would probably have been more compelling—and even might have survived past its initial 18 episodes—if the plotlines weren't cut from the standard issue-of-the-week cloth and the characters and the dialogue weren't so mired in clichés. Though the folks at Lifetime were fond of touting the series' "relevance" and newness, *Variety* was closer to the mark when its reviewer described *For the People* as "about as fresh as day-old bread."

Four Square Court

ABC: March 16–June 29, 1952. Produced by David Lown and Albert T. Knudsen. Cast: Norman Brokenshire (Himself).

This obscure ABC panel show predated the syndicated "reality" series *Parole* (q.v.) by nearly five years, featuring genuine parolees who discussed the crimes that had landed them in prison and the rehabiliation process they'd undergone since their release. To protect their identities, the ex-convicts wore masks and never gave their names. Actual state parole board officials also appeared on this New York–based effort, which was moderated by Canadian–born announcer Norman Brokenshire, whose reputation as one of radio's best and cleverest ad-libbers undoubtedly came in handy whenever one of the parolees said something that didn't quite measure up to broadcast-decency standards. Also known as *Foursquare Court*, the series was seen on Sundays, then as now the graveyard of low-rated public affairs programs.

Gary the Rat

TNN/Spike TV: June 26–Dec. 11, 2003. BlitzDS/Grammnet Productions/Viacom International. Executive Producers: Mark Cullen, Rob Cullen, Kelsey Grammer. Produced by James Fino and Joanne

Weiss. Voices: Kelsey Grammer (Gary Milford Andrews); Rob Cullen (Johnny Horatio Bugz); Spencer Garret (Truman Theodore Pinksdale); Billy Gardell (Jackson Buford Harrison); Vance DeGeneres (Gary's Therapist); Susan Savage (Boots the Cat); and Rob Paulsen.

When the travelling-salesman hero of Franz Kafka's *The Metamorphosis* awoke one morning to find himself transformed into a "monstrous vermin," it ruined his day. When the lawyer hero of the animated TV series *Gary the Rat* awoke one morning to find himself transformed into a six-foot rat, he put on his expensive suit and went to work as if nothing had changed. Perhaps nothing had.

As of this writing one of only two cartoon series with a lawyer protagonist (see also *Harvey Birdman, Attorney at Law*), *Gary the Rat* featured the voice of Kelsey Grammer (*Frasier*) as sleazy, amoral, thoroughly unscrupulous Manhattan defense attorney Gary Andrews, whom we had to take on faith had once been "normal," since we never saw him in human form. His courtroom dexterity slowed down not one whit by his new appearance, Gary continued to successfully defend such odious but extremely wealthy clients as Big Tobacco and Organized Crime, using every dirty legal trick at his disposal to sway judge and jury to his warped way of thinking. Of course, being a giant rodent did have its disadvantages: when visiting the Hamptons, he was refused service everywhere he went; at cocktail parties, his long tail would pop out of his trousers just as he was trying to impress a girl; he couldn't get through the night without suffering horrific, Bosch–like dreams; his neighbor Truman Pinksdale spent every waking hour trying to get Gary evicted from his luxury apartment (there was nothing in the lease that said a rat couldn't live in the building so long as he paid his rent); and most irksome of all, a lunatic exterminator named Johnny Bugz never let a day go by without trying to kill Gary in some grotesque manner—usually wiping out a bunch of innocent bystanders in the process. On the plus side, Gary's ethically challenged partner J.B. Harrison wasn't unduly stressed over having a rat in the office (maybe he considered it an improvement), while doped-up delivery boy Bud (who thought Gary was a dog) plied our hero with tons and tons of exotic foreign cheeses.

Given movement via the cost-cutting "flash animation" computer process, *Gary the Rat* originated in 2000 as an interactive cartoon on the mediatrip.com website: on November 17th of that year, Gary's mutation from human to rat was unveiled to a nationwide audience on *The Tonight Show with Jay Leno*. The character then began appearing as an interstitial (showing up between programs and during commercial breaks) on a variety of cable cartoon shows before graduating to his own series, one of three animated efforts — the others were *Ren & Stimpy's Adult Party Cartoon* and *Stripperella*— premiering the same night over the male-centric TNN cable network (soon to be renamed Spike TV). In addition to Kelsey Grammer, the series attracted such stellar voice talent as Robert Goulet, Joe Pantoliano, Mary Stuart Masterson, Michael Keaton, Brooke Shields and Betty White, to say nothing of Grammer's *Cheers* colleague Ted Danson and his *Frasier* costars David Hyde Pierce and John Mahoney. Though it won two major awards at the U.S. Comedy Arts Festival in Aspen, Colorado, *Gary the Rat* earned as many "boos" as "bravos," with *Entertainment Weekly*'s Ken Tucker summing up his opinion in three little words: "Well, it stinks." Despite respectable ratings, the series was flushed down the sewer after thirteen half-hour episodes.

girls club

Fox: October 21–28, 2002. David E. Kelley Productions/20th Century–Fox Television. Executive Producer: David E. Kelley. Theme song: "Wild" by Poe. Cast: Gretchen Mol (Lynne Camden); Kathleen Robertson (Jeannie Falls); Chyler Leigh (Sarah Mickle); Giancarlo Esposito (Nicholas Hahn); Lisa Banes (Meredith Holt); Christina Chang (Rhanda Clifford); Sam Jaeger (Kevin O'Neal); Donovan Leitch (Michael Harrod); Brian Markinson (Spencer Lewis); Eddie Shin (Mitchell Walton); Shane Brown (Meredith's Assistant); Armin Shimerman (Edmund Graves).

Producer David E. Kelley's immediate followup to the cancelled *Ally McBeal* (q.v.), *girls club* (the e.e.cummings-esque lower case is accurate) was set in San Francisco rather than Kelley's beloved Boston because the producer wanted a more "cosmopolitan" setting. Gretchen Mol, Kathleen Robinson and Chyler Leigh were respectively top-billed as Lynne, Jeannie and Sarah, three Stanford Law School graduates who shared the same loft apartment and worked as junior associates at the law firm of Myers, Berry, Cherry & Fitch. Lynne was brilliant but hopelessly naïve; Jeannie was vulnerable and completely out of her depth; and Sarah took an immediate dislike to her back-stabbing associate Rhanda (Christina Chang). Other members of the firm included the hard-shelled, implicitly lesbian Meredith (Lisa Banes), who in fine David E. Kelley tradition was given a derogatory nickname, "The Praying Mantis"; likeable Mitchell (Eddie Shin); and not-so-likeable Spencer (Brian Markinson).

Though the ladies' bullying boss Nicholas Hahn (Giancarlo Esposito) was the series' nominal villain, the actual "heavy" was not a Him but an It: Myers, Berry, Cherry & Fitch. Not only did the firm encourage its employees to be as nasty and ruthless as possible, but it also savagely pursued legal action where none was warranted, solely for the purpose of fattening the firm's bank account. (Courtroom–drama historian Michael Asimow cited this aspect as being influenced by the novels of John Grisham. Others might argue it simply held up a mirror to reality.)

Heavily promoted by the Fox network as David E. Kelley's next big hit, *girls club* ended up ranking 82nd out of 131 network shows. The series could be regarded as a success only in that it succeeded in taking all the negative aspects of *Ally McBeal* and multiplying them tenfold. Writing in the *Washington Post*, Tom Shales commented that the premiere episode "comes off as if some lesser writer were doing a cheap imitation of a David Kelley show." Of the 13 planned hour-long episodes of *girls club*, only nine were filmed — and only two were shown.

Grand Jury

Syndicated: officially debuted October 1959. Desilu/NTA. Produced by Mort Briskin. Cast: Lyle

Bettger (Harry Driscoll); Harold J. Stone (John Kennedy).

With America's newspapers full of headlines about the ever-mounting legal action against Organized Crime in 1958, producer Mort Briskin decided to strike while the iron was hot by assembling a semi-anthology TV series about two Federal Grand Jury investigators. Since the function of the Grand Jury was not to prosecute criminals but to determine if there were sufficient grounds for indictment, the series' courtroom scenes would be kept to a minimum, while the actual investigation of criminal activities and gathering of evidence would be given first priority.

Grand Jury was conceived in such haste that only three half-hour episodes were completed before the series' targeted syndication debut in the fall of 1958. Thus, most local markets did not unveil the series' full 39-episode manifest until October of 1959, by which time the property had toted up a million dollars' worth of prerelease sales (an impressive figure at the time). Though no further episodes were produced beyond the initial 39, the series was popular enough to remain in first-run syndication well into 1961.

Like most syndies of its period, *Grand Jury* opened with a Voice-of-God announcer establishing the premise: "The formwork of liberty, protecting the inalienable rights of free people. Serving unstintingly and without prejudice to maintain the laws of our land ... the GRAND JURY." To underline the toughness of its Federal investigative team, the series starred two actors normally cast in villainous roles, Lyle Bettger and Harold J. Stone, as chief investigators Harry Driscoll and John Kennedy (sic!). These boys *had* to be quick with their fists and fast on the trigger: the series' villain roster included corrupt prison officials, homicidal arsonists, a nationwide narcotics ring, a vicious baby-adoption racket, a crooked transit-company owner who masterminded fatal but profitable "accidents," the staff of a seedy nursing home specializing in bumping off sweet old ladies for the insurance, and a scurrilous crime boss who was willing to kill a crusading minister to keep his operation afloat. Ironically, at

a time when certain politicians were calling for a Congressional crackdown on the three TV networks for the excessive violence on detective and western shows, *Grand Jury* took advantage of its non-network status to dole out even *more* murder and mayhem than what was regarded as "normal."

Grand Jury was filmed at the Desilu facilities in Hollywood and Culver City, utilizing most of the standing sets left behind by the studio's previous occupant, RKO Radio Pictures. With such action-packed series as *Whirlybirds, Man with a Camera* and *The Untouchables* going full blast, Desilu may not have been the busiest studio in Southern California, but it was certainly the noisiest.

The Great Defender

Fox: March 5, 1995; July 10–31, 1995. Cardea Productions/Warner Bros. Television. Executive Producers: Frank Cardea, George Schenck. Cast: Michael Rispoli (Lou Frischetti); Kelly Rutherford (Frankie Collett); Richard Kiley (Jason De Witt); Carlos Sanz (Asst. DA Jerry Perez); Peter Krause (Crosby Caulfield III); Rhoda Gemignani (Pearl Frischetti).

Set in Boston but actually filmed in Toronto, the 60-minute legal drama *The Great Defender* starred Michael Rispoli as Lou Frischetti, a blue-collar attorney who'd earned the titular nickname by advertising on TV, with a tacky chorus singing new lyrics to the old Platters standard "The Great Pretender." To call Lou a storefront lawyer would be a bit pretentious: Lou actually worked out of his apartment, with his mom Pearl (Rhoda Gemignani) screening his calls. Through a stroke of luck, our liberal-minded hero was hired — at a deferred fee — to oppose the snootily conservative Beacon Hill law firm of Osbourne, Merrill & Dewitt in an accident case. What happened next was duly noted by the TV reviewer in *Variety*: "Lou takes the leap that the 1970–71 *Storefront Lawyers* series [q.v.] took a full season to accomplish: In an unlikely move at the end of the episode, he's enfolded into the stuffy firm headed by Jason DeWitt ([Richard] Kiley), who admits to liking his style and his pro bono attitude."

Lou was actually hired as a tutor of sorts

for DeWitt's arrogant grandson and junior associate Crosby Caulfied III (Peter Krause)—and have you ever noticed that spoiled, aristocratic young TV characters are *always* "The III" and never "The VI" or "The VIII"? Also, it turned out that Lou was a package deal: he saw to it that DeWitt retained the services of his faithful investigator Frankie Collett (Kelly Rutherford).

For reasons that probably made sense at the time, Fox chose to premiere *The Great Defender* in the Sunday–night timeslot opposite CBS' indefatigable *60 Minutes*. The new series' calamitously low ratings have traditionally been blamed for its swift cancellation after a single episode (four of the eight completed episodes were "burned off" in the summer of 1995). But series star Michael Rispoli (most fondly remembered as the ill-fated Jackie Aprile on *The Sopranos*) offered another explanation in a 2004 interview with *BackStage* magazine's Simi Horowitz: "I've done 17 pilots, seven of which got on the air.... A turning point was getting the lead in a TV show, *Great Defender*. Seven episodes were ordered, but when one of the top [Fox] executives left and a new one was brought in, all of the projects which had been identified with that first executive got swept out, too, including *Great Defender*. But the experience put me inside the world of TV deals."

The Guardian

CBS: Sept. 25, 2001–May 4, 2004. David Hollander Productions/Gran Via/Rosecrans Productions/Sony Pictures Television. Executive Producers: David Hollander, Mark Johnson, Michael Pressman. Theme music: "The Empire in My Mind" by the Wallflowers. Cast: Simon Baker (Nick Fallin); Dabney Coleman (Burton Fallin); Alan Rosenberg (Alvin Masterson); Charles Malik Whitfield (James Mooney); Kathleen Chalfont (Laurie Solt); Wendy Moniz (Louise "Lulu" Archer); Erica Leerhsen (Amanda Bowles); Raphael Sbarge (Jake Strata); Johnny Sneed (Brian Olsen); Amanda Michalka (Shannon Gressler).

According to creator David Hollander, CBS' *The Guardian* was inspired by the career of Hollander's brother, a child advocate lawyer. Initially, the series was planned as a theatrical feature film about a disgraced attorney who

found redemption by working with troubled children, but during the development process Hollander determined that a mere 90- or 120-minute running time would not be sufficient to allow the main character to grow, mature and ultimately "see the light."

Doing a masterful job of hiding his native Australian accent, Simon Baker starred in the TV version of *The Guardian* as Nick Fallin, a successful but seriously flawed young corporate attorney working for his father's prestigious Pittsburgh law firm, Fallin & Associates. A lifelong substance abuser, Fallin was arrested on a cocaine-possession charge in the very first episode. Facing disbarment, Nick was offered a reprieve in the form of a suspended sentence, a $1000 fine, and 1500 hours of community service, forcing him to use his legal skills as a child advocate for Children's Legal Service (CLS), a financially strapped nonprofit organization. At the same time, he kept his job with Fallin & Associates, though his father Burton (Dabney Coleman) clearly disapproved of the direction his son's life had taken, especially since the younger Fallin was expected to one day assume control of the firm. In most legal dramas, the schism between the lofty world of corporate law and the lowly world of public-service advocacy is represented by casting one actor as the arrogant "big" lawyer and another as the compassionate "little" one, then focusing on their conflicts of character and viewpoint. Since Nick Fallin was both these contradictory stereotypes rolled into one, the major conflicts on *The Guardian* roiled within Nick himself— and as a result, *Entertainment Weekly*'s Ken Tucker was able to describe the series as "a dandy schizophrenic potboiler."

Avoiding the obvious ploy of having Nick Fallin react to his reduced circumstances by bitterly lashing out at his CLS coworkers and charges, *The Guardian* conveyed Nick's selfishness and insensitivity by having him adopt an aura of brooding detachment. He went through the motions professionally enough, but no one really knew if he gave a damn about the physically battered and emotionally scarred children whose causes he championed in the courtroom. In fact, one would be hard pressed to remember if he ever so much as smiled during the earliest episodes. Only as the series progressed did the viewer realize Nick was holding back his true feelings — and avoiding eye contact with his youthful clients — because of the damage done to his own psyche by being the product of a broken home and a workaholic father who never had time to tell him that he was loved — not even by disciplining him when he needed it most. Remarkably, even when the viewer knew full well what was eating away at Nick and it was obvious that he was matriculating into a human being in spite of himself, star Simon Baker resisted all efforts to lighten up the character, retaining and even amplifying all of Nick's rough edges — especially in the courtroom, where Nick used his "compassion" for the lost souls who passed through the CLS as an excuse to pull every rotten and underhanded legal trick in the book! As presciently noted in *Media Life* a few weeks after the series' debut: "Viewers will admire Fallin's talent as a lawyer, they'll admire his results, but they won't admire his character. At least not for a while."

As for the outwardly callous Burton Fallin, he too revealed a hitherto untapped depth of emotion by struggling to make up for lost time with Nick, "proving" that he honestly cared about his son by imposing impossibly high and rigid standards upon him. In an even more extreme effort to compensate for past failings, Burton at one point cut an under-the-counter legal deal on Nick's behalf, an act that nearly wrecked both their careers. Ultimately, Burton Fallin resigned from his own firm, not because his son had finally proven capable of succeeding him (in fact, Nick had petulantly joined a rival firm) but because he wanted to be a full-time grandfather to Nick's newborn son. Clearly, the road to redemption on *The Guardian* was heavily travelled by both generations of Fallins. (As a footnote without further comment, the series' theme song was sung by Jakob Dylan — son of Bob.)

In addition to Simon Baker and Dabney Coleman, *The Guardian* boasted one of the best ensemble casts on television. Alan Rosenberg (late of *L.A. Law*) was seen as Nick's down-to-earth CLS boss Alvin Masterson, who waged an ongoing war with Burton Fallin over the monopolization of Nick's services. Charles Malik Whitfield played Nick's fellow advocate James Mooney, whose duties at CLS forced him to neglect a problem closer to home with his drug-addict sister; when Whitfield left the show at the end of its second season, James Mooney was abruptly killed in a drive-by shooting. Wendy Moniz appeared as the abrasive Luise "Lulu" Archer, another advocate, with whom Nick enjoyed an off-and-on romance: "off" when she briefly wed another guy (Johnny Sneed) and was seriously injured in a car accident in Season Two, "on" when she bore Nick's child in Season Three (similar to Nick, Lulu had serious relationship issues with her own mother, played by Rita Moreno). Other regulars included Erica Leerhsen (first season only) as Amanda, Nick's assistant at Fallin & Associates; Kathleen Chalfont as Laurie, a social worker; Raphael Sbarge as Jake, who became Nick's junior partner when Burton changed the name of the firm to Fallin & Fallin; and Amanda Michalka as Shannon, the daughter of a junkie whose death was unfairly pinned upon Nick, thus opening up a fascinating story arc in which Nick's dad fell in love with Shannon's embittered grandmother Mary — a recurring role which earned guest star Farrah Fawcett an Emmy nomination. Several other prominent actors made brief but memorable appearances, notably Lolita Davidovich as social worker Victoria Little, whom Alvin Masterson had planned to marry shortly before being diagnosed with ALS in Season Three; and Will Ferrell, capriciously billed as "Phil Reston," in the role of a hotshot attorney who had the misfortune to be with James Mooney when he was gunned down in the street.

Hammocked between *JAG* and *Judging Amy* (both q.v.) to create a "legal night" on CBS' Tuesday–evening schedule, *The Guardian* was expected to perform adequately during its first year on the air; instead, it performed magnificently, closing out the 2000–2001 season as network television's highest-rated new show. The series faltered during its second season, a problem David Hollander attributed to shifting focus away from Nick Fallin in favor of the secondary characters. Entering Season Three as a companion piece to the *JAG* spinoff *Navy NCIs, The Guardian* was unable to sustain its early momentum and ended its run after 67 episodes. Right up to the end, viewers who disliked the series could not understand its fans' devotion to so aloof and unpleasant a leading character as Nick Fallin; but those fans knew better, agreeing with critic Ken Tucker's analysis of the series' appeal: "*The Guardian* is about a good guy who behaves like a bad guy; an ideal of both dramatic worlds."

Guilty or Innocent

Syndicated: debuted September 1984. Genesis-Colbert Productions. Produced by Mickey Grant. Cast: Melvin Belli (Host); John Shearin (Jury Moderator); Lee Ritchey (Counsel); John Wells (Announcer).

Even if TV fans of the early 1980s had not been aware of Melvin Belli's distinguished reputation as the attorney for such high-profile clients as Errol Flynn, Muhammad Ali, Sirhan Sirhan and Lana Turner, they certainly would have recognized the publicity-hungry lawyer from his portrayal of the diabolically manipulative Gorgan in the 1968 *Star Trek* episode "And the Children Shall Lead." Lauded as "The King of Torts" and denigrated as "Melvin Bellicose," Belli was as much a celebrity as anyone he'd ever represented in court, and thus eminently deserving of his own starring TV series. Like him or not, he unquestionably deserved far better than the series he ended up with.

Debuting in daily strip-syndication in the fall of 1984, *Guilty or Innocent* was a ludicrous mixture of courtroom drama and game show. Each half-hour episode was divided into two segments. In the opener, an actual court case was re-enacted on a performance level several notches below your average junior-high production of *The Boy Friend*. In the concluding segment, a "jury" of contestants was instructed to guess the outcome of the case. The winners received fabulous cash prizes, while the losers could at least brag to all their friends that they'd been on television. Belli hosted the proceedings, but actor John Shearin (better known as Lt.

Ambrose Finn on *Hunter*) did most of the heavy lifting as moderator of the quiz portion.

Among Melvin Belli's most famous quotes is the self-aggrandizing, "There may be better lawyers than I, but so far I haven't come across any of them in court." Commenting upon the failure of *Guilty or Innocent*, the best Belli could come up with was that his show was "a hell of a lot better" than rival lawyer F. Lee Bailey's equally unsuccessful syndicated daily *Lie Detector* (q.v.). That's like saying that bursitis is a hell of a lot better than sciatica.

Hardcastle and McCormick

ABC: Sept. 18, 1983–July 23, 1986. Stephen J. Cannell Productions. Executive Producers: Stephen J. Cannell, Patrick Hasburgh. Cast: Brian Keith (Judge Milton G. Hardcastle); Daniel Hugh-Kelly (Mark "Skid" McCormick); Mary Jackson (Sarah Wicks); John Hancock (Lt. Michael Delaney); Joe Santos (Lt. Frank Harper).

Producer Stephen J. Cannell specialized in lighthearted action series with offbeat, maverick heroes, as witness *The A-Team* and *The Greatest American Hero*. With *Hardcastle & McCormick*, Cannell pulled off a neat hat-trick: making the concept of Vigilante Justice entertaining, exciting and amusing all at once.

Hardcastle & McCormick qualifies as a "legal" series principally because one of its two protagonists was a judge—actually a retired judge, Milton C. Hardcastle (Brian Keith), known to colleagues and defendants alike as "Hardcase" because of his tough sentences and his throw-away-the-key attitude towards scofflaws. (Among the many mottos emblazoned on his custom-made T-shirts was the pithy NO PLEA BARGAINING IN HEAVEN. Judge Judy, move over.) Though no longer on the bench, the 65-year-old Hardcastle continued administering his own brand of justice by going after some 200 criminals who had used technicalities and loopholes to walk free from his courtroom over the past three decades. In this pursuit, Hardcastle arranged to have the last defendant to stand before him released in his custody: young, good-looking Mark "Skid" McCormick, a two-time loser who'd been charged — unfairly, as it turned out — with Grand Theft

Auto. It so happened that Skid was a champion race-car driver, and it was this skill that appealed most to Hardcastle, who liked nothing better than to literally chase down the bad guys. Besides, grumbled the crusty ex-jurist, "it takes one to catch one": since Skid was a veteran of the penal system, who better to trap a crook than a crook himself—even one who was falsely accused? As for Skid, he wasn't what one could call enthusiastic about being Hardcastle's sidekick, but considering the alternative he decided it was best to grin and bear it. (In a neat role-reversal situation, Hardcastle was the "loose cannon" of the duo, while McCormick was the more cautious member of the team.)

Despite the combined charisma of Brian Keith (who won the role over such candidates as Jack Warden, Keenan Wynn and even Fred Astaire!) and Daniel Hugh-Kelly (the latter was being groomed by his handlers as "the new Tom Selleck," though with his wry comic know-how he emerged more like the new James Garner), for many fans the *real* star of the series was "The Coyote," Hardcastle's souped-up custom vehicle, which boasted a Volkswagen chassis, a VW-Porsche 914 engine, and a vanity license plate emblazoned with the letters DE JUDGE. The Coyote was added to the show not as a balm to action fans, but simply because series producer Patrick Hasburgh was an unabashed race-car enthusiast.

Scrambling about to grab a few crumbs of screen time away from the series' two dynamic stars were supporting players Mary Jackson as Sara the housekeeper, who worked for Judge Hardcastle at Gulls Valley, the baronial state he'd inherited from his late wife; John Hancock as Lt. Michael Delaney, the Judge's police contact during the series' first season; and Joe Santos as Delaney's successor, Lt. Frank Harper. Acting as the series' legal consultant was Lawrence Waddington, a former California ADA and State Attorney General who from 1980 to 1995 served as an LA Superior Court judge. Though Waddington specialized in insurance claims, real-estate law and commercial disputes rather than criminal cases, he did an admirable job making certain that *Hardcastle and McCormick* never ran afoul of the ACLU,

seeing to it that all the criminals who found themselves in the heroes' crosshairs were being pursued for their *current* crimes rather than their past indiscretions. (Oddly enough, Lawrence Waddington felt no need to mention *Hardcastle and McCormick* on his post-series resume.)

Introduced with a movie-length debut episode in the fall of 1983, *Hardcastle and McCormick* zipped and zapped along the ABC highways and byways for three years and 67 hour-long episodes.

Harrigan and Son

ABC: Oct. 14, 1960–Sept. 29, 1961. Desilu Productions. Creator/Executive Producer: Cy Howard. Theme song: "H-A-double R-I-G-A-N," by George M. Cohan. Cast: Pat O'Brien (James Harrigan Sr.); Roger Perry (James Harrigan Jr.); Georgine Darcy (Gypsy); Helen Kleeb (Miss Claridge).

In 1960, ABC briefly experimented with a new TV-series format, which bore no official name at the time but would later be designated as the "dramedy": a half-hour program (later expanded to an hour) that wasn't funny enough to be a sitcom nor serious enough to qualify as a drama, but fell somewhere in between. *Harrigan and Son* and *The Law and Mr. Jones*, the first two ABC efforts to meet the requirements of this new format, had a couple of other things in common: both series featured lawyer protagonists, and both dispensed with a recorded laughtrack.

Following the alphabet, we'll deal first with *Harrigan and Son*. Created by comedy veteran Cy Howard (*My Friend Irma, Life with Luigi*) and filmed at Desilu studios, the series was primarily a vehicle for 61-year-old Pat O'Brien, formerly a mainstay of the Warner Bros. "stock company." Though not as dynamic a performer as his collegues James Cagney and Humphrey Bogart, O'Brien was a solid, dependable screen presence, versatile enough to play everything from cops to priests, and from newspaper reporters to football coaches (who could forget his portrayal of Notre Dame's immortal Knute Rockne?). By the late 1950s O'Brien's film stardom was a thing of the past, but he remained extremely active as a stage performer, TV guest star and even nightclub entertainer, describing his song-and-snappy patter act as an "Irish Myron Cohen" (a popular Jewish monologist of the day). With *Harrigan and Son*, Pat O'Brien enjoyed his first top billing in years, though the feisty old trouper would have bristled if anyone described the series as a comeback vehicle. ("I'm damn good and I know it!" snapped O'Brien to a reporter who had the temerity to suggest that the actor was "underappreciated" in Hollywood.)

In a lighthearted variation of a format later treated with more sobriety on *The Defenders* (q.v.), Pat O'Brien was cast as criminal attorney James Harrigan, senior partner of a father-son legal firm in New York City, while Roger Perry costarred as Jim Harrigan Jr., who joined his dad's firm straight out of Harvard Law School. Having practiced law long before his son was even a gleam in his eye, Harrigan Sr. relied more on gut instinct than conventional legal procedure, following his heart as much as his head as he slyly maneuvered the system in order to provide his clients with the best possible defense. In contrast, Harrigan Jr. went strictly by the book and played scrupulously within the rules, a philosophy that of course frequently brought him into conflict with his free-wheeling dad; indeed, in several episodes the two Harrigans found themselves as the opposing counsels. Back in the office, widower Harrigan and his bachelor son were matched with secretaries ideally suited to their personalities, if not their ages: Harrigan Sr.'s gal Friday was a sexy, wisecracking young blonde named Gypsy (played by Georgine Darcy, best remembered as the man-crazy dancer "Miss Torso" in Alfred Hitchcock's 1954 classic *Rear Window*), while Harrigan Jr.'s secretary was prim, businesslike, middle-aged Miss Claridge (Helen Kleeb).

"I like Harrigan," Pat O'Brien told *TV Guide* in a 1960 interview. "There's a little bit of everybody in him—Darrow, Fallon, Leibowitz, Welch." But unlike those worthies, "Harrigan won't take any murder cases because this show is played for comedy, and murders aren't particularly funny. He won't take a divorce

case, either. He explains this by saying, 'I wasn't there when they were joined together and I don't want to be there when they're pulled apart.'" Perhaps anticipating comparisons to TV's most popular lawyer show *Perry Mason*, O'Brien added, "Occasionally Harrigan & Son lose a case, right on the air. Makes things more believable, more human."

Like many other shows of its era, *Harrigan and Son* occasionally lapsed into banal and obvious storylines, such as the ancient wheeze about a big-city lawyer (in this case Harrigan Jr.) who tries to beat a traffic ticket in a small town, only to find out that the sheriff, the judge, the prosecutor, and the jury members are all cousins. And with "professional Irishman" Pat O'Brien on the premises, the scripts sometimes leaned too heavily on the Auld Blarney, reminding one of critic James Agee's reaction to watching a film by another Son of Erin named John Ford: "There is enough Irish comedy to make me wish that Cromwell had done a more thorough job."

But for the most part, the series went to great and admirable lengths to seek out fresh story material. Two episodes in particular have lingered in the memory of this writer. In "The Magnificent Borough," the elder Harrigan defended a pugnacious cab driver (Charles Cantor) on a charge of blackening the eye of a snobbish Uptown customer (Hayden Rorke) who made disparaging remarks about Brooklyn. In the end, the plaintiff's regal-looking wife (Barbara Drew) revealed that she herself was born in Brooklyn but never had the heart to tell her husband, whereupon all charges were dropped, husband and wife embraced, and everyone (including Harrigan Jr., who'd been representing the plaintiff) shared a good laugh. And in the series' best episode "100 Proof," ex-alcoholic Patrick O'Toole (played by another member of Hollywood's "Irish Mafia," Frank McHugh) was arrested for drunk driving, despite insisting that he hadn't touched a drop in years. In a virtuoso courtroom climax, Harrigan Sr. proved conclusively that the hapless O'Toole suffered from an unusual yeast infection, and inadvertently became intoxicated by eating a combination of grain and sugar products—his freakish condition transforming the poor fellow into a "human still"!

Counterprogrammed against the CBS western *Rawhide* on Friday evenings, *Harrigan and Son* had the feel of a hit at the outset, or so claimed Pat O'Brien in his autobiography: "Everyone was enthusiastic about it. I loved the series and everyone associated with it. But after thirty-six weeks of what we thought was a most successful project, the Ivy-league suited gods of Madison Avenue thought otherwise. It was replaced with a chimpanzee show, half ape, half human. I never ascertained whether those chimps passed the legal bar to practice." Hopefully, O'Brien took solace in the fact that the infamous "half ape, half human" sitcom *The Hathaways* (starring Peggy Cass, Jack Weston and the Marquis Chimps) was even less successful than *Harrigan and Son*.

Harvey Birdman, Attorney at Law

Cartoon Network: debuted Sept.30, 2001 (one-shot preview, Dec. 30, 2000). Williams Street Productions/Turner Productions. Executive Producers: Keith Crofford, Michael Lazzo, Khaki Jones, Michael Ouwleen, Erik Richter, Linda Simensky, Matthew Charde. Voices: Gary Cole (Harvey Birdman/Judge Hiram Mightor); Stephen Colbert (Myron Reducto Esq./Phil Ken Sebben); Pamelyn Ferdin (Sallie); Josh Albee (Bobby); Bill Callaway (Sparks); John Michael Higgins (Judge Mentok the Mind-Taker); Thomas Allen (Peanut); Peter MacNicol (X the Eliminator); Paget Brewster (Birdgirl, aka Judy Sebben); Chris Edgerly (Peter Potamus); and many others, including Lewis Black, Frank Welker, Maurice LaMarche, Laraine Newman, Billy West, Joe Alaskey, Grey Delisle, Steve Landesberg, Tress MacNeille, Michael McKean, Michael Bell, and Phil LaMarr.

The more successful of TV's two animated lawyer shows (see also *Gary the Rat*), *Harvey Birdman, Attorney at Law* was an indirect offshoot of Cartoon Network's earlier "retro" series *Space Ghost Coast-to-Coast*. In that effort, footage from the obscure 1966 Hanna-Barbera cartoon opus *Space Ghost* was re-edited and recomposited over new backgrounds, resulting not only in a devastating satire of H-B's rather dismal array of 1960s Saturday-morning superheroes but also a lampoon of the late-night talkshow format. Originally, the emcee of *Space Ghost Coast-to-Coast* was to have been *another*

caped-and-masked animated obscurity, the title character from H-B's 1967 opus *Birdman and the Galaxy Trio*. Created by the brilliant comic book artist Alex Toth, the original *Birdman* revolved around the exploits of police investigator Ray Randall, who having been gifted with super powers by the Egyptian sun god Ra was able to capture criminals by soaring in the air and swooping down on them (hence his name), and by zapping them into helplessness with the solar energy beams pulsating through his knuckles. Birdman had a sidekick named Birdboy, and was also accompanied by his pet eagle Avenger. Dispatching Birdman on his missions to "spread the light of justice into the darkest recesses of the human soul" was his superior Falcon 7, who sported a dashing black eyepatch.

Birdman might have been forgotten by all but the most anal-retentive of cartoon buffs had he not been revived for satirical purposes by Williams Street Productions, the same animation factory responsible for such exercises in cartoon irreverence as *The Brak Show* and *Aqua Teen Hunger Force*. With *Coast-to-Coast* already consigned to Space Ghost, the animators toyed with the idea of using Birdman in a similar chat-show spoof, but then decided upon a more original concept: transforming a "third-rate superhero" into a "third-rate lawyer." Even though there was no evidence that he'd ever actually studied law, Harvey Birdman (his "Ray Randall" alter ego completely buried) at least knew how to talk like a lawyer (courtesy of versatile film and TV actor Gary Cole), and except for those awkwardly protruding wings and that ever-present domino mask even *looked* like a lawyer in his form-fitting business suits. Besides, the law firm of Sebben & Sebben was obliged to honor its Affirmative Action requirements by hiring at least one token superhero.

Harvey's new boss was actually his old boss Falcon 7, who had adjusted his moniker to the more socially acceptable Phil Ken Sebben (voice provided by Stephen Colbert of *Colbert Report* fame). His personal assistant was still the eagle Avenger, who'd learned some rudimentary typing skills so he could prepare legal briefs; Birdboy was also on the payroll as a teenage paralegal named Peanut, and even

Birdgirl, aka Phil's daughter Judy Sebben, showed up before the series ended. Because he was low man on Sebben & Sebben's totem pole, Harvey was limited to representing other washed-up Hanna-Barbera stars, whose "guest shots" were largely accomplished by lifting extreme poses and closeups from the studio's earlier animated triumphs. Our hero's most formidable courtroom opponents included neurotic prosecutor Myron Reducto, who without the "Myron" had originated as a mad-scientist villain on *Birdman and the Galaxy Trio*, and who after all these years still wielded a mean "shrink-gun"; Judge Mightor, a prehistoric superhero introduced on H-B's 1967 opus *Moby Dick and the Mighty Mightor*, who administered justice at the end of a fossilized club; and Judge Mentok the Mind-Taker, another villainous refugee from the original *Birdman*, who lived up to his name whenever he couldn't sway the jury to his side by more orthodox methods. Also seen from time to time was X the Eliminator, whose mission on the earlier *Birdman* to rob the hero of his powers had devolved into his futile efforts to "do lunch" with the elusive Harvey; and Peter Potamus, the one "funny animal" who was given any sort of worthwhile job at Sebben & Sebben.

Harvey Birdman's first client was Race Bannon of *Jonny Quest* fame, who was locked in a battle with Dr. Benton Quest over custody of Jonny and Hadji. Subsequent litigation involved Yogi Bear's sidekick Boo-Boo, who led a double life as the mad bomber "Unabooboo"; Scooby-Doo and Shaggy, hauled in on a drug charge because the arresting officer misinterpreted their goofy behavior and nonstop giggling; Captain Caveman, suing a local school board for refusing to teach Evolution; Dynomutt, whom Harvey was accused of murdering just to spite his new legal rival The Blue Falcon; Magilla Gorilla, forcibly liberated from Mr. Peebles' pet shop by a group of animal activists; Apache Chief, the Native American member of the Superfriends, who filed a damage suit after spilling hot coffee on his crotch and mutating into a monstrosity; Grape Ape, suspected of using steroids to win the Laff-a-Lympics; and giant genie Shazzan, pressing

charges against the Mizwa of Muffay Tah for imprisoning him in a vase in Phil's office. The one episode that generated the highest level of water-cooler chatter was "The Dabba Don," a wild *Sopranos* takeoff in which "Freddy" Flintstone was accused of using his job at the Gravel Pit as a cover for his activities as a Mafia chieftan (who could forget the episode's *Godfather*-inspired highlight involving the severed head of Quick Draw McGraw?). The series ended with "The Death of Harvey," a takeoff of the famously all-inclusive *Seinfeld* finale, in which Harvey was forced to bring back all his former clients and re-argue all his old cases before Judge Mentok went on vacation.

Best appreciated by aging Baby Boomers who needed to get outside more often, *Harvey Birdman, Attorney at Law* was telecast as a fifteen-minute component of Cartoon Network's weekend *Adult Swim* omnibus.

Hawkins

CBS: Oct. 2, 1973–Sept. 3, 1974. Arena Productions/MGM Television. Executive Producer: Norman Felton. Cast: James Stewart (Billy Jim Hawkins); Strother Martin (R. J. Hawkins).

Burned by the failure of his eponymous 1971 NBC sitcom, venerable film star James Stewart would be mighty careful if he ever ventured into the world of series TV again. Never again would he subject himself to the grind of a traditional weekly-series schedule, which to his way of thinking yielded minimum results from a maximum of hard labor. If he should ever agree to star in a series again, it would be on a less frequent, perhaps monthly basis, just like Peter Falk on *Columbo* or Rock Hudson on *McMillan and Wife*—both components of NBC's successful rotating anthology *The NBC Mystery Movie*. And instead of debasing his talents in the role of a bumbling sitcom dad, Stewart would settle for nothing less than a multifaceted characterization perfectly tailored to his established screen persona. Thus, when Norman Felton of Arena Productions approached Stewart with a series about a shrewd West Virginia criminal lawyer who craftily adopted an "aw, shucks" country-bumpkin demeanor in order to exonerate his clients and entrap the real culprits, the actor jumped at the opportunity.

Test-marketed on March 13, 1973, with the made-for-TV movie *Hawkins on Murder* (later syndicated as *Death and the Maiden*), the triweekly *Hawkins* formally debuted October 2, 1973, as one of three rotating series offered by CBS on Thursday evenings from 9:30 to 11 P.M. EST (the other two were a neutered version of the movie property *Shaft*, with Richard Roundtree repeating his film role, and *The New CBS Thursday Night Movies*). Stewart was definitely in his element as Billy Jim Hawkins, a former deputy district attorney who'd resigned his post to set up a private legal practice in the tranquil little burg of Beauville, West Virginia. Though he preferred to handle only small cases for local clients, Hawkins' reputation as one of America's foremost specialists in murder cases drew the Rich and Famous to Beauville like ants to a picnic. As a result, Billy Jim spent a great deal of time away from home, defending a movie star's husband accused of murder in Hollywood, a young hothead charged with killing a girl in a Manhattan highrise, and so on. Hawkins' modus operandi was to ramble and shamble around, making bucolic statements like, "The law's kinda a funny critter, Mr. Harrelson," telling longwinded stories about his kinfolk (13 siblings, 52 nieces and nephews, 174 first cousins, etc.) and in general playing the role of the backwoods buffoon — enabling him to surreptitiously gather evidence overlooked by the authorities, lead wise-guy DAs down the wrong path, and lull suspects into a false sense of security so they would make the One Little Slip that would prove them guilty of murder. "Columbo in the courtroom" was how one observer described *Hawkins*, though in fact the series owed more to Stewart's film portrayal of deceptively "simple" Michigan lawyer Paul Biegler in the 1959 courtroom drama *Anatomy of a Murder*. While the series boasted a substantial guest star lineup (Sheree North, Cameron Mitchell, Julie Harris, Lew Ayres, Teresa Wright), the only other regular besides Stewart was Strother Martin, decked out in baggy white suit and handlebar mustache as Billy Jim's cousin, assistant and legman R.J. Hawkins.

I have chosen to ignore most of the contemporary reviews for *Hawkins* because of the critics' irritating tendency to translate their thoughts into Jimmy Stewart-ese, beginning each sentence with "Waaaal" and dispensing such bon mots as, "You ain't really foolin' nobody, Billy Jim Hawkins." The fact that Stewart won a Golden Globe for his portrayal of Billy Jim is sufficient proof that the Hollywood Foreign Press Association liked what they saw. Unfortunately, CBS' trio of rotating series was telecast opposite ABC's ratings magnet *Marcus Welby, M.D.*, consigning *Hawkins* to an early demise after seven 90-minute episodes. The premise of a dumb-like-a-fox Southern defense attorney would, however, prove immensely successful when it was resurrected on behalf of Andy Griffith and *Matlock* (q.v.)

Head Cases

Fox: Sept 14–21, 2005. 20th Century–Fox Television. Creator/Executive Producer: Bill Chais. Cast: Chris O'Donnell (Jason Payne); Adam Goldberg (Russell Shultz); Richard Kind (Lou Albertini); Krista Allen (Laurie Payne); Jake Cherry (Ryan Payne); Rockmond Dunbar (Dr. Robinson).

So you think *Head Cases* is an offensive title for a Fox Network series about a couple of mental patients? Maybe you'd prefer the series' working title: *Crazy Lawyers*.

Chris O'Donnell starred in this 60-minute dramedy as corporate attorney Jason Payne, who after being laid low by the one-two punch of a nasty divorce and a nervous breakdown ended up spending three months under treatment at a wellness center. Motivated either by an epiphany or by the stark, raw fear that he'd never work again, Jason jump-started his career with a private practice, specializing in underdog cases in which he could stick it to the Rich and Powerful. As a condition of his release, our hero was forced to team up with a fellow wellness-center alumnus: Civil Rights attorney Russell Shultz (Adam Goldberg), an unabashed ambulance chaser who suffered from a disorder that caused him to burst into rage at the most awkward moments — and to punch out anyone who happened to be in his general vicinity. Just to prove they were equal-opportunity employers, Jason and Russell took upon themselves a paralegal named Lou Albertini (Richard Kind), a former bank robber. (Lou was more or less a substitute for another recovering "head case" named Kate, played by Rachael Leigh Cook, who had the good luck — or good sense?— to be written out of the series after the pilot episode.)

Created by former attorney Bill Chais, who in happier times had worked on the popular legal drama *The Practice* (q.v), *Head Cases* earned the distinction of being the first new series of the 2005–2006 TV season to be cancelled, with only two of its six completed episodes ever seeing the light of day.

Headlines on Trial

Syndicated: debuted September 1987. Orbis Communications. Cast: Arthur R. Miller (Moderator).

Taped at the studios of Washington DC's NBC affiliate WRC-TV, *Headlines on Trial* was the second syndicated series hosted by celebrated Harvard Law School professor Arthur R. Miller (for more on his career, see the entry for *Miller's Court*). Less a legal series than an "issues" program, *Headlines* was a weekly round-robin discussion of provocative points of law and the hot-button topics of the day, moderated by Professor Miller and hashed out by a panel of legal experts. Usually telecast on weekends, the series remained in active production from 1987 through 1989.

Hidden Faces

NBC: Dec. 30, 1968–June 27, 1969. Created by Irving Vendig. Produced by Charles Fisher. Cast: Conrad Fowkes (Arthur Adams); Gretchen Walther (Dr. Katherine "Kate" Logan); Stephen Joyce (Mark Utley); Rita Gam (Elinor Jaffee); Tony LoBianco (Nick Capello Turner); Nat Polen (Earl Harriman); John Karlen (Sharkey Primrose); Joseph Daly (Sen. Robert Jaffe); Linda Blair (Allyn Jaffe) (Jeannet Sloan); Louise Shaffer (Martha Logan); John Towey (Wilbur Ensley); Ludi Claire (Grace Ensley).

Debuting the last week of 1968, the half-hour daytime drama *Hidden Faces* was NBC's bid to compete against the popular early-afternoon CBS soaper *As the World Turns*. Eschewing the domestic and sexual travails that dominated the CBS series, NBC counterprogrammed with a serial trafficking in crime and suspense, featuring a lawyer as the protagonist — a formula that had proven enormously successful for another CBS daytimer, *The Edge of Night* (q.v.). To this end, NBC went so far as to hire former *Edge of Night* producer Charles Fisher and head writer Irving Vendig to develop their new serial, and also cast Conrad Fowkes, previously seen on *Edge of Night* in the role of Steve Prentiss, as *Hidden Faces'* erudite, pipe-smoking attorney hero Arthur Adams.

The show's introductory story arc found Adams and his handsome-hunk associate Nick Capello Turner (Tony LoBianco) working together on behalf of Adams' client, gorgeous surgeon Katherine "Kate" Logan (Gretchen Walther), who had been accused of murder. Ultimately proving Kate's innocence, Adams also fell in love with her. Along the way, the attorney encountered several secondary characters played by such New York–based pros as Rita Gam and John Karlen, and also a cute little girl portrayed by an eight-year-old photographer's model named Linda Blair, in her first credited acting appearance.

According to daytime-drama historian Mark Faulkner, *Hidden Faces* earned a great deal of praise from the critics, with *Variety's* TV reviewer comparing the series favorably to ABC's popular Gothic serial *Dark Shadows*; the new NBC show also developed a near-fanatical fan following. Unfortunately, this wasn't enough to make a dent in the ratings of *As the World Turns*, and in June of 1969 *Hidden Faces* shut down production after only six months on the air.

His Honor, Homer Bell

Syndicated: debuted 1955. Galahad Productions/ NBC Films. Executive Producer: Himan Brown. Cast: Gene Lockhart (Judge Homer Bell); Mary Lee Deering (Casey Bell); Jane Moutrie (Maude).

Syndicated by NBC Films in the Spring of 1955, the half-hour series *His Honor, Homer Bell* is included in these pages on the off-chance that by *not* including it we would be committing an error of omission. Here's what we know about this maddeningly obscure property: it starred veteran stage and screen character actor Gene Lockhart as Judge Homer Bell, who lived and worked in the Western town of Spring City. A widower, Homer had a spunky daughter named Casey (Mary Lee Dearing), and an all-knowing housekeeper named Maude (Jane Moutrie). When wearing his judicial robes, the avuncular Homer Bell exhibited a boundless faith in human nature, handing down decisions that were determined as much by good old-fashioned common sense as by standard legal protocol. The series was produced on a million-dollar budget by legendary radio pioneer Himan Brown (*Inner Sanctum Mysteries, The CBS Radio Mystery Theater*) and filmed at the then-new NBC facilities in Brooklyn, which formerly housed the Vitagraph motion picture studio.

And that, ladies and gentlemen, is all we know for sure about *His Honor, Homer Bell.* The series was withdrawn from distribution in 1958 or thereabouts, and its 39 episodes are currently unavailable for reappraisal. Earlier published reports that the series was a Western would suggest the stories took place in the late 19th century, a la *Judge Roy Bean* (q.v.); how, then, can one explain the contemporary-sounding synopses in the pages of *TV Guide,* wherein Judge Bell tries to score tickets for a sold-out football game, delivers a speech before the town's traffic commission, and is unwittingly entered in a contest held by a movie star's fan club? Also, some TV history books (and those ubiquitous *TV Guide* listings!) have categorized *Homer Bell* as a situation comedy, and there is certainly sufficient evidence to support this theory: one of the series' most frequent writers was Si Rose of *Bachelor Father* and *McHale's Navy* fame; and among the plotlines were such familiar sitcom chestnuts as Homer Bell entering a "mother-daughter" dressmaking contest on behalf of his motherless daughter Cassie, teaching a lesson to a mooching relative, and reluctantly serving as an emergency baby-sitter.

At the same time, there are several clues suggesting that *Homer Bell* also qualified as a "legal drama." Other members of the writing staff were *Dragnet*'s Michael Cramoy and *Hawaii Five-O*'s Jerome Coopersmith; and the storylines included one in which Judge Bell triumphantly digs up a technicality to prevent a "marriage of convenience," another in which he must honor the terms of a will by deciding which of two heirs is "most deserving" of a huge legacy, and still another in which the Judge agrees to take charge of a troublesome dog that has nearly ruined a marriage in order to bring the couple back together.

On the basis of the last-mentioned evidence, we'll take a chance and include *His Honor Homer Bell* in this volume — with the understanding (and expectation!) that someone may someday come riding out of the hinterlands to prove us wrong.

The Home Court

NBC: Sept. 30, 1995–June 22, 1996. Paramount Network Television. Creators/Executive Producers: Sy Dukan, Denise Moss. Cast: Pamela Reed (Judge Sydney J. Solomon); Breckin Meyer (Mike Solomon); Meghann Haldeman (Neal Solomon); Robert Gorman (Marshall Solomon); Phillip Glenn Van Dyke (Ellis Solomon); Charles Rocket (Judge Gil Fitzgerald); Meagan Fay (Greer).

The NBC sitcom *The Home Court* was tailored to the talents of actress Pamela Reed, who had established herself in such tough-cookie, "Don't f* with me!" roles as Bonnie Dummar in *Melvin and Howard* (1980) and Detective O'Hara in Arnold Schwarzenegger's *Kindergarten Cop* (1990). (She was still successfully projecting this "hardcase" persona as late as 2006 in the apocalyptic TV series *Jericho*.) Reed starred as Sydney J. Solomon, a Chicago family-court judge notorious for her no-nonsense approach to justice and her harsh treatment of anyone who told anything less than the whole truth in her courtroom (and no, she *wasn't* based on Judge Judy Sheindlin, whose syndicated series hadn't yet made its debut). At home, single-mom Sydney was slightly easier to get along with — but only slightly. In the opening episode, she kicked her slacker son Mike (Breckin Meyer) out of the house after he dropped out of college, only to relent and give the kid a second chance when everyone convinced her that she was allowing the pressures of her job to cloud her judgment. The rest of the Solomon brood — 16-year-old Neal (Meghann Haldeman) 13-year-old Marshal (Robert Gorman) and 11-year-old Ellis (Phillip Glenn Van Dyke) — likewise managed occasionally to maneuver Sydney into more sensitive and understanding behavior. In addition to her children, Sydney had to contend with her self-indulgent sister Greer (Meagan Fay) and her immature colleague, Judge Gil Fitzgerald (Charles Rocket).

Beyond offering Pamela Reed in a leading role, *The Home Court* had precious little to recommend it. Scheduled opposite the infinitely more appealing *Touched by an Angel* on Saturday evenings, the series barely survived its first and only season, toting up a mere 20 episodes.

I'll Fly Away

NBC: Oct. 7, 1991–Feb. 5, 1993, Apr. 11, 1993; PBS (reruns): 1993–1994. NBC Productions/Lorimar/ Warner Bros. TV Productions. Creators/Executive Producers: John Falsey, Joshua Brand. Cast: Sam Waterston (Forrest Bedford); Regina Taylor (Lilly Harper); Jeremy London (Nathaniel "Nathan" Bedford); Ashlee Levitch (Francie Bedford); John Aaron Bennett (John Morgan Bedford); Kathryn Harrold (Christina LeKatzis); Peter Simmons (Paul Slocum); Mary Alice (Marguerite Peck); Bill Cobbs (Lewis Coleman); Roger Aaron Brown (Rev. Henry); Rae'Ven Larrymore Kelly (Adlaine Harper); Brad Sullivan (Zollicofer Weed).

It was no coincidence that the critically acclaimed legal drama *I'll Fly Away* bore a striking resemblance to Harper Lee's imperishable novel *To Kill a Mockingbird*. John Falsey and Joshua Brand, the team responsible for the whimsical dramedy *Northern Exposure*, were longtime fans of the 1962 film version of *Mockingbird*, in which small-town lawyer Atticus Finch (Gregory Peck) defended a black man against a charge of raping a white woman in the Jim Crow South of the 1930s. Falsey and Brand were especially fascinated by the minor character of Finch's black housekeeper Calpurnia, who functioned as surrogate mother for the widowed lawyer's children Scout and Jem. Though the story was told from the viewpoint of Harper Lee's alter ego Scout, it was fairly easy for the audience to follow the thought processes of the Lincolnesque Atticus Finch. But it was virtually impossible to determine what was going on in the mind of Calpurnia — particularly since the actress portraying the character, Estelle Evans, brilliantly conveyed the vocal and facial "neutrality" that African Americans had been forced to adopt in the white-dominated Southern society of the Depression Era. To betray such emotions as anger, frustration or ethnic pride could prove downright dangerous for black people at a time when bigoted whites were apt to organize a lynch mob at the first sign of "uppity" behavior, and Calpurnia clearly regarded "keeping her place" as the wisest course of action. With this in mind, Falsey and Brand elected to tell the story of *I'll Fly Away* from the perspective of the series' "Calpurnia" counterpart, rechristened Lilly Harper — a name obviously chosen because of its similarity to Harper Lee.

Filmed on location in Atlanta, the weekly, hour-long series took place sometime between the late 1950s and early 1960s in Bryland County, located in an unspecified Southern state. Future *Law & Order* mainstay Sam Waterston starred as Bryland's district attorney Forrest Bedford, a kindly, decent man who was firmly committed to equal justice for all — but who was also a product of his times, raised in the belief that the black man was inherently inferior to the white man and that segregation was an inviolate fact of life. Forrest's wife Gwen (Deborah Hedwall) had been institutionalized after a nervous breakdown, leaving him to care for their three children: rebellious 16-year-old Nathan (Jeremy London), precocious 13-year-old Francie (Ashlee Levitch) and contemplative 6-year-old John Morgan (John Aaron Bennett). Finding it difficult to balance his professional and domestic responsibilites, Forrest hired a black housekeeper named Lilly Harper (Regina Taylor), a quietly efficient woman possessed of an innate intelligence that belied her lack of formal education. Stoically suffering through an unhappy marriage, Lilly not only looked after Forrest's kids but also pulled extra duty as single parent for her own daughter Adlaine (Rae'Ven Larrymore Kelly). Each episode was narrated by Lilly in flashback, as if reading an entry from her diary.

Though hardly what one would call a revolutionary — she was more concerned with
holding her own family together than saving
the world — Lilly found herself inexorably
drawn into the burgeoning Civil Rights movement, beginning with the fourth episode in
which she attended a voting-rights meeting. As
the series progressed Lilly would become more
radicalized, especially after being brutally
beaten during a sit-in demonstration; but she
never forced her opinions on her white employer, always behaving respectfully and
adopting the aforementioned "neutral gaze"
when interracting with Forrest and his children.
While youngest boy John Morgan readily
bonded with Lilly, peppering her with questions about the differences between blacks and
whites, Forrest was initially resistant to the fact
that his housekeeper was awakening his family to the social upheavals occuring in the world
outside their cozy little house. It wasn't that
Forrest resented the Civil Rights advocates,
but he didn't exactly welcome them with open
arms either: "I want things to change, and I
want things to stay the same."

In spite of himself, Forrest gradually became a staunch opponent of racial prejudice,
especially in the courtroom, where he was confronted with such volatile cases as a a black
man charged with assaulting a white cop — an
incident which not only sparked a murder investigation, but also a lengthy story arc involving the Bryland branch of the Ku Klux Klan.
Complicating matters for Forrest were his political ambitions: he spent the first season running for the office of State Attorney General,
and the second season contemplating a promotion to U.S. Attorney. As he became less committed to the Southern status quo and more
convinced that the system must change and
people must change along with it, Forrest was
frequently forced to place his politics and his
inbred racial beliefs on hold — and he was none
too comfortable in the role of crusader, especially when it alienated him from people he
had always regarded as his friends.

I'll Fly Away made a point of avoiding
standard TV clichés in its recreation of the
racially divided South. Most of the white char-

acters were not cross-burning hooligans but
essentially good-hearted, well-intentioned individuals, bewildered and a bit frightened by
the sweeping changes in their society. In a similar vein, not all of the black characters were on
the same page as Lilly and her fellow activists:
indeed, Lilly's own father frequently admonished her to "go slow" when pressing for equality. Reflecting the realities and complexities of
the "Eyes on the Prize" years, there were no
easy answers or quick fixes on *I'll Fly Away*:
People chose their words and their battles carefully, and took plenty of time arriving at the solutions to the problems at hand — sometimes
never finding "closure" at all.

The most prominent of the series' supporting players were Kathryn Harrold as local
defense attorney Christina LeKatzis, who developed a romantic interest in Forrest despite
his marital status; Brad Sullivan as Nathan's
win-at-all-costs wrestling coach, Zollicofer
Weed; Peter Simmons as Nathan's "white trash"
pal Paul Slocum, whose screen time increased
during Season Two when he was accused of
rape by his pregnant girlfriend Parkie Sasser
(Amy Ryan); Bill Cobbs as Lilly's dad Lewis
Coleman, who basked in his past glory as a
baseball player in the Negro Leagues; and Mary
Alice, who won an Emmy award for her recurring portrayal of Marguerite Peck, the mother
of a black murder victim.

It goes without saying that *I'll Fly Away*
was not typical light entertainment, and the
NBC executives were only too aware that its
chances for success were slim to none. Sure
enough, the series ranked no higher than 40th
during its first week on the air, and quickly
plummeted to 55th in the weeks that followed.
Most of the network honchos were all for dropping the series immediately, but Les Moonves,
then in charge of Warner Bros. Television, virtually shamed NBC entertainment president
Warren Littlefield into standing behind the
show: "You don't want your epitaph to say
Punky Brewster." Despite its anemic ratings and
NBC's apparent efforts to kill the show by constantly changing its timeslot, research indicated
that *I'll Fly Away* was a solid favorite with black
audiences and female viewers between the ages

of 25 and 54, prompting the network to keep the series alive as a "prestige" item — at least until its $900,000-per-episode budget became too prohibitive.

The mainstream critics circled their wagons around *I'll Fly Away*, which the *New York Times* heralded as, "The most important series in the history of television." Other reviewers were no less effusive, though many complained that the show was too slowly paced and the characters too ambivalent for their tastes — a curious assessment, since these were the same people who constantly berated other TV programs for their artificially accelerated plot resolutions and cookie-cutter characterizations. In response to these detractors, Ken Tucker offered a thoughtful assessment of *I'll Fly Away* in a lengthy 1991 *Entertainment Weekly* piece. After labelling the series a "beautifully crafted, tough-minded hour" and an "anomaly on the current TV landscape," Tucker got down to business: "The networks have pretty much given up on programming that has any sense of narrative or emotional complexity, with any sense of history…. In their quest for instant ratings and immediate audience loyalty, the networks routinely yank shows off or all around their prime-time schedules and cancel series that need time for their ideal audience to find them. A good, knotty drama will always have the hardest time establishing itself in viewers' minds, and so they've always been the most obvious victims of the networks' fidgety, fast-buck moves. And good 'n' knotty is what *I'll Fly Away* is."

The only element Tucker found fault with was the character of Christina LeKatzis: "If *I'll Fly Away* has a major weakness, it's not in Kathryn Harrold's performance but in the conception of her role, a stumper on a couple of levels. For one thing, her character seems far too much of a modern feminist to have ingratiated herself with this deeply conservative town. Then too, her receptiveness to Forrest's advances renders both of these people more than a bit weasely: Forrest isn't just a married man — he's a married man whose wife has recently been institutionalized. Don't the producers think we're going to find a dalliance between Forrest and Harrold's Christina LeKatzis a cruel betrayal? This is a very odd way to establish these two characters as heroes to root for every week." (Footnote: After a few awkward episodes in which oldest son Nathan expressed disapproval over his dad's relationship with Christina, the writers "solved" the problem by having Forrest's wife conveniently pass away.)

In its brief two-year life span, *I'll Fly Away* earned three Emmies, two Golden Globes, five NAACP Image Awards and the Humanitas Award — but never managed to even crack the Top 30 network TV programs. Despite a letter-writing campaign from the series' most devoted fans, coupled with the suggestion from an organization called Viewers for Quality Television that the show be "exempt" from ratings consideration, *I'll Fly Away* was cancelled after 38 episodes in April of 1993. Six months later, reruns of the series were picked up by PBS, the second time in TV history a *succès d'estime* legal drama crossed over from commercial to public television (the first was *The Paper Chase* [q.v.]). "I believe the reason we're getting a second chance on PBS has a lot to do with TV-watcher activism," noted series star Sam Waterston to *TV Guide*'s Deborah Starr. "I think the noise that they made was heard by PBS and had a lot to do with its decision to pick up the show. The audience ought to know it, so they can flex their muscles more often in the future." Added co-producer John Falsey, "In retrospect, *I'll Fly Away* should have been on PBS to begin with. You're talking about a period piece that deals with racism — a show that's very painful for some people to watch."

Though no new hour-long episodes were filmed for the PBS run, a special TV-movie sequel was produced to tie up the plot strands left dangling by the series' cancellation. Telecast October 11, 1993, *I'll Fly Away: Then and Now* was narrated from the vantage point of the the Present by the 60-year-old Lilly Harper, who just before making a pilgrimage back to Bryland recalled the details of a 1962 murder case (clearly inspired by the 1955 murder of Emmett Till) to her 13-year-old grandson. With the notable exception of Kathryn Harrold's Christina

LeKatzis, most of the series' principal characters appeared in this 90-minute reunion film — though Jeremy London, then busy on another project, was replaced by his twin brother Jason in the role of Nathan Bedford.

InJustice

ABC: Jan. 1–March 31, 2006. Touchstone Television/Buena Vista Television. Creators/Executive Producers: Michelle and Robert King. Executive Producer: Stu Bloomberg, Jeff Melvoin. Cast: Kyle MacLachlan (David Swayne); Jason O'Mara (Charles Conti); Marisol Nichols (Sonya Quintano); Daniel Cosgrove (Jon Lemonick); Constance Zimmer (Brianna).

Premiering the first day of 2006, the hour-long ABC procedural drama *InJustice* was literally "torn from the headlines." The series was inspired by several recent instances in which such legal advocacy groups as the California Commission on the Fair Administration of Justice, using new DNA evidence, had exonerated innocent people who had been sentenced to death or life imprisonment (by 2007, more than 200 convicts had been freed through this process). Kyle MacLachlan starred as David Swayne, a high-powered corporate attorney who gave up his $650-per-hour practice (this part might *not* have been torn from the headlines) and used $5 million of his own money to create the National Justice Project. It was Swayne's avowed purpose to reopen closed cases and reassess old verdicts in order to provide a chance for freedom to those unfortunates whom he suspected had been railroaded into prison.

Each episode began in *Law & Order* (q.v.) fashion with a stern offscreen announcer intoning, "Every trial results in a verdict, but not every verdict results in the truth. This is what the jury believed...." At this point, the action flashed back to the crime at hand, dramatizing what was *assumed* to have happened. Step by step, the National Justice Project would provide new evidence and testimony that would topple those assumptions like so many sand castles, until at the end of the episode enough proof had been accumulated to free the innocent party. Taking matters even farther than

most real-life advocacy groups, the Project often solved the original crime on behalf of the authorities: in the very first episode, Swayne and his colleagues not only came to the defense of an ex-junkie who'd been convicted of murdering her father back in 1995, but also utilized fresh DNA evidence to track down and capture the real killer.

Swayne's team was headed by former cop Charles Conti (Jason O'Mara), who disapproved of his boss' idiosyncratic behavior but was willing to get past personalities in pursuit of the Truth. Among the team members were Sonya Quintano (Marisol Nichols), who was beholden to Swayne for releasing her own brother from prison; Jon Lemonick (Daniel Cosgrove), the resident computer wonk; and Brianna (Constance Zimmer), the youngest and most idealistic member of the Project.

If good intentions translated to good ratings, *InJustice* would still be on the air. As it was, the series failed to find an audience and was cancelled after 13 episodes. But before its condemnation to television's equivalent of Death Row, *InJustice* received an impassioned defense from Debra Watson of the *World Socialist Web Site*. "By virtue of its underlying premise that many people are in prison as the result of official malfeasance," wrote Watson on February 22, 2006, "the program is at odds with every police drama on U.S. television today." She went on to congratulate the series for not being a "glorification of the draconian law-and-order culture in the US," then launched into a jeremiad about the wrongly accused and the wrongheaded accusers, concluding, "It is the police, judges and prosecutors who are guilty of negligence, fraud and even murder. Thus, instead of encouraging trust in US criminal justice institutions, this program helps to educate its viewers in its endemic failures." Whew.

JAG

NBC: Sept. 23, 1995–May 22, 1996; CBS: Jan. 3, 1997–Apr. 29, 2005. Belisarius Productions/Paramount Network Television Productions. Creator/Exeuitve Producer: Donald P. Bellisario. Cast: David James Elliott (Lt./Lt.Cmdr. Harmon "Harm" Rabb); Catherine Bell (Maj./Lt. Col. Sarah "Mac" MacKenzie); Tracey Needham (Lt. j.g. Meg Austin); Patrick

Labyorteaux (Lt. j.g./Lt. Bud Roberts); Karri Turner (Ensign/Lt. j.g. Harriet Sims Roberts); Scott Lawrence (Cmdr. Sturgis Turner); David Andrews (Maj. General Gordon "Biff" Creswell); Zoe McLellan (Petty Officer Jennifer Coates); W.K. Stratton (Cmdr. Terry Lindsey); Andrea Thompson (Cmdr. Allison Krennick); John M. Jackson (Adm. Albert Jethro Chegwidden); Chuck Carrington (Petty Officer Jason Tiner); Mark Metcalf (Capt. Matthew Pike); Randy Vasquez (Gunnery Sgt. Victor Galindez); Nanci Chambers (Lt. Loren Singer); Cindy Ambuehl (Renee Peterson); Cynthia Sikes (Dr. Sydney Walden); Tamlyn Tomita (Lt. Cmdr. Tracy Manetti); Isabella Hoffman (Meredith Cavanaugh); Steven Culp (Agent Clayton Webb); Rayle Hollitt (Cassie Fuller); Chris Beetem (Lt. Gregory Vukovic).

The title of this extraordinarily popular legal series is a military acronym for the Judge Advocate General Corps, an elite wing of officers trained as lawyers who investigate cases involving military personnel, then act as either prosecutor or defender depending on the circumstances. If you didn't know what JAG stood for before the series' 1995 debut, you were in good company: up until the early 1990s, series creator Donald P. Bellisario had never heard of the real-life JAG either.

After a four-year hitch with the Marine Corps, Bellisario landed his first important network television job as story editor for the WW2 series *Baa Baa Black Sheep*. Since that time he has espoused a strongly pro-military stance in his TV work; indeed, most of the principal characters in his own series are either current or former members of the Armed Forces. For many years, Bellisario rowed against the current of mainstream show-business thinking, in which "anti-war" was all too often synonymous with "anti-soldier." At a time when the egregious "Crazed Vietnam Veteran" stereotype was still commonplace on Prime Time TV, Bellisario created the series *Magnum P.I.*, in which all three main protagonists had not only served in Vietnam, but were also well-adjusted human beings with nary a hint of mental aberration.

The seed for *JAG* was first planted in Bellisario's subconscious when he read a newspaper story about the introduction of women on U.S. Navy carriers. This served as inspiration for a script for a proposed theatrical feature, in which the main female character, a Naval officer, died under mysterious circumstances. While doing research to find out which military organization would be assigned to investigate and prosecute this case, he came across information concerning the Naval Criminal Investigative Services — which led inevitably to the JAG division. Bellisario was especially attracted to this branch of service because of its flexibility: "What a great franchise," he commented six years into the run of the subsequent *JAG* series. "Unlike most law shows, I've got a detective, a prosecutor and a defender."

Bellisario's one-shot screenplay evolved into a weekly, hour-long series, debuting on NBC in the fall of 1995. David James Elliott headed the cast as Lieutenant (and later Lieutenant Commander) Harmon "Harm" Rabb Jr., a former Navy flyer who had suffered from night blindness ever since his F14D Tomcat fighter jet crashed on the deck of a carrier. Though grounded, Harm wanted to keep his commission and continue serving his country; thus he transferred to Navy JAG, occasionally returning to flying for the purpose of gathering evidence — and then only in the daytime hours. In the pilot episode, Harm's legal partner was Lt. Caitlin Pike, played by Andrea Parker, but when the actress left to costar in another series, she was replaced by Tracey Needham as Lt. j.g. Meg Austin, a lawyer and computer/weapons specialist. W.K. Stratton costarred in the first few episodes as JAG's chief prosecutor Theodore Lindsey, who was clearly unsuited to the task. Though quickly replaced by Cmdr. Allison Krennick (Andrea Thompson), Lindsey proved to be the proverbial "bad penny," sporadically resurfacing to cause trouble for the people we cared about; in one later story arc, Lindsey mounted a campaign to destroy the career of Admiral A.J. Chegwidden (John M. Jackson), the tough-but-supportive head of JAG in Washington DC, who'd joined the cast in the middle of Season One. Also seen from time to time during the series' freshman year was Patrick Labyorteaux as Ensign (later Lieutenant j.g.) Bud Roberts, public affairs officer for the U.S.S. *Seahawk*.

Though Bellisario had hoped to secure the cooperation of the U.S. Navy in the production of *JAG*, he was turned down: the Navy was not only nervous about the series' investigation/prosecution/defense premise, but had also been burned by recent TV shows and movies that cast a negative light on men and women in uniform. Thus, Bellisario had to content himself with the props and costumes available from the Paramount studio warehouse; he also relied heavily on stock footage from such films as *Top Gun* and *Hunt for Red October*.

While the series' miserable ratings (it never ranked higher than 77th) would seem to be the principal reason that NBC cancelled *JAG* after a single season, Bellisario later told an interviewer that the show was considered "uncool" in Hollywood circles because people who'd never seen it assumed, "it's got to be jingoistic." *TV Guide*'s Jeff Jarvis, who *had* seen the series, dismissed it on the same basic grounds, insisting that the show was not about military justice but rather "macho swaggering, muscle-flexing, cigar-chomping military men who can't stop whining about having to share their Navy with sailors in skirts." For all that, NBC might have actually renewed *JAG* if Bellisario had agreed to downplay the legal aspects in favor of more violence and "hardware."

Coming to the rescue was CBS executive Les Moonves, who brought *JAG* to his network as a midseason replacement in hopes of attracting disenfranchised older viewers. From a ratings low of 68th place, *JAG* soared far beyond Moonves' wildest dreams, closing the 1996–97 season as Prime Time's 17th highest-rated series. By December of 1998 it was averaging 15.4 million viewers a week—not merely oldsters, but also young fans who adored the series' stunningly attractive leading players. *JAG* continued riding high until it was knocked off its perch by NBC's *Who Wants to be a Millionaire* in 2000, only to experience another spectacular ratings surge after the harrowing events of September 11, 2001, sparked a resurgence of interest in—and unqualified admiration for—the American Armed Forces. Small wonder that *Entertain-*

ment Weekly has described *JAG* as "the stealth bomber of prime time."

Tracey Needham and Andrea Thompson had both left *JAG* before the series joined CBS. Added to the cast was Harm's new Marine partner, Major (later Lieutenant Colonel) Sarah "Mac" MacKenzie, played by Catherine Bell. (The actress had originally been cast during Season One in what was to have been a recurring role as Harm's former girlfriend; when the network decided to pull the plug, Harm's ex-sweetie was summarily killed off, and as a result Bell made her only appearance as a corpse in a body bag!) Preferring to follow standard military procedure to the letter—"I'm a Marine first, a lawyer second"—Mac provided a fascinating contrast to Harm, who relied more on gut instincts. But Donald Bellisario was careful to elevate Mac beyond mere stereotype, fleshing out the character in a variety of intriguing ways. As the series developed, we learned that Mac's rules-are-rules stance, born of her determination to persevere in what had formerly been an "All Boys Club," could be modified if she genuinely believed that the spirit of the law would better serve her client. We also found out that Mac was a recovering alcoholic, the product of an absentee mom and an abusive dad; and that the memory of her deranged ex-husband had cast a pall over all future romantic entanglements. The fact that Mac was able to maintain her cool professionalism and rock-solid commitment to "Duty, Honor, Country" under the most trying of circumstances—both professional and personal—made her a most admirable and appealing character.

Promoted to "regular" status in the move from NBC to CBS was Bud Roberts, newly assigned to JAG as Harm's law clerk. In a sense, Bud was the embodiment of every idealistic young recruit who has ever aspired to serve his country faithfully and honorably no matter what obstacles and setbacks came his way. Inspired by Harm's example, Bud began training to become an attorney himself, finally taking on his own cases during the series' fourth season. After 9/11, Bud was assigned to a six-month deployment on the U.S.S. *Seahawk*, acquitting himself nobly throughout. At the tail end of Season Seven, he was seriously injured by an Afghan mine, ultimately suffering the loss of his leg. Though the period of adjustment to his prosthesis was not an easy one, and despite the prospect of never receiving another promotion due to his disability, Bud refused to even consider giving up his position with the JAG unit, and in so doing became something of a role model for thousands of genuine wartime amputees. Standing loyally and lovingly by his side through all his troubles was Bud's wife, Lt. j.g. Harriet Sims (Karri Turner), whose courage in the face of the worst kind of adversity won the admiration of scores of real-life Navy wives.

By the time *JAG* entered its third season in the fall of 1997, the U.S. Navy was sufficiently comfortable with the series to extend full cooperation to the producers, offering hardware, technical advice, and even an honest-to-goodness aircraft carrier for location shooting. The Navy also began to allow its personnel to appear in extra roles — on condition that, if these extras happened to be off-duty, their salaries would be donated to the Morale Welfare and Recreation Fund. Finally, the Navy accepted the dramatic necessity of occasionally featuring military men and women as villains: "We're fine with that as long as the bad guys are caught and punished, and the institution of the Navy is not the bad guy," noted Cmdr. Bob Anderson, Los Angeles Navy public information officer, in a *TV Guide* interview. No one was happier with this new arrangement than Donald Bellisario: "I've got all the big-boy toys to play with: Navy jets and Marine helos." In reciprocation for the Navy's assistance, Bellisario agreed to remain as much a stickler for accuracy as he'd always been: "I prefer to tell stories that cannot be told on any other television show dealing with the law or the police: stories that can only happen within the military environment or the military justice system. That's the hardest part of bringing in new writers. They have to learn military protocol." (So did the viewers: the series' various Internet fansites have regularly issued glossaries of the military terms and acronyms

the characters bandied about with reckless abandon.)

Bellisario had always made a habit of basing his scripts on such actual events as the Tailhook scandal and scattered reports of military gay-bashing. Before long the series was not only keeping apace of late-breaking events, but sometimes anticipating them. "We dealt with a dirty bomb in the final episode of the [2001-2002] season long before this stuff ever broke," noted Bellisario to *TV Guide*. "We take from the headlines, but we also precede reality." Keeping himself apprised of the latest military developments by poring over the *Navy Times*, *The Marine Corps Times* and similar house organs, the producer also hired a research team to root out stories from other countries.

Additionally, Bellisario became adept at the sort of instant diplomacy vital to the successful maintenance of civilian-military relations. Though he now had the Navy on board, the U.S. Marines were not always keen on *JAG*'s use of controversial subjects, in one instance holding off a planned ad campaign that was to be launched during an episode about the alleged misuse of nerve gas. To mollify the Marines, Bellisario suggested that their ad be placed at a point in the story when a character had just finished making a staunchly pro-military statement, simultaneously reprimanding the Media for unfairly persecuting the armed services. This last-mentioned insertion was obviously heartfelt on Bellisario's part, since the mainstream press had never entirely warmed up to *JAG*. Even after viewership increased in the wake of 9/11, *Time* magazine's James Poniewozik wrote off the show as "TV's least cool hit," adding, "Rip the epaulets and you've got one more lawyer/cop show, with flat characters and dialogue and extra rations of melodrama." Bellisario as usual had an answer: "In Hollywood and in most of the media, the military was spoken of in perjorative terms. Now people have a lot of curiosity about the military. It always changes when the country is in trouble and we need someone to protect our ass."

In fairness to *Time*, there was some truth in Poniewozik's charge of melodramatic excess. Bellisario was the first to admit that the series took a great deal of artistic license — so much so that in normal circumstances, that license would have probably been revoked. For one thing, while real-life JAG officers seldom leave their offices, Harm and Mac hopscotched all over the world in search of hidden clues and elusive villains. Also, no genuine JAG unit would have implicitly permitted Harm to take time off from his various investigations to search for his father, a Navy pilot who'd been M.I.A. since Vietnam. The coincidences and contrivances which placed Harm on his dad's trail strained credibility to the snapping point, especially during a lengthy story arc in which evidence arose that the senior Rabb may have conspired with the KGB — leading Harm to a rendezvous with his Russian half-brother Sergei (Jade Carter), about whom more later. Finally, while it was perfectly acceptable to find a legitimate means of postponing the wedding between Mac and her Australian boyfriend Mic Brumby (Trevor Goddard), having Harm crash-land in the ocean and subsequently suffer from amnesia may have been a bit too extreme.

Nowhere did *JAG* stray further from real life than when it contrived to have the lovely Mac fall into the clutches of kidnappers. Though star Catherine Bell insisted that real JAGs were envious of her fabricated adventures, we beg leave to doubt that genuine female military lawyers harbor dreams of being locked in a warehouse with a gun-wielding lunatic. This was *JAG*'s one glaring concession to Standard TV Formula, which since time immemorial has dictated the most surefire way to stir up audience empathy is to place a pretty girl, a small child, or a dog in jeopardy. And since there were no children or dogs in the JAG corps, guess who ended up getting tied to a chair?

One melodramatic machination that reaped long-range benefits occurred during the 2002–2003 season. While most of the new characters introduced in the course of the series' 227 episodes were essentially likeable — Randy Vasquez as Gunnery Sergeant Galindez, Cindy Ambuehl as Harm's movie-producer girlfriend Renee, Steven Culp as CIA officer Clayton Webb, Zoe McLellan as Admiral Chegwiggen's assistant

Jennifer Coates, Scott Lawrence as Harm's friendly JAG rival Sturgis Turner — Lieutenant Loren Singer rhymed with witch from her first appearance onward. Introduced in 1999 and reaching her villainous pinnacle as temporary replacement for the wounded Bud Thomas, Lt. Singer (ironically played by Nanci Chambers, the wife of series star David James Elliott) was as ethically challenged as any attorney on *Boston Legal* (q.v.), willing to step on and crush anyone who stood in the way of her ambitions, and positively swilling in her own nastiness. Shortly after taking pregnancy leave in the middle of Season Eight, Singer was found murdered, whereupon her ex-lover, Harm's half-brother Sergei (we *told* you we'd get back to him) was suspected of the crime. This plot twist ended up killing two birds with one stone when the series' perennial nemesis Teddy Lindsey was exposed as the murderer. More importantly, the investigation of Singer's murder served to introduce Mark Harmon as Jethro Gibbs, special agent for the Naval Criminal Investigative Service, a role he would carry over into the popular *JAG* spinoff series *Navy NCIs*.

While melodrama was undeniably a key ingredient of the *JAG* recipe, the series scored its highest points with its believable characterizations, provocative but credible courtroom scenes (especially when dealing with terrorists), and up-to-date story developments, with several post–9/11 episodes taking place in such international trouble spots as Afghanistan, Uzbeckistan and North Korea. But though viewers were happy, they weren't entirely satisfied, as witness a 2003 poll that revealed that 69 percent of the series' fans demanded a romance develop between leading characters Harm and Mac. Well, why not? Practically every other main character had something going — even old Admiral Chegwidden, who'd popped the question to lovely college professor Meredith Cavanaugh (Isabella Hoffman) just before retiring from service (his replacement was Marine Major General Gordon Cresswell, played by David Andrews). Unfortunately, by the time things finally heated up between Harm and Mac, the series was on a ratings downslide, largely due to a timeslot shift from Tuesday to

Friday. Also, rumors were flying that the show would be cancelled at the end of the 2003–2004 season because Bellisario and CBS had not finalized the renewal agreement.

As it happened, *JAG* was back in September of 2004, but things weren't quite the same; the ardor between Harm and Mac had cooled, and it appeared as though new cast member Chris Beetem was being groomed as Harm's replacement in the role of Lt. Gregory Vukovic. To an outsider, these developments were puzzling indeed, but it didn't take a tearful courtroom confession to explain what was happening. CBS was on another of its periodic austerity kicks, and had ordered Bellisario to make deep cuts in the *JAG* budget; at the same time, star David James Elliott was negotiating for a more lucrative contract. Before long, Elliott was issuing statements that he was being "forced" to leave the show, while other reports indicated he'd decided to leave on his own accord. This didn't stop Bellasario from optimistically predicting the series would be renewed for an eleventh season, promising a "younger, hipper *JAG*" in response to the news that that most of the 10 million regular viewers were past the age of 40. With the signing of the aforementioned Chris Beetem, and Bellisario's assurance that Catherine Bell would remain on the series, it looked like the handwriting was on the wall for David James Elliott. Thus, the Harm-Mac romance was wrapped up as quickly as possible, whereupon Harm was offered the position of Force Judge Advocate in London while Mac was offered a commander's post at Joint Legal Services S.W., headquartered in San Diego.

That's All, Folks? Not quite. Ever the showman, Bellisario still had one ace up his sleeve. In the climactic moment of the season finale, Harm turned to Mac and proposed marriage … and with that, the ten-season saga of *JAG* came to a close.

Jake and the Fatman

CBS: Sept. 26, 1987–Sept. 7, 1988; March 15, 1989–Sept. 12, 1992. Dean Hargrove Productions/Strathmore Productions/Columbia Pictures Television. Executive Producers/Creators: Dean Hargrove, Joel Steiger. Executive Producers: Fred Silverman, Philip Saltzman, Ed Waters, David Moessinger, Joel Steiger. Cast: William Conrad (Jason Lochinvar "Fatman" McCabe); Joe Penny (Jake Styles); Alan Campbell (Derek Mitchell); Lu Leonard (Gertrude); Jack Hogan (Judge Smithwood); George O'Hanlon Jr. (Sgt. Rafferty); Olga Russell (Lisbeth Berkeley-Smythe); Melody Anderson (Sgt. Capshaw).

The producing team of Dean Hargrove and Fred Silverman has made a cottage industry out of fashioning popular mystery series around the talents of fading TV stars, not only generating huge ratings and viewer goodwill, but also revitalizing several near-dormant careers. The Hargrove-Silverman factory had already turned out such hits as the *Perry Mason* TV movies starring Raymond Burr (see separate entry), and the Andy Griffith starrer *Matlock* (q.v.) when, in 1987, the producers hoped to make magic happen again by developing a vehicle for William Conrad, who hadn't been seen too much since the cancellation of *Cannon* back in 1976.

The Conrad property was tentatively tried out as a two-part episode of the legal drama *Matlock*, "The Don," which aired October 28 and November 6, 1986. In this one, wily Atlanta attorney Ben Matlock (Andy Griffith), dragooned into taking on the defense of dying Mafioso Nick Baron (Jose Ferrer), matched wits with porcine prosecutor James L. McShane, played by Conrad; also featured were Joe Penny as Nick Baron's son Paul, and Alan Campbell as McShane's assistant Palmer. Audience response was positive, and the following year the three aforementioned guest stars were spun off into their own series, albeit with different character names — and in the cases of Penny and Campbell, markedly different characterizations.

Originally set in "a large Southern California city" (presumably not Azuza), *Jake and the Fatman* starred Conrad as hard-nosed DA Jason Lochinvar McCabe, known for obvious reasons as "The Fatman." Crude, rude and relentless, The Fatman brought scores of criminals to heel mostly by doodling around the edges of accepted legal procedure, rather than coloring strictly within the lines — which made him extremely popular with public and the bane of defense attorneys and bad guys everywhere. McCabe's trusty aides were his studdish investigator Jake Styles (Joe Penny), who lived high on the hog when not on the job and frequently went undercover during working hours, and nerdy-but-dedicated assistant DA Derek Mitchell (Alan Campbell). Also seen during the first season were The Fatman's equally hefty secretary Gertrude (Lu Leonard) and his equally surly bulldog Max.

Unlike most other Hargrove-Silverman concoctions, *Jake and the Fatman* posted only lukewarm ratings for CBS, and the show went on hiatus at the end of Season One. Still, the network figured that the services of William Conrad were worth retaining, so the series was renewed for a second season, with a virtually complete face-lift and several new producers. Now retired from the DA's office, The Fatman bade farewell to California and moved to Hawaii, enabling the production company to assume command of the recently vacated *Magnum, P.I.* studio facilities. Working as an undercover agent for the Honolulu prosecutor's office (though how Bill Conrad could ever go undercover without immediately being recognized—or even fitting into the costumes—is beyond us), McCabe continued to be aided and abetted by Jake Styles (now his full partner) and Derek Mitchell (now his legman). The series remained in Honolulu for its third season, by which time McCabe was once again a official prosecuting attorney, and Melody Anderson had joined the cast as policewoman Neely Capshaw (a character added to hype the ratings amongst male viewers).

Also brought onto the project for Season Three was an influx of Young Blood, including producers David Moessinger and Jeri Taylor and story editor/producer J. Michael Straczynski, soon to be the guiding force of *Babylon 5*. Moessinger, Taylor and Strazynski quit the series en masse over creative issues at the end of the season; the following year McCabe, Jake, Derek and Lisbeth left Honolulu and relocated to California, where McCabe had been requested to root out corruption in the DA's office by the Mayor of Los Angeles. This done, McCabe was appointed DA again, with

Jake as his chief assistant. Throughout all these scenery shifts and job transfers, the series' ratings continued to be underwhelming. Indeed, in 1990 CBS was poised to pull the plug on *Jake and the Fatman* in favor of a new legal series produced by Dick Wolf, but at the last moment the Hargrove-Silverman property was renewed for a fourth season (with yet *another* new production team at the helm), and Wolf was compelled to take *Law & Order* (q.v.) to NBC. The only thing worth mentioning about *Jake and the Fatman*'s final two seasons on the air is that the episode of March 20, 1991, "It Never Entered My Mind," served as the pilot for the Dick Van Dyke vehicle *Diagnosis: Murder*.

Since leaving the air, *Jake and the Fatman* has joined the ranks of such obscure series as *Treasury Men in Action, Colt. 45* and *The Joey Bishop Show* as one of those aberrations that few people were excited over when they were run on the networks, but which inexplicably remained in production for multiple seasons. Quoted in the *Thrilling Detective* website, Lee Goldberg, who as one of scriptwriters for *Diagnosis Murder* knew whereof he spoke, dismissed *Jake and the Fatman* thusly: "This longrunning detective show ... was so bad that it has never been rerun domestically since it was cancelled. The cliché-ridden show is notable only for spinning off a series that was far more successful." Accordingly, it has been completely forgotten — by everyone but comedian Jerry Seinfeld, who when asked by Barbara Walters what he planned to do when his own series ended its run in 1997, pondered a moment and quipped, "revive *Jake and the Fatman*."

The Jean Arthur Show

CBS: Sept. 12–Dec. 5, 1966. Universal Television. Produced by Richard Quine. Cast: Jean Arthur (Patricia Marshall); Ron Harper (Paul Marshall); Leonard Stone (Morton); Richard Conte (Richie Wells).

Many movie buffs regard Jean Arthur as the finest and most appealing screen actress of the 1930s and 1940s. Blessed with an endearingly raspy, cracked voice and a sublimely oddball personality, Arthur excelled at both drama and comedy, especially the latter; when she collaborated with a director capable of harnessing her unique talents for ultimate effect — a Frank Capra (*Mr. Deeds Goes to Town, Mr. Smith Goes to Washington*) or a George Stevens (*The More the Merrier, Talk of the Town*) — the results were miraculous. Ironically, this most vivacious and ebullient of screen performers was in real life a morbidly shy and withdrawn person, who was known to become violently ill just before each take, and to react with extreme anger and terror whenever anyone invaded her space. As her star ascended her phobias intensified, and by 1953 she had completely given up show business, retiring to her lavish estate in Carmel, California. In the early 1960s, Arthur's close friend Lucille Ball urged her to return before the cameras with a guest shot on *The Lucy Show*, but the actress' new business manager Eddie Dukoff nixed this idea, convincing Arthur that if she was determined to make a comeback, it should not be in someone else's sitcom but as the star of her own TV show. Ultimately the sixtysomething actress subjected herself to a "trial run" on a *Gunsmoke* episode, which proved so satisfying to herself and her fans that she agreed to fully emerge from her shell in a series of her own.

Produced by prolific film director Richard Quine (*The Solid Gold Cadillac, Bell Book and Candle*), *The Jean Arthur Show* cast the eponymous star as attractive widow Patricia Marshall, a successful defense attorney famous for her garrulous, mile-a-minute summations to the jury. Patricia's legal partner was her 25-year-old son Paul Marshall (Ron Harper), fresh out of law school and eager to make his mark in the profession. A by-the-book type, Paul was taken aback by his mom's unorthodox approach to jurisprudence and her free-and-easy rapport with some of her more unsavory clients, notably soft-hearted former gangster Richie Wells (Richard Conte). With Patricia chattering away and Paul doing his darnedest to get a word in edgewise, the Marshalls' taciturn chauffeur Morton (Leonard Stone) came off about as loquacious as Harpo Marx.

As always, Jean Arthur was the thorough professional, throwing herself into the scriptwriters' clichéd machinations with the same zest and zeal she'd displayed in her best film appearances; she even sang the praises (yes, *sang* the praises!) of the sponsor's product in the closing cast commercials. It's too bad that *The Jean Arthur Show* was a hopeless wallow in the mire of mediocrity, the sort of comedy series in which the people on the canned-laughter track sound as if they're being forced to react at gunpoint. Jean Arthur herself knew she was trapped in a lemon, as she noted philosophically in a *TV Guide* interview: "I'm terribly disappointed, of course, because I love the work so much. But I wouldn't want to face another one of those scripts…. But here is the other side — a bunch of fellows at Universal who have a budget. These people have to do something, and then how are they going to change it? They have a budget. There wasn't anyone brave enough to say, 'Let's quit until we can find a way.'"

Ranking 65th out of 90 network series, *The Jean Arthur Show* was already a dead duck six weeks into its run, and was cancelled after its 12th episode. Surprisingly, this traumatic experience did *not* drive Jean Arthur back into seclusion; instead, it had the opposite effect, and before long the actress was contentedly back at work, teaching drama classes at Vassar College and North Carolina School of the Arts.

Jones and Jury

Syndicated: Sept. 1994–Sept. 1995. Lighthearted Entertainment/Group W. Executive Producer: Howard Schultz. Cast: Star Jones (Herself).

Long before her tempestuous tenure on the ABC morning gabfest *The View*, Star Jones came to TV prominence in the daily courtoom-talkshow hybrid *Jones and Jury*. A graduate of the University of Houston Law Center, Starlet Marie Jones (it really shouldn't be necessary to confirm at this late date that, yes, that *is* her real name) began her legal career in the General Trial and Homicide bureaus of the Kings County DA's office in Brooklyn, then in 1991 was promoted to senior district attorney. In this capacity, she prosecuted such volatile felony cases as the "Crown Heights" incident, in which a Hasidic motorist was charged with accidentally striking and killing a black child, a tragedy which culminated in a three-day race riot. Jones' subsequent on-the-air commentary for Court TV's coverage of the William Kennedy Smith trial led to a similar stint on NBC's *Today Show*— and ultimately, at age 32, her own syndicated half-hour series.

In the tradition of *People's Court* and its ilk, the set on *Jones and Jury* was a simulated courtroom. "Judge" Jones presided over actual small-claims cases from the various legal dockets of Southern California. The litigants— among them a brother suing a sister over her credit card bills and an incensed bird owner whose pet tweetie had been eaten by the neighbor's dog— were questioned not only by Jones but also by a "jury" comprised of audience members. Once the arguments were presented, the jury cast its votes and Jones pronounced the final, binding verdict.

Although *Jones and Jury* was only in syndication for a single season, it was enough to convince Star Jones that the time was ripe for a career change. Abandoning the legal world, she plunged headlong into Show Business, acting in such series as *Strong Medicine, Soul Food* and *Less Than Perfect*, trading quips with Barbara Walters et al. during a ten-year stint on *The View* (during which she proudly shed 160 pounds virtually before our eyes), and basking in the glow of Total Celebrity when her lavish 2004 wedding was treated as a media event commensurate with the marriage of Charles and Di. In 2007, Star Jones more or less came full circle as the host of an eponymously titled talk show on Court TV.

Judd for the Defense

ABC: Sept. 8, 1967–Sept. 19, 1969. 20th Century–Fox Television. Executive Producer: Paul Monash. Produced by Harold Gast and Charles Russell. Cast: Carl Betz (Clinton Judd); Stephen Young (Ben Caldwell).

Most of his Hollywood contemporaries recognized Carl Betz as one of the best and most reliable actors in the business, but it wasn't until he emerged from behind the shadow of his longtime sitcom costar Donna Reed that he was finally rewarded with a TV series of his own. Set in Houston (but largely lensed on the Fox backlot), *Judd for the Defense* cast Betz as flamboyant attorney Clinton Judd, who showed up in court in custom-made boots, a string tie and a cowboy hat, and who piloted his own private jet from one assignment to another. A lot of observers assumed that Judd was based on the equally colorful F. Lee Bailey; but in fact the character owed more to legendary Houston criminal-defense lawyer Percy Foreman, who during his sixty-year career defended more than 1000 accused murderers, of whom 75 percent were acquitted and only 53 sent to prison — and of those 53, only one was executed. Ironically, Foreman despised killers and preferred to accept high-profile divorce cases, but took on murder cases because he was opposed to capital punishment. A millionaire many times over, Foreman was willing to accept property rather than cash if his client was financially strapped — and if the client had no property, Foreman worked pro-bono just for the thrill of it all. Unlike Betz's crisply organized portrayal of Clinton Judd, Foreman enjoyed throwing prosecutors and juries off-guard by adopting a shambling, disheveled courtroom persona — rather like Peter Falk's Lt. Columbo, only older and taller. After *Judd for the Defense* was cancelled, Percy Foreman achieved a measure of notoriety as defense attorney for Martin Luther King's assassin James Earl Ray, advising his client to plead guilty in order to avoid execution — a strategy that has inadvertently stoked the flames for hundreds of future conspiracy theorists.

Many of the *Judd* episodes were based on recent headlines: in one, a catatonic 10-year-old girl was the only witness to a murder at sea; in another, a psychic was brought in to determine if a schizophrenic young man had killed three girls in a ritual slaying; and in still another, William Daniels played an Average Joe whose credit rating, and life, were destroyed by a computer error. Judd's more controversial clients included a Southern sheriff accused of violating the civil rights of a murdered Northern journalist, a Cesar Chavez-like activist driven to desperation by his futile attempts to organize a migrant's union, an anti-war lawyer charged with prodding a draft dodger to set himself afire, and a transplant surgeon suspected of removing an organ from a still-living donor. Taking advantage of the ever-relaxing network and sponsorial standards of the late 1960s, several episodes dealt with subject matter that might have been taboo a few years earlier: a professor fired for giving a job to a black militant, a blacklisted writer trying to wrest free from the ironclad Hollywood contract that kept him from earning a living, a mentally challenged woman coerced into pleading guilty on a murder charge. One remarkable episode, "To Love and Stand Mute," costarred Tom Troupe and Loretta Leversee as a blind and deaf couple who wanted to adopt a child: in the interests of authenticity, both actors took extensive lessons from sign-language expert Mrs. Faye Wilkie.

To maintain the series' fresh, contemporary slant, producer Harold Gast hired several new and untested writers, with generally positive results (except, perhaps, for the preponderance of badly dated "mod" episodes featuring overaged hippie types and nausea-inducing psychedelic camerawork). In the same vein, a number of youthful actors on the way up were prominently featured, notably Stephen Young in the regular role of Judd's novice assistant Ben Caldwell, and such guest stars as Karen Black, Richard Dreyfuss, Susan Anspach, Tyne Daly and Georg Stanford Brown.

Though *Judd for the Defense* was a slow starter ratings-wise, both star Betz and producer Gast had faith in the series, redoubling their efforts to win new viewers. At one point, the series participated in the publicity-savvy crossover stunt of providing the concluding episode to a two-part cliffhanger introduced on another 20th Century–Fox series, *The Felony Squad.* Ultimately, Carl Betz was honored for his labors with a Golden Globe award — which, unfortunately, was bestowed after the cancellation of *Judd for the Defense* in 1969.

The Judge

Syndicated: 1986–1988. Genesis-Colbert. Executive producers: Sandy Frank, Gary Ganaway, Barry Cahn. Cast: Bob Shield (Judge Robert Franklin).

The Judge began as a local program on WBNS-TV in Columbus, Ohio, a station which since the 1950s had garnered several local awards for such simulated-hearing efforts as *Traffic Court* and *Juvenile Court* (these were billed as "public affairs" programs). Most of the re-enacted cases on *The Judge* were domestic disputes, generally involving child custody: at one point in development the series was titled *Custody Court,* and reportedly the original intention was to televise *actual* custody hearings. But though each episode was based on fact, no "real" people appeared on the carefully scripted and rehearsed series. Even the title character, Judge Robert Franklin, was a ringer, portrayed by actor Bob Shield (There *was* an actual family-court judge named Robert Franklin in Toledo, Ohio, but he had no connection with the series). Highlighted by the sort of hair-tearing histrionics not seen since the days of *Ten Nights in a Bar-Room, The Judge* was one of two daily, half-hour "legal" series (see also *Lie Detector*) packaged and syndicated by the entrepreneurial Sandy Frank, who normally specialized in such game shows as *Treasure Hunt, Liar's Club* and *Name That Tune.*

Judge Alex

Syndicated: debuted Sept. 12, 2005. Twentieth Television. Executive Producers: Karen Melamed, Michelle Mazur, Burt Wheeler, Sharon Sussman. Cast: Judge Alex E. Ferrer (Himself).

The daily syndicated courtroom show *Judge Alex* was the first such series to feature a Hispanic jurist at the bench. Born in Cuba in 1960, Alex Ferrer was one year old when his family escaped to Miami. A lifelong law-and-order buff, Ferrer joined the Coral Gables police department at age 19, working his way up the ladder from patrolman to detective to undercover operative to SWAT team member. While still in uniform, he began attending law school, and upon retiring from the police force he became a Miami–based attorney, specializing in Civil Rights litigation. At 34, Ferrer became the youngest circuit court judge ever to serve in the Criminal and Family Divisions of Florida's 11th Judiciary, Miami-Dade County — and just before launching his own series at age

44, he served as the associate administrative judge of the 11th's Criminal Division. (Fine, Alex, but what *else* have you done?)

Generally avoiding the sort of domestic issues common to other shows of its type, *Judge Alex* focused on landlord disputes, worker-employee conflicts, consumer complaints and the like. Though the good-humored Judge Ferrer preferred to keep things light and noncombative, he refused to put up with any B.S. in his courtroom: his standard warning to lying litigants was, "Tell the truth or take a walk!" Sometimes he took his own advice, "taking a walk" outside the courtroom to follow up testimony and evidence in the streets.

Debuting in the fall of 2005, *Judge Alex* was primarily seen on Fox–affiliated stations, thanks to a huge pre-release sale to the network's owned-and-operated outlets. During its first weeks on the air, the series posted the highest ratings of any new half-hour syndie in its time period, and was still riding high when renewed for a third season in 2007.

Judge and Jury

MSNBC: Dec. 21, 1998–June 1999. Cast: Burton Katz (Himself).

Spurred by the popularity of CNN's *Burden of Proof* and CNBC's *Rivera Live* (both q.v.), the MSNBC cable service entered the daily legal-show field with the 60-minute *Judge and Jury*. Staged in a courtroom setting, each episode took the form of a debate, with teams of legal experts arguing the fine points of recent civil and criminal cases. *Judge and Jury* was moderated by former Los Angeles Superior Court judge Burton Katz, whose career extended back to the Charles Manson trial of the 1970s, and who in 1997 memorably appeared on *Rivera Live* presiding over a mock trial of the still-unsolved JonBenet Ramsey murder case.

Judge David Young

Syndicated: debuted Sept. 10, 2007. Sony Pictures Television. Executive Producers: Michael Rourke, Rich Goldman. Cast: Judge David Young (Himself); Tawya Young (Bailiff).

Sony Television's official press release for

the daily, half-hour syndicated courtroom series *Judge David Young* did a fairly thorough job itemizing the credentials of the series' boyish, bespectacled star, and ticking off the reasons viewers should tune in. The release dutifully noted that Judge Young was a native of Miami and a graduate of Tulane University and the University of Miami College of Law; that he had served as Miami-Dade County's assistant state attorney under Janet Reno, and had thrice been elected as the county's Circuit Court judge; and that he had "captured the nation's attention" in May of 2005 when he jailed two America West airplane pilots for attempting to fly while intoxicated — one of several reasons he'd received multiple honors from such organizations as Mothers Against Drunk Driving. The Sony scriveners also observed that Judge Young was known for his "sharp sense of humor," his "provocative sentencing" (such as imposing a 30-year jail term on a remorseless octogenarian woman who'd spent the past three decades on the lam for a murder rap), his "strong sense of responsibilty" coupled with "a real concern for people" as manifested in his probation monitoring system for people with addiction problems, and most intriguingly for his "occasional bursts of song from the bench."

Sony did a fairly thorough job promoting *Judge David Young,* except that they curiously omitted one small detail — a detail that was not only chronicled by virtually every other media report on the new series, but was also cheerfully acknowledged by the Judge himself. *Judge David Young,* you see, was the first syndicated courtroom series presided over by an openly gay man.

Asked by *TV Guide* if being gay had any effect on his job, Young chuckled, "I dress better than most other judges." On a more serious note, he added, "Being a victim of bias and prejudice, I'm very attuned to individuals who aren't as enlightened as they should be. It's important that I can show people, one case at a time, that I am what I am. I'm Judge David Young and you're going to want to watch me because you like me. You don't care with whom I sleep! You just like my rulings, you're entertained, you're going to be educated. It's going to be fun." As for his campy habit of singing show tunes from the bench (a chronic gambler was regaled with a rendition of "Luck Be a Lady," for example), one might be tempted to chalk this up as a byproduct of his "orientation"— if one wasn't worried about feeding into a tired and antiquated cliché.

With his warm smile and easygoing demeanor, David Young was a refreshing alternative to certain other TV judges who got their jollies by scowling, snarling and belittling. Equally likeable was the series' bailiff Tawya Young (no relation), a onetime lieutenant at Brooklyn Civil-Small Claims Court. Though *Judge David Young* lagged behind its court-show brethren in terms of ratings and audience support, the series was still popular enough to grab choice timeslots in most local markets, and also earned the Judge a Daytime Emmy nomination.

Judge for Yourself

Syndicated: Sept. 12, 1994–Apr. 7, 1995. Faded Denim Productions/Touchstone Television/Buena Vista Television. Executive Producer: Jerry Jaskulski. Cast: Bill Handel (Host).

Syndicated in the fall of 1994, the daily, hour-long *Judge for Yourself* was the Disney Corporation's contribution to the courtroom-TV field — and surely no one was more expert in the realm of litigation than the terrifyingly efficient legal team at Disney! Hosted by Bill Handel, a Los Angeles lawyer specializing in surrogate-parenthood cases, the series combined the the trial-drama and talk-show formats, with each episode set in an ersatz courtroom and twelve members of the studio audience serving on the jury. Once the "testimony" portion of the program had ended, the jurors withdrew to deliberate in full view of the cameras while the rest of the audience discussed the possible verdict, and viewers at home were encouraged to dial a 900 number to help sway the outcome. In most episodes, the jury foremen were portrayed by celebrities — and sometimes the legal issues invoked during testimony had direct relation to those celebrities' private lives! Unremarkable for the most part, *Judge for Yourself* enjoyed a brief burst of no-

toriety when, in an episode dealing with the as-yet-unresolved O.J. Simpson trial, the "jury" voted that Simpson was innocent.

Judge Hatchett

Syndicated: debuted Sept. 2000; in production until 2008. Sony Television. Executive Producer: Richard S. Goldman. Legal Producer: David Getachew-Smith. Cast: Judge Glenda Hatchett (Herself); Tom O'Riordan (Bailiff).

A graduate of the Emory University School of Law, Glenda Hatchett clerked in various U.S. Federal Courts before being hired as an executive in the legal and P.R. departments of Delta Airlines. Leaving Delta to accept a judicial appointment with the Fulton County (Georgia) Juvenile Court, Hatchett ultimately became the first African-American chief presiding judge of the Georgia State Court. As the founder of Atlanta's Truancy Invention Project, Judge Hatchett achieved nationwide fame with her "scared straight" approach of showing juvenile offenders what the future might hold by having them visit jail cells and homeless shelters; she also mentored young offenders after their release. "It won't be enough to hit the gavel and make a judgment," was Hatchett's credo. "It's more important that they understand the life lessons after the judgment ends."

Taped at the Metropolis Studios in New York City, the daily, half-hour syndicated series *Judge Hatchett* was staged in a family-court setting, with emphasis on juvenile cases. To keep apace of the many other "judge" shows on the air, Hatchett additionally handed down decisions to adult litigants, and also presided over small-claims cases ($3000 was the limit for cash awards). Distinctive elements of Hatchett's series included the Judge's predilection for "creative sentencing," such as forcing a deadbeat dad to wear a 30-pound weight around his abdomen so he'd know what it feels like to be pregnant; the daily "Intervention Segment," wherein Hatchett travelled all over the country to make miscreants take responsibility for their actions; and the generally optimistic "Aftercare" segments, following up previously televised cases and verdicts. When critics denigrated the series as tasteless and exploitational,

Judge Hatchett replied that she regarded her show as "educational," and its more lurid highlights as "reality checks."

The series' resident bailiff was Tom O'Riordan, who'd spent six years as a Court Officer for the city of New York — and who landed his TV job by answering a classified newspaper ad. The winner of the 2003 Prism Award for Best Unscripted Non-Fiction Series or Special for Television (this honor bestowed for an episode titled "Carrie's Out of Rehab"), *Judge Hatchett* retained its popularity for nearly a decade, especially with the crucial African-American demographic that constitutes the majority of daytime-TV fans. Even after production shut down in 2008, the series continued to be seen in rerun form in many markets.

Judge Jeanine Pirro

CW: Sept. 22, 2008. Telepictures/Warner Bros. Domestic Television. Executive Producer: Bo Banks. Executive consultant: Greg Mathis. Cast: Judge Jeanine Pirro (Herself).

Taped in Chicago, the daily, 60-minute *Judge Jeanine Pirro* was a vehicle for the former DA of Westchester Country, New York (the first female to hold that position in the County's history), who had risen to TV prominence as legal commentator for the Fox News Channel. Inasmuch as Judge Pirro was a Conservative who'd had the temerity to run against Hillary Clinton in a 2006 Senatorial election, the Liberal press was primed to savage the Judge's series the moment it made its first appearance in September of 2008. Her detractors spared her nothing, dredging up her well-publicized marital difficulties (culminating in a legal imbroglio when she was accused of planting an electronic bug on her husband), and emphasizing the fact that she had already tried and failed to achieve judge-show stardom as host of the never-released *Celebrity Jury* (see the entry for *Jury Duty* for a detailed account of this project's demise). But Jeanine Pirro was never one to wilt in the face of adversity, and when her new eponymous series finally aired, she more than lived up to the words of Telepictures' Hilary Estey McLoughlin, who described the Judge as "a powerful and dynamic television presence with a distinct point of view, and depth of professional life and experience."

Unfortunately, *Judge Jeanine Pirro* itself was not altogether worthy of its star's talent. Most of the series' small-claims cases were lurid in the extreme, with many of the litigants coming off like *Jerry Springer Show* rejects. The fact that during her years on the bench Pirro had specialized in domestic-abuse and sex-offense cases did not seem to prepare her for the shocking revelations made in her TV courtroom. In the very first episode, the Judge looked positively dumbstruck when a female plaintiff revealed that both she and her cousin had been raped by the same friend of the family — information that was provided as virtually an afterthought in an otherwise run-of-the-mill case about an unpaid rent bill. Small wonder the judge spent much of her time shouting "Let's back up a minute!" as the various litigants popped one surprise after another upon her. To her credit, Pirro diligently followed up on these startling revelations, demanding that previously closed cases be reopened (though not on *her* show) so that the guilty would not remain unpunished.

Despite the "surprise" factor inherent in *Judge Jeanine Pirro*, the series frequently smacked of contrivance. Whenever a plaintiff would make a particularly jaw-dropping statement, there would be an all-too-slick cutaway to Judge Pirro, whose response often sounded rehearsed: at times it appeared Pirro's scenes had been taped separately, rather than during the actual testimony. While this is only conjecture (the producers insisted that the show was totally unrehearsed), it is worth noting that Pirro seemed a lot more spontaneous in her Fox News Channel appearances than in her own series.

On the plus side, *Judge Jeanine Pirro* occasionally went "backstage," as it were, with Pirro taking the plaintiffs out of the courtroom and into her chambers so they could calmly hash out their differences in a less stressful setting. Also, in each of the three to four cases presented per episode, the viewers were provided with a detailed post-show followup, assuring them the Judge's actions had proven beneficial in the long run. And at episode's end, a "Pirro's Principles" segment allowed the Judge to offer savvy insights on the law in general and current high-profile cases in particular.

Unlike most other daytime courtroom shows of the period, *Judge Jeanine Pirro* was not syndicated, but instead telecast nationally in a 3–4 P.M. timeslot by the CW network. The series was part of a two-hour afternoon CW block, which also included reruns of the Prime Time sitcoms *The Jamie Foxx Show and The Wayans Brothers*. Given the fact that a goodly portion of the network's core audience consisted of young African Americans, it's a bit surprising to discover that during its first few months on the air, *Judge Jeanine Pirro* scored higher ratings than *Jamie Foxx* and *Wayans Brothers* combined!

Judge Joe Brown

Syndicated: debuted Sept. 14, 1998. Big Ticket Television/Worldvision (later CBS Paramount Television). Executive Producer: John Terenzio. Cast: Judge Joe Brown (Himself); "Miss Holly" Evans (Bailiff, 1998–2006); Sonia Montejano (Bailiff, 2006–); Jacque Kessler (Court Reporter).

Produced by the same team responsible for *Judge Judy* (q.v.)—and taped right across the hall from Judy Sheindlin's popular LA-based courtroom show—the daily, half-hour *Judge Joe Brown* was a vehicle for the "non-traditional, no-nonsense, no-holds-barred" jurist whose meteoric rise from the crime-ridden dregs of South Central LA is the stuff of which legends are made. Likening his childhood to the movie *Boyz in the Hood*, Joe Brown managed to resist gangland peer pressure, work his way through UCLA, land a respectable position as a member of the Equal Employment Opportunity Commission (Tennessee Division) in 1974, and enter private law practice at age 31 in 1978. He went on to become the first African-American prosecutor in the City of Memphis, and later was elected judge of the Shebly County State Criminal Courts, Division Nine. It was during this period he made national headlines (and raised a few national hackles) when he presided over the retrial of Martin Luther King's accused assassin James Earl

Ray at the behest of the King family, who was determined to prove the slain Civil Rights leader had actually been the victim of a government conspiracy. Making no secret of his belief that Ray had been set up, Brown was ultimately removed from the retrial petition amidst accusations of bias.

Joe Brown was still a sitting judge when, in 1998, he was tapped to star in his own daily, half-hour syndicated legal series, which focused primarily on small-claims cases (with a jurisdictional award limit of $5000). Once again Brown came under pressure, this time to follow the example of other "TV judges" by retiring from the bench to avoid any conflict-of-interest issues. Reluctantly admitting he'd been spending more time before the cameras than at his "day job," Brown finally agreed to step down in April of 2000; around the same time, he permanently relocated to Los Angeles (where the series was taped) to end the exhausting weekly shuttle to and from his home in New Mexico.

The dapper, mustachioed Brown tended to be a bit full of himself when presiding over his video courtroom, but none could deny his skill, intelligence and dedication. It was also hard to deny that he relished the role of "maverick judge," handing down unconventional sentences that were guaranteed to draw attention to himself: in the most famous example, he allowed a burglary victim to square accounts by robbing the burglar's house! Brown also took great personal pride in following up his previous rulings (especially when the culprit had managed to turn over a new leaf), and in helping troubled teenagers chart the proper course in life. For several seasons, *Judge Joe Brown* was the second most popular courtroom show in syndication (*Judge Judy* was the first), and Brown took full advantage of his celebrity by commiserating with other celebrities, notably rap artist Coolio—who appeared in one episode as a litigant—and actor Wesley Snipes, for whom Brown acted as a character reference in Snipes' tax-evasion trial.

Former El Paso (Texas) TV anchorperson Jacque Kessler was the series' court reporter, while appearing as bailiff were "Miss Holly" Evans and later Sonja Montejano. Like many another show of its kind, *Judge Joe Brown* was frequently telecast in an hour-long block consisting of one new episode and one repeat.

Judge Judy

Syndicated: debuted Sept. 16, 1996. Big Ticket Television/Worldvision (later CBS Paramount Television). Executive Producers: Peter Brennan, Randy Dothit, Timothy Regler. Cast: Judge Judy Sheindlin (Herself); Petri Hawkins-Byrd (Bailiff).

It could be said that two people were responsible for the recent resurgence of interest in syndicated courtroom shows: O.J. Simpson and Judge Judy Sheindlin. With the O.J. trial riveting viewers to their seats in 1995, Larry Lyttle, president of Big Ticket Productions, felt it was the right time to relaunch the daily "judge show" genre that had been more or less dormant since the cancellation of the original *People's Court* (q.v.) in 1991. What was needed for this undertaking was the right judge — and with Joseph Wapner having all but retired and Judge Lance Ito utterly disinterested in pursuing a TV career, Lyttle would have to look elsewhere.

He found what he wanted in the form of Brooklyn–born Judy Sheindlin (*née* Blum), who from 1972 until her retirement in 1996 had served as a prosecutor in New York's Juvenile and Family Court system. Violently opposed to the weeping-willow mindset that all youthful offenders were helpless victims of society who had no control over their actions, Sheindlin summed up her philosophy in the title of her first autobiography, *Don't Pee on My Leg and Tell Me It's Raining*. Though the civil-liberties crowd expressed outrage over Sheindlin's abrasive and insulting courtroom demeanor and her harsh sentencing, the Judge had scores of loyal supporters, many of them adult members of ethnic minorities who were fed up with being held in thrall by teenage thugs who'd twisted the System around their little fingers by whining, "It's not *my* fault!" and getting away with it. (Assuming that they'd even read Judge Judy's book, most of her detractors evidently hadn't bothered to get past the first few vitriolic chapters, else they would have realized Sheindlin ended her tome on a posi-

tive note, pointing with pride to the many youngsters whose lives she had turned around and who were now productive members of society.) Apprised of Sheindlin's existence by an in-depth 1993 *Los Angeles Times* article and her subsequent appearance on CBS' *60 Minutes*, Larry Lyttle was instantly captivated by the diminutive dynamo: "She reminded me of my mother. But with an edge."

"Justice with an Attitude," promised Big Ticket to potential buyers when *Judge Judy* entered syndication in September of 1996. The series trafficked in small-claims cases that had originally been filed in real courtrooms. For a typical two-week program cycle, the show's staff contacted 500 litigants by phone, from which 40 were brought into the studio for interviews. If their stories held up and they were still confident they could win, they appeared on the show — forfeiting the right to appeal, agreeing to a jurisdictional award limit of $5000, and going along with the show's policy of using the money to establish a "fund," with either the winner or the loser getting the bigger half depending on Judge Judy's whim. They also went in with eyes wide open regarding the Judge's rigid insistence that the litigants were never allowed to interrupt her no matter how often she interrupted them — and with full awareness of her Hellenic temper, volcanic outbursts, and ceaseless invocation of such caustic catchphrases as "Is 'stupid' written on my forehead?" Though the taping sessions were long and grueling, Judge Judy took it all in stride, observing that she heard fewer cases on TV per week than she had in an actual court on a single day. (Even so, she admitted that it took several taping sessions for her to feel "100 percent comfortable" with her new job.)

In light of the series' enduring popularity, it's difficult to remember that *Judge Judy* was not regarded as a sure-fire success when it first came out. Many of the major local stations took a pass on the property, writing off Sheindlin as an "unknown" and the court-show format as passé. So Big Ticket Productions all but gave the show away, cutting prices to the bone while retaining only three 30-second ad slots in each episode, and shipping it to second-echelon stations in smaller markets. Although the budget for a week's worth of *Judge Judy* was half the cost of a single network sitcom episode, the show lost money during its initial season on the air. But thanks to word of mouth, *Judge Judy* slowly but steadily built up a following. Its ascension can be charted by a perusal of New York City's TV listings during the ten-year period beginning in 1996: starting as a virtual throwaway on UPN affiliate WWOR-TV, the series was picked up for a more advantageous timeslot by CBS flagship WCBS, eventually settling into a choice 4–to–5 P.M. daily berth (two episodes run back-to-back) on NBC–owned WNBC — amassing huge ratings before returning to WCBS in 2006. Three years into her run, Judge Judy was generating $75 million in revenue for Big Ticket, her ratings doubling with each successive year. When she signed a new four-year contract in 2005, Sheindlin's $100 million paycheck was second only to Oprah Winfrey as the highest amount ever paid a woman in American daytime television. A 1999 *Fortune* magazine article on *Judge Judy* told the whole story in its title: "The Little Judge Who Kicked Oprah's Butt: Daytime Television's Hottest Property."

Not long before (temporarily) losing *Judge Judy* to WNBC, WCBS' vice president and station manager Steve Friedman summed up Sheindlin's appeal for *New York Times* correspondent James Barron: "She does a little bit mother, a little bit teacher, a little bit crazy aunt, and there is a lot of don't-mess-with-me, Bill. It's the kind of thing where she has the center and the magnetism to make you believe you better listen to her. Viewers, in this time of political correctness and nobody taking a stand, like people with something to say." Sheindlin herself has offered her own assessment of her TV persona. "I may be wrong, but you're not going to misconstrue what I said," she told Joel Stein of *Time* magazine. "Why do I have to use polysyllabic explanations when a single syllable will do it — 'No,' 'Wrong'? If I have to I use two — 'Stupid.'"

Though it goes without saying that the success of *Judge Judy* has begotten a plethora of robe-and-gavel shows with the likes of Joe Brown, Glenda Hatchett, Greg Mathis, Alex Ferrer, Maria Lopez, David Young and Jeanine Pirro, we'll say it anyway. It can also be assumed that Judge Judy's popularity caused the rival series *People's Court* to bump its resident judge, former New York governor Ed Koch, in favor of Judy's husband Jerry Sheindlin, a one-time Bronx judge. "I want him to become accustomed to the fact that No. 2 is very good," Judy coyly remarked on the topic of her husband's temporary rise to TV stardom. For the record, Ed Koch, who ironically had appointed Judy to NYC Family Court in 1982, was no great fan of his protégée's series — nor was Koch's *People's Court* predecessor Joseph Wapner, who castigated Judy as "insulting in Capital Letters." And despite her series' enormous following (enormous enough to insure that the show will remain in production at least until 2012), there remains to this day a loud and virulent faction, among them such heavyweight law professors as Harvard's Alan Dershowitz and USC's Erwin Chemerinsky, who'd prefer that Judy cease and desist her high-pitched histrionics, her litany of personal insults and her P.C.–unfriendly tirades against "liberal morons," and retire to her home in Greenwich, Connecticut.

One group of observers who have always loved Judy — or at the very least loved to hate her — is the fraternity of TV Comedy Writers, who have inserted so many mean-spirited "Judge Judy" soundalikes into their various sitcoms and sketch shows that it is impossible to catalogue them all. The best known of the Judy Sheindlin knockoffs was *Saturday Night Live* regular Cheri Oteri, whose spot-on impersonation of the Judge enlivened many an otherwise mediocre comedy bit in the years between 1996 and 2000. In the tradition established by Sammy Davis Jr. and Janet Reno, and more recently upheld by Hillary Clinton, Judy Sheindlin brought down the house by confronting her *SNL* alter-ego Oteri on the episode of October 17, 1998. Incidentally, Judge Judy (who claimed to find Oteri "adorable") was disappointed that her cameo appearance was scripted in advance and not the "surprise" that she wanted it to be. Evidently, the Judge was too preoccupied with her own show to understand that there are *never* any surprises on *Saturday Night Live*.

Judge Karen

Syndicated: debuted Sept. 8, 2008. Sony Pictures Television. Executive Producers: Richard Goldman, Susan Sobocinski-Puchert. Cast: Judge Karen Mills-Francis (Herself); Christopher Gallo (Bailiff).

One of three new daily "judge" shows unveiled in September of 2008, *Judge Karen* was produced by Rich Goldman, who at that time was supervising several similar series, including *Judge David Young* (q.v.). In fact, it was Judge Young, who hailed from the same Miami, Florida, jurisdiction as Judge Karen Mills-Francis, who recommended her to Goldman as the next rising judicial star.

Formerly a public defender for underprivileged adults and minors, Karen Mills-Francis was in 2000 appointed an administrative judge, the second African American woman ever seated on the bench in Miami-Dade County. Passionately devoted to children's causes and child-advocacy programs, Judge Karen was not only a foster mother, but also a mentor to dozens of at-risk youngsters; in addition, she regularly called upon the lawyers who appeared in her courtroom to act as guardians for kids from abusive families. For all that, Karen was known to be quite harsh with youthful offenders who evinced unwillingness to take responibility for their actions. As often as she could, she made certain that these wayward kids did not use their first courtroom experience as the launching pad for an "endless cycle of dependency" by seeing to it that her rulings, no matter how severe, would ultimately have a positive effect on their lives. "I'm like the mother everyone should have had," proclaimed Judge Karen in a 2008 interview.

Each half-hour episode of *Judge Karen* began with the words, "Justice isn't always black and white." Indeed, there weren't many courtroom series around that could lay claim to being more "colorful": A black judge, with blonde hair, garbed in a red robe, seated before a sky-blue backdrop! While several of the small-claims cases brought before Mills-Francis en-

abled her to plead the cause of children's rights, many others dealt with women who felt they'd been mistreated or betrayed by the men in their lives. Surprisingly, the series' mood was not entirely somber and serious: Judge Karen had a very lively sense of humor, and was given to such amusingly sarcastic homilies as "God protects babies and fools—and you're no baby." And whenever a litigant tried to out-shout the judge or take over the proceedings, Karen quickly put the upstart in his place by declaring, "Stay in your lane—I can drive!"

Judge Karen introduced several innovations to the judge-show genre. For example, the witnesses were sequestered so that one person couldn't play off another's testimony, as occasionally happened on such rival series as *Divorce Court*. Also, the series was the first courtroom show to feature telestrators, those devices used by sportscasters to draw football plays on the TV screen: on *Judge Karen*, the telestrator was deployed to help the litigants visually chart their testimony in order to clarify matters. And at the end of each episode, an "Ask Judge Karen" segment allowed the Judge to answer legal questions sent in on video by the viewers.

Taped at the Metropolis Studios in New York City, *Judge Karen* was not only "stripped" in syndication during its inaugural season, but also telecast daily over the Black Entertainment (BET) cable service.

Judge Maria Lopez

Syndicated: Sept. 11, 2006–Sept. 6, 2008. Sony Television. Executive Producers: Richard S. Goldman, Michael Rourke. Cast: Judge Maria Lopez (Herself); Pete Rodriguez (Bailiff)

One of a handful of Hispanic judges enjoying American TV prominence in the first decade of the 21st century (see also *Cristina's Court* and *Judge Alex*), Maria Lopez was like her contemporary Judge Alex Ferrer a refugee of Castro's Cuba, arriving in the U.S. at the age eight and learning to speak fluent English within three months. After graduation from Boston Law School, Lopez established herself as a champion of the underdog in a variety of high-profile civil cases. Having served as the assistant to the Massachussetts attorney general, Lopez became in 1988 the first Latina appointed to the Massachussetts bench, and two years later the first person of Latin origin on the state's Supreme Court. In the light of her Liberal political leanings, it was ironic that Lopez was forced to resign the bench by the "P.C. Police" for refusing to apologize for alleged judicial misconduct after convicting a transgender defendant of sexual assault.

In 2006, Lopez inaugurated her own half-hour syndicated courtroom series, from the same production staff responsible for the long-running *Judge Hatchett* (q.v.). In true judge show tradition, Lopez was match with a good-looking and personable bailiff, in this case Pete Rodriguez, a New York City fireman who had previously worked as a tech-crew member on several other Manhattan–based series.

Remaining undaunted by the threat of protests from special-interest groups, Maria Lopez continued to advocate for the rights of the oppressed — even those who'd been oppressed by people on the "correct" (read: left-of-center) side of the argument. "If you can't stand the heat, get out of the courtroom!" was Judge Lopez' motto, and it must have struck a chord with viewers: within a month of its debut, *Judge Maria Lopez* was earning higher ratings than any other new syndicated offering. Unfortunately, the series was unable to sustain this early momentum, and was cancelled after two seasons.

Judge Mathis [*aka* Judge Greg Mathis]

Syndicated: debuted Sept. 29, 1999. Syndicated Productions/TelePictures/Warner Bros. Television. Executive Producers: Gus L. Blackmon, Alonzo Brown, Vicangelo Bulluck, Bo Banks. Cast: Judge Greg Mathis (Himself).

Judge Greg Mathis brought a unique perspective to his daily, hour-long courtroom series: Not only was he able to appreciate the Justice system from the bench, but he could also fully understand (if not necessarily condone) the viewpoint of the defendant — having so often been on the wrong side of the law in his youth.

As a gang member and heroin dealer in the

mean streets of Detroit, Mathis had done plenty of time in Juvenile Detention before he was 17, and for all he knew was destined for more of the same as an adult. All this changed abruptly when Wayne County circuit court judge Charles Kaufman gave young Greg a choice: get a G.E.D. or go to jail. At the same time, Mathis found out that his mother was dying of cancer; rushing to her side, he made a solemn promise to turn his life around. With the help of Affirmative Action he launched his academic career at Eastern Michigan University, where he got his first taste of social activism by heading a "Free South Africa" voter registration. In 1983 he graduated from the University of Detroit Law School, but at first was barred from practicing law because of his criminal record. He did, however, work on the staff of Detroit City Councilman Clyde Cleveland, and in 1988 became manager of the Detroit Neighborhood City Halls, a position he held through the auspices of Mayor Coleman A. Young. During this same period, Mathis and his wife established Young Adults Asserting Themselves, a non-profit agency designed to help young people ages 17 through 25 find good jobs and pursue career opportunities.

At age 35, Greg Mathis was elected judge of Detroit's 36th district, an event that received coast-to-coast coverage. "When I ran for judge in '95," he recalled in a 2000 TV interview with Larry King, "my opponents brought up my youth criminal background, my juvenile delinquency background in an effort to sidetrack my election, and it backfired. And I was swept into office beating a 20-year incumbent by 10,000 votes. And it made national headlines that day. The headlines read 'Former Street Youth Becomes Youngest Judge in Michigan.' And Hollywood came calling. They thought it was something that the rest of the country needed to know." Hollywood, in the form of Warner Bros. Television, did indeed approach Mathis to star in his own syndicated TV series, a professional move that required him to quit the bench because Michigan law did not allow a judge to pursue a second vocation.

Referring to the lineup of stations carrying his Chicago–based program as "The Empowerment Network," Judge Greg Mathis promised to deliver "real people with real disputes before a real judge delivering real justice." Each episode featured two to four small-claims cases, with the litigants shown arguing both before the bench and in chambers. There was no nonsense and no bull tolerated in Mathis' TV courtroom: street-smart and street-tough, he had little patience for the lame excuses of people who held everyone liable for their problems but themselves. Withal, he was a strong believer in rehabilitation, even if it meant verbally beating the daylights out of those in need of rehabilitating. And despite the harsh and cutting words dispensed by Mathis in the course of events, he was possessed with a quick and ready wit, frequently breaking the tension in his courtroom with a joke that instantly brought everything into perspective. The end of each episode featured a "Mathis Moment," in which the Judge proselytized about a responsibility-related issue.

Playing to good if not spectacular ratings, *Judge Mathis* has been a steady performer since its debut in the fall of 1999. In 2002, the same year Mathis hosted Detroit's Self-Empowerment Expo, the series earned a Prism Commendation for a special, non-court-related episode about teenagers and drunk driving. Greg Mathis' remarkable life story — and for once the adjective is not casually applied — has been chronicled in his own book *Inner City Miracle*, the outgrowth of a biographical musical written in 1997 by playright Ron Miller.

Judge Mills Lane

Syndicated: Aug. 17, 1998–Aug. 31, 2001. Hurricane Entertainment Corporation/Rysher Entertainment (later Paramount Television). Executive Producers: John Tomlin, Bob Young. Cast: Mills Lane (Himself).

Descended from a long line of Georgia bankers, Mills Lane forsook the family business to become a Marine rifle instructor, a welterweight boxer, an Olympic contender, a lawyer, and ultimately a Nevada district court judge headquartered in Reno. It wasn't until 1979 that the 43-year-old Lane — aka "Maximum Mills" because of the severity of his sentences —

refereed his first world championship boxing match, and then strictly as a hobby to relieve the monotony of his legal activities. Eighteen more years would pass before the gravel-voiced, pitbull-faced Lane achieved national prominence as referee of the infamous rematch between world heavyweight champ Evander Holyfield and challenger Mike Tyson — better known as "The Bite Fite" because of Tyson's strenuous efforts to chew off Holyfield's ear, earning him a disqualification from the unamused Mr. Lane. It wasn't long before the producers of the satirical MTV animated series *Celebrity Deathmatch* asked Lane to provide the voice for his clay-model namesake, who ref'ed such bogus wrestling matches as "Monica Lewinsky vs. Hillary Clinton" and "Chris Rock vs. Adam Sandler." Other byproducts of Lane's overnight fame included his autobiography *Let's Get It On* (the phrase he always used to begin a professional bout), and, perhaps inevitably, his own daily TV courtroom series. It was because of this last-named venture that Lane stepped down from the bench in Nevada — not because of "conflict of interest," as is often the case in such matters, but because he didn't think he had enough time to pursue both a judicial and a showbiz career.

Taped at WPIX-TV in New York, *Judge Mills Lane* was in many respects a typical example of its genre, with Lane presiding over small-claims cases for which a $3000 jurisdictional limit had been imposed. What set it apart from the rest of the courtroom shows was Mills Lane himself, who, though he claimed not to be as "strict" as rival TV jurist Judy Sheindlin (see *Judge Judy*) was nonetheless a fierce and frightening presence — even more so to the home viewers, especially when the camera zoomed in on Maximum Mills' mug as he chewed out a particularly irksome litigant. Whenever Lane began shaking his gavel at a plaintiff or defendant, you could be sure all hell would soon break loose; indeed, on more than one occasion, the bailiff would be forced to clear the courtoom in the roughneck manner of a nightclub bouncer. Lane would sometimes let loose with so rapid a verbal barrage that no one had any idea what he was talking about — but they sure knew what he *meant*.

Ratings for *Judge Mills Lane* were never anything to brag about, but somehow the series managed to hang on for three years; reportedly, the only reason it was cancelled was because viewers were repelled by the new Season Three theme song. Reruns of *Judge Mills Lane* later popped up on the testosterone-soaked cable service TNN (now known as Spike TV).

Judge Roy Bean

Syndicated: debuted Oct. 1955. Quintet Productions/Screencraft Pictures. Executive Producers: Russell Hayden, Fred Franks. Theme song by Hal Hopper. Cast: Edgar Buchanan (Judge Roy Bean); Jack Beutel (Jeff Taggert); Jackie Loughery (Letty Bean); Russell Hayden (Steve, a Texas Ranger).

So much nonsense has been reported about the estimable "Judge" Phantly Roy Bean — and so many of the more famous aprocryphal yarns can no longer be repeated because of their viciously racist content (the Chinese and the Jews get a particularly rough time of it) — that it is worth recounting a few of the known facts before discussing the syndicated 1955 TV series inspired by Bean's rather spotty judicial career.

Born in Kentucky in 1825, the rapscallion Roy Bean cut quite a swath through the midwest before settling in Texas in 1862, where he opened the first of several saloons designed to service the many railroad companies which were then expanding Westward. Although the TV version of *Judge Roy Bean* was set in the town of Langtry, Texas, in the 1870s, the real Bean wasn't appointed Justice of the Peace of Pecos County until 1882, and he didn't establish the town of Langtry (named for world-renowned British actress Lily Langtry, whom Bean had never seen in person but nonetheless worshipped from afar) until 1883. It was here that he set up the Jersey Lily, a combination saloon and courthouse — and it was here that Bean began referring to himself as "The Law West of the Pecos." Armed with a single tattered, outdated lawbook, the self-styled Judge picked his juries from his regular customers, who were obliged to buy drinks during every

recess. There was no jail in Langtry: those whom Bean found guilty were heavily fined and chased out of town, but contrary to legend few of the defendants were hanged (even horse thieves were able to escape the noose if they returned their ill-gotten gains and greased the Judge's palm). Re-elected Justice of the Peace several times, Bean was finally voted out of office in 1896, though he defiantly continued to levy sentences and extort fines all up and down the Tex-Mex border. He died in 1903, allegedly suffering a heart attack while playing pool but actually while sleeping off an inordinately lengthy binge. Among the actors who have portrayed Judge Roy Bean in films are Walter Brennan and Paul Newman, neither of whom was inclined to allow the facts to get in the way of a good story.

Edgar Buchanan, who in his pre–*Petticoat Junction* days spent most of his time in Walter Brennan-ish character roles, landed the part of TV's Judge Roy Bean not long after wrapping up a lengthy tenure as Hopalong Cassidy's sidekick Red Connors. Since Buchanan's series was geared as much for children as for adults, Judge Bean was cleansed and sanitized from top to toe. He was still the self-appointed "Law West of the Pecos," but now operated out of a general store rather than a gin mill, and his verdicts were dictated by a sincere love of justice rather than dreamed up on the spot to fatten the Judge's coffers. In short, Bean went from rascally reprobate to lovable old codger in 39 easy episodes. Adding to the softening process was the presence of the Judge's pretty and virtuous niece Letty, a character created from whole cloth and played by former Miss U.S.A. (and future Mrs. Jack Webb) Jackie Loughery. And though Judge Bean was an excellent rider and sure shot (as proven in the opening title sequence of each episode), he was given a strapping young assistant in the form of Jeff Taggert, played by Jack Beutel with the same stolid stoicism he'd brought to the role of Billy the Kid in the much-ballyhooed Jane Russell "bosom western" *The Outlaw* (1941).

Occasionally appearing as a Texas Ranger named Steve was B–western leading man Russell Hayden, another former Hopalong Cassidy sidekick who served as *Judge Roy Bean*'s executive producer. The series was lensed some 130 miles east of Hollywood in Pioneertown, California, an all-purpose facility for Western filmmakers. The area boasted several permanent exterior sets, which doubled as living quarters for actors and crew members filming on location; among the other series produced at Pioneertown were *The Cisco Kid, The Gene Autry Show* and *Annie Oakley*. The 35 acres homesteaded by Russell Hayden for *Judge Roy Bean* were purchased outright by the producer and renamed the Hayden Ranch, which in later years became a tourist attraction.

With TV western fans of the 1950s preferring deeds over words, the legal aspects of *Judge Roy Bean* understandably played second fiddle to the hard-ridin', fast-shootin' action scenes. Though cheaply assembled, the series remained in circulation well into the 1960s thanks to the producers' foresighted decision to film the episodes in color.

Judging Amy

CBS: Sept. 19, 1999–May 3, 2005. Barbara Hall-Joseph Stern Productions/ CBS Productions/ 20th Century–Fox Television. Executive Producers: Amy Brenneman, Barbara Hall, Joseph Stern, Connie Tavel, James Frawley, Karen Hall, Carol Barbee, Alex Taub, Hart Hanson, Richard Kramer. Cast: Amy Brenneman (Amy Gray); Tyne Daly (Maxine Gray); Karle Warren (Lauren Cassidy); Dan Futterman (Vincent Gray); Richard T. Jones (Bruce Van Exel); Jessica Tuck (Gillian Gray); Marcus Giamatti (Peter Gray); Jillian Armenante (Donna Kozlowski-Pant); Brent Sexton (Oscar Ray Pant) Wendy Makkena (Susan Nixon); Samantha Shelton (Evie Martell); Jeana Lavardera (Lisa Matthews); Timothy Osmundson (Sean Potter); Sara Mornell (Carole Toby); Richard Crenna (Jared Duff); Reed Diamond (Stuart Collins); Alice Dodd (Kimberly Fallon); Kevin Rham (Kyle McCarty); Nia Long (Andrea Salamon); Chris Sarandon (Judge Barry Krumble); Kathryne Dora Brown (Zola Knox); Kristin Lehman (Dr. Lily Reddicker); Sarah Danielle Madison (Dr. Heather Labonte); Inny Clemons (Robert Clifton); Gregory Harrison (Tom Gillette); Blake Bashoff (Eric Black); Columbus Short (Thomas McNab); Cheech Marin (Ignacio Messina); Adrian Pasdar (David McClaren); Jay R. Ferguson (Todd Hooper); Crawford Wilson (Victor McClaren); Jennifer Esposito (Crystal Turner).

The CBS legal drama *Judging Amy* was very much a labor of love for its star Amy Brenneman; the series was inspired by the career of the actress' mother, trailblazing Connecticut judge Frederica Shoenfield Brenneman. After obtaining her undergraduate degree at Radcliffe, Frederica became a member of the first Harvard Law School class to admit women, graduating in 1953. Fourteen years later, she was the second woman in Connecticut history to be appointed a state judge, specializing in juvenile law. It was truly a case of on-the-job training: she'd had no appellate or trial experience, nor even a working knowledge of the juvenile court system, but managed to land the appointment through her college credentials and her connections. And though not required to wear judicial robes in courtroom, she did so anyway to project the same air of authority as her not-always-supportive male colleagues. As it happened, Frederica's 32-year career as a judge began at precisely the same time the U.S. Supreme Court ruled that juveniles were entitled to Constitutional due process, thus doubling the caseload of the Connecticut juvenile court and keeping "Judge Freddy" (as she was known to her staff) busy on a near round-the-clock basis. Though she could not be described as a Bleeding Heart — indeed, her impatience with habitual offenders and her acerbic "gallows humor" became legendary in courtroom circles — Frederica was a tireless advocate for the rights of children and families, unafraid to take up arms against a rule-bound, politically-motivated System that didn't seem to truly care about its charges and allowed too many unfortunate youngsters to slip through the cracks. When the Connecticut state trial courts merged in 1978, Frederica was promoted to Superior Court judge, using her position to broaden legal protections for kids caught up in abuse and neglect cases, and to train other judges and caseworkers, especially young women interested in law enforcement.

Because all Juvenile Court sessions were closed, Amy Brenneman never had a chance to see her mother in action while she was growing up, though she thought her mom's job was "cool" despite its long hours (Amy decided to become an actress in order to work evenings so that her mother had time to watch!). The idea for *Judging Amy* began to take root when Amy and her husband Brad Silberling made a video of the Hartford, Connecticut, courthouse on the occasion of Frederica's 70th birthday in 1996, interviewing many of her mom's colleagues. Having long since left her starmaking role as Officer Janice Licalsi on the ABC cop drama *NYPD Blue,* Amy was eager to return to television in a series that would emulate the best of *Blue* with a huge cast of interactive characters and with multiple story arcs rather than single linear plotlines; she also wanted a series that would serve as a tribute to her mother. Though warned by industry friends that a show dwelling upon the serious and tragic aspects of Juvenile Court might be "too depressing," Amy argued that the meat of the concept was the "tough moral choices made one step at a time, and the wins along the way ... a good soup." Even so, when the initial pilot script for *Judging Amy* proved to be a bit too intense for her tastes, Amy hired Barbara Hall, former co-executive producer of the legal drama *I'll Fly Away* (q.v.), to lighten up the concept and make the characters less iconic. "[Hall] has a twisted point of view that I love," Amy informed *TV Guide.* "She writes characters that are too smart for themselves. They talk, talk, talk—'Oh, we're sooo smart'—but they're totally messed up." Coincidentally, around the same time that *Judging Amy* was in development, Judge Frederica S. Brenneman was working on a TV concept of her own involving a Child Protective Services team led by "an aging, foul-mouthed lady judge of about 70." She mailed her idea to daughter Amy—who promptly lost it.

Judging Amy took considerable dramatic license with the events of Frederica Brenneman's life and career, ladling on huge dollops of soap opera-like twists and turns that bore scant relation to the facts. Amy Brenneman conceded this, adding "I play my mother's job, not my mother." Also, it was hard not to notice the similarities between *Judging Amy* and the popular NBC multigenerational drama *Providence,* which revolved around a female plastic surgeon who forsook her lucrative Hol- lywood practice and returned to her Rhode Island home town to work at a free clinic and also take charge of her large, dysfunctional family. Though no one could accuse Amy Brenneman of plagiarism—her series was already in the planning stages when *Providence* premiered in early 1999—*Entertainment Weekly*'s Ken Tucker was on the money when he observed that were it not for the surprise success of the NBC series, *Judging Amy* may not have gotten on the air. "I say this because prior to *Providence,* prime time hadn't had an hour-long, female driven drama that was a ratings success since *Dr. Quinn, Medicine Woman,* and the hour drama format was heavily titled toward either newsmagazines or male-dominated ensemble dramas such as *Law & Order* [q.v.] and *ER.*"

Brenneman was cast as 34-year-old Amy Madison Gray, a New York corporate attorney. After the bitter breakup of her marriage to Michael Cassidy (played first by John Slattery and later by Richard Burgi), Amy gave up her practice, moving herself and her 6-year-old daughter Lauren (Karle Warren) back to her home town of Hartford, Connecticut. There she signed on as a Family Court judge, where after a rocky start she quickly proved ideally suited to her new position. Faced with a stunning array of extraordinary cases—in one early episode, a man demanded custody of his child because his ex-wife had become a Wiccan— Amy admirably maintained her equanimity and common sense, seldom allowing her emotions (amply on display *outside* the courtroom) to overwhelm her judgment, but still exhibiting more compassion towards her charges than was customary in the sterile, soulless Juvenile Law system. She also never let anyone forget who was in charge in her courtroom: if, say, a lawyer wanted to use her to intimidate a youthful witness into revealing the name of the man who had taken pornographic pictures of him, Amy refused to do it merely at the lawyer's behest— even though she agreed that the pornographer was a scumbag—but cooperated only after a careful examination of all the facts at hand, and then in a way that was perfectly legal and above-board and did not further scar the young witness' psyche. But though she played scrupulously by the rules, Amy could be a bit unorthodox on occasion, as when she performed a joyous "unmarriage" ceremony to assure the children of a divorcing couple that their parents still loved them. To be sure, there were times of duress when Amy allowed her personal problems to seep into her work, especially at the outset of the series' second season when she blamed herself for the suicide of a woman to whom Amy had denied custody of her child, and in Season Five when, after suffering a miscarriage, she took a morbidly obsessive interest in a troubled young girl. But these were aberrations in the overall scheme of *Judging Amy*.

When she relocated to Hartford, Amy moved into the home of her pugnacious, iron-willed mother Maxine (Tyne Daly), a unreconstructed 1960s activist employed as a social worker at the Department of Children and Family Services (CFS), where she regularly butted heads with her traditionalist boss Sean Potter (Timothy Osmundson). Judge Frederica Brenneman has gone on record stating that she herself more closely resembled Maxine Gray than Amy Gray—this despite the barbs of TV critics who condemned Maxine as "an infantilizing rag with superior air," perpetuating "stereotypical ageist sexism." True, Maxine ran roughshod over daughter Amy and granddaughter Lauren, forever passing harsh judgments on their failings and eternally griping about having two more mouths to feed. Still, one sensed that she secretly enjoyed her "Earth Mother" status, forever providing comfort and shelter for the various physically battered and emotionally bruised youngsters with whom she came in contact at CFS, and intensely feeling their pain when the System failed them.

Among the "strays" taken under Maxine's wing at one time or another were Amy's two brothers, Peter (Marcus Giamati) and Vincent (Dan Futterman). A successful attorney, Peter was married to the high-strung Gillian (Jessica Tuck), whose strident behavior was dictated by her inability to bear children. Early on, Peter and Gillian agreed to take care of pregnant teenager Eve (Samantha Shelton) in exchange for permission to adopt the girl's baby—and

this being a legal series, viewers knew well in advance that Eve would wind up battling with the couple over the child's custody. Later, Gillian nearly died giving birth to a son who turned out to have a serious mental disorder—namely a genius-level IQ that made it impossible for the kid to get along with his peers. The pressures of this situation, coupled with an extramarital dalliance, ended up driving Gillian and Peter apart, whereupon Peter moved back in with dear old mom Maxine, much to her (alleged) dismay. Younger brother Vincent could be described as the black sheep of the family, an erstwhile writer who, though he managed to pen a Pulitzer Prize–winning novel, was never really able to fit in anywhere. Vincent also precipitated many of the series' most devastating dilemmas, not only by loving several ladies neither wisely nor well (though he finally found happiness with Crystal Turner [Jennifer Esposito], his coworker at an outreach program for troubled teens), but also by almost getting killed on two separate occasions—most spectacularly when he was nearly blown to bits in Amy's courtroom by a disgruntled ex-friend.

Another lost soul to whom Maxine opened her doors was her substance-abusing nephew Kyle (Kevin Rham), who during his clean-and-sober periods tried to follow in his aunt's Public Service footsteps, first by working at the Teen Harbor halfway house for youngsters with addiction problems, then as a resident at financially strapped St. Michael's Hospital, where he met and fell in love with Dr. Heather Labonte (Sarah Danielle Madison), who bore him a child. Also taken in by Maxine was Eric Black (Blake Bashoff), a troubled gay youth who ended up saving Amy's life by killing a dangerous stalker (Jack Noseworthy).

As can be seen, *Judging Amy* was certainly not lacking for melodrama, and at times the characters' emotional turmoil so overshadowed the legal aspects that one wondered if the series' title referred to Amy's "judging" profession or to how her actions were *being* judged by everyone around her (especially Maxine). An awful lot of screen time was devoted to Amy's tempestuous love life, beginning with her acrimonious relationship with her ex-husband

Michael, who on no fewer than two occasions made her life miserable by demanding full custody of daughter Lauren. (He finally backed off when his *second* wife walked out on him, and who could blame her?) Amy's later gentlemen callers included Tom Gillette (Gregory Harrison), who dropped out of her life almost as quickly as he dropped in, and attorney Stuart Collins (Reed Diamond), whose political beliefs were the polar opposite of Amy's, but who nonetheless won her heart when he agreed to defend Eric pro-bono during the boy's murder trial. Ultimately Amy and Stuart became engaged, but she left him standing at the altar as the result of last-minute attack of nerves, exacerbated by Stuart's confession that children (especially Lauren) made him nervous. Amy's final beau was Assistant DA David McClaren (Adrian Pasdar), a widower with two children who was unable to reciprocate Amy's affections until he achieved closure regarding the murder of his first wife. Accepting David's proposal upon discovering that she was pregnant with his child, Amy drifted away from him after suffering a miscarriage, and the series ended before their problems could be resolved.

While all this was going on, Amy's coworkers likewise had their own tribulations with affairs of the heart. The Judge's loyal and supportive Court Services Officer Bruce (Richard T. Jones) lost his job defending the honor of his ex-junkie girlfriend Andrea (Nia Long), then regained it through the efforts of feisty activist attorney Zola (Kathryne Dora Brown) with whom he subsequently fell in love — only to fall out of love when he and Zola clashed over their contradictory views of the Justice System. Meanwhile, starry-eyed court clerk Donna (Jillian Armenante) impulsively wed a jailed felon named Oscar (Bruce Sexton), subsequently bearing him a child (given the astronomical birth rate on *Judging Amy*, it's hard to believe that Hartford is only the 45th largest city in the nation). The flighty Donna can be regarded as the comedy relief flip-side of Amy: she spent so much of the series procrastinating over giving up her clerical job to study law that Amy finally forced the issue by firing the girl and ordering her to start shopping for a good school.

Some of the more fascinating of *Judging Amy*'s romantic entanglements involved the series' *most* fascinating character, Amy's widowed mom Maxine. Allegedly, the producers decided halfway through the first season to soften the hard edges of Maxine's personality by giving her a boyfriend (an allegation, incidentally, that Tyne Daly has always denied). Her first serious relationship was with Jared Duff (Richard Crenna), who, knowing Maxine's instinctive hatred of the Rich and Powerful, tried to cover up the fact that he was a multimillionaire businessman. Even after the cat was out of the bag, Jared refused to reveal that he was the mysterious benefactor who financed Maxine's pet project "Sanctuary House," which brought social workers, investigators and child therapists together under one roof. Overcoming her aversion to Big Money, Maxine finally agreed to marry Jared, but 48 hours before the wedding she received word that he had suddenly died while wrestling with a business crisis in Asia (this plot twist was written to acknowledge the real-life death of Richard Crenna on January 17, 2003). As she struggled to pick up the pieces of her shattered happiness, Maxine was drawn to philosphical landscaper Ignacio Messina (Cheech Marin). Alas, this romance came to a screeching halt when Maxine found out that Ignacio hadn't bothered to divorce his estranged wife. The shock of this discovery, combined with the disappearance of a young client who'd been betrayed by the System, led to Maxine having a nervous breakdown, followed by two heart attacks — thus providing Maxine a brand-new reason for sniping at her children when they began treating her as a helpless invalid rather than a still-worthwhile individual.

Getting back to the legal elements of *Judging Amy* (and about time!), the series' writing staff began de-emphasizing the characters' personal travails during the 2001–2002 season, when *Amy* was scheduled opposite the grittier ABC courtroom drama *Philly* (q.v.) The producers promised that this season would be "darker" and more realistic, as manifested in the political squabbles between Amy and attorney Stuart Collins. This, however, did not mean the series would degenerate into an issue-of-the-week tract: Amy Brenneman still remained less interested in specific legal topics, and more in "the human toil this work takes, day after day." Most of the series' actors were anxious to explore issues of social significance, but Amy warned, "it's a slippery slope when it becomes about 'messages.' I'm an actor, not a soapbox girl. I'm interested in portraying what goes on [in juvenile court] with dignity, intelligence and fairness...." Be that as it may, the series became increasingly politicized during the 2003–2004 season, when Judge Amy Grey was transferred from Juvenile to Criminal Court — a move apparently motivated by NBC's decision to counterprogram *Judging Amy* with *Law & Order: Special Victims Unit* (q.v.) During this season, Amy was constantly at odds with her judicial superiors, who, governed more by political pressure than the interests of justice, began imposing inflexible rules and limitations on both lawyers and litigants. In particular, Amy was appalled by the procrustean "Welfare to Work" program, which made it all but impossible for single parents to hold their families together and seek out employment at the same time.

Back in Juvenile Court for the series' sixth and final season, Amy took up a new cause: blocking the passage of retrogressive laws that would result in treating children like adults in the courtroom, subjecting even the youngest defendants to disproportionately severe penalties. To this end, and despite the by-now-obligatory bitching and moaning of mom Maxine, Amy stepped down from the bench and ran for the U.S. Senate.

Mixed reviews notwithstanding, *Judging Amy* was an instant hit when it premiered in the fall of 1999, ending its freshman season as CBS' highest-rated new dramatic series. Its cancellation after a six-year run in 2005 was due not to a loss of popularity, but because CBS felt the show was "skewing old" and therefore not attracting younger viewers. The series continued chalking up respectable numbers in reruns on cable's TNT service until 2007, but a much-anticipated DVD release has reportedly been held up indefinitely because of royalty issues involving the musical selections heard on the series' soundtrack.

The Jury

Fox: June 8–Aug. 6, 2004. 20th Century–Fox Television. Creators/Executive Producers: Barry Levinson, Tom Fontana, James Yoshimura. Music by Blue Man Group. Cast: Billy Burke (John Ranguso); Jeff Hephner (Keenan O'Brien); Shalom Harlow (Melissa Greenfield); Anna Friel (Megan Delaney); Adam Busch (Steve Dixon); Cote de Pablo (Marguerita Cisneros); Patrice O'Neal (Adam Walker); Barry Levinson (Judge Horatio Hawthorn).

Fox's *The Jury* was a fascinating but flawed effort from Barry Levinson and Tom Fontana, the same team responsible for such landmark TV series as *Homicide: Life on the Street* and *Oz*. Set in a New York City courtroom, each hour-long episode was devoted to a single homicide trial. Though the actors playing the attorneys, the judge, the bailiff and the other courtroom personnel remained the same each week, a different 12-person jury was chosen for every trial. For the most part, the action unfolded from the jury's point of view, with the events leading up to the trial dramatized in flashback form as the jurors recalled the testimony of the litigants and witnesses. However, the viewers at home were permitted to watch the lawyers plea-bargaining and conferring with the judge behind closed doors; and once the verdict was announced, a final flashback showed what had *really* happened, allowing the viewers to decide for themselves if justice had been served. No wonder that critics described *The Jury* as a combination of *Twelve Angry Men*, *Law & Order* and *Rashomon*.

The prosecution team consisted of John Ranguso (Billy Burke) and Keenan O'Brien (Jeff Hephner), while acting for the defense were Melissa Greenfield (Shalom Harlow) and Megan Delaney (Anna Friel). Their characterizations were straightforward and non-declamatory; conversely, legal intern Marguerite Cisneros (Cote de Pablo) came off as a naïve chatterbox, while bailiff Steve Dixon (Adam Busch) was a mass of quirks and eccentricities. Originally, the part of Judge Horatio Hawthorne was to have been played by film director Sidney Lumet, no stranger to the courtoom-drama *ouevre* as witness *Twelve Angry Men*, *The Verdict*, *Find Me Guilty* and the TV series *100 Centre Street* (q.v.) But when Lumet was incapacitated by slipping on a piece of ice, he was replaced by series producer Barry Levinson, who appeared *sans* screen credit. And in classic Hollywood tradition, the weekly "guest jury" was jam-packed with the sort of familiar TV and movie character actors whom audiences invariably recognize by face if not always by name. One episode, "Last Rites," reunited a dozen cast members from the Levinson-Fontana prison series *Oz* as the jurors in the trial of a convict accused of murdering a jailhouse priest.

The series did a commendable job showing how the attitudes brought in by the individual jurors might affect the outcome of the trial: one woman can't bring herself to send a teenager to jail as an adult, another has a built-in animosity for juvenile offenders, and so on. Also, *The Jury* demonstrated how extraordinary events both inside and outside the courtroom determine the verdict, notably in the episode "Bangers," wherein the Judge requests the attorneys to arrive at a quick plea agreement upon learning that gang members have intimidated the jurors in the trial of two drug dealers. Where the show frequently fell apart was in its script contrivances, with some trials following a narrative structure that would be nearly impossible to adhere to in real life, and others simply getting things wrong: the episode "Lamentation of a Reservation," about a murder solicited in the casino of an Indian reservation, holds its trial in a State court, even though the law stipulates that such a case must be handled in Federal Court or by the reservation's tribal council. Equally detrimental was the corny, ham-handed expositional dialogue: when one juror demands to know which of her peers has stolen her deodorant, another juror snaps, "We've got a bigger mystery than that to solve. The murder of Craig Sheridan." (Cue the organ!). Finally, the series was plagued with distractingly shaky camerawork, which was evidently supposed to intensify the drama but ended up giving some viewers a bad case of *mal-de-mer.*

As part of their preshow promotion for *The Jury*, the Fox network came up with a state-of-the-art audience participation gimmick. A few minutes before the end of each episode, viewers with wireless phones were invited to call in their verdict, then keep watching to see how the rest of America voted. "Viewer Verdict" turned out to be a singularly pointless exercise, since the disclaimer baldly stated that it was merely "an opinion poll for entertainment purposes only and has no effect on the outcome of the television program." What *did* have an effect the program's ultimate outcome was widespread viewer indifference. Initially airing on Tuesday evenings with a rerun on Friday, *The Jury* performed so dismally in the ratings that the weekly telecasts were quickly trimmed from two to one — and after ten episodes, from one to none.

Jury Duty

Syndicated: debuted Sept. 17, 2007. Radar Entertainment/Foster-Tailwind Entertainment. Creator/Executive Producer: Vincent Dymon. Executive Producers: Linda Dymon, Susan Winston, Diana Foster, Buddy Fisher. Cast: Bruce Cutler (The Judge).

The daily, syndicated *Jury Duty* might well have been your basic, garden-variety courtroom series, replete with an acerbic judge presiding over genuine small-claim cases with a $5000 award ceiling, but for a unique twist: the jurors on the series were all show-business celebrities — and their verdicts were binding.

Technically speaking, the show's judge wasn't really a judge but instead a prominent New York criminal lawyer: Bruce Cutler, who had risen to fame with his defense of "Teflon Don" John Gotti, and who had relinquished the opportunity to represent music mogul Phil Spector in his murder trial to appear on *Jury Duty.* (Cutler was an eleventh-hour replacement for Amy Dean, likewise no judge but instead an author and labor leader.) Each half-hour episode featured three celebrities in the jury box, all of whom had actually been subpoenaed to appear (we were shown videotaped proof of this, though their reactions when the process servers "surprised" them at the doorstep *did* seem a wee bit rehearsed). "Don't even try to get out of this one!" was the series' catchphrase, and indeed the celebrities never tried to duck their civic duty by claiming they were too busy to serve on a jury. Avoiding the churlish observation that some of the luminaries chosen for *Jury Duty* had very little to keep them busy, we respectfully submit a partial list of the stellar jury pool: Debbie Reynolds, Bernie Kopell, Phyllis Diller, Todd Bridges, Paula Poundstone, Dick Van Patten, Bruce Jenner, Tiffany, Kevin Sorbo, Charlene Tilton, Shadoe Stevens and Lee Meriwether.

As "Judge" Cutler listened to testimony from plaintiff and defendant, he continually hectored them with kidding-on-the-square comments about their problems, never passing up an opportunity to take a swipe at the overall mindset of Southern California. As the testimony proceeded, the camera kept cutting away to the jurors, who reacted in pop-eyed astonishment or valiantly attempted to stifle giggles. Then it was the jury's turn to ask the litigants a few questions, which they did with courtesy and restraint, though a few of the professional comedians on the panel could not be prevented from having their little jokes. In the final third of the episode, we saw the jurors deliberating behind closed doors, for the most part taking their responsibilities very seriously. Once the case had been settled, the three celebs made their final comments to the camera, sometimes verbally, sometimes with amusing sound effects or exaggerated facial expressions. Throughout the episode, an off-camera announcer would invite the home viewers to compete for fabulous prizes by answering some trivia questions about the celebrities-of-the-day. The viewers were also encouraged to submit their own small-claims cases for consideration, the better to experience the rush of getting "Justice with a Hollywood Ending."

The most surprising aspect of *Jury Duty* was that the show wasn't as bad at it sounded: Ray Richmond of *The Hollywood Reporter* spoke for many when he described the program as "shockingly watchable." Apart from that, the series' most fascinating aspect was its gestation. *Jury Duty* was the brainchild of wealthy casino entrepreneur Vincent Dymon, whose Radar Entertainment production firm was headquar-

tered in the basement of the Four Seasons Hotel in Westlake Village, California. Despite his prominence in gambling and hotel circles, Dymon had virtually no show business contacts—though his wife Linda, a nurse, actually did have one whole friend in the Industry. Shopping *Jury Duty* all by himself, Dymon was unable to get anyone in Hollywood to answer his calls until word leaked out that he'd sunk $8 million of his own money into the pilot episode: only after the first sale was made did Dymon discover that it was usually the distributor and not the producer who ponied up the preshow cash.

By January of 2007 *Jury Duty* was a "go"— and then Dymon encountered an obstacle that might have stopped the project in its tracks. It seemed that Warner Bros. Television was preparing its own first-run syndicated courtroom series called *Celebrity Jury*, in which a panel of *five* famous jurors would deliberate small-claims litigation on a daily basis (this series' "judge" was to have been former New York DA Jeanine Pirro, who went on to host her own daily courtroom series in 2008 [see separate entry]). Instantly, Dymon filed suit against Warner Bros., claiming the studio had swiped his idea, further alleging that Bruce Cutler had been approached to preside over *Celebrity Jury* after the Warners executives had seen him as a judge on the pilot of "another" reality TV program that had been floating around. Warners also reportedly assured Cutler that the rival series would "never get distribution" and that the studio "was not afraid" of the other show's producers. "David" Dymon's response to "Goliath" Warners was: "Be afraid … be *very* afraid." In an amazing turn of events—all the more so because the "little guy" seldom prevails against the Hollywood Studio Oligarchy—Warners withdrew *Celebrity Jury* from the marketplace, leaving the field free and clear for *Jury Duty*.

So whether you loved the show or hated it, it was impossible not to respect the symbolic significance of *Jury Duty*: a rare and gratifying example in which a tiny mom-and-pop operation ultimately triumphed over the big-box superstores.

Just Cause

PAX: Sept. 15, 2002–May 18, 2003. Minds Eye Pictures. Creator/Executive Producers: Gail Morgan Hickman, Jacqueline Zambrano. Executive producer: Kevin DeWalt. Cast: Lisa [Elizabeth] Lackey (Alexandra "Alex" DeMonaco); Richard Thomas (Hamilton Whitney III); Shaun Benson (Patrick Heller); Roger R. Cross (C.J. Leon); Mark Hildreth (Ted Kasselbaum); Khaira Ledeyo (Peggy Tran); Jason Schombing (DA Kaplan).

Thought you could slip one over on us, didn't you? Alexandra "Alex" DeMonaco, the heroine of the 60-minute legal drama *Just Cause*, was *really* Erin Brockovich with a brunette wig, more modest clothes and a criminal record.

Played by Australian actress Lisa Lackey (who as Elizabeth Lackey later costarred on the NBC sci-fi opus *Heroes*), Alex DeMonaco had spent five years in prison after her husband, a crooked attorney, framed her for embezzling $5 million, then vanished along with her two-year-old daughter Mia. Vowing never to be anyone's victim ever again, Alex studied law while behind bars, and upon her release headed to San Francisco (which looked a lot like Vancouver), where she applied for a job with the firm headed by successful, well-connected and highly regarded civil lawyer Hamilton Whitney III (Richard Thomas, a long *long* way from John-Boy Walton), whose prior career consisted of helping the Rich and Powerful stay that way. A Harvard–educated stuffed shirt whose priggish conservatism had already cost him three wives, Whitney was initially turned off by the brash and outspoken Alex, but was won over by her thorough knowledge of the law and her intense empathy for the Underdog. Though as a convicted felon Alex was not allowed to practice law herself (and in fact had to work nights at a cleaning service to supplement her income), nothing stopped her from giving Whitney the benefit of her vast expertise — nor was Whitney able to prevent Alex from dragging him into controversial cases he would otherwise have avoided like the plague.

In the two-part opener, Alex coerced Whitney into going after an Enron–like megacorporation whose CEO might have committed murder in order to cover up his illegal business practices. Later cases found Alex accusing the U.S. Army of chemical posioning during Operation Desert Storm, representing a grieving mother who claimed an experimental medicine developed by a major drug firm had rendered her son comatose, tackling a powerful college administration that was covering up the facts behind a football star's "suicide," and targeting the sleazy promoter of a potentially deadly reducing formula. As Whitney continued working in tandem with Alex, he found himself more and more drawn to the cause of judicial activism, reawakening a sense of moral outrage that had lain dormant since he'd given up the practice of criminal law several years earlier. Similarly, Alex proved to be an inspiration to her hard-bitten parole officer C.J. Leon (Roger R. Cross), a former public defender who'd lost the calling after a crushing professional setback. As for Whitley's associate Patrick Heller (Shaun Benson), he enjoyed keeping company with Alex because she was so gosh-darned cute. This fact may have also motivated ex-computer hacker Kasselbaum (Mark Hildreth) to step up his efforts to locate Alex's beloved missing daughter, and to accumulate enough new evidence to clear her name. Conversely, Alex aroused nothing but hatred in the heart of DA Kaplan (Jason Schombing), the incompetent clod who'd sent her up five years earlier and who was eager to find an excuse to throw her back in the slammer for keeps.

Though *Just Cause* could not be mistaken for reality television — this much was obvious in the opener, which implied Alex was the only white person in the entire prison system — the 22-episode series was an agreeable way to spend an hour each week. Filmed in Canada, where the pilot episode aired exactly one year before the series' official debut, *Just Cause* was seen in the United States over the PAX Network, which before its metamorphosis into the infomercial-dominated ION network was briefly committed to providing original, family-oriented entertainment.

Just Legal

WB: Sept. 19–Oct. 3, 2005, Aug. 20–Sept. 10, 2006. Jerry Bruckheimer Television/Warner Bros. Television. Creator/Executive Producer: Jonathan Shapiro. Executive Producers: Jerry Bruckheimer, Jonathan Littman. Cast: Don Johnson (Grant Cooper); Jay Baruchel (David "Skip" Ross); Jaime Lee Kirchner (Dulcinea "Dee" Real); Susan Ward (Kate Manat).

For every TV–series hit like *CSI* or *Cold Case*, producer Jerry Bruckheimer had a multitude of misses like the WB network's *Just Legal*. The failure of this effort must have been particularly galling for Bruckheimer in that he'd shelled out good coin to engage the services of two hot leading actors: Don Johnson, a proven TV commodity vis-à-vis *Miami Vice* and *Nash Bridges*, and Jay Baruchel, who'd soared to stardom as the mentally challenged wannabe prizefighter Danger Bartch in the Oscar–winning *Million Dollar Baby* (2004).

Top billing went to Johnson as Grant Cooper, once a prominent and powerful attorney, who after bollixing up a high-profile case and losing his client to Death Row rapidly spiralled downhill, drowning his sorrows in liquor and ending up in a crummy little office in one of the seedier corners of Venice, California. Embittered over his inability to get any work other than as a court-appointed attorney, Cooper might have sunk further into drunken obscurity had he not taken on a new junior partner in the form of David Ross (Baruchel), a 19-year-old legal prodigy (no, this isn't an inaccuracy: believe it or not, the cutoff age for lawyers in California is eighteen). Nicknamed "Skip" because he'd bypassed so many classes in high school and college, Ross still lived at home with his parents (Julie Warner and Rafael Sbarge) and his average-brained brother Tom (Reily McClendon), and thus was ignorant enough in the ways of the Outside World to retain his childlike idealism. Unable to land a position with any of the major legal firms because of his age and inexperience, Skip Ross made a beeline to his lifelong idol Grant Cooper, who agreed to hire him only because he worked cheap. Following TV's factory-tested, doctor-approved "redemption" pattern, Cooper was revitalized by Skip's boundless enthusiasm, and together the two partners devoted themselves to representing the downhearted and downtrodden in what the bigger legal firms had dismissed as hopeless cases. Also in the cast were Jaime Lee Kirchner as Cooper's explosively unpredictable ex-client Dee Real, a recent parolee who still wore her electronic ankle bracelet and who labored away as Cooper's secretary to pay off her legal costs; and Susan Ward as Kate Manat, a onetime law-school classmate of Skip's who still had a crush on him even though she was now employed by a rival firm.

Just Legal was created by longtime Bruckheimer associate Jonathan Shapiro, a genuine lawyer and court reporter with 10 years' practical experience, whose other legal-show credits included *The Practice* and *Boston Legal* (both q.v.) Describing his new series to *TV Guide*, Shapiro said, "You go into law thinking it's all guys like Gregory Peck in *To Kill a Mockingbird* and you get Paul Newman in *The Verdict*." The requisite eight hour-long "trial" episodes were commissioned by WB, but only three had aired when the network realized the series was attracting "too old" an audience: the average viewer was 50 or older, not at all the sort of demographic desired by a network trafficking in hip black sitcoms and teen-angst dramas. Also, the WB brass declared there were too many "geeky" characters on *Just Legal* to suit their tastes. The upshot of all these negative vibes was a quick cancellation, with the remaining five episodes discreetly burned off in the dog days of August.

Justice

NBC: Apr. 8, 1954–March 26, 1956. Talent Associates-John Rust Productions. Produced by Herbert Brodkin and David Susskind. Cast: Dane Clark (Richard Adams, 1954–55); William Prince (Richard Adams, 1955–56); Gary Merrill (Jason Tyler); Westbrook Van Voorhis (Narrator).

Dramatizing actual cases from the files of the National Legal Aid Society, the NBC series *Justice* was co-produced by a young David Susskind (see *Witness*) and future *Defenders* (q.v.) guiding hand Herbert Brodkin. The property was introduced as a one-off episode of *ABC Album*, a series comprised of unsold pilots, on April 12, 1953: featured in the cast was actress Lee Grant, in one of her last TV appearances before she was blacklisted.

Launching its NBC run as a spring replacement for *The Life of Riley* in 1954, *Justice* was in its early months a straight anthology, with no regular cast and a different case and/or lawyer per week. The stories, which focused on the work done by the Legal Aid Society to provide proper representation for those unable to afford attorneys, generally dealt with civil and criminal cases. One of the few existing kinescopes of *Justice* comes from this era: originally telecast live on May 20, 1954, the episode "An Eye for an Eye" stars Richard Kiley as an ex–POW who goes off the deep end when the fellow prisoner (Skip Homeier) whom he believes betrayed him to their Korean captors is defended in court by the Society and completely exonerated. Vowing to administer his own brand of justice, Kiley spends the rest of the episode dogging the trail of the alleged turncoat. Though it is presumptuous to judge an entire series based on a single example, "An Eye for an Eye" (directed by Daniel Petrie) indicates that *Justice* was a slick and polished effort, seamlessly combining its live interior scenes with filmed exterior shots and effectively harnessing the subjective-camera technique for full dramatic impact (specifically in a lengthy monologue delivered directly to the camera by supporting player Harvey Lembeck).

Eventually, the anthology format was dropped and Gary Merrill joined the cast as Legal Aid Society attorney Jason Tyler, with Dane Clark as fellow attorney Richard Adams. Later on Merrill left the series and Richard Adams became the principal character, with William Prince replacing Dane Clark in the role (ironically, Merrill and Prince had previously costarred on another legal series, ABC's *The Mask* [q.v.]). The offscreen narration was provided by Westbrook Van Voorhis, legendary "voice of doom" in the radio and newsreel versions of *The March of Time*.

Like most other New York–based drama series of the period, *Justice* was a showcase for the talents of future stars and established character actors. George Grizzard once told TV historian Jeff Kisseloff, "I lived off David Susskind. On *Justice*, Bill Prince would get me out of jail at least once a month." Others appearing on the series included E.G. Marshall, Ed Begley Sr., Ben Gazzara, Jack Klugman, Rod Steiger, Bob Cummings, June Lockhart, Eva Gabor and Jackie Cooper. Singer Gisele MacKenzie guested in the single most famous *Justice* episode, "Hard to Get," originally telecast May 12, 1955. In this exposé of crooked song pluggers, Ms. MacKenzie performed the episode's title song, which spawned a best-selling record thanks to her repeated renditions of the tune on her own NBC series, *Your Hit Parade*.

Justice

Fox: Aug. 30–Dec. 22, 2006; Dec. 21, 2007. Jerry Bruckheimer Television/Warner Bros. Television. Executive Producers: Jerry Bruckheimer, Jonathan Littman, David McNally, Jonathan Shapiro, Tyler Bensinger. Cast: Victor Garber (Ron Trott); Kerr Smith (Tom Nicholson); Eamonn Walker (Luther Graves); Rebecca Mader (Alden Tuller).

As a successful purveyor of procedural TV cop dramas like *CSI, Cold Case* and *Without a Trace*, producer Jerry Bruckheimer took second place to no one. It was, however, a different story with the lawyer-drama genre: Bruckheimer's 2005 series *Just Legal* (q.v.) had been one of the producer's most conspicuous and embarrassing failures, getting the ax after a scant three episodes. It was hoped that his next courtroom effort, *Justice*, would eradicate the bitter aftertaste of *Just Legal*— and when the new series premiered over the Fox Network on August 30, 2006, it appeared at first that Bruckheimer had scored one of his biggest-ever hits. At first.

Created by former attorney and court reporter Jonathan Shapiro, *Justice* (working title: *American Crime*) zeroed in on the California law firm of Trott-Nicholson-Tuller, better known in legal circles as "T.N.T." The Dream Team to end all Dream Teams, T.N.T. specialized in high-profile cases involving celebrity clients. Senior partner Ron Trott (Victor Garber) was in fact a celebrity in his own right, a near-permanent fixture on the TV talk show circuit. Second-in-comand Tom Nicholson (Kerr

Smith), a sly and subtle shark who assumed a self-effacing, regular-fellow attitude in the courtroom, was a past master at charming and manipulating juries and jurists alike: it was not for nothing that Nicholson was nicknamed "The American Face of Not Guilty." Finally there was Alden Tuller (Rebecca Mader), young, female, drop-dead gorgeous, and a brilliant researcher and cross-examiner. And though his name was omitted from the "T.N.T." escutcheon, the firm's chief advisor, ex-prosecutor Luther Graves (Eamonn Walker), was an invaluable member of the team, not only because of his valuable connections, but also because of his uncanny ability to examine each case from the viewpoint of both defense and prosecution, enabling his colleagues to anticipate and second-guess the prosecution's strategy to the tiniest detail.

Utilizing a technique previously touched upon by the short-lived Barry Levinson/Tom Fontana courtroom drama *The Jury* (q.v.), each episode was highlighted by a high-tech, dizzily paced reconstruction of the crime, as envisioned by T.N.T. in order to map out their defense. Only after the verdict had been registered did the viewer see how the crime *really* happened — and sometimes T.N.T.'s theory turned out to be dead wrong, and it was not unknown for the Guilty to escape unscathed. In keeping with the breathless pace of these weekly reenactments, the rest of *Justice* zipped along with the same helter-skelter rapidity, swooshing the viewers from one location to another in the blink of an eye, and representing the passage of time by having the characters' wardrobe suddenly change in mid-scene! One critic accurately summed up the series as "*CSI* at warp speed."

Bruckheimer's confidence in *Justice* would seem to have been rewarded when, on the occasion of its premiere, the series posted the evening's highest ratings, commanding the attention of 8.9 million viewers. Alas, what happened next can be encapsulated in a line from Shakespeare's *Hamlet*: "How all occasions do inform against me." Production on *Justice* was briefly held up by an industry strike, providing Fox with virtually nothing to follow up its spectacular debut. Also, the series was forced

on hiatus by the network's coverage of the 2006 World Series, and when it returned, most viewers had gotten over the excitement of that inaugural episode. Add to this the fact that the series' level of writing began to deteriorate the moment production resumed, with the lawyers' unbroken winning streak and the climactic "what really happened" gimmick quickly growing repetitive and predictable, and you have a classic example of an early front-runner that simply couldn't finish the race. A desperate effort to save the series by moving it from Wednesday to Fox's successful Monday-night lineup succeeded only in losing the audience built up by its lead-in *Prison Break*. Habitually ranking a poor fourth in the nightly network ratings, *Justice* was finally removed from view after 12 episodes, with a 13th unexpectedly popping up in December of 2007, one year after the series' cancellation.

Kate McShane

CBS: Sept. 10–Nov. 12, 1975 (previewed as a 90-minute TV movie on April 11, 1975). Paramount Television. Executive Producer: E. Jack Neuman. Cast: Anne Meara (Kate McShane); Sean McClory (Pat McShane); Charles Haid (Father Ed McShane).

With a title like *Kate McShane* and with actress Anne Meara in the lead, we know one thing for sure: this CBS TV series was not about a Lithuanian family. What it *was*, was the first prime time network drama series to top-bill a female lawyer. Not that total parity with the male legal establishment had been achieved: though Meara's Kate McShane was proudly single, fiercely independent, and a two-fisted courtroom adversary, she still betrayed a soft-and-sentimental streak for her more unfortunate clients, and tended to become emotionally invested in their problems. (Ain't that just like a dame?) The product of a boisterous Irish family, Kate could always rely upon her father Pat (Sean McClory), a retired cop, to do her legwork and round up recalcitrant witnesses, and her brother Ed (Charles Haid), a Jesuit priest and law professor, to handle the legal research and guide her through the more problematic points of law and ethics.

Typical of its era, *Kate McShane* drew

much of its story material from current events: in the opening episode, Kate defended a former antiwar activist who'd gotten mixed up with a band of young urban terrorists, most of whom turned out to be confused, misguided and willing to change their ways (a startling attitude, considering that the series debuted not long after the Patty Hearst kidnapping). But for all its topicality, and despite the novelty of a woman in the leading role, the series was dismissed by critics as just another lawyer drama, and a slow and plodding one at that. Viewers didn't care one way or another, since most of them were tuned into the rival ABC adventure series *Starsky and Hutch*. But before CBS pulled the plug on *Kate McShane* after a meager seven hour-long episodes (with two remaining unshown), the series managed to earn Anne Meara an Emmy nomination, and also featured the acting debut of 10-year-old Ben Stiller, son of Meara and her comedian husband Jerry Stiller.

Kaz

CBS: Sept.10, 1978–Apr. 22, 1979 (pilot episode telecast Apr. 14, 1978). Lorimar Productions. Created by Ron Leibman and Don Carlos Dunaway. Executive Producers: Lee Rich, Sam Rolfe, Marc Merson. Cast: Ron Leibman (Martin "Kaz" Kazinski); Patrick O'Neal (Samuel Bennett); Linda Carlson (Katie McKenna); Gloria LeRoy (Mary Parnell); Mark Withers (Peter Colcourt); Edith Atwater (Illsa Fogel); George Wyner (DA. Frank Revko); Dick O'Neill (Malloy).

The first of TV's "jailhouse lawyers," Martin "Kaz" Kazinski (Ron Leibman) earned his law degree while serving prison time for stealing furs, jewels, and cars. (You *really* didn't think an American TV series in 1978 would feature as its main character a convicted murderer or pedophile, did you?) Despite the fact that most states will not allow ex-convicts to practice law, Kaz was hired as a junior partner by Samuel Bennett (Patrick O'Neal), senior partner of the prestigious L.A. firm of Bennett, Rheinhart and Quist — at a rock-bottom salary, of course. Though Kaz cut an unprepossessing figure with his baggy clothes, unkempt hair and business cards printed in smeary wet ink, and while his hotheaded personality brought him dangerously close to contempt charges

every time he stepped into a courtroom, the onetime jailbird had the advantage of thoroughly understanding and appreciating the defendant's point of view, serving him well in "underdog" cases in which the cards seemed stacked against his client. Also, his years behind bars had taught Kaz some valuable survival skills, notably the ability to manipulate and hoodwink such stern authority figures as judges and prosecutors, with persnickety DA Frank Revko (George Wyner) his most frequent opponent in the latter category. (Though a stereotypical character, Revko was an honest and ethical one, at one point actually siding with Kaz when both of them came up against a mentally deranged judge.) Other continuing characters included Kaz' court-reporter girlfriend Katie McKenna (Linda Carlson); his old pal Mary Parnell (Gloria LeRoy), owner of the nightclub that housed Kaz' second-floor apartment, and where Kaz sometimes sat in as a jazz drummer (star Leibman was quite skilled with the "sticks"); Malloy (Dick O'Neill), the series' "Huggie Bear"-style street informant; Mrs. Ilsa Fogel (Edith Atwater), Samuel Bennett's no-nonsense chief secretary; and Peter Calcourt (Mark Withers), the law firm's conformist junior partner.

Though *Kaz* represented Ron Leibman's second-only series TV assignment (he'd recently turned down *Starsky and Hutch*), the actor wielded considerable clout on the set. After all, he'd co-created the show and co-written the pilot with Don Carlos Dunaway, and could even lay exclusive claim to the protagonist's name: Leibman had previously portrayed a cop named Kaz in the made-for-TV movie *A Question of Guilt*. An admitted control freak with a reputation for being difficult, Leibman hoped his all-consuming dedication to *Kaz* would result in big numbers for the series. In an interview with *TV Guide*'s Dick Russell, Leibman said, "The reason I hope the show is appealing is because we all do come from different prisons — marriages we don't like, jobs we don't like, schools we don't like. Yet people don't really want to change their lives. Well, this guy Kaz *does*. I hope people will root for him." Then, as if to explain away his *own* bursts of temperament, Leibman added, "Kaz is a man who, had he not been angry, never would've gotten in the trouble he did originally…. Kaz's story is that of an angry young man who changed, not by magic but by rehabilitation. And so did I."

While *Kaz* was undoubtedly a source of personal satisfaction for Ron Leibman, who earned an Emmy nomination for his performance, it was unable to overcome the stiff competition of ABC's *Sunday Night Movie*, and after 22 hour-long episodes — and a brief shift to an equally untenable timeslot opposite *Vega$*— the series was cancelled.

Kevin Hill

UPN: Sept. 29, 2004–May 19, 2005. Icon Productions/Touchstone Television. Creator/Executive Producer, Jorge Reyes. Executive Producers: Bruce Davey, Alex Taub, Nancy Cotton. Cast: Taye Diggs (Kevin Hill); Jon Seda (Dame Butler); Patrick Breen (George Weiss); Christina Hendricks (Nicolette Raye); Kate Levering (Veronica Carter); Michael Michele (Jessie Grey).

A vehicle for dazzlingly handsome African-American leading man Taye Diggs (previously of *Ally McBeal*, and more recently a regular on *Grey's Anatomy* and its spinoff *Private Practice*), the 60-minute legal dramedy *Kevin Hill* cast the star as a hip, high-powered 28-year-old entertainment lawyer with oodles of disposable income and a roving eye for the ladies, who worked for the New York megafirm of Davis, Dugan & Kelly. Kevin Hill's affluent bachelor lifestyle was permanently altered after his no-account cousin was electrocuted while trying to illegally tap into cable TV. Feeling that he'd betrayed his late cousin by not offering financial support, Kevin was now obliged to look after the dead man's 10-month-old daughter Sarah. The pressures of instant daddyhood — diapers, bottles et al. — not only put a crimp in Kevin's social life, but also cut deeply into his business activities. Desperate for enough flex time to properly care for baby Sarah, Kevin quit Davis, Dugan & Kelly and ended up at the much smaller firm of Grey & Associates, controlled and staffed entirely by women. Naturally, his hedonistic chauvinism ran counter to the atti-

tudes of his new boss Jessie Gray (Michael Michele), herself a single mom, and his two coworkers: Nicolette Raye (Christina Hendricks), whose brilliant and incisive legal mind was hidden behind a deceptively demure veneer, and Veronica Carter (Kate Levering), the firm's resident "drama queen," who happened to be one of Kevin's ex-lovers. Kevin's first case with Grey & Associates was a challenge to his cool ladykiller status: he was assigned to go up against his old firm by prosecuting a baseball superstar accused of sexual assault. Much to everyone's surprise, including his own, Kevin's courtroom summation read in part, "How do you get a guy — a privileged, protected guy — who attacks women to change his behavior? You make it so expensive for him that he doesn't have a choice."

This new, enlightened Kevin Hill was a boundless source of amusement and disdain for his wisecracking best bud Dame Butler (Jon Seda), who considered Kevin a fool for placing his family duties above his personal desires. Additionally, both Kevin and Dame were forced to keep their homophobic impulses in check due to the presence of little Sarah's imperious gay nanny George Weiss (Patrick Breen).

While the series was in development, *Kevin Hill* was planned as the story of a white Jewish lawyer. The hiring of Taye Diggs, coupled with the UPN network's ongoing pursuit of a young and predominantly black viewer demographic, dictated the ethnic change. Once on board, Diggs was enthusiastic about *Kevin Hill*, though he worried that the concentration on baby Sarah would dull the series' edginess; the actor felt the show took an upward swing with an intense story arc wherein Kevin was challenged for the custody of Sarah. Other potent plotlines involved a woman being sued by her mom for making trust-fund investments on the advice of a psychic; another woman charging her boss with sex discrimination for firing her, only to find that he'd actually wanted to marry her but was terrified that *he'd* be fired as a result; and most memorably, Kevin's coworker Veronica suing a contraceptive manufacturer after becoming pregnant by Kevin's pal Dame Butler! (Incidentally, the resolution of each case was chock full of last-minute surprises — which to some observers suggested that for all their legal acumen, Kevin and his associates hadn't really done their homework.)

Though not an enormous hit by traditional network–TV standards, *Kevin Hill* chalked up some of UPN's best ratings, with network president Dawn Ostroff boasting "double- and triple-digit time period gains in our core demographics" (having *America's Top Model* as a lead-in didn't hurt matters). Hopes ran high that the series would be renewed after its first 22-episode season, but such was not to be. Designed to appeal to the 18 to 34-year-old crowd, *Kevin Hill* was dropped in the spring of 2005, during a period of reorganization at UPN when its execs decided to start angling for a wider audience. The faint possibility that *Kevin Hill* would be revived during the 2006–2007 season was scuttled when UPN and rival hookup WB were folded into the new CW network.

Kid's Court

Nickelodeon: Sept. 1988–Jan. 1994. Cast: Paul Provenza (Moderator); Nobi Nakanishi, Asha Canalos (Sketch Artists); Nicola Stewart (Prefect).

A component of Nickelodeon's "Cable in the Classroom" service, *Kid's Court* offered a crash course in judicial procedure to a target audience of 8- to 13-year olds. Chosen from the members of the studio audience, the preteen plaintiffs and defendants engaged in simulated trials and hearings, precipitated by disagreements with parents and friends (on one episode, a youngster sued his mom for throwing away his *TV Guide* collection, which to our way of thinking is a hanging offense). Although comic actor Paul Provenza served as moderator, the actual arbiter of the cases at hand was a bewigged mechanical device, "Presiding Honorable Judge O. Meter," which arrived at a verdict by gauging the amount of applause earned by the plaintiffs. Throughout the proceedings, a pair of sketch artists drew funny pictures of the participants, and the audience was quizzed about the real-life legal system. Each half-hour episode of *Kid's Court* ended with a "Sound-Off," in which selected kids would relate instances of perceived injustice in their own lives, while the audience let loose with loud and lusty choruses of "Unfair! Unfair!"

L.A. Law

NBC: Oct. 3, 1986–May 19, 1994. Steven Bochco Productions/20th Century–Fox Television. Created by Steven Bochco and Terry Louise Fisher. Executive Producers: Steven Bochco, David E. Kelley, John Masius, John Tinker, William M. Finkelstein, Rick Wallace. Cast: Richard A. Dysart (Leland McKenzie); Alan Rachins (Douglas Brackman Jr.); Corbin Bernsen (Arnie Becker); Jill Eikenberry (Ann Kelsey); Michael Tucker (Stuart Markowitz); Harry Hamlin (Michael Kuzak); Susan Dey (Grace Van Owen); Jimmy Smits (Victor Sifuentes); Michele Greene (Abby Perkins); Susan Ruttan (Roxanne Melman); Joanna Frank (Sheila Brackman); Ellen Drake (Elizabeth Brand); Patricia Huston (Hilda Brunschwager); Cynthia Harris (Iris Hubbard); Blair Underwood (Jonathan Rollins); Larry Drake (Benny Stulwicz); Dann Florek (David Meyer); Joyce Hyser (Alison Gottlieb); Nancy Vawter (Dorothy Wyler); Diana Muldaur (Rosalind Shays); Jennifer Hetrick (Corrine Hammond); Dennis Arndt (Jack Sollers); Vincent Gardenia (Murray Melman); Renée Jones (Diane Moses); John Spencer (Tommy Mullaney); Amanda Donohoe (Cara Jean "C.J." Lamb); Sheila Kelley (Gwen Taylor); Cecil Hoffmann (Zoey Clemmons); Conchata Ferrell (Susan Bloom); Tom Verica (Billy Castroverti); Michael Cumpsty (Frank Kittredge); Anthony DeSando (Alex DePalma); A Martinez (Daniel Morales); Lisa Zane (Melinda Paros); Alan Rosenberg (Eli Levinson); Debi Mazar (Denise Ianello); Alexandra Powers (Jane Halliday); Kathleen Wilhoite (Rosalie).

Halfway through the first year of what would ultimately be an eight-year run for the NBC legal drama *L.A. Law*, Abner Mikva, esteemed Federal judge and former Democratic U.S. Representative, had this to say in an article for the *Chicago Daily Law Bulletin*: "The genre of courthouse television is varied and venerable. Some syndicated series, like *Divorce Court* and *People's Court*, try to replicate an actual case, with only a few deviations. While these shows have loyal followings, their formats don't allow for much character or plot development. Other TV productions, like the *Perry Mason* specials, are generally disdained by judges and lawyers because the courtroom is distorted for dramatic purposes ... *L.A. Law*, on the other hand, tries to bridge the gap, and it succeeds — sometimes. The lawyers in this weekly series act like lawyers — sometimes. The tensions and conflicts of the story lines can be quite real — sometimes. The series, for all of its emphasis on the supercharged sex lives of its younger characters, may fall short of providing its viewers with realistic and insightful glimpses into the everyday problems of law firms and their clients; yet, it comes a lot closer than its predecessors, probably as close as the one-hour format permits."

L.A. Law was the first successful dramatic legal series to focus in depth on the personal rather than professional lives of attorneys; the first to fully detail the inner workings of a typical upscale law firm, warts and all; the first to show without rancor or ridicule that lawyers aren't infallible, that there are as many negative as positive facets to their personalities, that sometimes they are forced to put ethics on hold, that they frequently lose cases they should have won and vice versa, and that they can't always prevent their private demons from worming their way into the courtroom; the first non-sitcom courtroom show to feature an abundance of female attorneys, each of whom was respected and paid as much as their male associates; and the first to shamelessly glorify a materialistic lifestyle for high-priced lawyers who were worth every penny they earned. (It was no accident that each episode opened with the closeup of a vanity license plate, the title of the show embossed in large letters, firmly fastened upon the gleaming bumper of a brand-new Jaguar.)

Two years before L.A. Law made its network debut, producer Steven Bochco read a newspaper interview with Grant Tinker, wherein the NBC chairman claimed he'd asked Bochco to develop a series based on the 1982 film The Verdict, in which Paul Newman played a burned-out lawyer who found redemption by taking on an "underdog" case. When a surprised Bochco asked him about this, Tinker didn't remember making any such statement, but admitted that a legal series was a damn fine idea. Upon being dismissed from his long-running cop drama Hill Street Blues, Bochco opted

to fulfill a prior commitment to NBC programming chief Brandon Tartikoff, who told the producer he could do whatever series he wanted. Having already gotten the blessings of Grant Tinker, Bochco began work on L.A. Law, which he envisioned as Hill Street Blues with lawyers. Indeed, practically everything that had worked on Blues was slavishly duplicated for Law: the weekly pre-credit expositional scene, taking place in a room where most of the principals could logically congregate (a squad room on Blues, a conference room on Law); a large, intersecting ensemble cast; numerous separate storylines (as many as four an episode) playing out simultaneously, sometimes in arc form from one week to the next; and an evocative theme song by Mike Post (Law went Blues one better by tipping off viewers as to whether the episode would be a comic or serious one, depending upon the musical instruments used). The major difference between the two properties was that where Blues was dark and grimy, Law was bright and tidy, with better-educated and far better-paid protagonists.

"The law has always interested me," Bochco told TV Guide in 1987. "It has everything — good and bad, right and wrong…. The law is human behavior in the crucible of stress — emotional, financial, moral, sometimes even physical stress." Though Bochco himself had no hands-on legal experience, his L.A. Law cocreator Terry Louise Fisher had been a Los Angeles County deputy DA and a Hollywood entertainment lawyer before becoming producer of such TV series as Cagney and Lacey. "I spent most of my adult life trying to get out of a corporate law firm," Fisher noted at the time. "[And now] I get together with Steve and the first thing you know, I'm spending most of my time in one." In the pursuit of judicial verisimilitude, Bochco and Fisher hired a writing staff consisting entirely of people with legal backgrounds, one of whom was an ex-attorney named David E. Kelley.

The series' legal consultant was former law professor Chuck Rosenberg, who'd previously worked on The Paper Chase (q.v.). Recalling his duties during a 1996 seminar at Chicago's Museum of Broadcast Communications,

Rosenberg allowed that he'd had no real power to okay or veto a script ("At least at first!"); rather, his function was to "help the writer mine for better drama," translate the legal terminology into workable English, and to maintain "substantial balance" in the more controversial storylines. He recalled that one writer had wanted to do a story about improprieties in a 24-hour "supermax" prison; Rosenberg's advice was to approach this story from both a Conservative and a Liberal point of view, "put up a flag" dramatically to let the audience know that the problems within the prison were genuine rather than ideological — *then* include dialogue to explain the ultimate legal decision made, and assure the viewer that it was the correct decision by adding a scene in which an attorney was shown conferring with the opposing counsel's client. (There may be an official term for this sort of thoroughness, but the more popular acronym is "C.Y.A.")

While Fisher and her fellow legal eagles handled the accuracy end of the series, Bochco busied himself with creative matters by hiring several *Hill Street Blues* veterans for the production staff, and selecting key cast members from his own friends and colleagues. Alan Rachins, cast as attorney Douglas Brackman Jr., was Bochco's brother-in-law; Michael Tucker, who played attorney Stuart Markowitz, had graduated with Bochco from Carnegie Institute in 1966; and though the producer was unfamiliar with the work of Susan Dey before hiring her as prosecutor Grace Van Owen, he'd met the actress while her daughter and his son were attending the same private school!

L.A. Law largely took place in the luxuriously appointed offices of McKenzie, Brackman, Chaney & Kuzak, which during the series' run morphed into McKenzie, Brackman, Chaney & Becker before finally settling on McKenzie, Brackman, Kelsey, Markowitz & Morales. With his customary flair for melodrama, Bochco made sure viewers were welded to their seats from the outset by having the office invaded by the gun-wielding disgruntled husband of a former client, and more spectacularly with his introduction of law partner Norman Chaney — slumped over dead on his desk,

face down in a dish of beans. (Ever wonder who plays all those corpses on TV drama shows? In this case, the late Mr. Chaney was enacted by Loren Janes, one of Hollywood's best stunt coordinators.)

Of the enormous ensemble cast, the only members who appeared in the series from first season to last were Richard Dysart as the firm's distinguished senior partner and resident father figure Leland McKenzie; Alan Rachins as obnoxious, egotistical, money-mad managing partner Douglas Brackman Jr., forever trying but failing to live up to the reputation of his celebrated attorney father (even unto sleeping with his late dad's mistress); Corbin Bernsen as the gloriously slimy domestic-relations partner Arnie Becker, an insatiable womanizer and also a sucker for every get-rich-quick scheme that came his way; Michael Tucker as tax partner Stuart Markowitz, with the face of a nebbish and the heart of a *mensch*; and Tucker's real life wife Jill Eikenberry as idealistic litigation associate (and later full partner) Ann Kelsey, wooed by many but ultimately won by Markowitz. Almost as long-lasting was Susan Ruttan as Arnie's mousy secretary Roxanne Melman, who before leaving the series at the end of Season Seven astonished everyone by ending up in bed with her boss (and becoming pregnant by another lawyer); and two other regulars, introduced during Season Two and remaining with the show to the end: Blair Underwood as human dynamo Jonathan Rollins, the firm's first black attorney, and Larry Drake as mentally retarded office clerk Benny Stulwicz, a sensitively written and beautifully played role that won Drake one of the series' fifteen Emmy awards.

The most prominent players during the series' first five seasons included Harry Hamlin as savvy, sardonic young legal partner Michael Kuzak, described by staff writer Jacob Epstein as the "dark prince" of the ensemble because of his "potential for explosiveness"; Susan Dey as driving, driven Assistant DA Grace Van Owen, perennial courtroom opponent and off-and-on lover of the puckish Kuzak; Michele Greene as talented but insecure firm associate Abby Perkins, whose anxious jockeying for a partnership resulted in endless frustration and a brief defection to her own practice; and Jimmy Smits as associate Victor Sifuentes, the firm's only Hispanic member, who knew full well he'd been hired to satisfy a racial quota and worked twice as hard and twice as ruthlessly to prove his salt.

The sexual fireworks between Kuzak and Grace frequently dominated the action, to the extent that in one episode a female judge ordered the two attorneys into her chambers, told them everyone in L.A. knew what was going on between them, and ordered them to stop using the trial process to work out their domestic squabbles. There was a brief respite between romantic rounds when Grace was herself appointed a judge — an appointment which, according to Steven Bochco, actress Susan Dey came to regret because she was now confined to the traditional black robes and thus unable to display the fabulous wardrobe that had become her trademark. (Nor was she the only one with fashion issues: Michele Greene grew so weary of the drab outfits worn by her character Abby Perkins that when the *L.A. Law* set was rocked by a California earthquake, the actress looked heavenward and screamed, "*Please,* God, don't let me die in Abby clothes!") Grace ultimately stepped down from the bench and joined McKenzie, Brackman, Chaney & Kuzak, where she not only resumed her affair with Michael but also began canoodling with Victor Sifuentes — and as a result was unable to ascertain the identity of the father when she became pregnant.

The four above-mentioned actors were still commanding a great deal of attention when a new regular signed up during Season Three: Dann Florek (later a fixture of Dick Wolf's *Law & Order* franchise [q.v.]) as David Meyer, a feckless direct-mail entrepreneur who brightened any room he entered by leaving it. Though David was only around for two seasons, he left an indelible impression on the firm, first by inveigling Arnie Becker into producing a best-selling "Do-It-Yourself Divorce" video, then by briefly wedding Arnie's secretary Roxanne.

Steven Bochco relinquished his role as ex-

ecutive producer on *L.A. Law* at the end of Season Three, passing the torch to executive story editor David E. Kelley. Exhibiting the bravura showmanship that would become the trademark of his own legal series *Ally McBeal, The Practice* and *Boston Legal* (all q.v.), Kelley pumped new life into the old property with an abundance of extravagantly overblown story arcs, most memorably a case in which Sifuentes challenged the tasteless and demeaning sideshow attraction "dwarf-tossing." Featured in this storyline as 4'3" opposing counsel Hamilton Schuyler was dwarf actor David Rappaport, whose real-life suicide on May 2, 1990, embroiled *L.A. Law* in one of the few controversies for which its producers were wholly blameless. On the night of the tragedy, New York's WNBC-TV ran promotional teases for its 11 O'Clock newscast during *L.A. Law*'s commercial breaks, announcing the sudden death of one of the series' stars. The station did not reveal Rappaport's name until the very end of the newscast, angering several viewers who'd assumed that one of the show's regular actors had died. No one was more incensed than WNBC news anchor Chuck Scarborough, who complained to the media about his station's questionable promotion stunt — which of course was designed to hold *L.A. Law*'s audience straight through to the end of the newscast. Station management defended the gimmick by pointing out that the strategy had worked, and that Rappaport's demise was a "bonafide news story." So the next time *your* wroth is waxed by a local announcer artlessly linking a network drama show with a "late-breaking" news event ("You've seen the horror of gerbil smuggling on *CSI,* now see the *real* story on Action News Eleven!"), you'll know it all began with *L.A. Law.*

It was during the David E. Kelley era that *L.A. Law* introduced the first regular character who could be described as an out-and-out villain. Diana Muldaur joined the cast as brilliant but thoroughly unscrupulous attorney Rosalind Shays, who connived her way into a partnership and ultimately the highest position in the firm before being toppled from her throne and forced to resign by Leland McKenzie. This bitter power struggle nearly destroyed the firm when Rosalind sued the remaining partners for $2.1 million dollars. (As an observation without comment, when series cocreators Steven Bochco and Terry Louise Fisher acrimoniously dissolved their partnership at the end of 1987, she filed a $50 million lawsuit against Bochco.)

The addition of Rosalind Shays was a godsend to *L.A. Law*'s ratings, which had been sagging a bit before her arrival but which shot to the stratosphere when she began perpetrating her perfidy. It might have been that Rosalind was the sort of character viewers "loved to hate," like J.R. on *Dallas* or Alexis on *Dynasty*. Or it might have been, as some close to the series have insisted, that Rosalind was not universally regarded as a villain, but instead as an ambitious, self-reliant woman who'd managed to survive in a cutthroat profession dominated by male chauvinists. During a 1996 appearance at the Museum of Broadcast Communications, Diana Muldaur recalled that she had based Rosalind on certain prominent female CEOs who'd so firmly established themselves that they would never be answerable to any man. Muldaur further commented that whenever she felt a particular script was too "rough" on Rosalind or the other women on the show, Bochco obligingly handed that script to a female director; also, Muldaur herself was invited to make any suggestions or additions she found appropriate. This willingness to accept input from the actors was typical of Bochco: it was Harry Hamlin, for example, who came up with the idea of having his character show up in a gorilla suit to disrupt Grace Van Owen's wedding to the Wrong Guy in the unforgettable 1986 episode "Simian Enchanted Evening."

At this high point in the fortunes of *L.A. Law*, we pause to survey the opinions of real attorneys regarding the series. The aforementioned Abner Mikva liked the show's "willingness to take on at least a few tough legal subjects and try to depict them in a realistic manner," but with qualifications: "The cases, with few exceptions, do not sound like the bill of fare of any real law firm, even a Los Angeles law firm." Melvin Belli (see *Guilty or Innocent*) regarded *L.A. Law* as a "good public service," suggest-

ing it could have gone even farther in showing how a case actually reaches the courtroom: "...any time you have a show that gives you a look at what goes on in a court, the more you educate laymen, so that, for example, if they sit on a jury they have more understanding of how the system works." F. Lee Bailey (see *Lie Detector*) thought the series' various backstage intrigues were fundamentally realistic, but hard to swallow because they were so "chronic." Alan Dershowitz complained that the series made lawyers too likable: "Most lawyers are money-grubbing and selfish," Dershowitz insisted, also taking the show to task for not offering enough representation to "the most disadvantaged" (even though the characters regularly did pro-bono work). Gerry Spence, the colorful attorney at the center of the *Silkwood vs. Kerr-McGee* case, agreed with Dershowitz, further grousing that the main characters were too young to be credible attorneys, claiming one needs at least twenty years of "ugly hard work" to even reach the level of competence (Spence was 61 at the time of these comments).

If there was a universal gripe against *L.A. Law*, it was manifested in the character of the redoubtable Arnie Becker, who gave divorce lawyers a worse name than they already had. Attorney Raoul Lionel Fender, New York's "dean of divorce," took Arnie to task on the basis of his appearance, insisting, "divorce lawyers are usually skinny little guys, sitting in libraries, wearing green eye-shades." Celebrated "palimony" attorney Marvin Mitchelson also found fault with Becker, but only because of his lack of caution: Arnie's habit of deliberately stirring up discord between warring spouses in order to ensure a larger cash settlement could, Mitchelson argued, prove dangerous because of the "emotional unpredictability" of the plaintiffs.

Countering these complaints, the series' producers pointed to the incredible upsurge in law-school applications since *L.A. Law*'s debut. And to those who argued that the series steered clear of really important issues, the producers cited various episodes involving AIDs, persecution of homosexuals, sexual discrimination, "heroic" measures to extend the lives of terminal patients, Tourette's Syndrome, and the ongoing blight of judges who were allowed to remain on the bench even when they were obviously bigoted, incompetent or senile.

The series was also clear-eyed and credible on the issue of capital punishment. In the very first season, the lawyers faced the challenge of whether or not to condemn a prisoner already serving a life sentence for a murder committed behind bars. The solution, which even so adamant an anti-execution activist as Abner Mikva lauded as realistic and satisfying, was to use a "bifurcated" trial, with the jury first deliberating on the prisoner's guilt or innocence — and then, only after *that* decision was made, tackling the question of the death penalty. In this way, Mikva noted, the viewer was able to hear both sides of the argument compellingly expressed (though he took the writers to task for making things "easy" by characterizing the defendant as "absolutely ruthless and totally devoid of human feeling").

Using a similarly realistic (if slightly farfetched) set of circumstances, a later episode violated one of the once-sacred edicts of TV law shows: that a first-degree murderer must *always* be punished. In this one, Kuzak mounted a brilliant defense to acquit a young woman charged with murdering her older, wealthier husband. Only after he'd finished congratulating himself did Kuzak discover that his client, herself well versed in the law and in cahoots with a less ethical attorney, had manipulated the legal team into finding convenient "evidence" — and that she'd been guilty all along, and was now immune from further prosecution.

Of course, it did little good to argue about the fundamental authenticity of *L.A. Law* whenever a critic was determined to nitpick, so producer David E. Kelley composed a form letter for chronically complaining attorneys: "Dear___: If I were a good lawyer, I'd still be practicing law. Instead, I'm stuck in Hollywood, making 10 times as much money. I hope you are as conscientious about your clients as you are about our show."

Admittedly, most viewers couldn't have cared less about legal accuracy so long as *L.A. Law* continued serving up what passed for salacious sex in the 1990s: Stuart Markowitz persuading Ann Kelsey to marry him with a mysterious erotic procedure known as "The Venus Butterfly"; Arnie and Roxanne making love so ardently that they crashed through a ceiling; bitter enemies Leland McKenzie and Rosalind Shays winding up together under the covers (a response perhaps to actor Richard Dysart's wistful request for a few "steamy love scenes" so he could keep apace of his younger costars); and, in a scene that made international headlines, American Network Television's first lesbian kiss, shared by Abby Perkins and bisexual attorney C.J. Lamb (Amanda Donohoe), who joined the regular cast during Season Five.

There was one "water cooler" moment that had very little to do with things carnal. In the shocking finale of the ominously titled episode "Good to the Last Drop" (telecast March 21, 1991), the estimable Rosalind Shays carelessly stepped into an empty elevator shaft and plummeted to her death. Producer Steven Bochco was not alone in observing that the abrupt killing off of Rosalind opened a major gap on the series that no later episode was able to completely fill. Though many latter-day critics are hard-pressed to explain what exactly went wrong, most agree that this was the episode in which *L.A. Law* "jumped the shark." Jeff Jarvis of *TV Guide* said it best: "…it was the moment that the show lost us, when it went too far and broke the spell."

It didn't help that David E. Kelley left as executive producer at the end of the 1990–1991 season (though like Bochco, he stayed on in the lesser capacity of "creative consultant"). To quote a critique of the series in *Newsday*, "The difference between good and bad *L.A. Law* … was David Kelley." Just before his departure, Kelley had buttressed the property against the mass exodus of regulars Harry Hamlin, Jimmy Smits and Michele Greene by introducing three dynamic new characters: the aforementioned C.J. Lamb, the fiercely anti-establishment litigator Tommy Mullaney (John Spencer), and Tommy's peppery ex-wife, DA Zoey Clemmons (Cecil Hoffman). Though these Season Six newcomers did their best, they were hampered by some of the series' weakest storylines, several of them perilously close to self-parody.

The depletion of funds caused by the late Rosalind's lawsuit had forced the firm to share office space with sharkish entertainment lawyer Susan Bloom (Conchata Ferrell) and her handsome-hunk associate Billy Castroverti (Tom Verica), whose presence was (in context of the show) supposed to drum up new business by bringing in wealthy Hollywood clients and (in context of the ratings) attract new viewers: it failed on both counts. Nor did the firm's latest member, "killer" litigator Frank Kittridge (Michael Cumpsty), make any impression on the public. By season's end, all three of these newcomers were out — as, unfortunately, were the fascinating C.J. Lamb and hardy perennial Grace Van Owen.

Came Season Seven, and with it a new team of executive producers: John Masius and John Tinker, who promised to cut back on "issues" (which had been annoyingly topheavy in the previous season) and lighten up the scripts. Even so, the series spent a great deal of time focusing on the recent Rodney King riots, which proved dramatically viable in giving black attorney Jonathan Rollins the inspiration to run for City Council in hopes of rebuilding L.A., but also a major setback by dwelling on a tiresome subplot involving Stu Markowitz, who'd been beaten in the riots and had temporarily lost his sex drive. Another story arc that was played for far more than it was worth was the brief affair between Roxanne and Tommy Mullaney, resulting in another out-of-wedlock pregnancy. The one truly worthwhile "improvement" made by the Masius-Tinker team was the introduction of A Martinez as the firm's new Hispanic associate Daniel Morales. (Martinez had appeared in the previous season as a condemned murderer; considering that this story arc ended with his execution, the actor was understandably nonplussed when the series' producers called him back!) After years and years of randy sex and indiscriminate birthing, Morales was an L.A. Law novelty: a single dad who took his parental responsibilities seriously, spending as much time at home with his babies as in the courtroom (though like everything else this season, Morales' devotion to his family was overstressed to the point of irritation).

This character notwithstanding, it was clear that L.A. Law would continue its downward slide unless major changes were made behind the scenes. Halfway through Season Seven, Masius and Tinker were fired and William M. Finkelstein, who'd left the L.A. Law production staff to helm his own recently cancelled legal series Civil Wars (q.v.), was brought back into the fold to spearhead what TV Guide's Jeff Jarvis called a "rescue squad."

Finkelstein's first act was to carry over two of Civil Wars' most popular characters, former divorce attorney Eli Levinson (Alan Rosenberg) and his feisty secretary Denise Ianello (Debi Mazar), and add them to the established L.A. Law team. He also added the sort of appealing "marginal" character that had not been seen since the advent of the retarded Benny Stulwicz in Season Two. Alexandra Powers was cast as attorney Jane Halliday, who represented a type seldom seen on Prime Time television in the early 1990s: a deeply religious woman whose born-again Christianity was treated with respect rather than derision, and who was entirely sympathetic and non-judgmental toward those who didn't share her beliefs.

Those viewers who remained loyal during Season Eight were gratified that L.A. Law had improved substantially, with better and more compelling story arcs and a refreshing influx of humanity. Some industryites believed the show would have been renewed for a ninth season had not NBC needed to make room on its Thursday-night docket for the new medical series ER. As it turned out, L.A. Law came to an end with senior partner Leland McKenzie's retirement.

But though the weekly series had wrapped, the property wasn't dead just yet. Several plot strands left dangling in the spring of 1994 were knotted up on May 12, 2002, with the two-hour TV reunion special L.A. Law: The Movie. Of the original cast members, only Jimmy Smits declined to repeat his characterization of Victor Sifuentes. The movie wasted no time in bringing viewers up to date with the other regulars: McKenzie remained in retirement, leaving Brackman in charge of the old firm; Arnie Becker was being divorced by his trophy wife, with former colleague Abby Perkins spitefully representing Arnie's ex; Roxanne was now the firm's office manager and a single mom, trying to figure out some way to divorce the troublesome (and allegedly dying) David Meyer; Stu and Ann Markowitz were heavy into spiritualism, making them easy pickings for a phony guru; and Grace Van Owen, now promoted to DA, was dutifully trying to block the appeal of one of Kuzak's old capital cases. And Kuzak himself? Well, he'd given up the law after one of his acquitted clients committed a brutal sex crime, and was now running a trendy restaurant. Only sweet old Benny Stulwicz, still working as office clerk, remained unchanged.

L.A. Law: The Movie was timed for broadcast during NBC's 75th anniversary celebration; unfortunately, this lukewarm reunion film fared only marginally better in the ratings than NBC's equally ill-fated "retro" special of May 5, 2002, *Laverne & Shirley: Together Again.*

The Law and Mr. Jones

ABC: Oct. 7, 1960–Sept. 22, 1961; Apr. 19–Oct. 4, 1962. Four Star/Naxan Productions. Created and produced by Sy Gomberg. Cast: James Whitmore (Abraham Lincoln Jones); Janet DeGore (Marsha Spear); Conlan Carter (C.E. Carruthers).

One of two half-hour ABC legal series unveiled in the fall of 1960, *The Law and Mr. Jones* was like its companion piece *Harrigan and Son* (q.v.) what would now be termed a "dramedy": not funny enough to qualify as a sitcom, yet not serious enough to be pigeonholed as a drama. The property actually made its first appearance in 1959 as a pilot film called *Lincoln*, created, written and produced by Sy Gomberg for Bing Crosby Productions. James Whitmore starred as maverick lawyer Lincoln Jones, who on this occasion won an acquittal for an eccentric gentleman (Strother Martin) accused of setting a bomb. As was customary with pilot episodes, at the end of the show executive producer Bing Crosby appeared on camera to make a direct appeal to potential sponsors. By the time the property had been renamed *The Law and Mr. Jones* and sold to ABC, Crosby had dropped out of the picture and Dick Powell's Four Star Productions served

as distributor for Sy Gomberg and James Whitmore's Naxan Productions.

Whitmore remained with the series as the title character, now known more expansively as Abraham Lincoln Jones. Like his namesake, Abe Jones was honest to a fault — and he had *many* faults. His compassion for justice and fair play at all costs frequently led him to settle his differences with his fists, and many episodes opened with our breathless hero readjusting the furniture and removing the broken pictures and bric-a-brac from his tiny brownstone office after squaring accounts with his latest adversary. Though Jones could be warm and fuzzy with clients and colleagues, he also had a volcanic and unpredictable temper, and had lost a number of sensitive secretaries (he was never *really* yelling at them, merely at the situation) until pretty young Marsha Spear (Janet DeGore) elected to "tough it out" and give back as good as she got whenever her boss went on one of his *saeva indignatio* tirades. Similarly, Abe Jones' much-maligned young law clerk C.E. Carruthers (Conlan Carter) was willing to absorb a lot of verbal abuse simply for the privilege of working with a lawyer who'd memorized every writing and quotation of Oliver Wendell Holmes backward and forward. And because he took so many underdog cases on a pro-bono basis, Jones was forever a day late and several dollars short with his creditors — not that his chronic insolvency ever prevented him from expressing outrage that anyone would *dare* to bother him with mundane money matters when his client's well-being or very life was at stake. (The series' aptly chosen theme music was "When the Saints Go Marching In"!)

Many of Jones' cases involved fraud, embezzlement, extortion, land disputes, juvenile delinquency and child-custody battles. In addition to helping the poor and downtrodden, Abe was surrounded by an unusually high number of eccentric and downright certifiable clients and hangers-on: an old dowager who pretended to be crazy to scare off her greedy relatives, an elderly codger who called the cops to report illegal gambling after an ex-friend banned him from a weekly poker game, a curvaceous French dancer who encountered great difficulty keeping her clothes on despite pressure from the vice squad, an impoverished Shakespearean actor willing to commit bodily harm to prevent an ancient theater from being demolished, a punchdrunk boxer who Honest-to-God couldn't remember beating up his wife and a total stranger, three octogenarian courtroom buffs who perversely interfered with Jones' handling of a delicate divorce case, and so on. More credible were the episodes that found Jones returning to his home town to help his lawyer dad (Russ Brown) rout out a political hack, convincing a murder suspect to undergo a life-saving operation, persuading the son of a private-school headmaster to "do the right thing" and confess to a hit-and-run his family was trying to cover up, working side by side with an alcoholic attorney who insisted upon defending himself in a competency hearing, defiantly taking on a vicious gang of jukebox racketeers, rushing to the aid of a jazz musician framed on a narcotics charge, and endeavoring to prevent a martinet colonel (played by executive producer Dick Powell) from taking his son away from the boy's beloved stepfather. And just to keep Jones even more honest than usual, the lawyer was capable of making grievous errors in legal judgment, such as copping an insanity plea for an accused killer, only to be informed that the man committed another murder the day after his release.

Although *The Law and Mr. Jones* built up a faithful following, the series seemed destined to be cancelled at the end of its maiden season. Sponsor Procter & Gamble had every intention of renewing the show, but was told at the last minute by someone at ABC that *Law and Mr. Jones* had been "locked out"—the network had no available timeslot for it. Around the same time this decision had been reached, Bert Resnik, TV editor of the Long Beach (California) *Independent,* called producer Sy Gomberg on another matter. When Gomberg mentioned his series was on the road to oblivion, Resnik reported the news in his column, suggesting that fans of the series should write to the producer and tell him how much they liked it. Encouraged by the 400 positive letters that poured in, Gomberg contacted a TV critic in

a larger market, Terry Turner of the *Chicago Daily News,* to plead the show's cause. (Turner loved *Jones* because he found it a refreshing antidote to the violence of westerns and detective series—this despite Abe Jones' frequent slugfests!) The critic wrote several columns of support, resulting in an even bigger deluge of fan mail. Now Gomberg was ready to fight for the series' survival throughout the country: "And we fought like bears." Ultimately, so many columnists and viewers had taken up the cudgel that, in a rare gesture of concession, ABC brought *Law and Mr. Jones* back from the dead. Seven months after its official demise, the series resurfaced on the network's Thursday night schedule in April of 1962, with thirteen brand-new episodes.

But for all the grass-roots enthusiasm to keep *The Law and Mr. Jones* alive, the series failed to win its new timeslot; it was suggested that viewers unfamiliar with the series took one look at its title and assumed it was just another Western. With only 45 half-hours in its manifest, *The Law and Mr. Jones* was cancelled a second time in the fall of 1962. Over three decades later, the series was briefly recycled as part of a Saturday-night package of classic TV shows on cable's Nostalgia Channel. Though some of the dialogue and characterizations had not aged too well, and despite Abe Jones' tiresomely repetitious (and now politically incorrect) fistfights, the show impressed latter-day viewers with its ability to use actual legal precedents to make even the most far-fetched of plotlines seem credible. And besides, it was always a treat to see James Whitmore at the top of his form—right down to the unique opening credits, in which the actor was shown proudly signing his own name to a legal document.

Law & Order (The Franchise)

All series listed below were produced by Wolf Films in association with Universal Television, and were created and/or executive-produced by Dick Wolf. The music in all series was composed by Mike Post.

LAW & ORDER

NBC: debuted Sept. 13, 1990. Executive Producers: Arthur W. Forney, Edwin Sherin, Richard Sweren, Rene Balcer, Jeffrey L. Hayes, William N. Fordes,

Peter Jankowski, Matthew Penn, Lewis Gould, Eric Overmyer, Michael S. Chernuchin, Wendy Battles, Walon Green, Roz Weinman, Joseph Stern, Barry Schindel, Rick Eid, Nicholas Wootton, Ed Zuckerman, David Wilcox, Chris Levinson, Peter Guiliano, Robert Nathan, William M. Finkelstein, Arthur Penn. Principal Cast [in order of appearance]: Christopher Noth (Det. Mike Logan; 1990–1995); George Dzundza (Det. Sgt. Max Greevey; 1990–1991); Dann Florek (Capt. Donald Cragen; 1990–1993); Richard Brooks (Asst. D.A. Paul Robinette; 1990–1993); Michael Moriarty (Asst. D.A. Ben Stone; 1990–1994); Steve Hill (D.A. Adam Schiff; 1990–2000); Paul Sorvino (Det. Phil Cerreta; 1991–1998); Jerry Orbach (Det. Lennie Briscoe; 1992–2004); S. Epatha Merkerson (Police Lt. Anita Van Buren; 1993–); Jill Hennessy (Asst. D.A. Claire Kincaid; 1993–1996); Sam Waterston (Exec. Asst. D.A.[later D.A.] Jack McCoy; 1994–); Benjamin Bratt (Det. Reynaldo "Rey" Curtis; 1995–1999); Carey Lowell (Asst. D.A. Jamie Ross; 1995–1998); Angie Harmon (Asst. D.A. Abbie Carmichael; 1998–2001); Jesse L. Martin (Det. [later Senior Detective] Edward Green; 1999–2008); Dianne Wiest (D.A. Nora Lewin; 2000–2002); Elisabeth Röhm (Asst. D.A. Serena Southerlyn; 2001–2005); Andrea Navedo (Det. Ann Cordova 2001–2006); Fred Dalton Thompson (D.A. Arthur Branch; 2002–2006); Dennis Farina (Det. Joe Fontana; 2004–2006); Annie Parisse (Asst. D.A. Alexandra Borgia; 2005–2006); Milena Govich (Det. Nina Cassidy; 2006–2007); Alana de la Garza (Asst. D.A. Connie Rubirosa; 2006–); Jeremy Sisto (Det. Cyrus Lupo; 2008–); Linus Roache (Asst. D.A. Michael Cutter; 2008–); Anthony Anderson (Det. Kenneth Bernard; 2008–). [Leslie Hendrix as Dr. Elizabeth Rogers sporadically appeared on this and the subsequent *Law&Order: Criminal Intent*.]

LAW & ORDER: SPECIAL VICTIMS UNIT

NBC: debuted Sept. 20, 1999. Executive Producers: Ted Kotcheff, Neal Baer, Patrick Harbinson, Robert Nathan, Peter Jankowski, Dawn DeNoon, Amanda Green, Judy McCreary, Arthur W. Forney, Jonathan Greene, Robert Palm, David J. Burke, Tara Butters, Lisa Marie Petersen, Roz Weinman, Michael Fazekas. Cast: Christopher Meloni (Det. Elliott Stabler); Mariska Hargitay (Det. Olivia Benson); Richard Belzer (Det. John Munch); Dean Winters (Det. Brian Cassidy; 1999–2000); Ice-T (Det. Odafin "Fin" Tutuola; 2000–); B.D. Wong (Dr. George Huang; 2000–); Stephanie March (Asst. D.A. Alexandra Cabot; 2000–2003); Diane Neal (Assistant D.A. Casey Novak; 2003–); Tamara Tunie (M.E. Melinda Warner; 2005– ; previously recurring); Adam Beach (Det. Chester Lake; 2007–). [Mary Stuart Masterson briefly replaced B.D. Wong as Dr. Rebecca Hendrix in 2005.]

LAW & ORDER: CRIMINAL INTENT

NBC: debuted September 30, 2001 [first-run on NBC ended May 21, 2007]; moved to USA Network beginning October 4, 2007 [USA reruns later shown on NBC]. Executive producers: Rene Balcer, Arthur W. Forney, Peter Jankowski, Warren Leight, Fred Berner, Marlene Meyer, Stephanie Sengupta, Gerry Conway, Norberto Barba, Julie Martin, Therese Rebeck, Geoffrey Neigher. Cast: Vincent D'Onofrio (Det. Robert Goren); Kathryn Erbe (Det. Alexandra Eames); Courtney B. Vance (Asst. D.A. Ron Carver 2001–2006); Jamey Sheridan (Capt. James Deakins; 2001–2006); Eric Bogosian (Capt. Daniel Ross 2006–); Chris Noth (Det. Mike Logan 2005–2008); Annabella Sciorra (Det. Carolyn Barek; 2005–2006); Julianne Nicholson (Det. Megan Wheeler; 2006–2007; Alicia Witt (Det. Nola Falacci; 2007–); Jeff Goldblum (Det. Zach Nichols; 2008–).

LAW & ORDER: TRIAL BY JURY

NBC: March 3–Aug. 3, 2005. Executive Producers: Arthur W. Forney, Walon Green, Peter Jankowski, Richard Pearce. Cast: Bebe Neuwirth (A.D.A. Tracey Kibre); Amy Carlson (A.D.A. Kelly Gaffney); Kirk Acevedo (D.A. Investigator Hector Salazar); Scott Cohen (Det. Chris Ravell); Fred Dalton Thompson (D.A. Arthur Branch).

CONVICTION

NBC: March 3–May 19, 2006. Executive Producers: Rick Eid, Walon Green, Peter Jankowski, Nena Rodriguez, Constantine Makris. Theme music: "Destiny," performed by Syntax. Cast: Stephanie March (ADA Bureau Chief Alexandra Cabot); Eric Balfour (ADA Brian Peluso); Julianne Nicholson (ADA Sara Finn); Anson Mount (Deputy DA Jim Steele); Jordan Bridges (ADA Nick Potter); Milena Govich (ADA Jessica Rossi); J. August Richards (ADA Billy Desmond); Fred Dalton Thompson (D.A. Arthur Branch); David Zayas (Hernan); Rebecca Mader (Hannah).

As these words are being written, there is every likelihood that *Law & Order*, currently in its 18th year on NBC, will make it to its 21st season, thus surpassing *Gunsmoke* as the longest-running dramatic series in network TV history. When this milestone occurs, there is sure to be some nitpicker who will bleat, "Yeh, *Law & Order* may have been on longer — but *Gunsmoke* still holds the record for the most episodes." True enough: at the standard rate of 22 hour-long episodes per season, by the end of its 20th year on the air *Law & Order* will have toted up 459 episodes, falling far short of

Gunsmoke's 635. But we cavil: unless there's a world war or a polar ice melting in the meantime, *Law & Order* will still be going strong in the Summer of 2018, by which time it will have wound up its 28th season and will definitely have churned out the requisite 635 episodes. Thus, at the outset of Season 29, series producer Dick Wolf will be perfectly justified to break out champagne and shoot off fireworks to celebrate the record-shattering 636th time off-screen announcer Steve Zirnkilton intones the immortal opening statement: "*In the criminal justice system, the people are represented by two separate yet equally important groups: the police who investigate the crime, and the district attorneys who prosecute the offenders. These are their stories.*"

Like so many other successful TV properties, *Law & Order* wasn't given much of a chance for longevity in its early stages. In early 1990, Dick Wolf's newly formed Wolf Films was reeling from three back-to-back failures: *Gideon Oliver, Christine Cromwell* and *Nasty Boys*. He knew that his next project, for which he'd already filmed an expensive pilot, would be a make-or-break proposition, and he wasn't keen on being broken. A lifelong devotee of Jack Webb's *Dragnet*, Wolf hoped to revive the "procedural" format that had worked so well for Webb but had been pretty much abandoned by other producers since. Wolf also held a special place in his heart for the long-forgotten 1963 series *Arrest and Trial* (q.v.), a 90-minute weekly in which the first half was devoted to the pursuit and capture of a suspected criminal, and the second half given over to the suspect's courtroom trial. If he could combine the best elements of these two programs, Wolf reasoned, he might be able to breathe new life into the TV legal-drama genre. In a later interview for the Archive of American Television, Wolf disdainfully described the late-1980s network scene as "comedies, comedies, and more comedies," and it was his dream to break this stranglehold.

There was another reason why the two-tiered *Arrest & Trial* structure appealed to Wolf. At that time, most 60-minute network series were not selling well in off-network rerun syndication because local sponsors preferred to underwrite 30-minute comedy shows. But some series like *Fame* and *Fantasy Island* had managed to make off-net inroads by dividing each of their hour-long episodes into two half-hours for the purposes of daily "strip" syndication. By splitting his new show straight down the middle, thereby creating two separate-but-equal "Law" and "Order" segments, Wolf could not only syndicate the series with impunity, but also advertise the property as "two shows in one," with twice the profit potential.

The look, format and operating philosophy of the new series were all in place from the beginning, even though the title wasn't (Wolf wanted something along the lines of *Cops & Prosecutors*, overlooking the obvious *Law & Order* until just before the cameras rolled). For starters, Wolf insisted that the show would be filmed on location in New York City, and that it would be completely story-driven, with virtually no time devoted to the central characters' private lives; indeed, if the producer had followed through with his original plan, there wouldn't have been any central characters at all. The first half of each episode would detail the crime, the police investigation, the pursuit and the capture: this section would be filmed with a jittery handheld camera, harsh lighting and substandard sound quality to create the illusion we were watching authentic news footage rather than fictional events (in one of its first articles on *Law & Order*, *TV Guide* observed that even though it was filmed in color, "you'd swear that it was shot in grainy black and white"). The second half would concentrate on the trial, during which the camerawork would be more sedate and the lighting, color and sound recording more in line with typical TV fare. One non-typical aspect, at least in 1990, was that the prosecutors would be depicted as the heroes of the courtroom, rather than the defense attorneys who usually held the monopoly on heroics. Wolf later explained that the country's mood was becoming more conservative, and it was hard for audiences of the period to pull for an attorney who was defending a obvious felon. But even when the guilt of the suspect had been firmly established, the courtroom scenes would permit an exploration of what Matthew Penn, one of *Law & Order's* executive producers, described as the "social, moral and legal conundrum" of the case at hand. Difficult questions as to whether justice was truly being served (especially when sensitive political issues were involved) would be raised at the end of each episode — and sometimes the Good Guys would lose, just like in real life.

Throughout the entire hour, Wolf would keep viewers on their toes by avoiding the standard establishing shots, using seconds-long intertitles to tell us where and when the action was taking place, then jumping headlong into each new scene as if it was already in progress. Finally, everything that took place on screen — except for the opening discovery of the crime or the corpse, which was handled by bit players in a short "slice of life" vignette — would be shown from a "locked" point of view: the audience saw only what the cops and the DAs saw, and knew only what the cops and the DAs knew, period. So rigid and unswerving was this format that *Law & Order* star Jerry Orbach described it as "almost like Catholic High Mass."

Barry Diller of the Fox network, ever on the prowl for offbeat TV properties, expressed interest in Dick Wolf's new series, but ultimately took a pass. CBS commissioned a pilot episode, going so far as to reserve a slot in the network's 1990–91 fall season, but at the last minute decided to renew *Jake and the Fatman* (q.v) instead; besides, explained the CBS suits, *Law & Order* was so much an ensemble show that it lacked "star quality" and featured no "breakthrough" actors. But Universal, who'd bankrolled the pilot, had faith in the series and suggested that Wolf pitch it to NBC Entertainment executive Brandon Tartikoff. Though he was literally the only person at NBC who liked the show, Tartikoff was able to greenlight Wolf for six trial episodes, subsequently expanding to thirteen — and this despite his prediction to the producer: "You can't do this every week." In March of 1990, filming began; and in September of that year, NBC launched the series with a conspicuous lack of fanfare in a

Thursday night timeslot opposite ABC's top-rated *L.A. Law* (q.v.)

The episode viewers saw on opening night was actually the sixth to be filmed; the pilot, featuring Roy Thinnes in the DA role later assumed by Steven Hill, was withheld until later in the season. The cast lineup in those first few months included, on the "Law" side, Christopher Noth as Detective Mike Logan, George Dzundza as his partner Detective Sgt. Max Greevey, and Dann Florek (late of *L.A. Law*) as Captain Donald Cragen, all attached to New York's 27 Precinct; and on the "Order" side, Steven Hill as District Attorney Adam Schiff, Richard Brooks as ADA Paul Robinette and Michael Moriarty as ADA Ben Stone. From the very beginning, composer Mike Post used the sound-effect punctuation described by Wolf as "Ching-Ching!" as a transitional device between scenes; this effect was supposed to emulate a jail door slamming shut. To ward off complaints from pressure groups and network censors, each episode of a controversial nature opened with a warning, and each episode based on an actual event was preceded with a "no similarity is intended or implied" disclaimer. Setting a precedent that would be followed ever afterward, most of the Season One episodes were "ripped from today's headlines," fictionalizing such recent events as the Bernard Goetz subway killings, the Tawana Brawley hoax, the sexcapades of the "Mayflower Madam," the assisted suicides of Dr. Jack Kevorkian, and the ever-burgeoning AIDs crisis. (Referring to *The New York Post* as his "bible" for sensational news stories, Wolf frequently showed the cases on *Law & Order* being extensively covered by the *Post*-like fictional tabloid *The New York Ledger*.) Wolf's favorite first-season episode, likewise based on a true story, was "Life Choice," depicting the prosecution of a pro-life fanatic accused of bombing an abortion clinic. The episode ended up costing NBC $800,000 in sponsor pullouts, and would remain out of rerun circulation until 2001; even so, "Life Choice" was instrumental in attracting the series' first serious fan base.

Unexpectedly drawing good numbers throughout its first season, especially after moving to a more advantageous Tuesday night slot, *Law & Order* not only justified Wolf's cast-in-stone format but also Brandon Tartikoff's confidence in the property. Even after the show was an established hit, however, Tartikoff remained a voice in the wilderness at NBC, as recalled by Wolf in an interview for *Newsweek*: "Everybody at NBC hated it. Not mildly disliked it. Hated it. For the first five years, it was the chief topic of the first day of scheduling every year: What do we have to replace *Law & Order?*" Only after NBC struck a deal with cable's A&E service to sell reruns of the series at $159,000 per episode did the network's top brass change their minds about the show. (A&E was the first of several cable outlets that would make hay off of *Law & Order* reruns; ironically, for all of Dick Wolf's preparations to repackage the series in a half-hour format, it has never to this day been released for off-network syndication.)

One of the advantages of a story-driven series, in which no single character is any more important than another, is that cast changes can be made without upsetting the flow of the show or losing audience support. The first such personnel change occurred on the "Law" side in 1991, when George Dzundza left the series: his character Detective McGreevey was promptly killed in the line of duty, whereupon Detective Logan was given a new partner, Detective Phil Cerreta, played by Paul Sorvino. Halfway through Season Three Cerretta was seriously injured in a shootout and voluntarily took a desk job; his replacement was Detective Lennie Briscoe, played by Jerry Orbach.

Around this time an industry joke was in circulation, suggesting that any new partner of Detective Logan had about as much chance for survival as all those ill-fated fiancees of Little Joe Cartwright on *Bonanza*. But Briscoe managed to beat the curse and remained with *Law & Order* for twelve seasons; in fact, Briscoe ended up outlasting Logan, who in 1995 was pulled from Manhattan Homicide and exiled to the Staten Island Harbor Patrol after punching out a gay-bashing politician — a clever if not entirely successful ploy to obscure the fact that Christopher Noth had been fired from the se-

ries over a highly publicized salary dispute. Briscoe's new partner was Detective Rey Curtis (Benjamin Bratt), who in 1999 took early retirement to care for his MS-stricken wife Deborah; Curtis was quickly succeeded by Detective Ed Green, Briscoe's first black partner, played by Jesse L. Martin. (Both Martin and Jerry Orbach were Broadway musical-comedy headliners, and many a rehearsal was enlivened by the two actors' robust renditions of the songs they'd made famous on stage.) Green was still on the job when Briscoe left the force to join the DA's office and was replaced by Det. Joe Fontana (Dennis Farina); two years later Fontana retired, and his place on the show was briefly filled by Det. Nina Cassidy (Milena Govich). As for Green, he was shot on duty and temporarily spelled by Detective Nick Felco (Michael Imperioli), and not long after his return was promoted to senior detective (Martin left the series in 2008). More recent additions to the 27th Precinct roster have included Jeremy Sisto and Anthony Anderson, respectively cast as Detectives Cyrus Lupo and Kenneth Bernard.

Though there was less turnover amongst the series' prosecutors, the "Order" column of the cast list has also gone through a number of changes. Steven Hill, longest-lasting of the original cast members, was seen as curmudgeonly old-line liberal DA Adam Schiff from the series' debut in 1990 until the actor's retirement in 2000, when it was explained that Schiff had left the country to negotiate reparations for Holocaust survivors in Vienna. He was replaced on an interim basis by politically moderate DA Nora Lewin, played by Dianne Wiest. Lewin disappeared when the ultra-conservative Arthur Branch was elected to the DA post in the wake of the 9/11 terrorist attacks; this role was played with a singular air of authenticity by Fred Dalton Thompson, a former lawyer and Republican U.S. senator who had risen to prominence during the Watergate hearings, and who since 1987 had alternated his public-service activities with occasional acting roles in both TV (*Wiseguys*) and films (*Hunt for Red October*).

We'll have more on Thompson in a moment, but in the meantime let's take a look at the DA's staff. Richard Brooks was seen as Assistant DA Paul Robinette from 1990 until 1993, at which time he went into private practice as a defense attorney, specializing in civil rights cases. Also on board from the beginning was Michael Moriarty as Assistant DA Ben Stone, a role he might have played indefinitely had it not been for a real-life confrontation between Moriarty and U.S. Attorney General Janet Reno. During a 1993 dinner meeting with the *Law & Order* staff, Reno implicitly threatened government intervention unless the networks clamped down on television violence. In retrospect, it is clear that Reno was not only overreacting but also not in full possession of the facts, especially when she cited the scrupulously family-friendly *Murder, She Wrote* as one of the most violent shows on TV! Even so, Michael Moriarty's reaction to Reno went even further over the top: bitterly describing the meeting as a "kangaroo court," he threatened to personally sue Reno and took out trade ads attacking the government for stifling artistic expression. When Dick Wolf asked Moriarty to cool it, the actor went completely ballistic, publicly accusing NBC of being part of a vast conspiracy to deny his freedom of speech — and once the dust had settled, Moriarty either resigned from *Law & Order* or was forced off the show, depending on which source one believes. Brought in to replace Moriarty was Sam Waterston, who had impressed Wolf with his "un–Hamlet-like" performance as a Southern lawyer on *I'll Fly Away* (q.v.) as Executive ADA Jack McCoy. As of 2008, the year McCoy was promoted to District Attorney upon the sudden departure of Arthur Branch, Waterston had been on *Law & Order* longer than any other male actor. (McCoy's former duties were taken over by Linus Roache as ADA Michael Cutter.)

Note the qualification "male": conspicuous by their absence during the series' first few seasons were any regular female characters, save for recurring appearances by Carolyn McCormick as police psychiatrist Elizabeth Olivet and Leslie Hendrix as coroner Elizabeth Rogers. This was probably not a conscious decision by Wolf, but more likely grew from his reluctance to dwell upon the personal lives of the principals: bring in women and you bring in romantic entanglements, as demonstrated by *L.A. Law.* Though the series' ratings were strong among male viewers, as a group women did not enjoy the show, many of them insisting the jittery camerawork gave them headaches. But NBC executive Warren Littlefield sensed this was a subterfuge, concluding that *Law & Order* needed female stars to attract female fans. As Dick Wolf remembered it, Littlefield summoned him to his office in May of 1993 and said bluntly, "I'm giving you a year's notice that you're cancelled," unless certain gender-driven cast changes were immediately implemented. The result: Dann Florek, who since the series' inception had appeared as the 27th Precinct's Captain Donald Cragen, was summarily dismissed and replaced by actress S. Epatha Merkerson as Lt. Anita Van Buren (a role she was still playing as of 2008); and the aforementioned Richard Brooks relinquished the role of ADA Paul Robinette, to be succeeded by Jill Hennessy as ADA Claire Kincaid, a character strategically introduced during "Sweeps Week." Littlefield's edict to Wolf evidently paid off: the season after Merkerson and Hennessy joined the cast, *Law & Order* broke into the list of twenty top-rated programs for the very first time.

Claire Kincaid remained on the DA's staff until the end of the 1995–1996 season, when she was struck down by a drunk driver. Jill Hennessy had asked to be written out to pursue a movie career, and the original plan was to have Claire survive, albeit confined to a wheelchair; instead, it was revealed during the 1996–1997 season that Claire had died from her injuries (though disappointed that she could never return to *Law & Order*, Hennessy could take solace in the success of her subsequent starring series *Crossing Jordan*). In the meantime, Carey Lowell signed on as Claire's replacement, ADA Jamie Ross. She in turn resigned the DA's office in 1998 to devote more time to her family, though like Paul Robinette she would pop up in later seasons as a defense attorney at odds with her former colleagues. With the departure of Ross, ADA Abbie Carmichael, played by

Angie Harmon, joined the prosecution team. Arguably the most popular of the series' female characters, Harmon sent many a male viewer's heart into a tailspin when she decided to give up *Law & Order* to start a family with her new husband, NFL star Jason Sehorn. Brought in as Abbie Carmichael's replacement was Elisabeth Röhm as ADA Serena Southerlyn; for some reason this character never caught on with viewers and was written out in 2005, but not before a controversial scene in which Southerlyn asked DA Branch point-blank if she was being fired because she was a lesbian. (Up to this moment, there'd been absolutely no mention of her sexual orientation!) After Southerlyn came Annie Parisse as ADA Alexandra Borgia, who lasted only one season before she was kidnapped and brutally murdered—a rare example of Dick Wolf succumbing to a shopworn woman-in-jeopardy cliché in order to juice up ratings. Enjoying a longer term of office was Borgia's replacement ADA Connie Rubirosa (Alana de la Garza), who joined the show in 2006.

Though the faces continually changed onscreen, *Law & Order* slavishly adhered to its established format, and fans ate it up with a spoon. Unlike certain other legal programs we could name, the series was just as popular with real-life lawyers and law-enforcement officials as it was with the general public. In an article for the *Houston Chronicle*, Mary Flood observed that while attorneys were not entirely happy about all aspects of Dick Wolf's output—"Their No. 1 pet peeve: Unlike *Law & Order*'s Jack McCoy, real-life prosecutors don't hear defendants confess while their lawyers sit silently by like crash dummies"—most defense lawyers enjoyed the show "in part because sometimes the prosecution loses or looks overzealous." Others could not entirely explain why they liked the program, just that they *did*. Writing for *Picturing Justice* in April of 2006, UCLA law professor and coutroom-drama historian Michael Asimow said, "To me, the phenomenal commercial success of *Law & Order* is baffling: there's no sex or violence and the show never dumbs itself down or underestimates the viewers' intelligence. If you doze off

or get up to answer the phone or go to the bathroom, you're going to lose the thread of the story. As a result, *Law & Order* seems to flout every conventional rule for success in the world of mass media, yet old episodes play every day on cable while new episodes continue to appear on NBC. Although I can't account for its success, I appreciate the high quality of the show and I watch it religiously."

Certainly one of the key reasons for its continued success was that the series always kept abreast of current events with thinly disguised dramatizations of actual news stories. We've mentioned the headline-inspired episodes during the series' first season: since that time, *Law & Order* has tackled virtually *every* late-breaking event, sometimes a mere few weeks or even days after the first "special bulletin": the omnipresence of the Russian Mafia, the celebrity prosecutions of O.J. Simpson, Robert Blake and Michael Jackson, the Clinton impeachment, the voting discrepancies during the 2000 Presidential election, the "pedophile priest" scandal, the beating death of a Jewish motorist at the hands of angry African Americans in the Bronx, Geraldo Rivera's questionable reporting of military maneuvers in Iraq, the right-to-death Terry Schiavo controversy, the anti-immigration movement, stem cell research, internet bullying, the "gay gene" ethical imbroglio, Mel Gibson's anti-semitic tirade, the Andrea Yates filicide case, the Gitmo "Koran–flushing" allegation, the "claque" killings of rap stars, the influx of defective products from China, the Michael Vick dogfighting muddle, the collateral damage of using Blackwater–like private contractors in war zones, the sub-prime mortgage mess, and even the 2008 Hollywood writer's strike. To avoid libel suits, the series' writing staff was adept at folding two real stories into one fictional plotline, so that viewers who think they've figured out which news story is getting the *a clef* treatment at the beginning of the episode are surprised when the plot segues into *another* topic of current interest—and the ending turns out to have no bearing on either case (In one instance, a lawyer who thought he recognized one of his own cases in a *Law & Order* episode promo threatened to sue — until he actually saw the episode, in which the character based on his client was acquitted.) Still, *Law & Order*'s heavy reliance upon Real Life has gotten the show in trouble from time to time: "Sunday in the Park with Jorge," an episode inspired by the outbreak of violence at the 2000 Puerto Rican Day Parade, came under so much fire from pressure groups that it was immediately yanked from rerun rotation and never seen again on NBC.

In keeping with his philosophy that the series' characters should be seen only in performance of duty — "meat and potatoes" is how he described this austerity — Dick Wolf has continued resisting the impulse to dwell upon the principals' private lives, except when it has some bearing on their job performance: Carey Lowell's bitter child-custody battle, Rey Curtis' marital infidelity, Lennie Briscoe's drinking problem, Mike Logan's memories of childhood sexual abuse, etc. One of the few times Wolf broke his self-imposed rule resulted in the producer's least favorite episode: the one in which Detective Lennie Briscoe's druggie daughter was killed. Wolf later explained that he commissioned this episode to mollify actor Jerry Orbach, who had been begging for an "Emmy moment." We can be grateful that Wolf vetoed Orbach's first request, made upon the firing of costar Chris Noth, that a sobbing Briscoe be shown cradling the dead body of Mike Logan in his arms!

Still, many viewers have continued to demand Wolf provide them with more than a few furtive glimpses of his characters' human side. Rather than overhaul the story-driven *Law & Order* format to satisfy these fans, the producer developed two spinoff series, both of them decidedly character-driven — and in so doing, he launched a franchise that by 2005 would be generating an annual revenue of one billion dollars for NBC-Universal and its cable partners.

Since this is a book about legal shows, we won't expend too much prose on the first two *Law & Order* spinoffs: as Mary Fischer of CNN.com noted in a 2004 article, The "Law" part of *Law & Order* gets smaller and smaller in each sequel." Both *Law & Order: Special Victims Unit* (hereafter known as *SVU*) and *Law & Order: Criminal Intent* (hereafter known as *CI*) bear a surface resemblance to their parent show: a variation of the original's "In the criminal justice system…" opening announcement, the white-on-black credits and expository intertitles, the "Ching-Ching" sound effect punctuating each scene, the New York backdrop, and so on. Beyond that, both series are entirely separate entitities — and, as Mary Fischer points out, "The original gives the police and prosecutors a half-hour per episode. But the lawyers don't get nearly as much time in *SVU*, and they seem to be an afterthought in *CI*."

The titular "Special Victims Unit" on *SVU* referred to sex crimes. Heading this unit was a carryover from the original *Law & Order*, Dann Florek as Capt. Donald Cragen; Wolf had never wanted to drop this character from the parent series, and welcomed the opportunity to bring Florek back into the fold. The three main SVU detectives were Elliot Stabler (Christopher Meloni), Olivia Benson (Mariska Hargitay) and John Munch (Richard Belzer), the latter a character transferred from the recently cancelled cop series *Homicide: Life on the Street*. (Belzer was no stranger to Wolf's franchise, having appeared as Munch in several crossover episodes between *Law & Order* and *Homicide*. One might even say Richard Belzer has made John Munch his life's work: the character has also popped up on *Law & Order: Trial by Jury, The X-Files, The Beat, The Wire* and *Arrested Development*.) Added to the cast in 2000 was singer Ice-T as Detective Odafin "Fin" Tutuola, replacing first-season regular Dean Winters as Det. Brian Cassidy. Another important member of the unit was forensic psychiatrist George Huang, played by B.D. Wong. Chris Orbach, the son of *Law & Order*'s Jerry Orbach, made sporadic appearances as Ken Briscoe, the *nephew* of Lennie Briscoe.

Far more time was spent in discussing and dissecting the criminals' motives and tracking down fragmentary clues on *SVU* than on *Law & Order*, and unlike the earlier series the main characters' personalities often determined the direction and outcome of the case: Stabler's unblinking professionalism, Benson's outrage and overemotionalism (she deeply empathized with victims of sex offenders because she'd been one herself), Munch's obsession with sinister conspiracies (real or imagined), and Fin's insouciant wit. With most of the attention directed towards the SVU personnel, the series' lawyers had only a smattering of screen time, often appearing midway through the story in a desultory effort to put away the pervert-of-the-week, who generally remained free as a bird until episode's end. Making their few scenes count were Stephanie March as ADA Alexandra Cabot and Diane Neal as her successor, ADA Casey Novak, who had taken over after Cabot was presumed to have been killed by a Colombian drug cartel.

While the DA's office was given at least a fraction of screen time on SVU, there were virtually no courtroom scenes in *CI* at all: this second *Law & Order* spinoff dealt exclusively with crime-and-capture. In a further break from *Law & Order*, the stories on *CI* were told as much from the criminal's perspective as that of the detectives; many crucial scenes took place with nary a law enforcement official in sight. Also, *CI* was more of a whodunnit than its predecessor, with the detectives frequently following up misleading clues or concentrating on the wrong person while the guilty party moved about with impunity until ten minutes or so before the end, when the cards were finally on the table and the villain confessed — sometimes while holding one of the good guys at gunpoint.

CI was even more character-driven than *SVU*— or more accurately, personality-driven. Though it boasted a strong cast, *CI* was no ensemble effort, but rather a showcase for the peculiar talents of Vincent D'Onofrio as Detective Robert Goren, a character described by *TV Guide*'s Matt Roush as "a mix of Columbo, Sherlock Holmes and armchair Freud." Habitually wearing rubber gloves to avoid compromising the evidence, forever cocking his head sideways and leaning over everyone like a human Tower of Pisa, speaking in a voice so low that one almost needed special equipment to pick it up, playing mind games of genomic-like

complexity with the suspects, and interpreting the slightest change of vocal inflection as conclusive proof of guilt, Detective Goren was such a mass of idiosyncracies that at times the viewer expected him to drop his pose and reveal that *he* was the mysterious murderer. With Vincent D'Onofrio so dominating the proceedings, it's a tribute to the talent of Kathryn Erbe as his level-headed partner Det. Alexandra Eames that we can remember her presence at all.

The strongest link between *CI* and *Law & Order* was the presence of Christopher Noth, reprising his role as Det. Mike Logan. The initial break between Noth and producer Wolf in 1995 had not been a friendly one, with Wolf exacerbating the situation by announcing he planned to replace Noth with a "younger, cheaper actor" (the producer later admitted he might have been a shade more diplomatic). But in 1998 the two men buried the hatchet sufficiently to allow production of a spinoff TV movie, *Exiled*, chronicling Mike Logan's fortunes after he was taken off NYPD Homicide and banished to Staten Island. It was in fact the success of *Exiled* that inspired Wolf to expand the *Law & Order* franchise in the first place, thus it was probably out of gratitude that the producer hired Noth, just coming off a lengthy run on *Sex and the City*, to reprise the role of Logan in *CI*. The actor swallowed hard and accepted the offer, remaining on *CI* from 2005 until 2008, at which time he departed for greener pastures and was replaced by Jeff Goldblum as Detective Zach Nichols. (The fact that both Chris Noth and *SVU*'s Dann Florek were willing to forgive, forget and rejoin the *Law & Order* fold after being so coldly dismissed says something very special about Dick Wolf's ability to engender respect and loyalty amongst his employees.)

The combined success of *SVU* and *CI* encouraged Wolf to develop still another spinoff, this one designed to balance the topheavy concentration of the crime-and-capture aspects of American Justice on the two previous *Law & Order* derivations by focusing almost exclusively on the courtroom. Premiering in March of 2005, *Law & Order: Trial by Jury* was told from the P.O.V. of the prosecutors, the DA's police investigators, the judges and the defendants. Former *Cheers* regular Bebe Neuwirth topped the cast as Chief Proseutor Tracy Kibre, with Amy Carlson as her ADA colleague Kelly Gaffney. Representing the NYPD was Scott Cohen as Detective Chris Ravelli, and acting as liason between the two branches of law enforcement was Kirk Acevedo as DA's Investigator Hector Salazar.

In the two-part opener, it was established that Salazar was the partner of *Law & Order* alumnus Lennie Briscoe, who had become an investigator for the DA's office and was supposed to have been a *Trial by Jury* recurring character. Though Jerry Orbach was so debilitated by prostate cancer that he could barely speak above a whisper, he rose to the occasion with a spirited performance in his signature role. Alas, it turned out to be his final bow: Orbach died on December 28, 2004, three months before the new series' debut (on the show itself, Orbach's death was acknowledged in Episode Five as the other cast members returned from Briscoe's memorial service). In another crossover from *Law & Order*, Fred Dalton Thompson appeared in his familiar guise as no-nonsense District Attorney Arthur Branch, thereby joining that small and select fraternity of TV-show regulars (*Man from UNCLE*'s Leo G. Carroll, *The Six Million Dollar Man*'s Richard Anderson) who have appeared simultaneously on two different series in the same role. *Trial by Jury* also briefly crossed over with its sister program *SVU*, in a two-part episode based on the Michael Skakel murder trial and starring Angela Lansbury as a Kennedyesque matriarch.

Despite overwhelmingly positive press coverage, *Trial by Jury* was cancelled with startling suddenness after 12 hour-long episodes. "I was stunned," recalled Wolf to *Newsweek*'s Marc Peyser. "I had been told, 'Everyone loves it! It's coming back.' Then, the night before they announced the fall schedule—cancelled. It still pisses me off." Once he'd recovered from the blow, Wolf opted to make the best of the situation by re-using the expensive *Trial by Jury* sets for still another *Law & Order* spinoff, again concentrating on the "Order" part of the equation. Debuting exactly one year after the inau

gural episode of *Trial by Jury, Conviction* (the first of the spinoffs not to include *Law & Order* in its title, and the first with entirely different theme music) starred onetime *SVU* regular Stephanie March, literally returning from the dead as Chief Prosecutor Alexandra Cabot. It had already been explained on *SVU* that Cabot hadn't been bumped off by Colombian drug lords at all, but instead had disappeared into the Federal Witness Protection program. Now presumably immune from further attempts on her life, she was back on the job and in charge of a young prosecutorial team including Anson Mount as office manager Jim Steele, Eric Balfour as Brian Peluso, Julianne Nicholson (previously *CI*'s Detective Megan Wheeler) as Sara Finn, Jordan Bridges as Nick Potter, J. August Richards as Billy Desmond, and Milena Govich (later seen as *Law & Order*'s Detective Nina Cassidy) as Jessica Rossi. And just as in *Trial by Jury*, Fred Dalton Thompson was on hand as DA Arthur Branch, still putting in overtime on *Law & Order*. (Thompson would remain part of the Dick Wolf team until 2007, when he put his acting career on the back burner to make an unsuccessful bid for the U.S. Presidency.)

In a 2006 *Picturing Justice* article, Michael Asimow wrote, "In *Conviction*, Dick Wolf creates a prosecution office as different from *Law & Order* as it could possibly be." Asimow likened this latest series to such David E. Kelley lawyer dramas as *The Practice* and *Boston Legal*, citing the scripts' preoccupation with the private lives—or more bluntly, the sex lives—of its protagonists. Hoping to attract the 18- to 34-year-old audience demographic that had avoided *Trial by Jury*, Wolf uncharacteristically surrendered to the siren call of sensationalism, as if to say, "Okay, folks, you want prurience, you *got* prurience!" As Asimow dryly observed, "We see an ensemble of impossibly good looking and extremely horny 20- and 30- somethings on a crusade against crime and sleeping alone."

Nor did Wolf stop with mere carnality: virtually every one of *Conviction*'s principal characters brought a personal agenda or an inner demon to the workplace. Alexandra

Cabot pursued the philosophy of rejecting plea bargains and trying juvenile offenders as adults no matter what the circumstances in order to advance herself politically; "child of privilege" Nick Potter had given up a cushy job with a Wall Street firm so that he'd have the opportunity to actually meet with clients and try cases, never mind that he had neither the experience nor the maturity to do so; Brian Peluso was a compulsive gambler whose unavoidable association with underworld types threatened to compromise his effectiveness; Billy Desmond was a fatuous grandstander who jockeyed to be assigned only those cases he was certain to win; Christina Finn's yearning to prosecute a case after two years' confinement to paperwork was hobbled by her abject terror of everyone and everything; Jessica Rossi was tormented by a checkered past; and on it went.

Though the individual crosses borne by the main characters prevented the series from following the pure procedural throughline of *Law & Order*, *Conviction*— described by its producer as a "charactercedural"— was to be commended for its realistic depiction of the hopelessly overstocked and backed-up workload of a typical DA's office, forcing the principals to tackle anywhere from three to five cases per episode. But with this bid for authenticity, Wolf may have overreached himself: there were so many characters and so many subplots that the scriptwriters had no time to properly develop any one person or storyline. Unable to hold its audience for any more than a few minutes at a time, *Conviction* was axed after 13 episodes.

Though two of the franchise entries have died in infancy, the original *Law & Order* has continued to flourish, as have *SVU* and to a lesser extent *CI*, which moved to the USA Network in 2007 when NBC decided it only had room for the other two programs. And while we predicted at the outset of this entry *Law & Order* will still be a contender ten years hence, Dick Wolf, for all his success in ringing variations on a single basic theme, has refused to take anything for granted: "The wonderful thing about television is, it's kind of like life," Wolf philosophised in a 2006 interview. "Everything dies — they just don't give you the date of execution."

The Law Firm

NBC: July 28–August 3, 2005. Bravo: Sept. 6–Nov. 27, 2005. Executive Producer: David E. Kelley. Cast: Roy Black (Himself).

It had to happen: an NBC "reality" series combining the lawyer-show genre with competition-elimination format popularized by such programs as Donald Trump's *The Apprentice*. As originally conceived, *The Law Firm* was to have run for eight hour-long episodes, during which twelve young-and-hungry lawyers, broken down into teams of three or four, tried actual court cases with binding decisions, simulataneously competing for a grand prize consisting of a position in a prestigious law firm and $250,000. Overseeing the contest was attorney and legal analyst Roy Black, who provided running commentary and criticism for the contestants, at least one of whom was eliminated from the running in each episode. In the finale, Black would determine which of the two finalists was the weakest, sending that person on his/her way with the Donald Trump-like catchphrase, "The verdict is in. You're out."

Apparently series producer David E. Kelley, whose more laudatory efforts have included *Ally McBeal* and *The Practice* (both q.v.), had no one on his staff brave enough to talk him out of this fiasco. A bad idea was made worse by the series' deadeningly slow pace and crashingly dull courtroom sequences, though somehow Kelley managed to dredge up a few "money scenes": one in which a lawyer shocked his colleagues by loudly disagreeing with a verdict, another involving a lawsuit filed by a dominatrix. Only two of the planned eight episodes of *The Law Firm* were seen on parent network NBC; the remaining six were burned off on the Bravo cable channel.

Lie Detector

Syndicated: debuted January 1983. Sandy Frank Productions. Created by Ed Gelb. Cast: F. Lee Bailey (Himself, Host). Ed Gelb (Himself, Polygraph Expert).

Many people assume that the daily, syndicated 1983 legal series *Lie Detector* was famed criminal attorney F. Lee Bailey's first regular TV gig. Wrong. The man who became a courtroom legend by representing such controversial defendants as accused wife murderer Dr. Sam Sheppard, self-confessed "Boston Strangler" Albert DeSalvo, alleged My Lai Massacre participant Ernest Medina and kidnap victim-cum-fugitive Patty Hearst (and whose future held O.J. Simpson in store), had previously hosted the antiseptic 1967 ABC interview program *Good Company*, where Bailey traded pleasantries with everyone from Hugh Hefner to Senator Everett Dirksen. There are also those who assumed that the presence of F. Lee Bailey would actually make *Lie Detector* worth watching. Wrong again.

Actually, Bailey was only the series' front man: *Lie Detector* was created by Ed Gelb of the American Polygraph Association, who also appeared on the show (Gelb would enjoy a burst of prominence in the late 1990s when he conducted polygraph tests on the parents of 6-year-old murder victim Jon Benet Ramsey). Both Gelb and Bailey were careful to point out that the results of a lie detector are not admissable as evidence in any courtroom, with Bailey adding that a polygraph exam was "certainly not free of controversy," even though he regarded it as "the best tool science has" for determining lies. But so what? We've got Bailey, we've got Gelb, we've got that swell machine with all the bells and whistles; let's strap 'em in, flip the switch and have some fun!

Among the guest participants who willingly subjected themselves to Gelb's lie detector were gas station attendant Melvin Dummar, still determined to convince the world Howard Hughes really did bequeath him his entire fortune; daredevil Evel Knievel, anxious to quell rumors that he'd been frightened into pulling his parachute ripcord during his motorcycle leap over the Snake River; and Ronald Reagan's barber, ready to testify before God and County that he never, *never* put any coloring in Ronnie's hair. Even though the show's polygraph failed to prove anything conclusively other than the fact that some people can fool a machine and some can't, a steady parade of celebrities, semi-celebrities, wannabes and nobodies paraded through the studio and dutifully answered Yes and No to Gelb's queries. In retrospect, the most pathetic of the participants were those convicted criminals who hoped that by allowing themselves to be polygraphed they'd earn a few extra days' vacation from the hoosegow.

Despite its faults, the 170-episode *Lie Detector* made a lot of money when it was distributed domestically by the enterprising Sandy Frank (see also *The Judge*), once again demonstrating his uncanny knack for extracting the sweetest of lemonade from the sourest of lemons. A 2005 version of *Lie Detector*, co-hosted by Ed Gelb and news anchor Rolonda Watts, briefly showed up in Prime Time on the PAX network. Minus the presence of an F. Lee Bailey, this version didn't really qualify as a legal series and was not promoted as such.

Life's Work

ABC: September 17, 1996–July 29, 1997. Touchstone Television. Executive Producer: Warren Bell. Cast: Lisa Ann Walter (Lisa Hunter); Michael O'Keefe (Kevin Hunter); Alexa Vega (Tess Hunter); Luca and Cameron Weibel (Griffin Hunter); Molly Hagen (DeeDee Lucas); Andrew Lowery (Lyndon Knox); Larry Miller (Jerome Nash); Lightfield Lewis (Matt Youngster); Shashawnee Hall (Coach Brick); Lainie Kazan, Jenny O'Hara (Connie Minardi).

Every time a stand-up comedienne becomes even marginally famous, a flock of predatory TV producers swoop down in the hope of transforming the poor lady into a sitcom star. Sometimes the strategy works, yielding such worthwhile vehicles as *Roseanne*, *Ellen* and *Grace Under Fire*. Most often, however, we end up with things like *Life's Work*— the second sitcom effort by comedienne Lisa Ann Walter after the resounding failure of her first starring series, Fox's *My Wildest Dreams*.

This time out, Walter was cast as Lisa Hunter, newly matriculated from law school at Baltimore City College, and hired right out of the chute as Assistant State's Attorney. At the risk of making a sexist remark, Lisa was no babe in the woods: having put her ambitions on hold for seven years after marrying college basketball coach Kevin Hunter (Michael O'-Keefe) and bearing his two children Tess (Alexa Vega) and Griffin (Luca and Cameron Weibel), Lisa was finally fulfilling her life's dream of a legal career at the tender age of 33. Described by *Entertainment Weekly*'s Bruce Ferris as "a mouthy feminist trying to balance her job with her home life," the *über*-abrasive Lisa used her endless arsenal of barbs and wisecracks as a defense mechanism against the cruelties of the justice system and the sappy behavior of her coworkers: to her way of thinking, one had to act crazy to avoid going crazy. Larry Miller costarred as Lisa's grouchy, ulcerated boss Jerome; Molly Hagen was seen as Lisa's only female coworker and occasional verbal sparring partner, airhead conservative DeeDee; Andrew Lowery appeared as our heroine's most vicious on-the-job critic, unreconstructed chauvinist-pig Lyndon; and Lightfield Lewis was funky office helper Matt. Lainie Kazan was seen in the earliest episodes as Lisa's mom Connie, but it wasn't long before Jenny O'Hara stepped into the role.

Life's Work is a typical example of over-stocking a star vehicle with easy and convenient targets for the protagonist to shoot down — the laziest form of sitcomery. The only logical reason that Lisa's workplace was populated solely by goofballs and misfits was to give the heroine justification in tearing them apart (the pompous-ass Lyndon virtually had a bull's eye painted on his back). No genuine legal office would survive beyond painting the names on the door if it was really staffed by such losers. It was also hard to believe the series was produced in 1996 rather than 1976, or even earlier: *Variety*'s Ray Richmond noted that the show "alternates confusingly between 1950s and 1990s sensibilities.... Everyone reacts to Lisa as if she's the first intelligent and attractive woman they have ever seen in the workplace." Even more surprising, considering that star Walter cowrote the pilot episode with executive producer Warren Bell to emphasize her character's innate ability to juggle her duties at home and at the office, was that Lisa Hunter did not come off as being all that qualified for her work: indeed, she lost her first important

case by showing up late. Was the viewer supposed to sympathize with Lisa for being unable to cope with the pressures brought to bear against a mere woman by the insensitive Male Establishment? If so, she'd have been better off dropping her attitude, giving up law and resigning herself to a life of gingham aprons and dirty dishes.

And in case you think this writer is just another male sexist boor, consider the words of female legal author Christine Alice Corcos, who in an article for the *Picturing Justice* website dismissed the heroine in the mercifully short-lived *Life's Work* as "clumsy" and "clueless."

The Line-Up

Fox News Channel: debuted Oct. 15, 2005. Cast: Jamie Colby (Herself); Kimberly Guilfoyle (Herself).

An entry in the Fox News Channel's weekend "Crime Time Lineup," *The Line-Up* was a TV newsmagazine concentrating on current criminal and legal stories. In-depth reports and analyses were initially provided by host Jamie Colby, a lawyer and former CNN and CBS correspondent; later on, Colby was replaced by another lawyer-turned-TV personality, one-time Court TV anchor Kimberly Guilfoyle. In keeping with Fox News Channel's pursuit of the sensational, *The Line-Up* was advertised with the lurid tagline "Fugitives! Amber Alerts! Missing Persons! Captures! Tips!"; and though the series was supposed to feature several different stories per episode, the bulk of the airtime during the first season was dominated by the Natalie Holloway disappearance.

Lock-Up

Syndicated: 1959–1961. Ziv Productions. Produced by Jack Herzberg and Henry S. Kesler. Cast: Macdonald Carey (Herbert L. Maris); John Doucette (Lt. Jim Weston); Olive Carey (Casey).

The second of Ziv Productions' syndicated legal dramas (see also *Mr. District Attorney*), the weekly, half-hour *Lock-Up* was inspired by the career of Philadelphia defense attorney Herbert L. Maris, who during a long and distinguished career had saved some 300 wrongly convicted persons from life imprisonment or execution (the series went into production as *Philadelphia Lawyer*). Seventy-nine years old at the time the series debuted, the real-life Maris generously opened his voluminous files to Ziv executive Herbert Gordon, who'd read a magazine article about the veteran lawyer and sensed series potential. Maris did not object to the casting of the much-younger Macdonald Carey (previously starred on the Ziv medical drama *Dr. Christian*) as his TV counterpart, nor was he disturbed that the TV version of Herbert L. Maris was more an adventurer than in reality, forever putting his own life on the line in pursuit of justice and reaping such consequences as getting threatened, beaten up, kidnapped and shot at. The real Maris was even willing to accept the never-ending efforts by the "reel" Maris' mother-hen secretary Casey (played by Olive Carey, no relation to Macdonald but instead the widow of venerable cowboy hero Harry Carey) to find a suitable bride for her bachelor boss. Quoted in *TV Guide*, Maris dismissed these deviations from the facts by noting that *Lock-Up* was "well enough done that it won't — shall we say — do me any harm."

Most of the series' episodes opened with a pre-credits teaser, establishing that someone had been falsely accused of criminal activity and was thus in dire need of Maris' services. This was handled with considerable originality and panache: the debut episode's introductory scene dropped several enticing hints suggesting that a recently released convict was planning to bump off the man responsible for his arrest — and only at the story's halfway point was Maris convinced that the ex-con, by now his client, was being set up as a fall guy by the *real* villain. In another innovative touch, several of the episodes' epilogues were played in pantomime as the closing credits rolled.

To relieve the monotony of Maris' infallibility in choosing *only* innocent clients, occasionally the lawyer came up against an accused felon who was unmistakably guilty, at which point he would shift his attention to a secondary character who *could* be redeemed with his help. One of the many *Lock-Up* episodes in

current DVD circulation begins with Maris and his young assistant being taken hostage by a psychotic escaped prisoner. This guy turns out to be hopeless, so Maris makes it his mission to prevent the escapee's hero-worshipping kid brother from following in his sibling's footsteps. (Of course, in so doing Maris will be able to save his *own* neck, but this is treated as a minor consideration).

Outside of Macdonald Carey and Olive Carey, the only other *Lock-Up* actor who could be considered a regular was John Doucette as police lieutenant Weston, invariably expressing skepticism over Maris' gut instincts regarding guilt and innocence, but always generous with his praise for the lawyer's skill and dedication. Showing up in supporting roles were such celebrities-in-training as Leonard Nimoy, Dyan Cannon, Angie Dickinson, James Drury, Gavin MacLeod, Robert Conrad, and, in the role of a tremulous young damsel accused of smuggling microfilm behind the Iron Curtain, Mary Tyler Moore.

Despite the diminishing market for first-run syndicated drama series in the late 1950s, *Lock-Up* was reasonably successful, thanks in large part to Ziv's customarily aggressive promotional stategies. Hastily assembled to fulfill a September 1959 release commitment, the 78-episode series was still making new sales well into the mid–1960s.

Lyon's Den

NBC: Sept. 28–Nov. 30, 2003. Baby Owl Productions/ thinkfilm/ Brad Grey Television/20th Century–Fox Television. Executive Producers: Bernie Brillstein, Brad Grey, Rob Lowe, Daniel Sackheim, Kevin Falls. Executive Producer/Creator: Remi Aubuchon. Cast: Rob Lowe (John "Jack" Turner); Matt Craven (George Riley); Elizabeth Mitchell (Ariel Saxon); Kyle Chandler (Grant Rashton); James Pickens Jr. (Terrance Christianson); Frances Fisher (Brit Hanley); David Krumholtz (Jeff Fineman); Robert Picardo (Det. Nick Traub).

Having quit the popular political series *The West Wing* over a salary dispute, actor Rob Lowe soon found himself again standing before a Washington D.C. backdrop and playing an idealistic crusader in the same mold as his *West Wing* alter ego Sam Seaborn. The NBC legal drama *Lyon's Den* cast Lowe as attorney John "Jack" Turner, described in the series' official website as the "maverick scion from an American political dynasty"—that is, the son of a powerful U.S. Senator, played in the pilot episode by Rip Torn. (We won't belabor the obvious by explaining whom Jack Turner was *supposed* to be, except to repeat the five-word synopsis provided by an anonymous wit: "Mr. Kennedy Goes to Court.") Passionately plying his trade in a woefully underfunded D.C. pro-bono clinic, Jack was urged by his more pragmatic friend and colleague George Riley (Matt Craven) to accept a managing partnership with the clinic's sponsoring law firm, Lyon, LaCrosse & Levine. Jack finally agreed to do so, but only to save his clinic from going under.

Though we never saw the "Lyon" of Lyon, LaCrosse & Levine — except in the series' title, of course — we were given plenty of exposure to the firm's autocratic senior attorney, Terrance Christianson (James Pickens Jr.), and to the other members of the team: recovering alcoholic Ariel Saxon (Elizabeth Mitchell), picking up the pieces of a once-promising career; cynical, fiercely competitive Grant Rashton (Kyle Chandler), whom Jack disliked at first sight and vice versa; paralegal Jeff Fineman (David Krumholtz), who regarded the law as his own private playground rather than a serious business; and ambitious legal assistant Brit Hanley (Frances Fisher), willing to cut any and all available throats to advance herself. Each episode found Jack handling a different case of sociopolitical significance, questioning his own values and ethics while cast adrift in a sea of moral ambiguities.

According to Rob Lowe, *Lyon's Den* was originally concieved as a Western, in which he would play the new sheriff of a frontier town reeking with corruption and conspiracies. Asked by Mike Duffy of the *Detroit Free Press* to explain the format change, Lowe replied, "What's a western? It's a lone sheriff coming to town, the dusty, corrupt village. And everybody is out to get him...." It was series creator Remi Aubuchon who persuaded the actor that the show would play better (and presumably cost less) as a lawyer show. Nevertheless, many of the story elements retained the vestiges of traditional western plot devices. Jack Turner spent a great deal of his downtime helping police detective Nick Traub (Robert Picardo) learn the truth behind the highly suspicious suicide of Jack's legal mentor, all the while systematically exposing the sinister secrets, hidden intrigues and unholy alliances deep within the bowels of Lyon, Lacrosse & Levine. And in place of the "lawman-turned-outlaw" character so common to TV westerns, Steven Weber was cast in the recurring role of a former lawyer turned serial killer.

Quoted in *Entertainment Weekly*, Rob Lowe promised, "My stories are going to have meat every week. *The West Wing* was always fantastic, but it wasn't enough. Sam Seaborn was one of the great characters, but they just weren't giving him enough to do." Invoking the most successful courtroom series of the period, *Lyon's Den* creator Remi Aubuchon added, "[With *Law & Order*] Dick Wolf has done an amazing job developing a series about the legal aspects of how you prosecute a case. But I wanted to see how lawyers react to certain moralistic choices."

Harking back to the series' western roots, one could respond, "Them's mighty tall words, podner. Think you can back 'em up?" Well, the preshow buzz for *Lyon's Den* suggested that the series would be one of the biggest hits of the 2003–2004 season. Marc Berman, "programming insider" for the *MediaWeek1* website, had this to say on July 25, 2003: "The Scoop: Lucky Rob Lowe — six months ago the press was calling him the next potential McLean Stevenson for exiting a hit series. Now he's on a new show that looks potentially promising. The Reality: With the growing *Law & Order Criminal Intent* [q.v] as its lead-in and opposite ABC's re-tooled *The Practice* [q.v.], viewers bored with the David E. Kelley drama might find the legal-themed *The Lyon's Den* a satisfying alternative. Maybe Rob Lowe's departure from *The West Wing* wasn't such a bad career move after all."

As it happened, *Law & Order: Criminal Intent* was no help whatsoever as a lead-in, and viewers weren't *that* bored with *The Practice*.

Only ten weeks after its NBC debut, *Lyon's Den* rode off into the sunset. Rob Lowe went on to star in another flop called *Dr. Vegas* before reconciling himself to less screen time but more job security by returning to *West Wing*— accepting the salary he'd asked for when he ankled the series two seasons earlier.

The Mask

ABC: Jan. 10–May 16, 1954. Created and produced by Halsted Welles. Produced by Robert Stevens. Cast: Gary Merrill (Walter Guilfoyle); William Prince (Pete Guilfoyle).

No relation to the Jim Carrey film comedy (and subsequent animated-TV spinoff) of the same name, the ABC legal drama *The Mask* is historically significant in several respects. Telecast live from New York, it was the first prime time network series to be repeated twice a week, with the original broadcast seen on Sunday evening and a kinescope (a film taken directly off the TV screen) rerun the following Tuesday — or Wednesday, depending on the needs of the local stations. And it was the first hour-long dramatic series with a continuing cast of characters, forsaking the anthology format then common to 60-minute programs.

Gary Merrill and William Prince, both of whom later starred on the NBC lawyer show *Justice* (q.v.), were cast respectively as brothers Walter and Pete Guilfoyle, two attorneys who headed the firm of Guilfoyle & Guilfoyle. Knight errants in mufti, the Guilfoyles specialized in defending poor and persecuted clients against impossible odds, and also carried on a personal vendetta against organized crime. According to TV historian Robert Larka, "Each week's episode involved a homicide, and as the series progressed, the lawyer image was dropped and the Guilfoyle brothers became just another pair of private eyes."

In its very brief life span, *The Mask* boasted some impressive talent both behind and in front of the camera. The series was created by Halsted Welles, whose best-known screenwriting credit was the tense 1957 western *3:10 to Yuma*, and who later regularly contributed to such TV weeklies as *Bonanza, Mannix* and *Kojak*. A handful of episodes were adapted from the novels of mystery specialists Cornell Woolrich, Philip MacDonald and Frank Gruber, while legendary screenwriter/journalist Ben Hecht penned an original script for the program. And the guest-star roster included such seasoned character players as Luther Adler, George Macready and Jo Van Fleet, as well as such promising newcomers as Steven Hill, Cloris Leachman, Tom Tryon and, in the episode "The Party Night" (April 11, 1954), Paul Newman.

A number of factors doomed *The Mask* to an early demise: for one thing, the series was extremely expensive to produce; for another, its allegiance to the third-rank ABC network (which at the time had less than 20 primary affiliates in the country) made it impossible to attract a sponsor, forcing ABC to run it on a sustaining basis. After 15 one-hour episodes, *The Mask* (and no, I *don't* know the significance of the title) was cancelled — which in the long run turned out be beneficial to the careers of creator Halsted Welles and producer Robert Stevens, both of whom went on to lengthy associations with the popular suspense anthology *Alfred Hitchcock Presents*.

Matlock

NBC: Sept. 20, 1986–Sept 11, 1992; ABC: Nov. 5, 1992–Sept. 7, 1995. Dean Hargrove Productions/Fred Silverman Company/Matlock Company/Viacom. Executive Producers: Fred Silverman, Dean Hargrove, Andy Griffth, Joel Steiger. Cast: Andy Griffith (Benjamin L. Matlock); Linda Purl (Charlene Matlock); Kene Holliday (Tyler Hudson); Nancy Stafford (Michelle Thomas); Kari Lizer (Cassie Phillips); Julie Sommars (Asst. D.A. Julie March); David Froman (Lt. Bob Brooks); Clarence Gilyard Jr. (Conrad McMaster); Brynn Thayer (Leanne McIntyre); Daniel Roebuck (Cliff Lewis); Carol Huston (Jerri Stone).

Outside of Perry Mason, the best known, best loved and most enduring of TV's ficitional defense lawyers is Benjamin Layton Matlock. Beyond that, the two characters could not be more different. Mason is the stern, humorless, impeccably dressed, well-groomed, slightly distant father figure; Matlock is the affable, jocular, rumpled, tousle-haired, eminently approachable favorite uncle. In either case, if ever you were falsely accused of murder and needed the best defense money could buy, you needn't have gone any farther in the Yellow Pages than the letter "M."

Although he had undergone a professional eclipse in the years following his successful 1960–1968 CBS sitcom, Andy Griffith remained a welcome and profitable commodity on the local-rerun circuit; even those independent stations that had abandoned old black-and-white network programs in the daytime hours never removed *The Andy Griffith Show* from regular rotation — and, indeed, risked open rebellion from many viewers of a certain age if they ever had the audacity to drop the show. (This, by the way, is the only reference you'll find in this entry to Andy Griffith's superannuated fan base: There'll be no mention of Grandpa Simpson in *this* book). Aware of this phenomenon, and having seen Griffith play a defense attorney in the 1984 miniseries *Fatal Vision,* NBC Entertainment head Brandon Tartikoff felt the time had come for the actor to make his TV comeback in a lawyer show. Tartikoff asked Fred Silverman, himself a former network CEO and now an independent producer, to fashion a suitable vehicle for Griffith. With the success of the recently revived *Perry Mason* (q.v.) still fresh in his memory, Silverman and his *Mason* partner Dean Hargrove agreed that Griffith should shed his small-town-sheriff persona and be reborn as an unassuming but extremely shrewd Southern lawyer.

When ultimately the call was placed to Andy Griffith, it came at a time when the actor was convinced he would never work again. Already disheartened by his fading career and two failed marriages, Griffith was dealt a potentially fatal blow in 1983 when he came down with Guillain-Barre syndrome, which left him virtually paralyzed for seven months. Gratified that his services should still be in demand, and with his private life on the upswing with a third and lasting marriage, Griffith eagerly signed on to the project, willing to endure the first few months of filming on leg braces if it meant that he'd be back in business.

Introduced on March 3, 1986, with the two-hour pilot film *Diary of a Perfect Murder,*

Matlock began its regular weekly run the following September 20th. Though largely filmed in Griffith's home state of North Carolina (with pickup work at the former MGM facilities in Hollywood), the series was set in Atlanta. Even at this early date, the essential character of widowed attorney Ben Matlock was fairly well established. Despite his Harvard education, Ben assumed a shambling, shucking courtroom personality, using his bumptuous veneer to mask a brilliant and sometimes deadly legal mind, the better to throw his opponents completely off guard. Using every trick at his disposal to clear his client (need we add, the client was always charged with murder and always innocent) Matlock would blurt out statements that were obviously inadmissable as evidence, then, feigning wide-eyed innocence, sheepishly apologize to the judge and withdraw those statements — knowing full well that the jury would *not* disregard them as ordered. He also derived a childlike pleasure from trapping witnesses in their own web of lies, barely suppressing a chuckle or a whoop of triumph when he let drop the vital piece of evidence or zeroed in on the "one false step" that would simultaneously exonerate his client and place the noose around the actual culprit. Conversely, no one could display more slack-jawed astonishment and infantile outrage than Matlock whenever one of his clever strategies failed in court — or when opposing counsel used a similar strategy on *him.*

Unlike other series that featured colorful, eccentric defense attorneys who regularly trumped the more traditional prosecutorial establishment, *Matlock* was seldom subjected to criticism from actual lawyers. This may have been because each episode was carefully vetted for legal accuracy by Dennis Smith, a highly respected California attorney. Also, those dramatically compelling but laughably unrealistic courtroom confessions that were part and parcel of *Perry Mason* were not to be found on *Matlock*: when ol' Ben closed in for the kill and proved that one of the state's witnesses was the real murderer, the culprit seldom admitted his perfidy on the spot, but instead responded calmy, "I want an attorney" — or more often, said absolutely nothing at all.

While he charged a minimum $100,000 fee for his services, Matlock was endearingly tight with a dollar and carried on a lifestyle that could in no way be described as ostentatious. He lived in a roomy but unpretentious converted farmhouse in the pastoral Georgia town of Willow Springs, always wore a slightly seedy-looking light gray suit, drove an all-grey Ford Crown Victoria that had obviously seen better days, preferred to do his own cooking (hot dogs and fudge were his passions), and spent his spare time hunting, singing and banjo-plunking with his friends, raising chickens, and obsessively shining his shoes (including the bottoms!).

It has been suggested that the Matlock character was based on Daniel Cook, "Dean of the Georgia Criminal Defense Attorneys," who mostly handled high-profile personal injury cases. The bearded, deceptively bucolic Cook has appeared as a peripheral character (either under his own name or pseudonymously) in several noteworthy novels about Southern justice, including John Berendt's *Midnight in the Garden of Good and Evil;* more recently, he raised a few Georgian eyebrows with his fervent opposition of the Patriot Act. For his part, Andy Griffith told *TV Guide* that Ben Matlock's voice and gestures were inspired by his own father (Ben's dad Charley occasional showed up on the series, also played by Andy Griffith). The actor also claimed to have based certain of Ben's traits and idiosyncracies on himself: citing Matlock's preference for living the simple life despite his wealth, Griffith insisted it was "something I kinda share with this character. I've kept one foot in one part of the world [Hollywood] and one foot in another [North Carolina]." Two years after the series was cancelled, Griffith reaffirmed this point in another *TV Guide* interview, adding, "[Matlock's] a very bright, cagey guy. He's very vain and cheap. I always loved Jack Benny. And I do little takes of Jack Benny as Matlock."

There was even more to it than that. Never missing an opportunity to insist that in real life he bore scant resemblance to his *Andy Griffith Show* character Andy Taylor, the actor was also the first to admit that he had one of the shortest fuses in show business, and this was certainly invested in the personality of Ben Matlock, as were Griffith's actual likes, dislikes, personal prejudices (notably his distrust of people whom he felt were taking advantage of him) and occasional illnesses and infirmities (if Matlock was suffering from a bad head cold or a sore tooth, chances are that Griffith wasn't faking). Those who worked on *Matlock* confirmed that during the first two seasons, Griffith had a tendency to blow his top at the slightest provocation, just like Matlock, but that he was harder on himself than anyone else, just like Matlock, and as the years passed Griffith relaxed and mellowed, just like Matlock.

During the series' first season, Ben Matlock had a junior law partner in the form of his daughter Charlene, played by Lori Lethin in the pilot and Linda Purl thereafter; his young chief investigator was Tyler Hudson, a stock-market genius played by Kene Holliday. Charlene moved to Philadelphia to set up her own practice at the same time that Linda Purl left the series; her place in the firm was taken by Nancy Stafford as Michelle Thomas. Kene Holliday also left the show early on due to profound personal problems, to be replaced by Clarence Gilyard Jr. as ex-police deputy Conrad Mc-Masters. Another character, Matlock's office manager Cassie (Kari Lizer), was briefly added during this period, but it was Michelle and Conrad with whom Matlock would enjoy the most chemistry—especially Conrad, who frequently joined his boss during his weekend back-porch "musicales."

When *Matlock* moved from NBC to ABC in the fall of 1992, Nancy Stafford was replaced by Brynn Thayer as Matlock's heretofore barely mentioned *other* daughter Leann, likewise an attorney, who after breaking up with her husband joined her dad's law firm, which was promptly renamed Matlock & McIntyre. In 1993 Clarence Gilyard Jr. departed and Daniel Roebuck joined the cast as new legman Cliff Lewis, whom Matlock had begrudgingly hired to prevent his obnoxious cousin Billy Lewis (Warren Frost), Cliff's dad, from harping on the time Ben had broken the heart of Billy's sister. Finally, Carol Huston was added to the cast as another of Matlock's investigators, Jeri Stone, who like the departed Conrad McMasters enjoyed lifting her voice in song with her employer.

One of the most memorable of the supporting players appeared in only 42 of the series' 195 episodes: Julie Sommars as Julie March, the "wildest, most ruthless prosecutor" in Georgia. One of the few attorneys who relished the opportunity of facing down Matlock in the courtroom, Julie was not only Ben's most formidable legal opponent but also one of his closest and most trusted friends. There were hints aplenty that their relationship went far beyond the friendship stage, but both Ben and Julie were far too stubborn, independent and set in their ways to seriously entertain thoughts of matrimony.

Although Andy Griffith was adamant in his refusal to copy any of his familiar "Sheriff Taylor" mannerisms on *Matlock*, the series' guest-star pool included a number of former regulars from *The Andy Griffith Show*, notably Aneta "Helen Crump" Corsaut in the recurring role of a judge, Betty "Thelma Lou" Lynn as as woman named Sarah, Arlene "Millie Swanson" Golonka as *two* different women named Jackie, Jack "Howard Sprague" Dodson in the 1989 episode "The Cult," and best of all, Don Knotts, Barney Fife himself, as Les "Ace" Calhoun, aka "The King of Plastic." Introduced as one of Matlock's clients in the November 19, 1988, episode "The Lemon," Knotts showed up as Ace in several subsequent stories, rekindling the warm rapport he'd shared with Griffith back in the good old days. Finally, Sheldon Leonard, onetime executive producer of *The Andy Griffith Show*, made a unexpected appearance in the 1987 episode "The Gambler" reprising his unforgettable "Race Track Tout" character ("Hey, bud … c'mere a minute") from *The Jack Benny Show*.

While no one episode can be regarded as a classic, several of the *Matlock* entries are, within their own limitations, standouts. Three of the best were seen early in the series' run. The 1987 two-parter "The Network," all about the murder of a much-despised network executive, featured cameos by several then-current NBC series stars, including Corbin Bernsen, Rhea Perlman, Jason Bateman, Malcolm-Jamal Warner, Betty White, and even the title character from *ALF*. And on the occasion of the February 16, 1988 episode "The Hucksters," three different endings were filmed so that viewers could dial a 900 number to pick the murderer (this meant a different killer could theoretically pop up in each of the three major time zones, though in fact only one of the endings was actually shown). Finally, the 1986 episode "The Don" served as a dry run (if not an official pilot) for another Fred Silverman-Dean Hargrove lawyer show, *Jake and the Fatman* (q.v.)

During its years on NBC, *Matlock* was essentially a whodunnit, with the identity of the murderer withheld until just before the final commercial. After the series switched to ABC, the producers adopted the *Columbo* format of revealing the killer early in the proceedings, whereupon Matlock spent the rest of the episode trying to prove that the aforementioned killer, and not his own client, committed the foul deed. It has been theorized that the mystery angle was played down to accommodate Andy Griffith, whose screen time diminished as the series progressed due to health and age considerations. With the guest stars taking up the slack for a largely absent protagonist, there wasn't much time to trot out the traditional mystery-yarn trappings of red-herring characters, false leads, and the slow but steady accumulation of new evidence.

As early as the series' fifth season, Andy Griffith was telling the press that he was worn out and wanted to give up *Matlock*. The producers regarded this as the latest example of Griffith's traditional salary-negotiation process, citing the number of times that the actor threatened to leave *The Andy Griffith Show*, only to be lured back with more money and a better percentage deal. Ultimately, Griffith was persuaded to remain with the show at least until the end of the 1993–94 season. Officially on hiatus from that point onward, *Matlock* remained in play as a possible backup for one of ABC's cancelled shows, and in October of 1994 it served this function by rushing in to replace the unsuccessful cop drama *McKenna*. After *Matlock*'s final first-run episode in the spring of 1995, Andy Griffith reprised the character for "Murder: Two," a 1997 episode of Silverman and Hargrove's Dick Van Dyke vehicle *Diagnosis: Murder*. This appearance has given rise to the prevalent belief that Griffith and Van Dyke had made other crossovers when both of their series were up and running, though in fact Van Dyke's only *Matlock* appearance was in the role of a homicidal judge.

In the years since leaving Prime Time, *Matlock* has joined its spiritual predecessor *The Andy Griffith Show* as one of the permanent fixtures of the daytime-rerun field, playing to excellent response not only on hundreds of over-the-air stations but also on several cable services. And, just as in the case of *The Andy Griffith Show*, it ill behooves any station manager to even entertain the thought of removing *Matlock* from the daytime schedule, lest he incur the wrath of the series' legions of devotees.

Maximum Bob

ABC: Aug. 6–Sept. 15, 1998. Sonnenfeld Josephson Worldwide Entertainment/Warner Bros. Television. Executive Producer: Barry Sonnenfeld, Alex Gansa, Barry Josephson. Based on the novel by Elmore Leonard. Cast: Beau Bridges (Judge Bob Gibbs); Kiersten Warren (Leanne Lancaster); Liz Vassey (Kathy Baker); Sam Robards (Sheriff Gary Hammond); Rae'Ven Larrymore Kelly (Wanda Grace); Brent Briscoe (Elvin Crowe); Beth Grant (Inez Crowe); Peter Allen Vogt (Dirk Crowe); Paul Vogt (Bogart Crowe); William Sanderson (Dicky Crowe); T. Scott Cunningham (Hector Finch); Sam Trammell (Sonny Dupree); Garrett Dillahunt (Dep. Dawson Hayes).

ABC greenlighted the limited Summer series *Maximum Bob* principally because the property represented a collaboration between two very hot properties: author Elmore Leonard, upon whose 1991 novel the series was based, and producer-director Barry Sonnenfeld, then riding high on the contrails of his movie blockbuster *Men in Black*. Beau Bridges starred in the series as Judge Bob Isom Gibbs, the genially corrupt town boss of Deepwater, Florida. The ultra-right-wing, orchid-growing Gibbs was known as "Maximum Bob" because of the severity of his sentences, even for the most minor and insignificant of crimes (in one episode, he condemned a teenager to death for drinking a beer in public). Treating his courtroom as a pulpit, Maximum Bob brought an evangelical fervor to his calling; indeed, those brought before him were so enraptured by his fire-and-brimstone oratory that they all but shouted "Hallelujah!" and begged to be sent to jail for salvation! Saddled with a schizophrenic wife named Leanne (Kiersten Warren) who claimed to be possessed by the spirit of a 12-year-old black slave girl (Rae'Ven Larrymore Kelly), Bob demonstrated the full weight of his authority by first frightening Leanne into a divorce, then refusing to recuse himself from his ex-wife's trial on smuggling charges because he didn't feel like paying alimony.

The only person who seemed capable — or willing — to stand up to Bob's benevolent despotism was Miami–based public defender Kathy Diaz Baker (Liz Vassey), whom he'd earlier tossed in the slammer on contempt charges for refusing his sexual advances. Nonetheless, Kathy decided to stick around Deepwater, dividing her time between fighting Bob in the courtroom and helping him track down lawbreakers — which was just fine with Bob, who affectionately described Ms. Baker's backside as "two bulldogs in a bag." The supporting characters included ballroom-dancing sheriff Gary Hammond (Sam Robards), eternally-in-heat deputy Hayes (Garrett Dillahunt), and bird-watching crime boss Sonny Dupree (Sam Trammell). Other ingredients essential to the action were a soda-guzzling mermaid, a pistol-packing pizza guy, and an incestuous family whose members all wore identical eyeglasses.

Ostensibly, neither Barry Sonnenfeld nor ABC intended to extend *Maximum Bob* beyond its initial seven hour-long episodes. Even so, the series labored so strenuously to become a *Twin Peaks*-style cult item that it was fairly obvious *someone* was hoping viewers would rise up and demand its renewal for a full season. But despite a full complement of quirks and eccentricities, *Maximum Bob* failed to attract a following, and ABC did nothing to save the series when it ran its predetermined course in September, 1998.

McBride

Hallmark Channel: debuted Jan. 14, 2005. Alpine Medien Productions/Larry Levinson Productions/Hallmark Entertainment. Creator/Executive Producer: Dean Hargrove. Executive Producers: Robert Halmi Jr, Larry Levinson. Cast: John Larroquette (Mike McBride); Matt Lutz (Phil Newberry); Marta Dubois (Sgt. Roberta Hansen).

A star vehicle for John Larroquette, whose law-show credentials include *Night Court* and *Boston Legal* (both q.v.), *McBride* was not in the strictest sense a series, but instead a group of feature-length specials telecast on a monthly basis by cable's Hallmark Channel. Larroquette was cast as Mike McBride (the actor hadn't wanted the character to have a first name, preferring to be identified only as "M," but he was outvoted by the producers), a 12-year veteran of the Los Angeles Police Department. Upon his retirement as a detective in 1997, McBride remained in law enforcement as a public defender, setting up headquarters in a huge house owned by a previous client who was serving a life sentence (he also inherited the client's dog). Tagged as a "maverick" during his days as a cop, McBride carried over his unorthodox methods to the legal world, displeasing many with his questionable approach to his work, but invariably clearing his woebegone clients of murder charges. Inasmuch as the series was created by Dean Hargrove, the same man who pumped new blood into the *Perry Mason* (q.v.) franchise with a series of TV-movies in the 1980s and 1990s, lawyer McBride generally adhered to the *Mason* formula of trapping the real murderer during cross-examination, often backed up by key evidence uncovered at the last possible moment. Seen on a semi-regular basis were Marta Dubois as McBride's police contact and ex-lover Sgt. Roberta Hansen; and Matt Lutz as Phil Newberry, formerly an incompetent DA whom McBride had replaced in midtrial while serving on a jury (!), and latterly the protagonist's legman and "gadget guy," working for free because he was living off a sizeable trust fund.

McBride's most intriguing nuance occurred during the flashback sequences, in which Mike McBride inserted himself into the action, standing over the shoulders of both murderer and victim as he described what he believed had happened (occasionally he was wrong, requiring one or two revised flashbacks as the episode progressed). Several of the storylines were likewise off the beaten track: in one instance, McBride was called upon to clear his client of killing a woman who turned out to have several different identities — and who had allegedly been murdered several times before, resulting in an embarrassing superfluity of falsely accused prisoners. Otherwise, the show was standard stuff, right down to such musty old devices as a would-be killer shooting several holes into an already dead victim.

Beginning with "McBride: The Chamelon Murder" on January 14, 2005, *McBride* has been shown in rotation with a number of other movie-length Hallmark Channel whodunnits, including *Mystery Woman* with Kellie Martin, *Jane Doe* with Lea Thompson, and *Murder 101* with Dick Van Dyke.

Michael Hayes

CBS: Sept. 15, 1997–June 15, 1998. Baumgarten-Prophet Entertainment/New Regency Pictures/Trotwood Production / Columbia Tri-Star Television. Executive Producers/Creators-Developers: Nicholas Pileggi, Paul Haggis. Executive Producers: John Romano, Michael S. Chernuchin. Cast: David Caruso (Michael Hayes); Ruben Santiago-Hudson (Eddie Diaz); Mary B. Ward (Caitlin Hayes); Jimmy Galeota (Danny Hayes Jr.); David Cubitt (Danny Hayes); Hillary Danner (Jenny Nevins); Rebecca Rigg (Lindsay Straus); Peter Outerbridge (John Henry Manning); Jodi Long (Joan); Helen Slater (ADA Julie Siegel).

No sooner had actor David Caruso ascended to TV fame as Detective John Kelly on the ABC cop drama *NYPD Blue* than he decided he had outgrown the series, quitting the show cold in a bid for movie stardom. Even those critics and fans who thought Caruso was the greatest thing since sliced bread were worried that he was committing professional suicide; and when his first two starring films bombed, there was *Schadenfreude* to spare amongst the actor's envious and spiteful Hollywood colleagues. Bloody but unbowed, Caruso returned to television in 1997 with his own weekly, hour-long CBS legal series, *Michael Hayes*— no longer a "lowly" ensemble player as he'd been on *NYPD Blue*, but for all intents and purposes the Whole Shooting Match.

Cut from the same pattern as the intense, furrow-browed John Kelly, Caruso's Michael Hayes was an ex–NYPD cop turned Federal Prosecutor (he'd taken law courses in night school). When his boss was taken out of action by a would-be assassin, Hayes was thrust into the position of acting U.S. Attorney for the Southern District of New York. Tough, hard-driving and an unapologetic loose cannon, Hayes had no patience for politics or diplomacy: it was his job to put the Worst of the Worst behind bars, and no one was gonna stop him no matter how much they bellyached. Our hero not only bulldozed his way through the justice system but also ran roughshod over his harried but loyal staff: Chief Investigator Eddie Diaz (Ruben Santiago-Hudson); assistant U.S. Attorneys Jenny Nevins (Hillary Danner), Lindsay Strauss (Rebecca Rigg) and John Manning (Peter Outerbridge); and secretary Joan (Jodi Long). For all his ranting and railing, Hayes could be compassionate and even tender at times, especially when acting as surrogate patriarch for Caitlin (Mary B. Ward) and Danny Jr. (Jimmy Galeota), the wife and son of Michael's irresponsible ex-con brother Danny (David Cubitt), who further distanced himself from his family when he entered the Witness Protection Program. Early in the proceedings, Michael was humanized by the presence of a policewoman lover, but she was killed in the line of duty, allowing David Caruso ample opportunity for those mute, morose closeups he did so well — and so often.

Caruso came on so strong in the pilot episode that the CBS executives, worried that viewers unfamiliar with the actor's established bare-knuckle image might run screaming from their sets, insisted that a new "prequel" be hastily filmed, explaining and justifying Michael Hayes' combustible behavior. Brought in for this emergency operation was new executive producer Paul Haggis (late of *L.A. Law*, later of *Family Law* [both q.v.]), who received a "developed by" screen credit for his troubles. Noted Haggis in *TV Guide*, "Despite all the lawyer and cop shows on TV over the years, the [U.S.] attorney is one job that TV hasn't explored before. David Caruso is mesmerizing in this role. He can say more with the silences between the lines than most actors can say with all the words in the world." (Haggis was gone after eight episodes, reportedly because more than silence had passed between himself and the star.) Once the show was up and running, the tendency to heap the full weight of each plot crisis on Michael Hayes' shoulders at the exclusion of the supporting characters had the effect of making Hayes less an ordinary mortal and more a saint. As Salon.com's Joyce Milman noted in exasperation, "What the heck is this? *Jesus Christ, Federal Prosecutor?*"

Of course, most observers had a pretty good idea why this had been allowed to happen. The kindest remark that anyone made about David Caruso during this period was that he lacked objectivity about himself. The next-kindest remark was from a critic who wrote off the series as "an exhausting one-man show." After that, the critical commentary became pretty nasty, and it serves no purpose to go into further detail. Ah, but there's a happy Hollywood ending: four years after *Michael Hayes* wound up its sole season on the air, David Caruso got his groove back as star of Jerry Bruckheimer's *CSI: Miami* ... an ensemble show.

Miller's Court

Syndicated: 1982–1985. Metromedia. Cast: Arthur R. Miller (Himself).

Taped at the Boston studios of WCVB-TV, the weekly, half-hour *Miller's Court* was the first of two syndicated programs starring Harvard Law School professor Arthur R. Miller (see also *Headlines on Trial*). Celebrated in academic circles for his lively classroom sessions — notably his fondness for dressing up as famous historical characters to illustrate his lectures —

Professor Miller spent nearly two decades as legal editor for ABC's *Good Morning America* (more recently, he was an on-air contributor to cable's Court TV). In his own series, Miller presided over dramatizations of current legal issues, which were then discussed by a panel of prominent attorneys and a "layman" studio audience. Debuting as a local Boston program in 1982, *Miller's Court* entered syndication that same year, remaining in distribution for three seasons thereafter courtesy WCVB's then-parent company Metromedia.

Miss Match

NBC: Sept. 26–Dec. 15, 2003. 20th Century–Fox/ Darren Star Productions/Imagine Entertainment. Executive Producers: Brian Grazer, David Nevins, Jeff Rake, Darren Star. Cast: Alicia Silverstone (Kate Fox); Ryan O'Neal (Jerry Fox); David Conrad (Michael Mendelsohn): James Roday (Nick Paine); Lake Bell (Victoria); Jodi Long (Claire); David Alan Basche (Brian).

In her first Prime Time TV dramedy, an all-grown-up Alicia Silverstone (*Clueless, Batman & Robin*) starred as Kate Fox, divorce lawyer by day, matchmaker by night. Servicing her recently liberated clients, the good-hearted Kate made it her mission to link up these wandering souls with appropriate new mates, sometimes successfully, oftentimes less so. Looking askance at Kate's good works was her boss—and father—Jerald "Jerry" Fox (Ryan O'Neal), an attorney with no room for humanity on his schedule; likewise contemptuous of Kate's cupidity was her prickly co-counsel, Nick Paine (James Roday). And though fairly adept at mending broken hearts, Kate couldn't seem to connect with "Mr. Right" herself, though her erstwhile client Michael Mendelsohn (David Conrad) could have fit the bill had he not made all the wrong moves.

Miss Match was inspired by an actual person named Samantha Daniels—not an attorney, but instead the creator of the successful matchmaking service Samantha's Table. Daniels co-produced the series and made occasional guest-star appearances, admirably stifling the impulse to launch into an informercial whenever the action flagged. Shuffled off to NBC's

Friday-night "death slot," *Miss Match* managed to pick up a Golden Globe nomination before its cancellation after eight episodes (leaving the remaining five unshown to this day).

Miss Susan

NBC: March 12–Dec. 28, 1951. Main writer: William Kendall Clarke. Cast: Susan Peters (Susan Martin); Mark Roberts (Bill Carter); Helen Ray (Laura); Kathryn Gill (Mrs. Peck).

Having earned an Oscar nomination for her performance in *Random Harvest* (1942), Susan Peters was one of Hollywood's most promising young actresses when, in 1944, she suffered a severe spinal injury in a hunting accident. Though now confined to a wheelchair, the 23-year-old actress refused to give up her career and courageously attempted a comeback. Her first post-accident starring film was *Sign of the Ram* (1948), in which she played a vicious, conniving cripple who used her handicap to manipulate and control everyone around her. It was brilliant piece of work—but sadly, it was her last movie appearance.

Turning her attention to radio and television, Peters landed the lead role in *Miss Susan*, a daily, 15-minute NBC soap opera produced in the Philadelphia studios of WPTZ-TV (now known as KYW). The actress forsook her brief brush with villainy to portray Susan Martin, a wheelchair-bound attorney who'd left her big-city practice and returned to her home town of Martinsville, Ohio. No sooner had she put out her shingle than Susan found herself defending her housekeeper Laura (Helen Ray) on a theft charge. Later episodes focused on Susan's romance with Bill Carter (Mark Roberts), who wanted her to give up the Law so she could marry him and raise lots of kids (this being 1951, it was of course out of the question that she could handle a job and a family at the same time).

Throughout the series' nine-month run, Susan Peters was in very fragile health, but she valiantly soldiered on—while the producers, sensing they might lose their star at any moment, retitled the series *Martinsville U.S.A.* Ultimately, the show ended because producer

Procter & Gamble decided to hand the timeslot over to a new game show, *The Big Payoff.* (It has been suggested but not confirmed that P&G made this move because of Peters' increasing infirmities; the sponsor's representatives claimed that game shows were more profitable than soap operas, but that wasn't P&G's story in 1956 when they launched another lawyer serial, *The Edge of Night* [q.v.].) Susan Peters continued to pursue an acting career in such carefully selected stage vehicles as *The Barretts of Wimpole Street* until she died of complications related to kidney disease and pneumonia on October 23, 1952.

The Mississippi

CBS: March 25–May 6, 1983; Sept. 27, 1983–March 13, 1984. Hajeno Productions/Ralph Waite Productions/Warner Bros. Television. Executive Producer: Ralph Waite. Producers: Ed Waters, Liam O'Brien, Christopher Morgan, Robert Crais, Stanley Kallis. Cast: Ralph Waite (Ben Walker); Linda G. Miller (Stella McMullen); Stan Shaw (Lafayette "Lafe" Tate).

Actor Ralph Waite's first significant TV project after winding up his nine-year run as patriarch of *The Waltons* was the leisurely paced legal series *The Mississippi*, introduced by CBS on a tryout basis in the spring of 1983. Waite was cast as Ben Walker, a big-city attorney who forsook the Fast Lane to journey up and down the Mississippi river, purchasing his own sternwheeler riverboat for this purpose. Though officially retired, Walker was obliged to use his legal expertise at every port of call, most often by clearing someone falsely accused of a serious crime. His assistants and fellow travellers were garrulous former client Stella McMullen (Linda G. Miller) and troubled Vietnam veteran Lafe Tate (Stan Shaw), the latter an avowed pacifist who nevertheless got into at least one fistfight a week.

As executive producer of *The Mississippi*, Ralph Waite had a vested interest in the show's success, and he strived for realism in every aspect of the show, hoping to capture what *TV Guide*'s Robert MacKenzie described as "the look and character" of America's heartland. The series was filmed entirely on location along the

Mississippi in such river towns as LaCrosse and Prairie du Chien, Wisconsin, and Dubuque, Iowa. Whenever possible, Waite used genuine courtrooms for the trial scenes, hiring local nonprofessionals to play judges, bailiffs and court clerks — even when it meant innumerable retakes as these "real" people repeatedly blew their lines.

In the scripts themselves, Waite's Ben Walker was shown to be possessed of such integrity that he was willing to alienate old friends and turn conventional wisdom on its head in the interests of justice. In a typical episode — and one that would probably not get a pass from network executives of the early 21st century — a young military-academy cadet was accused of deliberately shooting a tormenting bully during a Civil War reenactment. Despite political pressure from the bully's father, who ran the academy, Walker revealed in open court that the victim had made homosexual advances to the defendant — and in order to clear his client, the old lawyer not only had to find out who had actually loaded the murder weapon, but also convince the jury that the defendant himself was neither gay nor acting out of fear of exposure. Perry Mason never had a day like this!

The Mississippi did well enough in its trial run for CBS to add the series to its fall 1983 lineup. Ralph Waite had wanted "realism," and he got it in spades, as the reality of the TV ratings system buried *The Mississippi* under the combined weight of NBC's *The A-Team* and ABC's *Happy Days*.

Mister District Attorney

Network version: ABC: Oct. 1, 1951–June 23, 1952. Created and produced by Edward A. Byron. Developed by Phillips H. Lord. Cast: Jay Jostyn (Mr. District Attorney); Len Doyle (Harrington); Vicki Vola (Miss Miller).

Syndicated version: originally released 1954–1955. Ziv Productions. Supervisors of Production: Jack Herzberg and Herbert L. Strock. Cast: David Brian (District Attorney Paul Garrett); Jackie Loughery (Miss Miller). [By contractual agreement, Edward A. Byron was not credited in the syndicated version, with sole credit for "development" going to Phillips H. Lord.]

Long before he was reduced to the punchline of a story about a premature post–Presidential election *Chicago Tribune* headline, Thomas E. Dewey was the fearless, crusading District Attorney of New York City. Dewey's relentless prosecution of such vicious gangsters as Lucky Luciano and Dutch Schultz earned him the nickname "The Gangbuster" and made him the *a clef* hero of films like 1937's *Marked Woman* (with Humphrey Bogart as a thinly-disguised Dewey) and 1938's *I Am the Law* (Edward G. Robinson, ditto). Dewey also served as the model for the title character in the long-running radio series *Mister District Attorney*, created by former law student Edward A. Byron and developed by Phillips H. Lord — whose best-known radio effort was, appropriately, *Gangbusters*.

Making its NBC debut April 3, 1939, as a replacement for the nightly comedy series *Amos 'N' Andy* (which had defected to CBS), *Mister District Attorney* began life as a 15-minute serial, with Dwight Weist as the title character, who throughout the series' network run was never identified by his given name, only as "Mister District Attorney," "Boss" or "Chief." This early version, of which a handful of episodes still exist, found the protagonist faithfully following in Dewey's footsteps by targetting a powerful extortion gang. To quote radio historian Jim Harmon, it was obvious that "justice could not be done in fifteen minutes," so the series was expanded to a full half-hour on October 1, 1939, eschewing serializations in favor of self-contained stories, and with Raymond Edward Johnson (best known as the unctuous host of the suspense anthology *Inner Sanctum*) taking over as star. By 1940, the series' most familiar format was firmly in place, with Jay Jostyn (previously heard on the series in minor roles and as the announcer) as Mr. District Attorney, Vicki Vola as his faithful, briskly efficient secretary Miss Edith Miller and Len Doyle as Mr. District Attorney's oafish but loyal chief investigator Harrington.

As the series progressed, Mr. District Attorney's activities expanded far beyond merely bashing organized crime, as the hero went after bank robbers, jewel thieves, kidnappers, con artists, juvenile deliquents, serial murderers (then known as "thrill killers"), corrupt public officials, and — once America entered World War II — Nazis, Fifth Columnists, saboteurs and crooked war profiteers. It was during this period that Ed Byron established a reputation for his encyclopedic knowledge and understanding of crime and criminals (he had a personal library of 5000 books on the subject), his fluency in current gangster jargon, and his close contacts with both law enforcement officials and minions of the Underworld. Something of a precursor to Dick Wolf's *Law & Order* franchise (q.v.), *Mister District Attorney* not only ripped its stories from current headlines, but often anticipated and predicted late-breaking news events. John Dunning, author of the all-inclusive radio history tome *On the Air*, records several examples of Byron beating the newspapers and even the Government to the punch with his eerily prescient stories of phony sanitariums, gasoline hijackers, and deranged war veterans: and when in 1942 *Mister District Attorney* featured a story about German submarines depositing spies on the Atlantic coast a scant few days before a group of real Nazi spies was captured under similar circumstances, Byron received a not-so-cordial visit from the FBI.

Like most crime shows of the period, *Mister District Attorney* boasted an aggressive rather than passive hero, with the title character often as not abandoning his office to personally hunt down lawbreakers, frequently using his own fists to administer on-site justice. Mocking this lapse of accuracy (not unique to Byron's series) in a 1963 *TV Guide* article, famed defense attorney Louis Nizer noted, "real district attorneys, living sedentary lives at their desks, must remove their silver-rimmed eyeglasses in amazement at this depiction." But Byron knew what the public wanted, and *Mister District Attorney* remained a network-radio fixture until June 13, 1952, by which time it had moved from NBC to ABC.

The radio version was still going strong when, in October of 1951, the TV arm of ABC introduced a video version of *Mister District*

Attorney, telecast live from New York and seen on an alternate-week basis with another legal drama, *The Amazing Mr. Malone* (q.v.). In an unusual move, the radio cast of *Mr. District Attorney* was transferred intact to television, with Jay Jostyn as the still-unnamed protagonist, Vicki Vola as Miss Miller and Len Doyle as Harrington. The TV version continued to draw its material from current events, and also retained the radio version's stirring theme music by Peter Van Steeden and opening signature, delivered with gusto by "Voice of the Law" Maurice Franklin: "*Mister District Attorney. Champion of the People! Guardian of the fundamental rights to life, liberty and the pursuit of happiness!*" And as ever, this intro was followed by the echo-chambered voice of Jay Jostyn: "*…and it shall be my duty as district attorney not only to prosecute to the limit of the law all persons accused of crimes perpetrated within this country but to defend with vigor the rights and privileges of all its citizens…*"

The radio edition of *Mister District Attorney* ended on June 13, 1952, with the TV spinoff following suit ten days later. That same year, independent producer Frederick Ziv acquired the rights to the property and produced a half-hour, 52-episode radio revival for local syndication. The Ziv version starred David Brian as the title character, now at last given a name, Paul Garrett. Brian also headlined Ziv's subsequent filmed, syndicated TV version of *Mister District Attorney*, replacing Jay Jostyn — who despite having essayed the role for 12 years on radio had not registered well on TV because the 50-year-old actor seemed too young for the role! (The shock of losing the part must have caused Jostyn to age overnight: by the time he appeared on the syndicated *Night Court* in 1958, he looked older than dirt.)

Debuting in most markets in the spring of 1954, Ziv's *Mister District Attorney*, like the producer's "canned" radio version, bore only a superficial resemblance to the network original. To be sure, the new version still began with that inspirational "Champion of the People!" opening narration, followed by the hero's promise to "prosecute to the limit of the law,"

etc. etc.; but Peter Van Steeden's theme music had been replaced by a less expensive stock theme ("Brutal Regiment") from the all-too-familiar Music For Television library. Also, by this time DA Paul Garrett had pretty much given up on organized crime and was focusing almost exclusively on isolated gangs and individual perpetrators. While Garrett's secretary Miss Miller, now played by former Miss USA Jackie Loughery (see also *Judge Roy Bean*), was still on the job, his radio sidekick Harrington had disappeared, and now Garrett worked in tandem with the regular police and the Feds. Virtually all pretense of reality had gone out the window, with Garrett seldom wasting time at his desk so long as he could don his trenchcoat and personally hunt down the bad guys: in one episode he jovially confessed, "There's nothing I like better than playing cops and robbers." (Incidentally, on those few occasions Garrett was actually in his office, he was headquartered at the Los Angeles City Hall building — the same edifice that stood in for *The Daily Planet* on TV's *The Adventures of Superman*.)

Though the depiction of Garrett was more fanciful than ever, the syndicated *Mister District Attorney* retained a veneer of authenticity by filming extensively on location in the streets of Los Angeles, and at various actual law-enforcement locales — courtrooms, forensic labs, interrogation chambers and the like. Ziv was careful to build up a strong relationship with local LA police officials, employing as technical advisor a 33-year-old police sergeant who ended up contributing several scripts to the series. Ultimately, this young man gave up police work to become a full-time screenwriter — and that, kiddies, is how the career of Gene Roddenberry was born.

One of Ziv's more popular 1950s efforts, *Mister District Attorney* yielded 78 episodes, 52 of which were lensed in color (but generally shown in black and white). The filmed series remained in active circulation well into the 1960s, by which time the original radio version of *Mister District Attorney* was nothing more than a title to all but the most dedicated of pop-culture buffs.

Moral Court

First-run syndication: Oct. 2, 2000–Sept. 28, 2001. ION (all reruns): April–June 2007. Edwards-Billett Productions/Warner Bros. Television. Executive Producer/Creator: Stu Billett. Cast: Larry Elder (Host/Judge); Vivian Guzman (Court Reporter); Russell Brown (Bailiff).

Most daily, syndicated courtroom shows have provided the public with practical and sometime painful examples of the Legal System at work, illustrating that it doesn't really matter who is right or wrong, but instead which of the two plaintiffs has presented the most persuasive argument before judge and jury. Created by Stu Billett of *People's Court* (q.v.) fame, *Moral Court* lived up to its title by ignoring the subtle complexities of American jurisprudence and offering verdicts that were morally rather than legally "right."

Billett had originally approached Conservative radio talkshow host Dennis Prager to appear as *Moral Court*'s ersatz judge, but Prager didn't test well with preview audiences and Libertarian radio chatmeister Larry Elder was hired instead. Throughout each hour-long episode, disclaimers repeatedly assured viewers that Larry Elder was not a real judge, that Russell Brown was not a real bailiff (he didn't even wear a uniform), that the cases heard were not actual small-claim litigations, and that the verdicts were non-binding. To further distance the series from genuine courtroom procedure, the litigants were billed not as Plaintiff and Defendant but instead as "Accuser" and "Accused"—and if you happened to be the "Accused," you entered Judge Elder's courtroom at your own peril. Clearly predisposed to take the Accuser's side when listening to tales of woe involving unpaid debts, unkept promises and personal betrayals, Elder levied fines against the Accused based on the degree of the "crime." If it was just a minor misunderstanding or conflict of interest, Elder awarded the Accuser $500; if the Judge found the moral issue "Offensive," the Accuser got $1000; and if the case *really* got Elder's goat, he'd scream "Outrageous!" and lavish the maximum $2000 upon the aggrieved party. (Of course, the Accused never had to pay a cent, since all financial transactions were handled by the producers; still, it was extremely embarrassing to be exposed as a Grade-A heel on national television.) The series' ad copy wasn't kidding when it proclaimed *Moral Court* as the show, "Where it PAYS to be right."

Throughout the proceedings, "court reporter" Vivian Guzman interviewed the plaintiffs and members of the studio audience, asking how they thought the show was going and soliciting predictions as to the outcome. There had been similar "vox populi" moments on Stu Billett's *The People's Court*, but none were as lively or opinionated. (We'd add that none were as much fun to watch, but that depends on your own definition of "fun.")

Because *Moral Court* was still in active syndication as late as 2006 (and also showed up on the ION Network's Prime Time schedule in the Spring of 2007) many people have assumed the series was in production for longer than a single season, though in fact its final first-run episode was seen in September of 2001. A radio version of *Moral Court* was also heard for a brief period, with attorney Gloria Allred—whose political ideology was as far from Larry Elder's as it was possible to get without leaving the galaxy—serving as host.

Morning Court *see* Day in Court

Murder One

ABC: Sept. 19, 1995–Apr. 23, 1996; Oct. 10, 1996–Jan. 23, 1997; May 25–29, 1997. Steven Bochco Productions/20th Century–Fox Television. Executive Producers: Steven Bochco, Charles H. Eglee, William M. Finkelstein, Michael Fresco. Cast: Daniel Benzali (Theodore Hoffman: 1995–96); Patricia Clarkson (Ann Hoffman: 1995–96); Michael Hayden (Chris Docknovich); J.C. MacKenzie (Arnold Spivak); Mary McCormack (Justine Appleton); Grace Phillips (Lisa Gillespie: 1995–96); Barbara Bosson (ADA Miriam Grasso); Dylan Baker (Det. Arthur Polson: 1995–96); Kevin Tighe (David Blalock: 1995–96); Gregory Itzen (DA Roger Garfield); Jason Gedrick (Neil Avedon: 1995–96); Stanley Tucci (Richard Cross: 1995–96); John Fleck (Louis Heinsbergen); Vanessa Williams (Lila Marquette: 1995–96); Anthony LaPaglia (James Wyler: 1996–97); David Bryan Woodside (Aaron Moseley: 1996–97); Missy Cryder (Sharon Rooney: 1996); Ralph Waite (Malcolm Dietrich: 1996–97); Clayton Rohner (Vince Biggio: 1996–97); Rick Worthy

(Rickey Latrell: 1996–97); Pruitt Taylor Vance (Clifford Banks: 1997).

The success of such series as *Hill Street Blues, St. Elsewhere* and *NYPD Blue* has enabled producer Steven Bochco to experiment with chancy and controversial projects, not so much to please the crowd as to please himself. Critics have applauded Bochco for the courage to try out such curiosities as *Cop Rock*—a bizarre and baffling blend of crime story and musical comedy—which, though they bombed with the public, at least proved that the producer was willing to go against the flow in pursuit of free artistic expression. (Of course, sometimes even his staunchest supporters turned their backs on Bochco, as witness his misbegotten animated political satire *Capitol Critters*.) Unlike many of his other experimental efforts, *Murder One*—Bochco's first legal series since the cancellation of *L.A. Law* (q.v.)—managed to please both critics and viewers … at least temporarily.

Anticipating a format that would later be adopted and refined to perfection on Fox's espionage thriller *24*, the first season of *Murder One* focused primarily on a single murder case throughout the entire first season, from discovery of the body to the final jury verdict. Bearing such titles as "Chapter One" and "Chapter Two," each episode began where the previous one left off, with viewers brought up to date via 90-second summaries provided by the fictional "Law TV" cable channel. Some observers suggested that Bochco had been inspired by the public's fascination with the recent O.J. Simpson trial, noting that Bochco's then-wife Barbara Bosson, cast as chief prosecutor Miriam Grasso, had researched her role by sitting in Judge Lance Ito's courtroom. There were also rumors that the first season's principal character, defense attorney Ted Hoffman (played by Daniel Benzali), was based on real-life attorney Howard Weitzman, who'd stepped down from the O.J. defense team just as the most damning evidence against Simpson began pouring in. Though Bochco dismissed these comparisons, the fact remained that Weitzman served as the series' technical advisor.

The case dominating the first 23 episodes of *Murder One* was set in motion by the murder of 15-year-old Jessica Costello, whose naked, bound and strangled corpse was found surrounded with narcotics and drug paraphernalia. The prime suspect in what the press called "The Goldilocks Murder Case" was wealthy businessman and philanthropist Richard Cross (Stanley Tucci), who owned the building where the killing took place and who had been having an affair with the victim's sister. Though Cross appeared to be guilty as hell, the legal team headed by tall, imposing, bald-domed, fundamentally honest but ethically ambivalent Ted Hoffman—described by actor Daniel Benzali as "the Philip Marlowe of the 1990s," doing his best to survive and maintain his integrity "in a world gone wrong"—was determined to give their client the best possible defense. Claiming that Bochco had never told him whether Cross was guilty or not, Stanley Tucci superbly played the character right down the middle, keeping viewers guessing as he charmed and manipulated everyone around him. And in a brilliant bit of Bochco sleight-of-hand, Cross was abruptly dropped as Number One Suspect in favor of spoiled young movie star Neil Avedon (Jason Gedrick).

The rest of the Season One cast list included Hoffman's associates Justine Appleton (Mary McCormack), Chris Docknovich (Michael Hayden), Arnold Spivak (J.C. MacKenzie) and Lisa Gillespie (Grace Phillips); sassy legal-office receptionist Lila Marquette (Vanessa Williams); obsequious officer manager Louis Heinsbergen (John Fleck); Hoffman's sympathetic, supportive wife Ann (Patricia Clarkson); DA Miriam Grasso's politically ambitious superior, DA Roger Garfield (Gregory Itzin); persistent police detective Arthur Polson (Dylan Baker in a role for which, ironically, Stanley Tucci was orginally considered); and Polson's veteran investigator David Blalock (Kevin Tighe), who was killed off early in the action.

To break up the monotony of the single-case throughline, the lawyers were shown dealing with other clients at the same time they were defending Richard Cross and Neil Avedon, most of the "outside" cases serving as object lessons and growth experiences for Hoffman and his team (these secondary stories had been developed by Bochco as backups in case the main plotline failed to click with viewers). Refreshingly, the prosecutors and defenders were not depicted as sworn enemies constantly at each other's throats, but as mature professionals willing to cooperate and compromise in the interest of Justice. Some of the series' best scenes were the businesslike conferences involving the judge, the prosecution and the defense as they discussed the day's upcoming court proceedings, carefully parsing the words to be used during questioning and cross-examination. Likewise entertaining and enlightening were the pithy dialogue exchanges indicating that if one lawyer did another a favor, payback was expected somewhere down the line — exchanges spoken completely without malice or contempt, but with calm, cleared-eyed pragmatism.

Murder One was heavily promoted in the weeks prior to its September 1995 debut, not only via a personal-appearance tour undertaken by star Daniel Benzali, but also with TV ads shown on 10,000 Delta Airlines flights and in 3000 Blockbuster video stores. Despite stiff competition from NBC's *ER* (ABC Entertainment president Ted Harbert explained, "Somebody has to go up Pork Chop Hill. We are sending our best soldier") — the series posted decent ratings and was enthusiastically received by both viewers and reviewers. But once the initial novelty wore off, ratings began to flag; and when in the final episode the murderer's identity was revealed, fans were sorely disappointed, many of them dashing off angry letters to Bochco stating that they couldn't care less whether the series was renewed. Some observers suggested that the viewer dropoff was due to audience fatigue over the O.J. coverage and recent revelations of big-city police and judicial corruption; others noted that the pendulum had swung away from Bochco's *L.A. Law* in the past few seasons, and prosecutors rather than defense attorneys were now regarded as the heroes of coutroom shows; and still others, dismayed over the series' lack of *L.A. Law*–style histrionics, perceived that Bochco had grown soft over the years.

The knee-jerk reaction from the production team was to alter the series' format and replace many of the principal actors. Abandoning the one-big-case-per-annum premise, the second season of *Murder One* (now in an even more precarious timeslot, opposite NBC's *Seinfeld*) featured two different cases. One of these, despite the perceived "audience fatigue" mentioned in the previous paragraph, was blatantly inspired by the Simpson trial: Rick Worthy appeared as Ricky Latrell, a basketball superstar charged with murdering a duplicitous team owner. The other case, which came first in the chronology, involved the murder of a controversial California governor and his paramour (sounds a bit Clintonesque, but try to get anyone to admit it). Gone from the regular cast were Daniel Benzali, Vanessa Williams, Patricia Clarkson, Grace Phillips, Dylan Baker, Kevin Tighe, Stanley Tucci and Jason Gedrick. New to the proceedings were Anthony LaPaglia as Theodore Hoffman's younger and less follicle-challenged replacement James "Jimmy" Wyler; David Bryan Woodside as Wyler's hotheaded associate Aaron Mosely, replacing Lisa Gillespie; Missy Crider as Sharon Rooney, principal suspect in the Governor's murder; and Ralph Waite as ruthless political power broker Malcolm Dietrich, in cahoots with first-season holdover Roger Garfield — who in his rise from DA to acting governor had gone completely over to the Dark Side. Those who'd enjoyed the comparative restraint and decorum of *Murder One*'s inaugural season were distressed by the *sturm und drang* that infested the series' second year on the air. And if viewers felt betrayed by the out-of-left-field resolution of Season One, they were positively apoplectic over the ludicrous finale of Season Two, in which the O.J. clone confessed to murder while on the stand — only to be acquitted because the jury was impressed by his honesty!

On this discordant note, *Murder One* was cancelled in mid-season, resulting in a hue and cry from those faithful few who still felt the show was worthwhile. The series briefly returned in May of 1997 with a final three-episode story arc, in which James Wyler defended serial killer Clifford Banks (Pruitt Taylor

Vance), better known as "The Street Sweeper" because he only bumped off violent criminals. While this climactic three-parter earned praise for its stylish camerawork, some viewers detected a whiff of sour grapes on the part of Steven Bochco, who, in a scene showing the Street Sweeper negotiating a movie deal with several disreputable-looking network suits, may well have been taking a swipe at ABC for insufficiently supporting his series.

Readers who would prefer to decide for themselves whether *Murder One* was a *succes d'estime* or a stinker are advised to punch up the series on the Internet; as of this writing, all 41 of the series' hour-long episodes are available on the hulu.com website.

The New Perry Mason *see* Perry Mason

Night Court

Syndicated: debuted Sept. 1958. Sandy Howard Productions/Banner Films. Created by Sandy Howard. Produced and directed by Sanford Spillman. Associate Producers: Jesse Corallo and Robert Collins. Technical advisor: Deputy Marshal Commissioner Ed Faulkner. Cast: Jay Jostyn (The Judge); Phil Tully (Court Clerk); Barney Viro (Public Defender).

No relation to the later NBC sitcom of the same name, the simulated-courtroom series *Night Court* first aired April 18, 1958, as a live local program from the Los Angeles studios of KTLA. Unlike many of the other LA–based courtroom series of the time, which were largely ad-libbed from "fact sheets" provided to the actors, *Night Court* was fully scripted, save for the extemporaneous comments from the show's "judge," actor Jay Jostyn. Previously the star of the *Mister District Attorney* radio series (see separate entry), Jostyn had been carried over to the 1951 TV version, only to be replaced by David Brian because he appeared too young for the role. But by the time *Night Court* rolled around, Jostyn seemed far older than his 57 years, thanks to his sour-pickle countenance and bellicose attitude.

Writer-producer Sandy Howard (see also *Congressional Investigator*) described *Night Court* as a "frank" show, noting that all cases

presented on the series were based on fact, with special emphasis on such no-punches-pulled topics as dope addiction, public intoxication, sexual assault, adultery, pornography, parental neglect and indecent exposure. Howard also insisted his show was a faithful replication of an actual night court, which processed as many as 30 cases per hour; when the series was seen in its original 60-minute format, Judge Jostyn heard up to a dozen cases per episode, each running between four to eight minutes. Carrying realism to the nth degree, Howard even intermingled amateur performers with professional actors in the earliest episodes, though this eventually proved impractical.

When *Night Court* was prepared for national syndication the fall of 1958, the series was whittled down to 30 minutes per episode; at the same time, the on-screen title was changed to *Night Court, U.S.A.*, indicating that its case load was no longer harvested merely from California courtroom records, but from the files of the "inferior courts" (the series' own description) throughout the nation. Filmed at Paramount Pictures' Sunset Studios in Hollywood, the 78 episodes were hastily shot on an assembly-line basis, using multiple cameras and lengthy, uninterrupted takes. While the actors playing the litigants and the attorneys were expected to learn their dialogue, Jay Jostyn was obviously making up most of his lines as he went along, using the "court documents" laid out before him as his continuity guide.

As in the original LA version, several different cases were processed during a single *Night Court* episode: the less serious and more frivolous cases were quickly disposed of, while the ones with the most dramatic or sensational value were played for all they were worth. Each defendant was given the opportunity to plead Guilty or Not Guilty: those taking the latter option would be offered the further choice of a getting an immediate decision from the bench or going to trial. That shameless old barnstormer Jay Jostyn was a joy to behold as he balefully chewed out the more contentious defendants, never missing the opportunity to rail against those "subversives" who complained about the American justice system. But for all his hamminess, Jostyn was given plenty of competition by the overloud, overgesticulating plaintiffs, many of whom behaved as if they were still auditioning for their roles. (Most of the actors were anonymous day players, none of them receiving screen credit: the only ones I have been able to firmly identify are Alvin Childress [onetime "Amos" on the *Amos 'n' Andy* TV show], David Hoffman and Frank Mills.)

Some of the series' best moments were also, in retrospect, the funniest (one public-domain DVD service has gone so far as to categorize *Night Court* as a comedy). In one episode, a woman sobbingly explained that she'd absent-mindedly stolen items from a department store because she was still giddy over her boyfriend's sudden marriage proposal: digesting this, Judge Jostyn agreed that it was a "beautiful story"— and had also been a beautiful story when the woman used the same excuse after being arrested for shoplifting ten years earlier. In another instance, Jostyn was clearly caught off guard when a man charged with stealing a duck from a local park was brought before him — along with the duck, demonstrably unhappy about making its TV debut. Then there was the episode in which a seedy-looking cowboy actor was arrested for allowing his stallion to run wild through the city streets, just to impress his girlfriend. Though he valiantly tried to maintain a straight face, Jostyn nearly lost it when the cowboy passionately proclaimed, "Ah'm gonna take after that gal like a honeysuckle takes after the front porch!"

Syndicated by Banner Films, *Night Court* was seen on a Monday-through-Friday basis in most markets. The series remained in distribution until the early 1960s, sometimes retitled *Courtroom U.S.A.*; and in the mid–1990s, it resurfaced on cable's Nostalgia Channel as *Vintage Court U.S.A.*

Night Court

NBC: Jan. 4, 1984–July 1, 1992. Starry Night Productions/ Warner Bros. Television. Creator/Executive Producer: Reinhold Weege. Executive Producers: Chris Cluess, Stu Kreisman. Cast: Harry Anderson (Judge Harry T. Stone); Karen Austin (Court Clerk Lana Wagner: 1984); Selma Diamond (Selma Hacker: 1984–85); Richard Moll (Bailiff Nostradamus "Bull" Shannon); John Larroquette (ADA Dan Fielding); Paula Kelly (Liz Williams: 1984); Charlie Robinson (Court Clerk Mac Robinson: 1984–1992); Ellen Foley (Billie Young: 1984–85); Markie Post (Christine Sullivan: 1985–92); Florence Halop (Florence Kleiner: 1985–86); Marsha Warfield (Roz Russell: 1986–92); S. Marc Jordan (Jack Griffin: 1990–91); Joleen Lutz (Lisette Hochstetter: 1990–92); Terry Kiser (Al Craven: 1984); William Utay (Phil/Will Sanders: 1985–86, 1989–92); Bumper Robinson (Leon: 1985–86); Mike Finneran (Art Fensterman: 1986–92); Denice Kumagel (Quon Lee Robinson: 1985–90); John Astin (Buddy Ryan: 1988–90); Mary Cadorette (Margaret Turner: 1990–91); Mel Torme (Mel Torme).

The most successful of the many courtroom-based sitcoms, *Night Court* rose from the ashes of an earlier, failed effort, *Sirota's Court* (q.v.). At the time of its 1976 debut, *Sirota's Court* was heralded as the first American comedy series to show judges and lawyers actually on the job and working, rather than dwelling upon their private lives. The series' *dramatis personae* included an eccentric night-court judge, a womanizing DA, an attractive female public defender, a wisecracking female court clerk, and an eccentric bailiff. Languishing in a death slot opposite CBS's *All in the Family* and *Alice*, *Sirota's Court* was cut short after 14 episodes, but the NBC higher-ups were certain that the series' premise would be worth reviving at a later date. Some six years later, writer-producer Reinhold Weege (whose previous courtroom sitcom *Park Place* [q.v.] had expired after four episodes) approached the network with a series concept that was virtually *Sirota's Court* redux, but with a typically 1980s emphasis on unbridled raunchiness and zany verbal slapstick. NBC liked Weege's *Night Court*— but they didn't like his leading man.

There is a prevalent belief that *Night Court* was created specifically for its star Harry Anderson, tailored to his special talents from the very beginning. To be sure, the protagonist of Weege's series was an irreverent judge named Harry, who behaved more like a confidence trickster and sleight-of-hand artist than a jurist. And it was certainly true that Harry Anderson had spent the bulk of his career as a professional magician and card manipulator who incorporated pickpocket techniques in his act — and likewise true that Anderson had risen to national prominence as light-fingered con man "Harry the Hat" on the NBC sitcom *Cheers*. But though his work on *Cheers* had brought Harry Anderson to the attention of Reinhold Weege, the producer initially rejected Anderson because he felt the 31-year-old actor was too young for the role of 34-year-old Judge Harry T. Stone. After Anderson read for the part, Weege changed his mind, and soon was so committed to having Anderson star in *Night Court* that he waged and won a bitter battle with NBC, who had initially insisted that Weege hire someone else. (A well-circulated report that Harry Anderson accepted the part only because he was desperate for work can be discredited in light of the fact that he was then pulling down $50,000 a year as a stage magician.)

Debuting in January, 1984, as a midseason replacement for *Cheers* (which had moved to an earlier Thursday-night timeslot), *Night Court* at first bore a strong resemblance to Reinhold Weege's previous effort *Barney Miller*: an ensemble comedy with a central urban setting (a courtroom rather than Barney Miller's police station house), featuring a seemingly endless parade of oddball litigants, lawyers and witnesses. The youngest and wackiest judge in New York City, Harry T. Stone, had been elevated to his position at the Manhattan Night Court simply because he happened to be home when the outgoing mayor was calling around making judicial appointments. Harry's cheeky irreverence and complete disregard of protocol drove his coworkers to distraction — in the opening episode, he determined a female defendant's sentence with the flip of a coin, then incited a riot when he ordered the defendant, her husband and their lawyer to go out to dinner together — but few could deny that his decisions were fair and just, yielding generally positive results. As the series progressed, Judge Stone's softhearted lenience towards the more downtrodden and deserving of defendants became so well known that the entire courtroom

would shout Harry's standard ruling in unison: "Fifty dollars and time served."

Of the series' many supporting players, the only two seen from beginning to end were John Larroquette (who'd originally read for the role of Harry Stone) as hard-drinking, skirt-chasing prosecutor Dan Fielding, and Richard Moll (his head shaved for a concurrent role in the theatrical feature *Maelstrom: The Destruction of Jared-Syn*) as tall, taciturn, trivia-obsessed bailiff Bull Shannon. Gail Strickland was cast in the first episode as combative public defender Sheila Gardner, but was replaced in Episode Two by Paula Kelly as Liz Williams. Kelly left at the end of the first season, whereupon Markie Post made her first appearance as legal-aid lawyer Christine Sullivan. For those who are now shouting, "Whoa! Markie Post didn't join the show until Season Three!," it must be noted that the producers had wanted Post to join *Night Court* upon Kelly's departure, but the actress was already contracted to another series, *The Fall Guy*, and wouldn't be free until the 1985–86 season. Thus, after her initial appearance, Post was immediately replaced by Ellen Foley as Billie Young (former *Charlie's Angels* star Shelley Hack, originally slated to play Billie, was unavailable). Returning upon Foley's exit in 1985, Post's Christine Sullivan remained with the series until its conclusion, briefly entering into a romantic relationship with Judge Stone after the breakup of her 1990 marriage to undercover cop Tony Guilano (Ray Abruzzo). Plot developments of this nature were indicative of the fact that the ensemble aspects of *Night Court* diminished and the focus on the principal characters increased as the series rolled on.

There was just as much turnover amongst Judge Stone's staff as in the public defender's office. During the first season, fabled "Golden Age" TV comedy writer Selma Diamond was seen as abrasive, chain-smoking court matron Selma Hacker. When Diamond died of lung cancer on May 13, 1985, she was replaced by veteran radio actress Florence Halop as the equally caustic, equally nicotine-addicted Florence Kleiner. Sadly, Halop had been on the series for only a year when she, too, passed away on July

15, 1986 (also of lung cancer). Filling the breach at the beginning of Season Four was future talk-show host Marsha Warfield as smart-n'-sassy matron Roz Russell, a role she would continue playing until *Night Court* ended its run.

Another first-season regular was Karen Austin as perky court clerk Lane Wagner, but she lasted for only ten episodes. Following an interim in which Lana was spelled by Mavis Tuttle (Alice Drummond) and Charly Tracy (D.D. Howard), Charlie Robinson signed on as the new court clerk, sweet-natured Vietnam vet Mac Robinson. Like his fellow "latecomers" Markie Post and Marsha Warfield, Robinson was still on the job during the series' final season — by which time two new regulars had been added, blind newsstand proprietor Jack Griffin (S. Marc Jordan) and airheaded court stenographer Lisette Hocheiser (Jolene Lutz). (Reinhold Weege has described the preponderance of female characters on *Night Court* as a reaction to the all-male ambience of *Barney Miller*.)

One "regular" who was more talked about than seen was real-life singer/composer Mel Torme, Harry Stone's personal idol. The common belief is that Harry Anderson incorporated his own idolatry of Mel Torme into the script: in truth, Judge Stone was "Torme-crazy" long before Anderson had even been cast. Once it was established that the Judge adored Mel and would walk through fire to be near him, Torme cheerfully agreed to make sporadic guest appearances and do anything the scripts required of him — and in a classic example of life imitating art, Mel Torme and Harry Anderson, who'd never met before shooting started, became close friends.

Throughout its nine-year run, *Night Court* managed to sustain a veneer of verisimilitude no matter how absurd the plotlines became. "We've had shows where some ventriloquist's dummy will leap out of the window to his death," noted Reinhold Weege in 1987. "In the context of *Night Court* logic, that makes sense." Weege qualified this statement by claiming to have witnessed even weirder events during actual night court sessions. Though he didn't offer any examples, one doubts that anything Weege happened to see in real life could

match the weirdness of such *Night Court* plaintiffs as the old guy who claimed to be God but refused to accept responsibility for creating Pia Zadora, the diaper-clad "New Year" impersonator, the preteen electronics geek who threatened to crash every computer in New York after an argument with his principal, the gypsy fortune-teller who placed a curse on the courtroom, the four pregnant defendants trapped along with Harry in the middle of a freak hurricane, or the indigent hillbilly family headed by future *Star Trek: The Next Generation* costar Brent Spiner. And for sheer volume, no one could top Judge Harry Stone during *Night Court*'s periodic "Day in the Life" episodes, in which the beleaguered Harry would be forced by some plot contrivance or other to preside over an impossibly high number of cases before a predetermined deadline.

Suprisingly, for all its comic exaggeration *Night Court* was a favorite among genuine attorneys. Articulating the series' appeal was famed defense lawyer Alan Dershowitz, who in a 1985 *TV Guide* article deriding the "myth" that "the law is always clear and certain" on TV legal shows, stated, "The judges who apply the law suffer from the same human qualities that affect the very human cops so passionately portrayed by the assortment of heroes who parade across the our TV screens. (In that respect, perhaps the most realistic TV show on 'justice' is a sitcom: *Night Court*)."

The series began losing momentum after the departure of creator Reinhold Weege, and there was serious talk about ending the show after its eighth season, with Harry Stone and Christine Sullivan getting married and the libidinous Dan Fielding doing a complete 180 by becoming a priest. But when *Night Court* was unexpectedly renewed for a ninth season, the writers went to work on an even more spectacular series finale. In the hour-long closing episode "Opportunity Knock-Knocks," Harry turned down the opportunity to manage Mel Torme's upcoming singing tour to remain on the bench; Dan gave up his serial womanizing and popped the question to Christine, who had been elected to Congress; court clerk Mac dropped out of law school to become a filmmaker after his video of Bull Shannon's wedding became a *Rocky Horror*-style cult item; and Bull himself, who was never quite of this world to begin with, departed for the planet Jupiter in the company of two tiny aliens!

The Nine Lives of Elfego Baca

ABC: originally telecast as a component of *Walt Disney Presents*. First episode: Oct. 3, 1958; final episode, March 25, 1960. Executive producer: Walt Disney. Produced by James Pratt. Cast: Robert Loggia (Elfego Baca).

The spectacular success of the three-part *Davy Crockett* serial on the original *Disneyland* TV anthology proved both a blessing and curse for producer Walt Disney. Though he'd originally intended *Disneyland* to offer a cornucopia of entertainment from a variety of genres, the ABC brass insisted that he try to replicate the popularity of *Davy Crockett* by churning out additional quasi-western yarns featuring famous historical characters. Disney reluctantly responded with several elaborately mounted miniseries, seen as components of both *Disneyland* and its successor *Walt Disney Presents*, chronicling the lives and times of frontiersmen Andy Burnett and Daniel Boone, Revolutionary War hero Francis "Swamp Fox" Marion, and western lawman Texas John Slaughter. Between 1958 and 1960, Disney also offered a multipart saga which sprang as much from the popularity of his *Zorro* TV series as from ABC's demand for "more westerns": *The Nine Lives of Elfego Baca*.

Though for dramatic purposes he exaggerated the heroic exploits of real-life sheriff and attorney Elfego Baca, Disney could easily have stuck to the facts, which were pretty dramatic in themselves. Born in New Mexico in 1865, Baca was raised in Topeka, Kansas, where his father served as a marshal. Ordering a badge from a mail-order company and stealing a few weapons, the 18-year-old Baca installed himself as sheriff of Socorro County, New Mexico. His reputation was established in 1884 when, in the village of San Francisco Plaza, he arrested a drunken cowboy for taking a few potshots at him. At least 80 of the cowboy's pals descended upon the small hut where he was being held prisoner, demanding the man's release. During a siege that lasted 33 hours, Baca single-handedly held off the crowd, killing four men and wounding eight others—but never getting shot himself, thereby creating the legend that he had "nine lives" like a cat. (Lucky for him the hut had a sunken floor!) After surviving what came to be known as "The Frisco War," Baca was appointed U.S. Marshal in 1888. Ten years later he was admitted to the Bar, joining a law firm in Socorro and practicing in both New Mexico and Texas during the first decade of the 20th century. He went on to serve as official U.S. representative of the Huerta government during the Mexican revolution of 1911–1916, beating a prison rap when he was cleared of a conspiracy charge arising from the escape of a prominent rebel general. He spent his final professional years as a private detective, a political investigator for the New Mexico senate, and a bouncer in a bordertown casino. Though extremely partial to wine, women and song, Baca lived to a ripe old age, relating his life story (and making several pungent comments on the current political scene) to historian Janet Smith as part of a WPA Federal Writers Project in 1936, and running for District Attorney on the Democratic ticket in 1944, the year before his death. On May 10, 1995, New Mexico governor Bill Richardson delivered a tribute to Elfego Baca on the floor of the U.S. House of Representatives.

Considering this resume, it is surprising that Elfego Baca's life had previously been filmed only once before Disney took a crack at it, as a half-hour episode of the 1957 ABC anthology *Telephone Time*. Starring in Uncle Walt's version of the story was Robert Loggia, an Actors' Studio alumnus who was hired because of his resemblance to the real Baca (though Loggia had a bit less hair), his rich theatrical voice, and his athletic prowess. The saga was launched on the October 3, 1958, episode of *Walt Disney Presents*, "The Nine Lives of Elfego Baca," a recreation of the 1884 Frisco War replete with such Disneyesque "humanizing" touches as showing Baca preparing dinner and coffee for himself and his prisoner between volleys. In Episode Two, "Four Down and Five Lives to Go" (October 17, 1958), Baca had just been installed as Sheriff of Socorro, and was seriously considering the study of law. By Episode Three, "Lawman or Gunman" (November 28, 1958), he was learning the legal trade under the tutelage of Santa Fe attorney J. Henry Newman (James Dunn), getting practical experience by butting heads with corrupt officials and and hired guns. Appointed Newman's partner in the fourth episode "Law and Order, Inc." (December 12, 1958), Baca came to the aid of beleagured rancher Don Esteban Miranda (Gilbert Roland), making a great show of avoiding gunplay in favor of proper legal procedure (quite a contrast with the real Baca, who was notoriously trigger-happy). Matriculating to "Elfego Baca: Attorney at Law" in Episode Five (February 2, 1959), our hero defended a former outlaw from a bank-robbery charge: this episode was graced by the presence of former Mouseketeer Annette Funicello, in her first dramatic role as the ex-outlaw's daughter. The sixth and last *Elfego Baca* installment of the 1958–59 season was "The Griswold Murder" (February 20, 1959), in which Baca squared off in a courtroom showdown with his former law partner.

Although *The Nine Lives of Elfego Baca* was as extensively publicized and promoted as any of Disney's other offerings of the period, the miniseries wasn't the big hit that the producer had hoped it would be: possibly the title worked against it, with millions of youthful Disney fans (this writer included!) expecting the famously bullet-proof Baca to literally rather than figuratively rise from the dead in each episode. Whatever the reason, the one *Walt Disney Presents* component that really clicked with viewers during the 1958–59 season was *Tales of Texas John Slaughter*, a straightforward western starring future novelist Tom Tryon. With this in mind, Disney stepped up the "western" elements of *Elfego Baca* during the following season, with the character's legal acumen taking second place to his shooting, fighting and riding skills.

Baca made his first appearance this season in a two-part story arc (November 15 and 20, 1959) involving a nomadic religious sect called the Mustangers, led by Shadrack O'Reilly (Brian Keith). Dedicating himself to protecting the Mustangers from evil land barons, Baca naturally found himself relying more on gunplay than litigation. Next up was "Friendly Enemies at Law" (March 18, 1960), wherein Baca faced down a formidable array of antagonists in his struggle to defend honest ranchers against a powerful town boss. The final *Elfego Baca* installment was "Gus Tomlin is Dead" (March 25, 1960), all about a fugitive gunslinger (played by Alan Hale Jr.) living peacefully with his family under an assumed name until his past was exposed, whereupon Baca dusted off his law books and his six-shooters to make certain that the reformed gunslinger got a fair trial.

Despite so-so viewer response, *The Nine Lives of Elfego Baca* was considered a valuable enough property to be revived in theatrical-feature form. But it wasn't so valuable that Disney was willing to shoot any new footage: thus, the first two TV episodes were spliced together and released overseas as a "movie" version of *Elfego Baca*, while the later episodes "Elfego Baca: Attorney at Law" and "The Griswold Murder" were combined into the 1962 theatrical release *Six Gun Law*.

On Trial

ABC: Nov. 22, 1948–Aug. 12, 1952. Cast: David Levitan (Moderator).

The first of three TV series bearing the title *On Trial* was a public-affairs program telecast live by ABC beginning in the fall of 1948. David Levitan, who was connected with both the United Nations and the Nassau County (New York) Board of Social Services, served as host and moderator for the series, in which debates on topical issues were conducted in the manner of a courtroom trial, with an actual judge presiding and experts from a variety of fields acting as opposing counsels. In a classic example of "*plus ça change, plus c'est la même chose*," the first episode, telecast November 22, 1948, debated the question, "Should Wiretapping Be Prohibited?" In later episodes the format was modified, with "affirmative" and "negative" arguments telecast on alternating weeks.

Designed as a stopgap series to fill empty half-hour timeslots on ABC's schedule (of which there were many at the time), *On Trial* hopscotched all over the network's Prime Time manifest until March of 1950, when it finally settled into a regular Monday-night berth. The series moved to Tuesdays for the 1951–1952 season, and was last telecast nationally on July 1, 1952, though it continued to be seen in New York and a handful of other east-coast markets until August.

On Trial [*aka* The Joseph Cotten Show]

NBC: Sept. 14, 1956–Sept. 13, 1957, June 14–Aug. 30, 1958; CBS: July 6–Sept. 21, 1959. Fordyce Enterprises/Revue Television. Cast: Joseph Cotten (Host).

Hosted by Joseph Cotten, *On Trial* originated as a filmed NBC anthology series dramatizing actual court cases, both famous and obscure, from all over the world. Cotten himself starred in the opening episode, recreating the 1865 prosecution of notorious wife poisoner Dr. Edward Pritchard. Later episodes focused on the trials of Colonel Blood (Michael Wilding), who in 1642 stole the Crown Jewels of England, and Mary Surratt (Virginia Gregg), who was hanged for her alleged complicity in the 1865 Lincoln assassination. Also covered were Emile Zola's legal woes after the publication of his 1898 broadside "J'Accuse," and future Secretary of State William Seward's precedent-setting 1846 defense of William Freeman, a mentally disturbed black man accused of murdering four whites. Other less familiar stories involved a former shoplifter who wanted to adopt a child, a group of gold prospectors charging their leader with the killing of their Indian guide, and an esteemed senator representing an elderly farmer in a damage suit over a dead dog. Among the series' guest stars were Keenan Wynn, Alexis Smith, Chuck Connors, June Lockhart, Diana Lynn, Hoagy Carmichael and Joan Fontaine.

On February 1, 1957, the series' title was changed to *The Joseph Cotten Show: On Trial*. The *On Trial* part was subsequently dropped when, in the summer of 1958, the series resurfaced on NBC as a summer replacement for *Your Hit Parade*. Though this revival consisted solely of rebroadcasts from the initial run, a handful of new episodes were filmed when *The Joseph Cotten Show* was brought back for a third and final time on rival network CBS, as a Summer 1959 replacement for *The Ann Sothern Show*. Also included during the CBS run were repeats from two other Revue Television anthologies, *General Electric Theater* and *Schlitz Playhouse*. The original *On Trial* identification would have been meaningless for this go-round, since few of the stories had anything to do with courtroom procedure: one conspicuous exception was the telecast of August 3, 1959, "Strange Witness," a rerun of a 1958 *General Electric Theater* playlet starring Joan Crawford as the central figure in a crime of passion.

On Trial

Syndicated: Sept. 1988–Sept. 1989. Reeves Entertainment/Republic Pictures. Executive Producer: Woody Fraser. Cast: Nick Clooney (Host).

Produced by reality-show pioneer Woody Fraser (*That's Incredible*), the daily syndicated half-hour *On Trial* represented the first significant national exposure for host Nick Clooney, who had spent the previous two decades emerging from the shadow of his famous sister Rosemary Clooney to become a popular Cincinnati TV personality in his own right. (Ironically, Nick Clooney's fame would later be eclipsed by his actor son George Clooney, but that's material for another book.) "Real People! Real Cases! Real Life!" trumpeted the ad copy for *On Trial*, which consisted in the main of genuine public-domain courtroom footage, culled from the archives of 40 states.

Not yet inundated by such slice-of-life efforts as MTV's *The Real World* and CBS' *Survivor!*, many contemporary critics found *On Trial* to be tasteless and exploitive. Citing the opening episode in which a pathetic-looking punk was sentenced to life imprisonment, John J. O'Connor of *The New York Times* remarked: "His life shattered, he now has the dubious pleasure of knowing that, for some 20 minutes, he was the star of a television reality entertainment." One positive aspect of the series was that it led to Nick Clooney's much lengthier association with the American Movie Classics cable network, courtesy of former *On Trial* staffer and later AMC honcho Jane Wallace.

100 Centre Street

A&E: Jan. 15, 2001–March 5, 2002. Jaffe/Braunstein Films. Executive Producer: Sidney Lumet. Cast: Alan Arkin (Judge Joe Rifkind); Phyllis Newman (Sarah Rifkind); LaTanya Richardson (Judge Attallah "Queenie" Sims); Paula Devicq (ADA Cynthia Bennington); Joseph Lyle Taylor (ADA Bobby Esposito); Bobby Cannavale (ADA Jeremiah "J.J." Jellinek); Manny Perez (Ramon Rodriguez); Michole White (Fatima Kelly); Margo Martindale (Michelle Grande); Chuck Cooper (Charlie the Bridgeman); Joel de la Fuente (Peter Davies).

100 Centre Street (the Manhattan address of the holding cells in which criminals cool their heels before their court appearances) represented a "return to the soil" for celebrated Hollywood filmmaker Sidney Lumet, who not only conceived the weekly, hour-long cable series but also served as executive producer and primary director. Lumet had established his reputation in the halcyon days of live television, helming such acclaimed New York–based anthologies as *You are There*, *Omnibus* and *Goodyear Playhouse*. His first feature-film directing assignment was the classic jury-room drama *Twelve Angry Men* (1957), and among his many later projects were the courtroom-based films *The Verdict* (1982) and *Guilty as Sin* (1993). Who better, then, to handle a provocative, high-quality legal series, largely filmed on location in the Big Apple ?

Explaining the series' gestation to *Entertainment Weekly*'s Ken Tucker, Lumet recalled that the idea came to him while researching his 1981 film *Prince of the City*: "I went to night court. I couldn't believe the sheer drama of it: the juxtaposition of the banal, with everybody [acting] bored, yet the fate of people's lives are being decided. I sat there and thought, My God, this is the most natural TV series I've ever seen. There are so many stories, with so many fixed, regular characters—the judges, and lawyers, and repeat offenders. And the cases

were so varied." When, seventeen years later, NBC asked him if he wanted to do a series for them, Lumet suggested the "night court" premise. "They gave me a lot of money to write a pilot and they turned it down, thank God, because I knew I would have problems with this on a network. Not because of [bad] language but because of the kinds of stories I wanted to do." Ultimately, Allen Sabinson, programming chief of the A&E cable service, saw the pilot, liked it, and asked Lumet to do a full season's worth of episodes.

Alan Arkin starred in *100 Centre Street* as Night Court Judge Joe Rifkind, an unreconstructed 1960s Liberal whose predilection for setting minor perpetrators free had earned him the nickname "Let 'Em Go Joe" (reportedly, the character was based on real-life NYC Supreme Court judge Bruce "Turn-'em-Loose" Wright). This being a typically stark-and-dark Lumet production, Judge Rifkind's bleeding heart landed him in hot water in the very first episode, wherein a petty criminal killed a policewoman a few hours after being released, stirring up a political maelstrom eagerly exploited by the candidates in an upcoming mayoral election. Worse still, the dead cop was the daughter of Rifkind's former law partner. This unfortunate chain of circumstances would torment "Let-'Em-Go-Joe" throughout the series' first season, which may explain why he began dallying with Zen Buddhism to alleviate the pressures of his job; in this, he was following the lead of his estranged daughter Rebecca (Amy Ryan), whose out-of-wedlock status was one of the plot points that would probably have been vetoed if NBC had chosen to go with the series.

Rifkind's polar opposite—and close friend—was ultraconservative Judge Attallah "Queenie" Sims (LaTanya Richardson), whose harsh sentences had earned her the soubriquet "Attallah the Hun." While Rifkind's comparative tolerance could be explained by his Jewish heritage, it was something of a novelty— at least by cable-TV standards—that Judge Sims, an African American, should take such a hard-line stance (it was explained that she had fought her way to the top from a hardscrabble existence on a Georgia farm, and had been raped as a child). She, too, had plenty to keep her occupied during Season One, including a concerted effort by her enemies to drive her off the bench, and a bid for the mayor's office which forced her to sacrifice some long-held principles.

Likewise on opposite ends of the spectrum, legally and socially, were young ADAs Cynthia Bennington (Paula DeVicq) and Bobby Esposito (Joseph Lyle Taylor). A child of privilege who had breezed through law school on a trust fund, Cynthia felt she could "make a difference" if she devoted her legal career to the downtrodden, while her society-lawyer father patiently awaited the day that Cynthia would bid Night Court goodbye and join his firm (he was still waiting when the series ended). Conversely, Brooklyn–born Bobby was an immigrant's son whose legal education was funded by his mob-connected family, in hopes that he would grant them special favors when the going got tough (fans of the series could not help but notice the similarities between Joseph Lyle Taylor's Bobby Esposito and Al Pacino's roughhewn but fundamentally ethical cop Serpico, from the 1973 Lumet film of the same name). Other first-season regulars included legal-aid lawyer Ramon Rodriguez (Manny Perez), whose Liberalism was less altruistic and more materialistic than Rifkind's, and who by his own admission chased "every bit of prosecutorial tail" who came into the courtroom; overworked ADA supervisor Michelle Grande (Margo Martindale); and Rifkind's far-from-patient wife Sarah (Phyllis Newman).

Pushing all the right social-issue buttons, the first-season *100 Centre Street* scripts highlighted such situations as a deadbeat dad being prosecuted for his son's death, a teenager on trial for gay-bashing, and a police officer putting his life and career on the line by providing testimony against dirty cops (there's that *Serpico* connection again). With Season Two came a few alterations. Bobby Esposito's checkered past came back to bite him when, caving in to family pressure, he tried to erase a previous arrest from the record of his drug-addicted

brother. Booted from office and stripped of his law license, Bobby set up a Legal Aid office with the assistance of his former colleague (and now erstwhile lover) Cynthia. His Night Court replacement was ambitious, ethically challenged ADA Jeremiah "J.J." Jellinek (played by Sidney Lumet's son-in-law Bobby Cannavale, late of *Ally McBeal* [q.v.]), who subsequently carried on a romance with legal aid Fatima Kelly (Michole White)—while Fatima herself tried to juggle her job duties with a secret drug habit. As Andrew Wallenstein noted in a 2001 *Media Life* article, "*Centre* concentrates on the limits of mercy. All of its characters struggle in their own way with handling their responsibilities without losing their humanity."

Sidney Lumet found *100 Centre Street* a particularly exhilarating experience, especially since it allowed him to work at his customary rapid-fire pace, a practice often discouraged when he was making theatrical films with high-priced actors who preferred a more leisurely and methodical work schedule. So determined was Lumet to get things done at top speed that he even continued filming location scenes in Lower Manhattan during the chaos of September 11, 2001—and *remained* on the job well into September 12, while most other New Yorkers were sitting at home recovering from the catastrophe.

Especially thrilling for Lumet was the opportunity to give the series an expensive "35 millimeter, three-camera look" on a modest television budget. *100 Centre Street* was the first TV series shot on high-definition, 24-frame videotape, utilizing multiple 24p CineAlta camcorders for both location and studio work (interiors were filmed at the venerable Kaufman Astoria facilities). The results were so impressive—not to mention 50 percent cheaper than comparable film costs—that Lumet redeployed the same High-Def process on his 2006 courtroom movie *Find Me Guilty*, with the same cinematographer (Ron Fortunato) and production designer (Chris Nowak).

Lasting 31 episodes, *100 Centre Street* garnered mixed but generally positive reviews. Typical was the assessment of *Time's* James Poniewozik: "The dialogue can be as heavy-handed as, well, a Sidney Lumet picture. But Alan Arkin is powerful yet subtle as a liberal judge…. Worth putting on your docket for a probationary period."

Orleans

CBS: Jan. 8–29, 1997; March 28–Apr. 11, 1997. Samoset Productions/Paramount Network Television. Executive Producers: John Sacret Young, Toni Graphia. Cast: Larry Hagman (Judge Luther Charbonnet); Brett Cullen (Det. Claude Charbonnet); Michael Reilly Burke (Deputy DA Jesse Charbonnet); Colleen Flynn (Paulette Charbonnet); Lynette Walden (Rene Doucette); Vanessa Bell Calloway (Rosalee Clark); Lara Grice (Gloria); Cotter Smith (DA Bill Brennecke); Richard Fancy (Vincent Carraze); Jerry Hardin (Leon Gillenwater); O'Neal Compton (Curtis Manzant); Charles Durning (Frank Vittelli); Melora Hardin (Gina Vittelli).

The lurid legal drama *Orleans* was designed as a comeback vehicle for longtime *Dallas* star Larry Hagman, who had recently survived a potentially fatal bout with cirrhosis and had undergone liver transplant surgery. That Hagman received some of the best reviews of his career was undoubtedly due to his acting skills rather than the fact that he'd overcome some very serious health issues. That *Orleans* itself was was not so enthusiastically received upon its January, 1997, debut suggested that while people were glad to see Hagman back on the job, they would have preferred a vehicle more worthy of his talents.

Like *Dallas, Orleans* was a continuing drama, each hour-long episode picking up where the previous one left off. Unlike *Dallas*, Larry Hagman was not cast as a lovably unscrupulous scoundrel, but instead as a highly respected and basically honest New Orleans judge named Luther Charbonnet. However, Charbonnet and J.R. Ewing had one thing in common: an extremely dysfunctional family. Luther's eldest son Claude (Brett Cullen) was an N.O.P.D. homicide detective whose taste for sluttish young women threatened to destroy both his reputation and career. The judge's daughter Paulette (Colleen Flynn) was the manager of the Lady Orleans Riverboat Casino, a friend and confidant of the Big Easy's gangster population, and a profoundly bitter

young woman who harbored a mysterious animosity for her father and everything he stood for. And his youngest son Jesse (Michael Reilly Burke) was an ambitious deputy DA, who, in a secondary plot (which stirred up a lot of controversy at the time) was carrying on a sexual relationship with his first cousin Rene Doucette (Lynette Walden). Even the distinguished Judge Charbonnet was not free of eyebrow-raising idiosyncracies: he was known to allow attorneys to place bets in his courtroom, and despite the still-prevalent racial barriers in New Orleans (to say nothing of "conflict of interest" issues), he was enjoying a secret romance with African-American defense attorney Rosalee Clark (Vanessa Bell Calloway).

In the movie-length opener, Jesse Charbonnet incurred the wrath of his sister Paulette when he prosecuted a man accused of killing a senator who was last seen leaving the Riverboat Casino. Later episodes found Luther himself running afoul of his sons when a gambling party at his home was busted, and Paulette being abducted and her casino hijacked, presumably on orders from the Judge's sworn enemy, mob boss Frank Vittelli (Charles Durning). Meanwhile, Luther's voluptuous secretary Gloria (Lara Grice) was virtually given a miniseries to herself as she juggled the affections of attorneys Leon Gillenwater (Jerry Hardin) and Curtis Manzant (O'Neal Compton). It was claimed at the time that *Orleans* was based on the real-life experiences of co-creator Toni Graphia, herself the daughter of a Louisiana judge; Graphia's subsequent credits included *Battlestar Gallactica*, which was set in Outer Space, a slightly less exotic locale than New Orleans.

Those observers who felt *Orleans* was overhyped due to Larry Hagman's presence were merciless in their criticism. After its first few episodes posted dismal ratings, the series went on hiatus until March of 1997: at that time, CBS relocated the property to Hagman's old *Dallas* timeslot on Friday evening, hoping to recapture the old magic. The strategy didn't work, and *Orleans* died after only seven episodes.

Owen Marshall: Counselor at Law

ABC: Sept. 16, 1971–Aug. 24, 1974 (two-hour pilot episode telecast on Sept. 12, 1971). Universal Television. Created by David Victor and Jerry McNeely. Executive Producer: David Victor. Produced by Douglas Benton. Cast: Arthur Hill (Owen Marshall); Lee Majors (Jess Brandon); Reni Santoni (Danny Paterno); David Soul (Ted Warrick); Joan Darling (Fried Krause); Christine Matchett (Melissa Marshall).

Co-created by University of Wisconsin law professor Jerry McNeely (whose other credits included, of all things, *The Man From U.N.C.L.E.*), the weekly, hour-long *Owen Marshall: Counselor at Law* starred Arthur Hill, a Canadian–born actor who had pursued a law degree at the University of British Columbia before his studies were interrupted by World War II. After serving in the RCAF, Hill gravitated to radio work because he was in desperate need of money. Though he went on to a stellar career in theater and films (he created the role of George in the original 1962 Broadway production of *Who's Afraid of Virginia Woolf*), he never forgot his legal roots, claiming to have based his characterization of Owen Marshall on his own father, a lawyer in the prairie community of Melfort, Saskatchewan.

Hanging out his shingle in Santa Barbara, California, Owen Marshall was a widowed defense attorney who lived with his twelve-year-old daughter Melissa (Christine Matchett). Marshall's dedication to his profession and his compassion for the Underdog were so strong that the series was honored with public-service awards from several genuine legal organizations. Inasmuch as the series took place in the early 1970s, several controversial issues were addressed in the course of its 69 episodes. Examples include the episode "Murder in the Abstract," inspired by the 1971 California Supreme Court decision holding accomplices responsible for any deaths occuring during the commission of a crime (Karen Valentine played the hapless defendant, involuntarily dragged into a life of crime by her unsavory boyfriend); and "Words of Summer," in which a lonely teenage girl accused her swimming coach (Meredith Baxter) of what *TV Guide* described as "lesbian seduction."

Joan Darling costarred as Owen's unflappable secretary Frieda Krause, while several handsome young Universal contractees were featured as the lawyer's junior associates. During the first season, Lee Majors costarred as Jess Brandon; when Majors was tapped to star in *The Six Million Dollar Man* in 1973, he was replaced by Reni Santoni as Danny Paterno; and upon Santoni's departure, future *Starsky and Hutch* costar David Soul signed on as Ted Warrick. The series' guest-star lineup included such promising young performers as Ed Begley Jr., Farrah Fawcett, Anson Williams, Lindsay Wagner, Mark Hamill, Tom Selleck, Robert Urich, Kathleen Quinlan, Louis Gossett Jr., and, in her professional acting debut, Arthur Hill's real-life daughter Jenny. Also seen were singers Rick Nelson and Peggy Lee, respectively cast as a vicious-but-charming rapist and a veteran entertainer swept up in a plagiarism suit. And in the crossover episodes "Men Who Care" (October 21, 1971) and "I've Promised You a Father" (March 9, 1974), the cast of *Owen Marshall: Counselor at Law* intermingled with the cast of the concurrently produced medical series *Marcus Welby, M.D.*

The Paper Chase

CBS: Sept. 19, 1978–July 17, 1979; Showtime: Apr. 15, 1983–Aug. 9, 1986. 20th Century–Fox Television. Executive Producer: Robert C. Thompson. Series developed by James Bridges, from the novel by John Jay Osborn Jr. Cast: John Houseman (Prof. Charles W. Kingsfield Jr.); James Stephens (James T. Hart); Tom Fitzsimmons (Franklin Ford III); Robert Ginty (Thomas Craig Anderson); James Keane (Willis Bell); Jonathan Segal (Jonathan Brooks); Francie Tacker (Elizabeth Logan); Jane Kaezmarek (Connie Lehman); Charles Halahan (Ernie); Carole Goldman (Carol); Betty Harford (Mrs.Nottingham); Jack Manning (Dean Rutherford); Jessica Salem (Mallison); Stanley De Santis (Gagarian); Michael Tucci (Gerald Golden); Claire Kirkconnell (Rita Harriman); Andra Millian (Laura); Penny Johnson (Vivian); Lainie Kazan (Rose Samuels); Peter Nelson (Tom Ford); Diana Douglas (Prof. Tyler).

Based on the autobiographical 1970 novel by John Jay Osborn Jr., the 1973 theatrical feature *The Paper Chase* starred Timothy Bottoms as James T. Hart, a wide-eyed Minnesota lad enrolled as a first-year student at Harvard Law

School. While struggling to balance his studies with his social life, Hart was held in thrall by his autocratic contract-law professor Charles W. Kingsfield Jr. Brooking no nonsense or laxity in his classroom, the stern, uncompromising Kingsfield was given to statements like, "You'll teach yourself law but I'll train your minds. You come in here with a skull full of mush and, if you survive, you'll leave thinking like a lawyer." He seemed to take particular delight in tormenting Hart, at one point handing the student a dime and murmuring, "Take it, call your mother, and tell her there is serious doubt about you ever becoming a lawyer." Withal, Kingsfield was not without compassion and a dry sense of humor. It was obvious he was being tough on Hart and the other students to force them to perform far beyond their expectations; and when, after being kicked out of class for a minor infraction, Hart angrily called Kingsfield a "son of a bitch," the gimlet-eyed professor replied, "Mr. Hart! That is the most intelligent thing you've said all day. You may take your seat."

Striking a responsive chord with college kids and recent grads, *The Paper Chase* proved a surprising financial success and also made an overnight star out of the man playing Professor Kingsfield, 71-year-old John Houseman. Best known to the public for his award-winning achievements as a stage and film producer—he had co-founded the legendary Mercury Theater with Orson Welles, while his movie credits included *The Bad and the Beautiful* (1952) and *Lust for Life* (1955)—the Rumanian-born Houseman was at the time of *The Paper Chase* the director of the drama division at the Juilliard School of Fine Arts, where his pupils included Robin Williams and Christopher Reeve. The last person in the world to call himself an actor, Houseman's film appearances had previously been confined to unbilled cameos in *Ill Met by Moonlight* (1957) and *Seven Days in May* (1964), both of which he accepted with reluctance. He took on *The Paper Chase* only because the original Professor Kingsfield, James Mason, had dropped out at the last minute, and such established actors as Edward G. Robinson, Henry Fonda, James Stewart,

Melvyn Douglas, John Gielgud and Paul Scofield proved unavailable; it was director James Bridges, a friend of Houseman, who submitted his name to the producers.

Though he won an Academy Award for his performance as Kingsfield, many of his associates and former students believed Houseman was merely playing his own irascible self; when asked by the producers if he would allow some of his own Juilliard pupils to play bit roles as law students, Houseman barked, "I won't let them out of school that long." For his part, Houseman insisted that he himself was far less tyrannical—and far less erudite—than Kingsfield, claiming he'd patterned the character after fabled property-law professor Edward Henry "Bull" Warren, who taught at Harvard Law School from 1904 until his death in 1945. Warren, recalled Houseman in a 1986 *TV Guide* article, was just as "sarcastic and sadistic" as Kingsfield, often even more so. Wrote James M. Landis in the October 1945 edition of *Harvard Law Review*, "Edward H. Warren will always remain a somewhat legendary figure in the history of the Harvard Law School.... His cutting, merciless wit in class, his impatience with stupidity, his insistence upon discipline of action as well as of thought—these qualities, irrespective of any others, made him a figure to be feared, if not admired." And in his introduction to the same article, *Law Review* editor W. Barton Leach presciently pointed out the one major trait "Bull" Warren and Charles Kingsfield had in common: "Why do young men go through months of an ascetic regime in order to strain and sweat through the ordeal of a four-mile boat race? Why do they climb mountains? Why do they fight for places on an Arctic expedition? Because these things are hard and young men like the feeling that they can lick whatever comes along. The Bull tossed the same type of challenge at the young men who came before him. It's tough, he said, and nobody's going to help you, but a good man can do it; are you a good man? This brought the right type of young man to Harvard Law School and still does."

When *The Paper Chase* was adapted as a weekly, hour-long CBS TV series in 1978, it was a given that John Houseman would be invited to recreate his role. Houseman not only accepted the invitation, but immediately began exercising control over the property to make certain it would maintain the same high standards as the 1973 film, collaborating with executive producer James Bridges on story ideas, treatments and scripts. Despite this input, the similarities between the film and TV versions of *The Paper Chase* were mostly superficial. While the film had concentrated primarily on the relationship between James Hart and Professor Kingsfield, the series was more an ensemble effort, giving equal attention to the other students. The series' setting was no longer specifically Harvard, but instead an unidentified "Eastern Law School." The film's important subplot involving a romance between Hart and the widowed Kingsfield's daughter Susan (Lindsay Wagner) was eliminated—as was Susan herself, except for the episode of November 28, 1978, in which she was played by Susan Howard. And with the exception of John Houseman, none of the actors from the movie version were carried over to the TV adaptation.

Newcomer James Stephens was cast as James T. Hart, with Tom Fitzsimmons as Hart's classmate and best friend Franklin Ford III, the son of a prominent attorney. James and Tom were part of a study group including fellow first-year students Willis Bell (James Keane), Elizabeth Logan (Francine Tacker), Tom Anderson (Robert Ginty), and Jonathan Brooks (Jonathan Segal), who was married to a girl named Asheley (Deka Beaudine, billed as a regular even though she appeared only in the first episode); Jonathan left the group—and the school—a few months into Season One when he was caught cheating. To make ends meet, James Hart worked at a local pizza parlor run by a guy named Ernie (Charles Halahan), where he met attractive waitress Carol (Carole Goldman), the series' nominal romantic interest. Also making occasional appearances was Kingsfield's strictly-business secretary Mrs. Nottingham (Betty Harford).

In standard late-1970s fashion, *The Paper Chase* served up a smattering of mildly contro-versial episodes involving Women's Rights, Affirmative Action, physical handicaps, organized crime and legal ethics. If the series seemed to be pulling its punches at times, it was because CBS in its infinite wisdom had chosen to schedule *The Paper Chase* during the 8– to 9 P.M. "Family Hour," where stories of a potentially volatile or offensive nature were strongly discouraged by the FCC. John Houseman was not the only cast member lodging a protest when, because of the show's timeslot, the producers vetoed a script involving a pregnant woman who accused a law student of being the father. And though the series was universally praised by critics for being "intelligent" and "mature," CBS could not resist trying to woo viewers away from the competing ABC shows *Laverne and Shirley* and *Happy Days* by demanding that *The Paper Chase* truckle to the Lowest Common Demoninator. When the network forced the producers to feature an episode in which Hart fell in love with a Russian gymnast who'd temporarily enrolled in the Law School, creator John Jay Osborn Jr. and the cast members refused, insisting that the plotline was ridiculous and illogical. But CBS held firm, whereupon both Osborn and John Houseman quit the series cold (the episode was filmed as ordered, with Houseman replaced by Pernell Roberts as a visiting professor).

Even if Houseman hadn't walked out on *The Paper Chase*, the series' cancellation after a single season was a foregone conclusion: CBS may have paid lip service to "prestige" programming, but the network wasn't about to continue underwriting a show that failed to attract the preferred audience demographic. Even so, CBS received 13,000 letters of protest when *The Paper Chase* was axed—the angriest response within the network executives' memory. The series' loyal fan base prompted the viewer-supported network PBS to rebroadcast the original 22 episodes in 1981, minus commercials. Encouraged by the overwhelmingly positive response to the PBS run, the Showtime pay-cable service commissioned a batch of brand-new episodes—the first time an over-the-air network dramatic series was ever revived on cable. Twentieth Century–Fox Television agreed to film seven episodes on a trial basis, to be aired once per month.

Retitled *The Paper Chase: The Second Year*, the property began its Showtime run in 1983, with five of the CBS version's cast members intact: James Stephens, Tom Fitzsimmons, James Keane, Betty Harford, and of course John Houseman: "I don't think there would have been a demand for *Paper Chase* today without John Houseman," noted costar Stephens. New to the cast were 18-year-old Jane Kaczmarek (later a full-fledged star on such series as *Equal Justice* [q.v.] and *Malcolm in the Middle*) in the role of first-year student Connie Lehman, who became Hart's girlfriend; Michael Tucci as Gerald Golden, Hart's editor at the school's Law Review; and Hart's second-year colleagues Rita (Clare Kirkconnell), Laura (Andra Millian) and Vivian (Penny Johnson), all of whom were added to show when the original seven-episode commitment was expanded to a full season's worth of 19 installments.

In contrast with the CBS brass, Showtime gave the series' writers a fairly free hand, allowing a far wider range of subject matter: the unintended consequences of bestowing tenure upon a senile professor, the pressures of academia causing a student to commit suicide, a single mother enrolling in Kingsfield's class, political duplicity in high Washington circles, Hart being assigned to defend the thug who had attacked Kingsfield, and so on. Also, the characters were permitted to use profanities from time to time, and to indulge in sexual activity; viewers couldn't help but notice Hart's single bed had suddenly grown into a double! As for John Houseman, he welcomed the opportunity to explore the human side of Kingsfield, though the old crustiness remained: obviously touched when his students gave him a birthday cake, Kingsfield nonetheless warned them that the 80 candles constituted a fire hazard.

In May of 1985, a block of 12 new episodes premiered on Showtime under the title *The Paper Chase: The Third Year*, with Lainie Kazan joining the cast as Rose Samuels, a fortysomething housewife who'd decided to enroll in law school (her fellow students ended up representing Rose in court when her chauvinistic hus-

band filed for divorce). Also new to the proceedings was Peter Nelson as Franklin Ford III's younger brother Tom Ford, dutifully following his sibling's footsteps by signing up for Kingsfield's course. By this time, James Hart had succeeded Gerald Golden as president of the Law Review, adding an additional burden to his academic career. Among the touchy legal issues addressed this season were Golden's defense of an anti-semitic client in a libel suit and Kingsfield's quixotic battle against Medicare bureaucracy.

Having won one of the first Cable ACE awards, the series concluded with a six-episode continuity in 1986, titled *The Paper Chase: The Graduation Year*. In the final episode, originally telecast August 9, 1986, James Hart finally graduated (evidently completing what was normally a three-year curriculum in a record eight years!) and joined a law firm after losing out on a teaching position. As the series reached its climax, John Houseman expressed gratitude for the opportunity not only to play Professor Kingsfield, but also to create the illusion that he was a legal expert—though he confessed that he really wasn't any smarter about the law than he'd been back in 1973. (Houseman died in 1988).

Park Place

CBS: Apr. 9–30, 1981. CBS Television Productions. Creator/Executive Producer: Reinhold Weege. Executive Producer: Tom Blomquist. Cast: Harold Gould (David Ross); David Clennon (Jeff O'Neill); Don Calfa (Howard "Howie" Beech); Cal Gibson (Ernie Rice); Alice Drummond (Frances Heine); Lionel Smith (Aaron "Mac" MacRae); Mary Elaine Monti (Joel "Jo" Keene); James Widdoes (Brad Lincoln).

Fresh from a lengthy hitch as a staff writer on the ensemble sitcom *Barney Miller*, Reinhold Weege set up his own production company and began work on a new comedy series, *Park Place*—which resembled *Barney Miller* in more ways than one.

Like its predecessor, *Park Place* took place in a single urban setting, this time the Legal Assistance Bureau of Manhattan rather than the 12th Police Precinct of Greenwich Village. The leading character was a long suffering father figure, Bureau director David Ross; he was played by Harold Gould, who bore a slight resemblance to *Barney Miller* star Hal Linden and even emulated Barney's habit of staring out of his office window while mulling over a problem (though unlike Miller, Ross was forever being distracted by the young lady who stood before the window of the adjacent building in a seemingly permanent state of *deshabille).*

The supporting characters were individualized by sharply-drawn character traits and eccentricities: Brilliant but naïve lawyer Jeff O'Neill (David Clennon); overambitious, status-seeking lawyer Howie Beach (Don Calfa); funky receptionist Ernie Rice (Cal Gibson), who disdainfully handed out numbers to the office's clients as if they were waiting to pick up their crescent rolls; born-again Christian secretary Frances Heine (Alice Drummond), who ended every conversation with "Jesus loves you"; wheelchair-bound associate Mac MacRae (Lionel Smith), a Vietnam vet who wasn't as cynical as he pretended to be; sexy militant-feminist associate Jo Keene (Mary Eaine Monti); and the newest member of the staff, shavetail Harvard Law School grad Brad Lincoln (James Widdoes). Finally, just as in *Barney Miller,* the central setting in *Park Place* accommodated a steady stream of bizarre and downright balmy clients, beginning with a befuddled woman (Florence Stanley) who walked into the Bureau and declared, "I murdered my husband"—and when asked why she didn't go to the Police, replied meekly, "I wasn't sure what was proper."

Described by the *New York Times* TV critic as "promising," *Park Place* debuted April 9, 1981, the same night as another CBS sitcom, *Checking In.* Four weeks and four episodes later, both *Park Place* and *Checking In* were cancelled. Disappointed but undaunted, Reinhold Weege returned to the drawing board to refine his concept of an ensemble legal sitcom — but it wasn't until he incorporated elements from another failed series, *Sirota's Court* (q.v.), that Weege finally struck gold with the long-running *Night Court* (q.v.)

Parole

Syndicated: 1958. Parole Productions/Telestar Films. Produced by Fred Becker.

Reminiscent of the 1952 ABC reality series *Four Square Court* (q.v.), the filmed, syndicated *Parole* featured actual prison inmates as they pleaded their cases before genuine parole boards. Each episode was filmed at a different house of correction, with a brief tour of the facilities featured during the opening credits. After the anonymous narrator explained the parole requirements of the state in which the action took place, the scene dissolved to the members of the parole board, all shown full-face and identified by name. Conversely, the prisoners were filmed from the back and identified only by number, with the offender's age, crime and sentence superimposed. As the would-be parolee stated his or her case, the camera continually cut away to the dispassionate faces of the board members, with occasional closeups of court records and depositions being shuffled about. Once the argument was ended, the prisoner would leave the room while the board discussed his or her situation. Surprisingly, even when the board decided not to grant an immediate parole, the prisoners avoided emotional outbursts, possibly intimidated by the presence of camera equipment and microphones. Generally speaking, the series' thrill content was on the level of a Monday-morning business conference, but *Parole* got by on its novelty value.

Though several existing prints of *Parole* bear a 1956 copyright date, the series was not placed into national syndication until 1958. Each half-hour episode featured two separate hearings, allowing local stations to decide for themselves whether to telecast *Parole* in a 15-minute or 30-minute timeslot.

The Paul Lynde Show

ABC: Sept. 13, 1972–Sept. 8, 1973. Ashmont Productions/Screen Gems Television. Creator/Executive Producer: William Asher. Executive Producer: Harry Ackerman. Music by Shorty Rogers. Cast: Paul Lynde (Paul Simms); Elizabeth Allen (Martha Simms); Jane Actman (Barbara Simms Dickerson); John Calvin (Howie Dickerson); Pamelyn Ferdin

(Sally Simms); Herb Voland (T.J. McNish); James Gregory (T.R. Scott); Allison McKay (Alice).

TV sitcoms featuring lawyers as protagonists have seldom taken advantage of the leading character's legal background for story material. In such programs as *Bachelor Father, Father of the Bride* and *Will & Grace*, the principal player is identified as a lawyer merely to explain why he wears expensive clothes, lives in a beautiful home, and has so much disposable income. In contrast, audiences were never allowed to forget that the star of *The Paul Lynde Show* was a lawyer: not only did several episodes revolve around legal issues, but the lawyer hero relentlessly harrassed and hectored his friends and loved ones, almost as if he was trying to get them to break down on the witness stand.

The Paul Lynde Show came into being because ABC wanted a new comedy series to replace its popular fantasy sitcom *Bewitched*, using the same production personnel. William Asher, longtime producer-director-writer of *Bewitched*, knew that comic actor Paul Lynde, who'd appeared on the series as puckish warlock Uncle Arthur, had been jockeying for his own starring comedy show for years. Tapping into his memory bank, Asher recalled that among Lynde's many unsold pilots was the 1962 effort *Howie*, in which the actor played a conservative lawyer whose daughter had married a beatnik. *Howie* had almost been picked up by CBS for the 1962–63 season, but was dropped when *The Dick Van Dyke Show* got a last-minute reprieve. Now ten years had passed, and beatniks had morphed into hippies, but the old "generation gap" premise was still alive and kicking, as witness the CBS blockbuster *All in the Family*. Asher reasoned that with few minor adjustments, *Howie* could easily be updated for the 1970s — and luck of luck, Paul Lynde was still available.

Lynde was cast as Paul Simms, a well-respected and tightly wound lawyer who lived in the suburban community of Ocean Grove, California, with his wife and two daughters. The star's close friend Elizabeth Allen landed the role of Paul Simms' wife Martha after negotiations with June Allyson fell through; seen as Paul's daughters Barbara and Sally were Jane

Actman and Pamelyn Ferdin, respectively. The Simms' peaceful household was thrown into turmoil when Barbara married Howie Dickerson (John Calvin), a shaggy-haired college student who, despite an I.Q. of 185, was constitutionally incapable of holding down a job. With no money and no home, Jane and Howie moved in with Jane's parents, giving the disgruntled Paul Simms ample opportunity to glare disdainfully at his slacker son-in-law and spew forth the sort of lip-curling, gritted-teeth sarcasm Paul Lynde's fans had come to know and love from his many appearances on *The Hollywood Squares*.

When not at home, Paul shared an office with his law partners T.J. McNish (Herb Voland) and T.R. Scott (James Gregory) and his secretary Alice (Allison McKay). The storylines regularly brought Paul's profession into play, as he was taken to court for unauthorized house renovations, butted heads with his pro-environmental children after taking on an oil executive as a client, entered into a noisy property dispute with a neighbor, forced Howie and Jane to submit to a "proper" wedding ceremony after digging up a technicality voiding their marriage, and tried to curry favor with an old law school buddy who was now a powerful politician. In a similar vein, liberal-minded Howie attempted to prove to right-leaning Paul that criminals could be rehabilitated by bringing a mugger home to dinner — whereupon the entire family was taken hostage. Other episodes were strictly from sitcom hunger, including the ancient wheeze about an angry letter to Paul's boss being delivered by mistake, and an inordinate amount of slapstick byplay surrounding the Simms family's swimming pool.

Unfortunately, *The Paul Lynde Show* merely confirmed what the comedian's closest associates had been telling him for years: though his brand of caustic humor was hilarious in small doses, it became awfully monotonous — and sometimes just plain awful — when stretched out over a thirty-minute timeframe. Nor was that the only problem: exercising his prerogative as star, Lynde insisted that he get all the funny lines while the rest of the cast played straight — and in many instances what passed for "funny" consisted of Lynde screaming full-decibel at everybody in sight. Even more deleterious was the plain and simple fact that Paul Lynde was thoroughly unconvincing as any sort of husband or father (even more so today, now that the particulars of his private life have become public knowledge). In fairness to the actor, however, ABC must take some of the blame for the series' failure: Lynde had begged the writers (including *I Love Lucy*'s Bob Carroll and Madelyn Davis and future *Alice* contributors Bob Fisher and Arthur Marx) to invest the show with the same satirical cutting edge as CBS' *All in the Family*, but ABC's standards-and-practices division blocked him at every turn, eviscerating the only passable fun in the series.

Despite a near-total lack of audience support (most viewers were preoccupied with CBS' *The Carol Burnett Show* and NBC's *Adam-12*), *The Paul Lynde Show* managed to survive for an entire season, proof positive that a powerful producer like William Asher could negotiate an ironclad 12-month commitment from one of the Big Three networks as late as 1972. But even if the series hadn't been cancelled in the fall of 1973, Paul Lynde had already torpedoed the project when, after losing the "Best Actor in a Comedy Series" award to *Sanford and Son*'s Redd Foxx at the Golden Globe Awards ceremony, the actor raised such an embarrassing row that he had to be physically removed from the auditorium.

The People's Court

1. Syndicated: Sept. 11, 1981–Sept. 7, 1993. Edwards-Billett Productions/Telepictures [Later Lorimar-Telepictures]. Executive Producers: Ralph Edwards, Stu Billett. Developed by John Masterson and Stu Billett. Cast: Judge Joseph A. Wapner (Himself); Doug Llewelyn (Himself); Rusty Burrell (Himself); Jack Harrell (Himself, narrator).

2. Syndicated: debuted Sept. 8, 1997. Edwards-Billett Productions/Warner Bros. Domestic Television Distribution. Executive Producers: Stu Billett, Harvey Levin. Cast: Ed Koch (Judge: 1997–1999); Jerry Sheindlin (Judge: 1999–2001); Marilyn Milian (Judge: 2001–); Carol Martin (Herself); Harvey Levin (Himself, host/legal reporter); Josephine Ann Longobardi (Bailiff: 1997–2001); Davey Jones (Bailiff: 2001); Douglas MacIntosh (Bailiff: 2001–); Curt Chaplin (Himself, in-court reporter/narrator).

The People's Court was the series that singlehandedly revived the daytime TV-courtroom genre that had flourished in the 1950s, withered in the 1960s and died on the vine in the early 1970s. It was also the first series in which genuine small-claims cases, rather than re-enacted or simulated litigations, were presented before a real judge on a Monday-through-Friday basis.

The series was the brainchild of John Masterson, a producer whose first brush with the "reality" format had been the daily televised-wedding opus *Bride and Groom* (1951–1958). Masterson began shopping *The People's Court* in 1975, a year in which TV's daytime hours were dominated by network reruns, Phil Donahue and Dinah Shore–style gabfests, and game shows. In the latter category was *Let's Make a Deal*, whose producers Monty Hall and Stephen Hatos optioned *The People's Court*, dispatching their associate Stu Billett to various California courtrooms to determine whether a series featuring actual cases would be saleable. In the course of his research, Billett found that only the Small Claims courts (representing 76,000 yearly cases in LA alone) captured his interest, noting that the courtrooms in which these cases were hashed out boasted a "constant audience" of fascinated onlookers.

Since existing California laws banned cameras from the courtroom, Billett arranged for the *The People's Court* pilot to be taped in the Hollywood studios of KTLA. But the next hurdle proved harder to clear: none of the three major networks was enthusastic about resurrecting a genre that had been moribund since the old *Divorce Court* (q.v.) ceased production in 1970. As Billett persisted, network resistance weakened, with NBC evincing interest in *The People's Court*—but only if a standup comedian like Nipsey Russell was hired to play the judge, and a comic actor like Charles Nelson Reilly was cast as the defense attorney. Another network was willing to consider *People's Court* on condition that a different "celebrity judge" be featured each week. Rejecting both of these "improvements," Billett elected to bypass the networks in favor of first-run syndication. He managed to convince actor-producer Ralph

Edwards, whose previous credits included *Truth or Consequences* and *This is Your Life*, that *The People's Court* had "smash hit" written all over it—and thus was born the prolific TV-show factory known as Edwards-Billett Productions.

Now came the icing on the cake: finding a genuine judge with built-in audience appeal. Several authentic jurists were auditioned, including LA Superior Court judge Christian Markey, a personal friend of Ralph Edwards. Markey liked the idea of *The People's Court* but balked at retiring from the bench, which would be required of the series' judge to avoid scheduling and conflict-of-interest problems. So Markey recommended his longtime colleague and tennis partner: 61-year-old Joseph A. Wapner, who *had* recently retired after two decades' worth of public service, including a lengthy tenure as presiding judge of LA Superior Court and a year-long term as head of the California Judge's Association. Though Wapner's only prior connection with show business occurred when he dated film star Lana Turner in his youth, Stu Billett liked the man's looks: and after a three-hour conversation with Billett's partner Ralph Edwards, followed by an "audition" in which he presided over a prickly domestic-abuse case, Wapner landed the job.

Now it was Wapner's turn to veto a ill-conceived decision on the part of Stu Billett. Just as the networks had wanted to "juice" the appeal of *The People's Court* by hiring a popular actor as judge, Billett had originally planned to pair Wapner with a sexy starlet as the bailiff. The Judge would have none of this, insisting that the producer hire real-life bailiff Rusty Burrell, who had worked with Wapner's lawyer father on the original *Divorce Court* back in the 1950s (and besides, Wapner and Burrell shared the same birthday: November 15). The regular cast was rounded out by court reporter Doug Llewelyn, a seasoned talkshow host and Washington DC news anchor who also produced a number of popular TV specials, among them the notorious *The Mystery of Al Capone's Vaults*.

After shelling out $150,000 for the half-hour pilot, Edwards-Billett settled down to a weekly budget of $75,000, or fifteen grand per episode (a full week's worth of shows were completed during each taping session). Once the new series was sold to the powerful ABC-owned station group, other local outlets followed suit, and by mid–1982 the show was being seen in 105 markets, a number that would increase dramatically in the seasons to come.

Though technically speaking the series featured private arbitrations that would normally be settled in the judge's chambers or a lawyer's office, *The People's Court* was staged in a courtroom setting for dramatic effect. Each episode began with Alan Stanley Tew's pulsating, bongo-dominated theme song "The Big One" as the plaintiff and defendant walked into camera range and offstage narrator Jack Harrell breathlessly detailed the case at hand, whereupon an encapsulation of Harrell's words was "typed" onto the screen. Then came the now-familiar signature: "*What you are witnessing is real. The participants are* not *actors. They are actual litigants with a case pending in a California municipal court. Both parties have agreed to dismiss their court cases and have their disputes settled here, in our forum:* The People's Court."

Generally, Judge Wapner heard two cases per 30-minute episode. A brace of researchers was retained by the production staff to prowl around 20 small-claims courts within a 60-mile-drive's radius of the KTLA studios. Those cases deemed the best and most interesting were chosen for the series, and those plaintiffs who could talk without mumbling and possessed good looks were among the first to appear on camera (of course, if the plaintiff was a beautiful woman, her speaking ability was a moot point). No lawyers were allowed in Wapner's court, and all participants had to agree to abide by the Judge's binding decision and not pursue the case any farther. The original maximum cash award was $750, which soon increased to $1500. Except for the series' producers, no one ever really lost any money: the defendant was automatically paid $25 for showing up, and if the Judge ruled in favor of that defendant, both litigants received $250. At the end of each case, Doug Llewelyn collared the litigants in the hallway outside the courtroom to solicit their opinions on the outcome; he then quickly herded them off-camera, ostensibly to "sign some papers," but actually to clear the deck for the next case or to give himself enough time to deliver his famous sign-off: "Remember, when you get mad, don't take the law in your own hands. Take 'em to court."

Although *The People's Court* was not essentially humorous in nature, the individual episodes bore such tongue-in-cheek titles as, "A Head with a Beer on It," "The Case of the Overdone Underthings" and "Snowball Fight — D.C. Style." Certain of the series' litigations have achieved a measure of permanence in the history of television, notably such oft-quoted examples as the stripper who was sued for not sufficiently disrobing during a stag party, the professional clown charged with terrifying the kiddies at a birthday celebration, the angry consumer who proved that the clock he'd purchased was defective by smashing the timepiece on Wapner's bench, or the poor old duffer who discovered that the car he'd just bought had a garden tiller where the steering wheel was supposed to be. In a 1982 *TV Guide* interview, producer Stu Billett readily admitted that the best episodes featured the angriest and most vengeful plaintiffs: "We're all getting ripped off every day, so people will finally take a stand. They say 'I'm mad as hell and I'm not going to take it any more.'"

Some of the series' critics argued that the plaintiffs wouldn't have been so hot under the collar if Judge Wapner hadn't goaded them into anger, charging Wapner with being unduly harsh with the participants, bullying and browbeating them at the slightest provocation. What these detractors didn't understand was that the Judge reserved his grouchiness only for those who came before him without adequately preparing their arguments ahead of time, or who refused to behave like grownups. (One classic episode found the Judge reading the riot act to an obnoxious plaintiff who'd actually *won* his case and was dancing around the courtroom taunting the defendant!) Though he recognized the value of putting on a good show, Wapner never displayed anger merely for effect, but for purpose. He took his duties on The

People's Court quite seriously, regarding the series as a public service by showing viewers how best to represent themselves under actual courtroom circumstances. "I always felt the public had the wrong perception about judges and courts," he explained to *TV Guide*'s Ellen Torgerson Shaw. "I want to help enlighten the public. I think this show is better for young people than to see cops and robbers." Wapner's colleagues, as a group, heartily approved of the series: "I see the humanity in it," declared retired Superior Court judge Harry Shafer. "I think people want realism, and it *is* real."

In retrospect, media historians have found Joseph Wapner to be a pillar of restraint and his series to be a textbook description of decorum, compared to the shouting, garment-rending and character assassination that proliferated on the TV courtroom shows of the 1990s and 2000s. In 1998, *Time* magazine's Joel Stein nostalgically recalled that *The People's Court*, "entertained with silliness, not heated conflict"; and in a 2003 piece for *The Daily Record*, Irwin Kramer summed up Wapner as "a model of judicial restraint," adding "I really miss [him].... Rarely losing his cool, Judge Wapner addressed litigants with respect, listening patiently as they presented their cases, asking thoughtful questions designed to test their credibility, and retiring to review the facts before rendering a reasoned decision."

As the series entered the home stretch, *The People's Court* modified its format to allow a $5000 ceiling on cash awards, and to feature outside legal experts and commentators like Harvey Levin and Steve Doocy contributing their observations and expertise. Also, Judge Wapner occasionally took his show on the road, taping selected episodes in local studios throughout the country. By the time the series wrapped up production in the summer of 1993, Wapner had taped 2,340 individual segments. After leaving syndication, *The People's Court* was rebroadcast on cable's USA Network from 1993 through 1995 — and at the time, that seemed to be that.

The unexpected success of *Judge Judy* (q.v.) revitalized the courtroom-show market in 1996, whereupon Edwards-Billett productions got to work on a new version of *The People's Court*, this one running a full hour each day. Fans of the original who eagerly anticipated a reunion with Judge Wapner were disappointed when he was not hired for the new *People's Court*— and indeed, had not even been asked to return. The official story was that the producers felt the 77-year-old Wapner was "too old," an argument that was ridiculous on its face considering that former New York City mayor Ed Koch, who'd been engaged as the series' "judge," was himself 71. Another rumor circulated that Wapner had excluded himself from consideration because of his adverse criticism of *Judge Judy* star Judy Sheindlin, of whom he said, "She is not portraying a judge as I view a judge should act. She's discourteous, and she's abrasive. She's not slightly insulting — she's insulting in capital letters!" This, claim the rumormongers, was perceived as a backhanded slap at the original production staff of *The People's Court*, several of whom were now working on *Judge Judy*. But even if Wapner had hurt some feelings by deriding Judy Sheindlin, surely the new *People's Court* packagers could have used a "judge vs. judge" feud to increase the viewership of their program.

It would appear from other evidence that Judge Joseph Wapner was frozen out of the *People's Court* revival because its producers were hoping to eventually take the property in the direction of the loud, confrontational *Judge Judy* school of video jurisprudence, and this was a school in which Wapner had never enrolled. Stu Billett has attributed Judy's success to, "a humiliation factor, and maybe people like to see people humiliated"— but not people like Judge Wapner. And despite his fondness for the good gray Judge, *Daily Record* correspondent Irwin Kramer admitted that Wapner's "style of deliberation failed to satisfy the public thirst for combative courtroom drama." (Incidentally, in 2000 Joseph Wapner joined former colleagues Doug Llewelyn and Rusty Burrell to make a guest appearance on the 3000th taping of *The People's Court*— for a fat fee, of course.)

The most popular elements of the original series were retained for the 1997 revival: the theme music, the post-verdict interviews,

the magnificently unfunny episode titles ("The Case of the Izuzu That Wasn't a Trooper"), and a distinct preference for the bizarre and offbeat in its selection of cases (one man claimed to have suffered whiplash from a topless dancer's breast, while another insisted that his chihuahua had been devoured by a python). The differences between the two *People's Courts* began with the new 60-minute length, which accommodated three separate cases of 20 minutes' duration in each episode. Production was now headquartered in New York rather than Hollywood; the plaintiffs were now permitted to collect damages as high as $7500; and throughout the hearings, viewers were apprised of the facts, trivia and minutiae pertaining to the plaintiffs and the case at hand by way of VH1-style superimposed "pop-ups." The proceedings were also telecast live outside the courtroom to a crowd of people gathered at the Manhattan Mall, who would offer their opinions of the testimony, ask questions relating to the finer points of the law, and hold straw polls to predict the outcome of the case. And during the first few seasons, viewers at home were invited to cast their votes on the testimony and possible verdict via the Internet: this, according to *Time*'s Joel Stein, was "almost as much fun to watch fluctuate as the Dow or Congressional voting on C-SPAN."

The number of regular cast members expanded along with the running time on the new *People's Court*: In addition to the Judge and Bailiff, there were now two in-court reporters, Carol Martin and Curt Chaplin, while series producer and legal consultant Harvey Levin handled the "man-in-the-mall" interviews (which became "man-in-the-street" when the series abandoned the Manhattan Mall in favor of Times Square). As mentioned, the first of the new series' resident judges was ex–New York City mayor Ed Koch, who as a partner in the law firm of Robinson, Silverman, Pearce, Aronsohn & Berman had as much right to wield a gavel on a TV legal series as anyone. Interestingly, for all the industry speculation that *The People's Court* would abandon its original no-frills austerity in favor of *Judge Judy*-like sensationalism, Koch went to great lengths to distance himself from his onetime protegee

Judy Sheindlin. "The reason she's No. 1 is her style, which is very confrontational, like a public scold," commented Koch in a *Time* magazine article. "I happen to know her. She's a very nice lady. Judge Judy has said she makes decisions on the basis of common sense. And I have said, 'That's not what the law is all about.' We do it on the law. This is not a court of compassion; this is a court of law."

Despite the fact that since stepping down as Mayor Koch had made so many TV and movie appearances that he'd been forced to join the Screen Actor's Guild, the new judge did not come across all that satisfactorily before the camera. He seemed ill at ease in his judicial robes, appeared vague and indecisive when rendering verdicts, and tended to allow the more combative litigants to run amok, apparently unable to impose the same control and discipline Joseph Wapner had exercised so well. Also, Koch had difficulty maintaining the pace of the program, especially when the plaintiffs began digressing at great and boring length: in the words of *Time*'s Joel Stein, "Unfortunately, the show runs for an hour, with only three cases, and no amount of cool graphics is going to make some guy's poor used-car purchase seem interesting for 20 minutes."

Finally, Koch had a habit of making the same sort of foot-in-mouth statements that had frequently gotten him in trouble during his political career. During one "behind the scenes" episode, he referred to Officer of the Court Josephine Ann Longobardi as "cute and Bailiff-Licious," prompting court reporter Carol Martin to step up to her microphone and demand that Koch apologize on the air. He refused on the grounds that he didn't believe in political correctness, effectively driving a wedge between himself and Martin that remained in place until she left the show at the end of the 1998–99 season. As for Ms. Longobardi, she apparently didn't sustain any permanent emotional damage from Koch's sexist ramblings and remained with *The People's Court* until 2001, when she was replaced first by Davey Jones (*not* the ex–Monkee!), and then by Douglas MacIntosh.

In 1999 Ed Koch himself was replaced, ironically by Judge Judy Sheindlin's husband Jerry Sheindlin — meaning that in several markets, Judy and Jerry were in direct competition. While few mourned Koch's exit, fewer still were pleased with Jerry Sheindlin, who many felt would never have gotten his own TV show if it hadn't been for his wife's popularity. Bill Haltom of the *Tennessee Law Journal* described the Sheindlins as "the Lucy and Ricky" of TV judges — and we need not have it 'splained to us which of the two Ricardos the fans of *I Love Lucy* preferred.

With the ratings of the new *People's Court* taking a noticeable dip, the producers bade Jerry Sheindlin farewell and brought in Marilyn Milian, a former Florida judge and deputy DA for the Dade County State's Attorney's office. By this time, most of the series' litigants had completely forsaken propriety in favor of loud and obstreperous name-calling. To her credit, Judge Milian did not surrender to this cacophony, but instead managed to restore at least the illusion of restraint and control upon her courtroom, her own performance bridging the gap between the abrasive showboating of Judy Sheindlin and the taciturn self-possession of Joseph Wapner. As a tribute to Milian, the series' long-standing opening signature was completely overhauled, each episode now beginning with, "*Everybody's talking about the honorable Marilyn Milian, the hottest judge on television. Real cases, real litigants. Here, in our forum:* The People's Court."

Never as highly rated as the original *The People's Court*, the newer version has nonetheless occasionally managed to match the numbers of its courtoom-TV brethren, especially with such "stunt" episodes as the one in which Judge Milian was asked to determine ownership of the baseball that scored Mark McGwire's 66th homer. And if anyone should question the extent of the property's popularity, please note that a British version of *The People's Court* was launched on ITV in 2005, hosted by *two* judges, Jerome Lynch and Rhonda Anderson. While this version did not survive, *The People's Court* has since its inception remained a rich and bountiful source of pop-culture references, ranging from autistic savant Dustin Hoffman's devotion to the series in the 1989 Oscar winner *Rain Man* to a 1995 episode of TV's *Sliders*, in which an alternate-universe Joseph Wapner presided over a Communist edition of *The People's Court*, spelled with the Cyrrilic "Я" (meaning, one supposes, that it should be pronounced "PEOPLE'S KOYAT").

The People's Court of Small Claims

Syndicated: 1959. ABC Films/Guild Films. Cast: Orrin B. Evans (Himself, Judge).

Occasionally cited as a precursor to the fabulously successful syndicated legal series *The People's Court* (q.v.), the obscure half-hour weekly *The People's Court of Small Claims* actually bore little resemblance to the later series beyond a similarity of titles. *The People's Court* featured actual, unscripted small-claims cases before a genuine judge: *The People's Court of Small Claims* was completely dramatized, and the "judge" was nothing of the kind.

First telecast February 11, 1959, as a local offering on Los Angeles' KCOP, the series was presided over by Orrin B. Evans, a USC law professor who went on to serve as his department's Dean from 1963 through 1968. Since there were rules prohibiting lawyers and judges from advertising themselves on TV, it was a good thing Professor Evans was not a member of the State Bar of California.

In each 30-minute episode, "Judge" Evans listened patiently to a steady stream of carefully scripted small-claims cases — some based on fact, others fabrications — with the actor-litigants adhering faithfully to the lines written for them, though they were permitted to improvise if they didn't stray too far from the scenario and the predetermined verdict. Several members of Orrin Evans' own family appeared as plaintiffs and defendants, including his teenage son David R. Evans, who grew up to become a Pasadena criminal defense lawyer. Since KCOP was one of the first local stations to install videotape equipment, it was possible for the producers to put several episodes in the can in the course of a single recording session. The series posted excellent prime time ratings in its original Wednesday-night timeslot, at one point attracting more viewers than the competing ABC western *Wagon Train*.

Advertised as "52 half-hours packed with humor and despair and dramatic insight into people and their troubles," *The People's Court of Small Claims* was nationally distributed beginning in the spring of 1959. The series performed best in those markets with independent TV outlets, such as Indianapolis and Minneapolis-St. Paul. Cities with only two or three stations tended to shunt the series to the early-afternoon or late-night hours, when viewership was at its lowest — or, if those stations lacked videotape facilities, the show was bypassed altogether.

Perry Mason

CBS: Sept. 21, 1957–Sept. 4, 1966. Paisano Productions/TCF Television Productions Inc. Created by Erle Stanley Gardner. Executive Producers: Gail Patrick Jackson, Cornwell Jackson. Cast: Raymond Burr (Perry Mason); Barbara Hale (Della Street); William Hopper (Paul Drake); William Talman (Hamilton Burger); Ray Collins (Lt. Arthur Tragg); Karl Held (David Gideon); Wesley Lau (Lt. Anderson); Richard Anderson (Lt. Steve Drumm); Lee Miller (Sgt. Brice); Dan Tobin (Terence Clay)

THE NEW PERRY MASON

CBS: Sept. 16, 1973–Jan. 27, 1974. 20th Century–Fox Television. Exeutive Producer: Cornwell Jackson. Produced by Ernie Frankel and Art Seid. Cast: Monte Markham (Perry Mason); Sharon Acker (Della Street); Albert Stratton (Paul Drake); Harry Guardino (Hamilton Burger); Dane Clark (Lt. Arthur Tragg); Brett Somers (Gertie).

Before discussing the most famous TV criminal lawyer of all time, we respectfully ask the indulgence of the court to reconstruct the case by chronicling the events leading up to the premiere of *Perry Mason* on September 21, 1957.

While holding down a successful law practice in Ventura, California, in the early 1920s, Erle Stanley Gardner began to write as a hobby, publishing several mystery stories in the legendary pulp magazine *The Black Mask*. Writing under a variety of pen names, Gardner churned out scores of fictional pieces, usually featuring a hard-boiled detective protagonist; he also wrote six *Black Mask* stories featuring a crusading defense lawyer named Ken Corning. Having more affinity for lawyers than detectives, he penned a brace of novels, each featuring a different attorney hero. Though neither novel sold, a publisher suggested that Gardner combine the best elements of both characters into a single person. The name he chose for his new protagonist was Perry Mason, an all-purpose moniker that Gardner recalled from his favorite boyhood magazine *The Youthful Companion*.

Published in 1933, *The Case of the Velvet Claws* was the first of 82 Perry Mason novels; the character also appeared in dozens of short stories. While several of these were straightforward mysteries in which Mason would clear a client (not always accused of murder) by unearthing evidence overlooked by the police, the bulk of the Mason novels were whodunnits. Sam White, a producer who worked on the *Perry Mason* TV series, explained to film historian David N. Bruskin, "a pure whodunnit is not a murder mystery, per se. It's a different type of writing. You set up characters and develop them in a certain way. We used a term, 'He needs killing.' When a character was killed, you figured that any one of them deserved to be killed. There were at least four or five red herrings in every story. You set them up first. Then you had your principal guy — in this instance, Perry Mason — uncover these people, and it wound up in the courtroom."

Of course, Perry didn't do it alone. He had a "Girl Friday" in the form of his beautiful and efficient secretary Della Street, and also benefitted from the input of private detective Paul Drake, who shared office space with Perry in the same building. (In the novels, Mason was but one of Paul's many clients: the close friendship depicted in the TV series was largely absent.) Gardner's version of Perry Mason was tougher and more blunt-spoken than on television, an extension perhaps of the author's famously antagonistic personality; similarly, District Attorney Hamilton Burger, Perry's perennial courtroom adversary, was more a villain in print than on the small screen, reflecting Gardner's intense dislike of prosecutors.

Outside of making lots and lots of money, Gardner's main motivation in creating Perry Mason stemmed from his feeling that defense attorneys had been dealt a bad hand in mystery fiction. All too often the criminal lawyer was depicted as an unethical shyster, and Gardner hope to clean up this tarnished image of the Defense. With this in mind, he couldn't have been too happy when Warner Bros.-First National chose Warren William, one of Hollywood's foremost purveyors of crooked lawyers, to star in the first film based on a Mason novel, 1934's *The Case of the Howling Dog*. Nor was Gardner pleased when the studio gave Mason an outlandishly huge suite of offices, then proceeded to glamorize Della Street (played in this film by the forgotten Helen Trenholme) with too much makeup and too many diamonds. Admittedly, *The Case of the Howling Dog* bore only a passing resemblance to its literary counterpart, but credit is due Warners for optioning the actual Gardner novels as the basis for their subsequent Perry Mason series, which was unusual for Hollywood at the time. (Most such series ignored the original books in favor of studio-hack scripts, or adaptations of novels from different authors: RKO's *The Falcon Takes Over* [1942], for example, was derived not from any of Michael Arlen's "Falcon" stories but instead from Raymond Chandler's *Farewell, My Lovely*.)

At the same time, it was clear that Warner Bros. didn't give a fig about Perry Mason as created by Erle Stanley Gardner: the studio shamelessly reshaped the property as a knockoff of MGM's *Thin Man* films, which starred William Powell and Myrna Loy as Nick and Nora Charles. Warren William's Perry Mason was transformed into a suave, wisecracking, martini-imbibing Nick Charles clone, just as Della Street — played by Clare Dodd in *The Case of the Curious Bride* (1935) and *The Case of the Velvet Claws* (1936) and Genevieve Tobin in *The Case of the Lucky Legs* (1936) — morphed into an imitation Nora Charles, wearing expensive evening gowns even while taking dictation. To further underline the Nick-and-Nora similarity, at the beginning of *Velvet Claws* Perry and Della were married and on their honeymoon, a far cry from the Gardner novels, in which Della repeatedly turned down Perry's proposals. As for the brainy, no-nonsense Paul Drake, he was dumbed down as Mason's comedy foil Spudsy Drake, played first by Allen Jenkins and then by Eddie Acuff. Of the Warren William films, the only one to rise above the level of mediocrity was *The Case of the Curious Bride*, which featured a young Errol Flynn in his first American role as the murder victim. Otherwise, *Howling Dog* is memorable only for its curiously amoral finale, in which Mason intimated that his client (Mary Astor) actually *did* commit one of the two murders for which she had been arrested; while *Lucky Legs* included a remarkable sequence in which Perry rescued a suspected killer (Patricia Ellis) from the clutches of the DA by smuggling her out of a hotel room and bundling her into an airplane. (The less said about *Velvet Claws*—Perry kidnapped on his wedding night, treating the whole ordeal as an uproarious joke — the better.)

After *Velvet Claws*, Warner Bros. lost interest in the Mason series and burned off the remaining Gardner novels they'd purchased as cheap B pictures. *The Case of the Black Cat* (1936) starred matinee idol Ricardo Cortez, no more convincing as Mason than he'd been as Sam Spade in the original 1931 version of *The Maltese Falcon*; and *The Case of the Stuttering Bishop* top-billed Donald Woods, somewhat closer to Gardner's Mason than Warren William but an undynamic screen presence. There was still no consistency in the casting of Della Street, with June Travis taking over the role in *Black Cat* (her marriage to Perry in the previous film unmentioned and forgotten), and Ann Dvorak seen as Della in *Stuttering Bishop*. Paul Drake was given back his "real" name, with Garry Owen giving a noncommittal performance in *Black Cat* and Joseph Crehan doing his standard Irish-cop characterization in *Velvet Claws*. New to the series was DA Hamilton Burger, reduced to a bit role and played by a pair of bit players, Guy Usher and Charles C. Wilson; apparently to avoid Gardner's capricious little pun — "Hamilton Burger" as in "Ham Burger"— the DA's last name was pro-

nounced "Ber-*jer*." The final Gardner novel in Warners' possession, *The Case of the Dangerous Dowager*, was filmed as a comedy western (!) in 1940 under the title *Granny Get Your Gun,* with Perry Mason written out of the story and character actress May Robson taking over as the amateur sleuth. Bad as it sounds, *Granny Get Your Gun* is actually a lot of fun on its own terms, though Erle Stanley Gardner always claimed that seeing the film caused him to burst out crying!

Slightly more satisfying to Gardner was the radio version of *Perry Mason*, sponsored by soap-opera specialists Procter & Gamble and heard as a daily, 15-minute CBS serial beginning October 18, 1943. At first Gardner didn't object to the radio show because it promoted his books and the producers allowed him a degree of control over the story material, but he grew to dislike the cliffhanger format and was frustrated over his inability to adapt his writing style to the rigors of a daily soap. Also, there was little time for mystery on this edition of *Perry Mason*: in the words of historian John T. Dunning, "On radio, [Perry] was as much detective as lawyer, at times swapping gunfire with criminals." But even though his control over the series diminished with each passing year, Gardner heartily approved of the contributions made by staff writer Irving Vendig, who fleshed out the character of Mason and actually made him more human and appealing than in the novels.

Bartlett Robinson, Santos Ortega, Donald Briggs and John Larkin were heard as Perry, with Gertrude Warner, Jan Miner and Joan Alexander as Della Street, and Matt Crowley and Chuck Webster as Paul Drake. Though Hamilton Burger was still a peripheral presence, Lt. Arthur Tragg, one of the many "official" police characters in the Mason novels, was prominently featured, played first by Mandel Kramer, then by Frank Dane (several of these actors later showed up in supporting roles in the TV *Mason*). In terms of longevity, radio's *Perry Mason* was the most successful adaptation of all, running twelve consecutive years before it was cancelled on December 30, 1955. Also during this period, a *Perry Mason* comic strip was nationally distributed by Universal Syndicate; seen from October 16, 1950, through June 21, 1952, the strip was drawn and written by Mel Keefer and Charles Lofgren.

The next logical step would seem to have been to transfer radio's *Perry Mason* to television. But Gardner had had a falling out with CBS, and never really warmed up to the daytime-drama format: besides, the author had gone into partnership with his agent Cornwell Jackson to form his own Paisano Productions (named after Gardner's ranch in Temecula, California) for the purpose of creating a weekly, prime time version of *Perry Mason*. Thus, Procter & Gamble contracted Irving Vendig to formulate a TV soap opera that would resemble but not slavishly copy the radio *Mason*—and the result was *The Edge of Night* (q.v.).

It was agreed that Gardner would have the lion's share of creative control over the proposed *Perry Mason* TV series, with complete story and casting approval; Cornwell Jackson's wife, former film actress Gail Patrick, would serve as executive producer, while Ben Brady and Sam White functioned as line producers. Studio facilities were secured at 20th Century–Fox, where White had previously worked on the TV-series version of *My Friend Flicka*. Though NBC expressed interest in *Perry Mason*, the network was outbid by CBS, who in a co-venture deal with Paisano Productions financed the series and assumed control of all future syndicated and worldwide distribution.

Reports vary as to how many big-name actors were considered for the role of Mason, but by early 1957, Efrem Zimbalist Jr. had made a firm commitment to star in the series. While auditioning actors for the other major roles, Erle Stanley Gardner reportedly offered the part of Della Street to his partner Gail Patrick Jackson, but she wasn't interested in resuming her acting career. Among those reading for the part of DA Hamilton Burger — at last promoted to "costar" status — was a bulky, bass-voiced actor named Raymond Burr, hitherto typecast as a villain in such films as *Tarzan and the She-Devil, A Cry in the Night* and Hitchcock's *Rear Window*; some sources indicate that Burr was actually cast as Burger during the preliminary auditions.

Even though Efrem Zimbalist Jr. was considered a shoe-in for Perry Mason, Raymond Burr begged the producers for the opportunity to read for Mason himself, promising to accept the part of Burger if he didn't make the grade. At this point, the legends conflict. The official story is that Erle Stanley Gardner took one look at Burr before the actor even said a word and exclaimed, "That's Perry Mason." But Sam White has insisted that he and the other producers sat through Burr's audition in shock and awe, then looked at each other and said, "What do we do now? This guy is far superior. He'll give this thing a dimension that Zimbalist could never give it"— and only then did they summon Gardner, who agreed Burr would be the better choice. Whatever the case, Raymond Burr was cast as Perry Mason, and Efrem Zimbalist Jr. went on to star in *77 Sunset Strip* and *The FBI*.

The rest of the casting proceeded smoothly and swiftly: Barbara Hale, an accomplished actress who'd been pursuing stardom with only modest success since 1943, was hired as Della Street; William Hopper, son of gossip columnist Hedda Hopper and best known for his portrayal of Natalie Wood's implicitly incestuous father in *Rebel Without a Cause* (1955), landed the role of Paul Drake; William Talman, like Raymond Burr previously limited to such villainous roles as the psycho title character in *The Hitchhiker* (1953), took over from Burr as Hamilton Burger; and Ray Collins, a 67-year-old veteran of stage, radio and Orson Welles movies, was signed as Lt. Arthur Tragg. In one of those rare instances of show-business serendipity, *Perry Mason* was blessed with a perfect cast, each actor fitting his or her role like a glove, and all of them getting along famously once the cameras stopped rolling.

A significant contributor to the success of *Perry Mason* was mystery writer and former lawyer Eugene Wang (pronounced "Vahng"), a past master at blending the whodunnit form with credible legal procedure. Wang and the other staff writers quickly adjusted to the rigors of *Perry Mason*, expertly adapting dozens of the original Erle Stanley Gardner novels and short stories to the hour-long form during the first few seasons; eventually all of the old Warner Bros. Perry Mason movies — even *Granny Get Your Gun* — would be refilmed for the series, and vastly improved in the process.

It was positively miraculous how the writers could wring so many variations from so "locked" a format. Each episode began by setting up a tense situation revolving around a disagreeable character who "needed killing." Once the murder was committed, several likely suspects were introduced, of which Lt. Tragg invariably arrested the wrong one — who of course was already a client of Perry Mason, or soon would be. Assisted by Paul and Della, Perry would nose around and interview the other suspects, keeping an eye peeled for any clues that might have escaped the less-than-eagle eye of Tragg and his fellow cops (CBS' original ad copy actually billed Mason as a "private investigator," ignoring his legal credentials entirely). The action would then usually move to the courtroom — not a murder trial, but a preliminary or evidentiary hearing, the better to avoid exceeding the $100,000-per-episode budget by hiring 12 expensive "business extras" as jurors. The hearing format also gave Mason a freer hand in cross-examining the witnesses, rebutting prosecution, and introducing new evidence, though this never prevented DA Burger from dismissing Mason's tactics as "irrelevant, incompetent and immaterial." Approximately five minutes before the final commercial, Mason would zero in on a heretefore completely unsuspected character — a witness, a friend of the family, a courtroom spectator — and, carefully framing his words in the form of speculation, force that person to break down and confess that he or she was the real murderer. Earl Stanley Gardner's most inviolate rule, and the one that caused the most frustration to the writing staff, was that Perry Mason *could never lose* (or almost never, as we shall see).

To be sure, there were a few deviations from the basic set-up. Occasionally Perry would be out of town on vacation, where he would somehow get tangled up in a local murder case and defend his client before a small-town prosecutor, often played by crusty old Paul Fix. Once in a while, Hamilton Burger would ask his "friendly enemy" Perry to help clear one of his own associates on a trumped-up charge, recusing himself from the subsequent hearing. Every so often, there'd be no hearing nor judge at all, with Perry extracting a confession in a police waiting room, his own office, or even a neighborhood tavern. And in one unique instance, all bets were off when Perry tried a case in a Communist–bloc courtroom, where presumption of innocence was just a bad joke. But by and large, the fundamental "crime/false accusation/court hearing/tearful confession" continuum was followed to the letter.

Though reviewers liked the series, they were not impressed by Raymond Burr, at least not at first. Audiences, however, responded to Burr's strong, solid, incisive, all-business, no-play portrayal of the cagey defense attorney, and before long thousands of letters were pouring into CBS, soliciting Burr's advice on all manner of real-life legal problems. One block of viewers who weren't entirely won over by *Perry Mason* were the prosecuting attorneys of America, who took the series to task for creating the false impression that defense lawyers were always right and DAs were always wrong — and worse, that a genuine courtroom experience would automatically be as exciting and suspenseful as what was seen on TV. More than one actual attorney griped that spectators and even jurors had approached him in mid-trial to complain about the dullness of the proceedings and demand that he get on the stick and force one of the witnesses to dissolve into tears. (Even defense attorneys hated those last-minute confessions, with F. Lee Bailey — ironically a close personal friend of Erle Stanley Gardner — speaking for his brethren by stating flatly, "It *never* happens.")

But lawyers don't determine TV ratings, and by the beginning of the 1959–60 season *Perry Mason* was securely in the Top 10, though it would slide a bit in the next few months thanks to its new Saturday-night NBC competition *Bonanza*. A sure sign of the series' success were the endless parodies and lampoons on such variety programs as *The Red Skelton Show* and *The Jack Benny Program*, all of them introducing their Perry Mason sketches with variations on the real series' jazzy theme music by Fred Steiner (official title: "Park Avenue Beat"). *Perry Mason* even survived a scandal that a few years earlier would have sent a lesser show hurtling to oblivion. In 1960, series costar William Talman was arrested for attending a "wild, nude" marijuana party. Though the charges were later dropped, CBS invoked the morals clause in Talman's contract, suspending him from *Perry Mason* on the grounds the viewers would reject the presence of a "tainted" performer. Not only did viewers fervently demand Talman's reinstatement, but his fellow cast members rallied around the actor and warned CBS that if he wasn't immediately allowed to return to the set, they might seriously reconsider signing contracts for the following season. Bill Talman did indeed come back, remaining with the show until its final episode, good-naturedly grousing about being a "perennial loser" straight down the line.

Virtually the only thing that could have killed *Perry Mason* during its first five seasons would have been the loss of Raymond Burr — and this dread calamity came very close to fruition on several occasions. Though unfailingly courteous and helpful to his fellow actors and the crew members, Burr was insisting as early as 1961 that the series had worn him out and he was on the verge of retiring. The actor wasn't just blowing smoke: he worked five days a week until 7:30 P.M., awakening at 3:30 A.M. on many occasions to learn his lines (he would later be permitted to use a teleprompter, but the dialogue was only a small part of the problem). To facilitate the hectic production schedule, Burr virtually lived in a furnished bungalow on the studio lot. "To live at the studio was the only way I could do the job," he recalled in a 1988 *TV Guide* interview. "I had no close friends. I had no life whatsoever, except the show. I was alone. We did 39 shows in our first season alone. If you went over six days on an episode ... [there would be] pressure from CBS. The tension to get it right would get to you." To alleviate that tension, he began playing practical jokes on the set, with Barbara Hale as his most frequent patsy (she never complained, and could still laugh about some of his pranks

nearly five decades later). This was quite a contrast with the taciturn Perry Mason, of whom Burr said, "Mason never once, in nine years, had a sense of humor."

The second time Burr made serious noises about leaving the show was during the 1962–63 season, when CBS had moved *Perry Mason* to Thursdays to accommodate Jackie Gleason's new Saturday-night extravaganza. Though the ratings weren't as strong as they'd been in past seasons, CBS knew that *Mason* was their strongest defense against ABC's Thursday-night sitcom lineup, and they were willing to mollify their star to the extent of allowing him his first extended leave of absence in six years. Informing the press that Burr had to undergo minor surgery (to this day the story has not been verified), CBS endeavored to stir up audience interest by hiring several "guest" attorneys to sub for Mason while he was ostensibly out of town "on business." Thus Bette Davis played the titular lady lawyer in the January 31, 1963, episode "The Case of the Constant Doyle"; Michael Rennie guested as law professor-turned-trial-attorney Edward Lindley in "The Case of the Libelous Locket" (February 7, 1963); Hugh O'Brian showed up as entertainment lawyer Bruce Jason in "The Case of the Two-Faced Turnabout" (February 14, 1963); and Walter Pidgeon spelled the missing Mason as Sherman Hatfield in "The Case of the Surplus Suitor" (February 28, 1963). A similar stunt was staged during the 1964–65 season, with Mason off on another convenient overseas trip while he was subbed by Mike Connors as Joe Kelly in "The Case of the Bullied Bowler" (November 5, 1964); on this occasion, it was leaked that Connors was being considered as Burr's replacement should he actually leave *Perry Mason*. (Needless to say, this didn't happen, and Connors would have to wait three more years before drawing a regular paycheck on *Mannix*.) Finally on January 14, 1965, Barry Sullivan as attorney Ken Kramer temporarily took over from Mason in "The Case of the Thermal Thief."

The big problem during the 1963–64 season was not the impending defection of Raymond Burr, but a sharp decline in ratings, not unusual for a program that had been running for seven years. On this occasion, the producers came up with a doozy of an audience-grabber, one that was heavily promoted in print and on the air in the weeks before its execution. On October 17, 1963, in the ominously titled "The Case of the Deadly Verdict," the unthinkable finally happened: Perry Mason lost a murder case, and his luckless client Janice Barton (Julie Adams) was sentenced to the gas chamber! After several minutes of navel-gazing on the part of Mason and his staff—and a commendable lack of hubris from Hamilton Burger, who despite having finally defeated his *bête noire* was no happier than Mason that poor Janice was doomed to die—the attorney decided to seek out new evidence to prove his client's innocence before the fatal pellets were dropped. Of course Janice was innocent, but she had signed her own death warrant by failing to reveal vital information that would have cleared her. Without giving away the ending, let us merely comment that fans of the old *Dobie Gillis* TV series were in for quite a jolt (and no, Bob Denver *wasn't* among the guest stars!).

Perry Mason ended up lasting nine seasons, spending its final year on Sunday evening opposite its former nemesis *Bonanza*—a move allegedly dictated by CBS to force the cancellation of a series that was by now attracting only the "wrong" (read "older") viewers. By this time the show was for the most part simply going through the motions, even remaking several of its earlier episodes under new titles: for example, 1965's "The Case of the Impetuous Imp" was a by-the-numbers rehash of 1958's "The Case of the Negligent Nymph." Also, the original cast was no longer intact: Ray Collins, aka Lt. Tragg, had retired due to ill health in 1964 (he died one year later); his replacements, Richard Anderson as Lt. Anderson and Wesley Lau as Lt. Drumm, were serviceable enough but never quite jelled with the other regulars.

Interestingly, *Perry Mason* suddenly rallied in the midst of its death throes with two of its most offbeat entries: "The Case of the Twice-Told Twist" (February 27, 1966), the series' only color episode, with Victor Buono as a modern-day Fagin; and "The Case of the

Dead Ringer" (April 17, 1966), featuring Raymond Burr in the dual role of Perry Mason and a homicidal cockney seaman named Grimes. And no one could ever claim that the series didn't have a wow finish: in the May 22, 1966, finale "The Case of the Final Fadeout," we were treated not only to two mysterious murders but also an impassioned curtain speech from Hamilton Burger, a "surprise killer" who was *really* a surprise, an abundance of inside jokes sprinkled throughout the script, and cameo appearances by members of the *Perry Mason* production staff—including, in the role of a judge, Erle Stanley Gardner.

Contrary to popular belief, the cancellation of *Perry Mason* was not due to Raymond Burr's disenchantment with the role: in fact, he had verbally agreed to do a tenth season, only to find out the show was cancelled when he read a newspaper story to that effect. No sooner had this occurred, however, than CBS began having second thoughts about pulling the plug on *Mason*. According to onetime CBS programming vice president Perry Lafferty, the network spent seven years after the series' cancellation looking for another "puzzle show" with a courtroom setting, shelling out "more than a million dollars in pilot films and scripts." One proposed *Mason* replacement was *Higher and Higher*, starring John McMartin and Sally Kellerman as a husband-wife defense team and an unknown Dustin Hoffman as the prosecutor. CBS wasted $500,000 on the pilot for this turkey, which gathered dust in the network's vaults until the overnight stardom of Dustin Hoffman vis-à-vis *The Graduate* resulted in a one-time-only telecast on September 8, 1968.

During this period, *Perry Mason* continued to spin money for Paisano Productions and CBS's domestic distribution arm (later known as Viacom), one of the few hour-long, black-and-white shows to retain its popularity in off-network syndication. What the series' new generation of fans didn't know was that only 195 of its 271 episodes were in active circulation. A number of episodes, including the "guest-lawyer" entries of 1963 and 1964, the various remakes of earlier shows, the balloon-busting "Case of the Deadly Verdict," the color-filmed

"Case of the Twice Told Twist" and the climactic "Case of the Final Fadeout" were withdrawn from view, remaining unseen until the late 1980s. (The unavailabilty of the aforementioned "Deadly Verdict" would give rise to an urban legend claiming that the one episode in which Mason lost a case was shown on a night when Good Friday and Passover coincided, guaranteeing that absolutely no one — except, presumably, atheists — would be watching!)

Unable to ignore the property's second wind in syndication, the folks at CBS decided that the only saleable substitute for *Perry Mason* would be a full-color revival of the series, along the lines of the resuscitated *Dragnet* of 1967–1970. Actually, former *Mason* production executive Cornwell Jackson had been pushing such a revival for years. Just before the death of Erle Stanley Gardner in 1970, Jackson had been beating the bushes promoting other Gardner literary properties for series consideration, including "David Selby" and "Lam and Cool." While making a "David Selby" pilot film for 20th Century–Fox, Jackson was urged to bring Perry Mason back to life, albeit as a a daytime serial. Picking up the ball and running with it, Jackson went to Fox's Willam Self, selling the executive on a full-fledged *Mason* revival in weekly prime time.

At the same time, Gail Patrick, who though now divorced from Jackson remained a partner in Paisano Productions, was in talks with *Dragnet* producer Jack Webb to restart *Perry Mason* as a Universal series. William Self beat her to the punch by shopping the new *Perry Mason* to CBS, who bought it — and then Gail Patrick chimed in with a *third* offer from her own Tomorrow Entertainment. The Paisano board of directors took a vote as to who should proceed with *Mason*, and Cornwell Jackson won, with his former wife retained as "executive consultant" and Erle Stanley Gardner's widow and sister-in-law signed as "script consultants." Jackson assembled the survivors from the original series' production team, including Ernest Frankel and Art Seid, and in early 1973 set the wheels in motion for *The New Perry Mason*.

But *not* with the original cast. Raymond Burr was then headlining another series, *Ironside*; Barbara Hale refused to consider appearing as Della without Burr; and William Hopper and William Talman had passed away. Screening 315 possible applicants for the role of Perry Mason, the producers chose Monte Markham, who despite having struck out with his two previous starring series *The Second Hundred Years* and *Mr. Deeds Goes to Town* enjoyed a solid reputation as a reliable character actor (ironically often cast as the Least Likely Suspect who ended up screaming "I did it! I did it! Hahahaha!"). Reportedly, the rest of the regulars were chosen on the basis of 15 ballots cast by Paisano and 20th Century–Fox each: Sharon Acker as Della Street, Albert Stratton as Paul Drake, Harry Guardino as Hamilton Burger and Dane Clark as Lt. Tragg. Perennial game-show panelist Brett Somers was also billed as a regular in the role of Mason's receptionist Gertie, who'd been played on a sporadic basis by Connie Cezon in the earlier series. Pooh-poohing the notion that Raymond Burr and his original costars were an impossible act to follow, William Self insisted, "the real strength of the show is the puzzle."

Scheduled on Sunday evening opposite ABC's *The FBI* and NBC's *Wonderful World of Disney*, *The New Perry Mason* was unveiled on September 16, 1973. Though a handful of scripts were unfilmed leftovers from the earlier *Mason*, most of the episodes strove to avoid the onus of being old-fashioned and outdated by focusing on contemporary issues — often obsessively so. The first episode dealt with a supercomputer; the second involved a "Jesus freak"; and in a plot twist that would not have even been considered in the old days, one of the "surprise" killers turned out to be an African American.

For all its up-to-date trappings, *The New Perry Mason* carried over the one constant from the original: Perry Mason was never permitted to lose. It was this element that came off as debilitatingly anachronistic to 1970s viewers, especially in the wake of such recent fallible TV attorneys as Carl Betz on *Judd for the Defense* and Arthur Hill on *Owen Marshall: Counselor at Law* (both q.v.). As *TV Guide* critic Cleveland Amory quipped, "If *The New Perry Mason Show* is new, we're Baby Snooks." Failing to make any impression at all on its target audience, *The New Perry Mason* folded after 15 hour-long episodes.

But by the mid–1980s, the baby boomers who'd grown up on the old *Perry Mason* reruns had reached the age when they were waxing nostalgic over the comfortable consistency of their TV-addicted youth, when they could rely upon Mason to triumph over adversity week after week — allowing them to live vicariously through Perry and briefly forget the disappointments and setbacks in their own lives. The unabated popularity of the syndicated *Mason* package proved that there was a sizeable if slightly superannuated core audience for such retrospective entertainment, and it was in this spirit that former NBC honcho Fred Silverman and veteran scripter Dean Hargrove joined forces on a two-hour, made-for-TV *Perry Mason* "reunion" film. And *this* time, they could count on the participation of surviving cast members Raymond Burr and Barbara Hale.

In the years since relinquishing the role of Mason, Burr had wound down his acting career and basically retired, living off his many business ventures and investments, including his own island in Fiji. Though he always claimed that the backbreaking production schedule of the old *Perry Mason* had robbed him of a personal life, the actor was coaxed back before the cameras on the promise that the TV movie would be filmed at a relaxed, leisurely pace — and also because Fred Silverman, Dean Hargrove and Viacom couldn't get the property sold without him (some sources suggest that Burr's more recent financial transactions had gone South and he desperately needed the money, but this has been unconfirmed). Barbara Hale had also been living in semi-retirement, but when she was told that Burr had signed on to the reunion film, she eagerly agreed to participate as well. The trick now was to inject credibility into the premise that a heavily bearded, overweight, 68-year old Perry Mason was still an active defense attorney. This was solved by explaining that Perry had left private practice to become a judge, but

had agreed to temporarily set his robe and gavel aside and act as counsel for his dear friend and former colleague Della Street, recently employed as executive secretary to billionaire Arthur Gordon (Patrick O'Neal) and now charged with Gordon's murder. In a sublime stroke of casting, the role of private eye Paul Drake Jr., now in charge of his late dad's business, was played by Barbara Hale's real-life son William Katt — inadvertently encouraging speculation that Della Street had *not* been the secret lover of Perry Mason as was often assumed by the more impressionable fans of the 1950s and 1960s, but that she had been carrying on an affair with Paul Drake Sr.!

First telecast by NBC on December 1, 1985, *Perry Mason Returns* succeeded beyond Silverman and Hargrove's wildest dreams, ending up as the week's top-rated network telecast. As proof this wasn't just a nostalgia-driven fluke, the second Mason TV movie, *Perry Mason: The Notorious Nun*, was likewise well received when it aired on May 25, 1986. In all, Burr and Hale costarred in 26 feature-length *Perry Mason* specials, which came out every three to six months over a period of eight years. Though just as formula-bound as the original series, the new Mason films benefited from an abundance of exciting action sequences (invariably handled by the younger members of the supporting cast, notably future *Baywatch* regulars David Hasselhoff and Alexandra Paul) and such appealing guest stars as Debbie Reynolds, Robert Stack, Mariette Hartley, Brian Keith, Teresa Wright, Tim Reid, Vanessa Williams, Robert Culp, Linda Blair, Jerry Orbach, Valerie Harper and Regis Philbin — the latter cast as a vicious talk-show producer who was bumped off by one of his many enemies (the producers manfully resisted the temptation to hire Kathie Lee Gifford … or even "Gelman"). Perhaps the most fondly remembered of the latter-day Mason efforts was 1987's *Perry Mason: The Case of the Lost Love*, in which the dour attorney shared his first on-screen kiss with guest star Jean Simmons — "a media event," in the words of F. Lee Bailey.

After the death of Raymond Burr on September 12, 1993, Hargrove and Silverman tried to keep the property alive by casting other actors as Perry's "colleagues," with mixed results. These post–Burr films were given the blanket title *A Perry Mason Mystery* and remained in production until 1995's *The Case of the Jealous Jokester*, in which sole survivor Barbara Hale costarred as Della Street opposite Hal Holbrook as flamboyant defense attorney/private detective Wild Bill McKenzie.

Petrocelli

NBC: Sept. 11, 1974–March 3, 1976. Miller-Milkis Productions/Paramount Television. Executive Producers: Thomas L. Miller, Edward J. Milkis, Leonard Katzman. Created by Sidney J. Furie and Harold Buchman. Developed for television by E. Jack Neuman. Cast: Barry Newman (Tony Petrocelli); Susan Howard (Maggie Petrocelli); Albert Salmi (Pete Ritter); David Huddleston (Lt. John Ponce).

The character of blunt-spoken, Harvard-educated defense attorney Tony Petrocelli was introduced in the 1970 theatrical feature *The Lawyer*. Directed by Sidney J. Furie from a script by Furie and Harold Buchman, the film was a thinly disguised dramatization of the Dr. Sam Sheppard murder case, which had previously inspired the popular TV series *The Fugitive*. Since Dr. Paul Harrison (William Sylvester) was clearly patterned after accused wife-murderer "Dr. Sam," it was a logical conclusion that Harrison's brash, stubborn, and cheerfully opportunistic lawyer Tony Petrocelli was inspired by F. Lee Bailey. In the absence of concrete evidence, Petrocelli based his defense of Harrison on "reasonable doubt," with three different flashbacks illustrating *Rashomon*-style what might have happened on the night of the murder. Barry Newman starred as Petrocelli in both *The Lawyer* and a TV-movie followup, *Night Games*, which aired on NBC on March 16, 1974.

In *Night Games*, Petrocelli moved from the Big City to the small Arizona farming community of San Remo (the film was actually shot in and around Tucson), accompanied by his wife Patsy, played by Susan Howard (in *The Lawyer*, his wife was named Ruth and was portrayed by Diana Muldaur). Assisting Tony in his defense of socialite Pauline Harrigan (Stefanie Powers), who had been accused of murdering her wealthy husband, was cowboy-turned-investigator Pete Toley, played by Albert Salmi. Though Mrs. Harrigan was an A-list client, Petrocelli let it be known that he preferred to cater to less affluent and more needy citizens, even offering to work pro-bono if it meant saving an innocent person from prison or worse — which is probably why Tony and Patsy were living in a low-cost trailer.

Night Games led to a weekly, hour-long *Petrocelli* series, commencing in September of 1974. The premise, locale and characters remained the same, though now Mrs. Petrocelli's first name was Maggie and Pete's last name was Ritter. New to the cast was David Huddleston as police lieutenant John Ponce, who maintained a friendly relationship with Petrocelli despite frequently butting heads with the lawyer in the courtroom. As in *The Lawyer* and *Night Games*, the crime-of-the-week was recalled in flashback from the various perspectives of the people involved, resulting in some truly baffling contradictions that Petrocelli would be forced to chew on before arriving at the truth.

Getting off to a shaky start, *Petrocelli* improved considerably when producer Leonard Katzman (*Hawaii 5-O*) was brought in to salvage the show. During filming, actor Barry Newman found himself getting into some curious Petrocelli-like situations. After a fan approached him on location to praise his acting, Newman was slapped with a subpoena: it turned out the fan was a juror under oath, and the DA wanted an exact account of the juror's conversation with the actor — forcing Newman to sing his own praises in a sworn deposition! In another instance, one of the crew members was arrested on an assault charge, whereupon Newman, anxious to play the role of attorney in real life, offered to defend the man in court. It didn't work: as Newman recalled, the judge "didn't like the idea of a TV star barging into a courtroom."

Before completing its two-season run, *Petrocelli* offered several young actors an opportunity for network exposure, including future *Star Wars* leading men Harrison Ford and Mark Hamill, and *Three's Company* star-to-be

John Ritter. The series was cancelled abruptly in March of 1976, its last four episodes remaining unshown until *Petrocelli* entered syndication.

Philly

ABC: Sept 18, 2001–Apr. 23, 2002; one leftover episode telecast May 28, 2002. Steven Bochco Productions/Paramount Television. Creators/Executive Producers: Steven Bochco, Alison Cross. Executive Producers: Kevin Hooks, Rick Wallace. Cast: Kim Delaney (Kathleen Maguire); Tom Everett Scott (Will Froman); Scott Leavenworth (Patrick Cavanaugh); Kyle Secor (ADA Daniel X. Cavanaugh); Rick Hoffman (ADA Terry Loomis); Diana-Maria Riva (Patricia); Kristianna Loken (Lisa Walensky); Scott Alan Smith (Jerry Bingham); Jamie Denton (Judge Augustus Ripley); Robert Harper (Judge Irwin Hawes); Dena Dietrich (Judge Ellen Armstrong).

Describing his newest legal drama *Philly* in a 2001 interview, producer Steven Bochco invoked the name of his then-most-successful cop series: "It's sort of a legal equivalent of *NYPD Blue*. It's a low-rent deal, the daily grind of the judicial system." Confirming the link with *NYPD Blue* was *Philly*'s leading lady Kim Delaney, who had just concluded a six-year run as *Blue*'s emotionally unstable, alcoholic police detective Diane Russell. Delaney had in fact agreed to star in *Philly* with the understanding that if the series failed, she would be allowed to return to *NYPD Blue*.

The actress was cast as Kathleen Maguire, a single mother and recent law-school graduate now working as a criminal attorney in the courtrooms of Philadelphia's City Hall. When not coping with her unsavory clients (rapists, murderers, pedophiles, pornographers and other such pillars of society), Kathleen waged an ongoing professional and personal war with her vengeful ex-husband, Assistant District Attorney Daniel X. Cavanaugh (Kyle Secor), who was systematically turning their 10-year-old son Patrick (Scott Leavenworth) against his mom. Viewers knew that Kathleen was in for some mighty rough sledding in the very first episode, in which her law partner was bundled off to a sanitarium after suffering a diet pill-related nervous breakdown in the middle of a court trial, leaving our heroine to deal with six arraignments, three pretrial conferences, four witness interviews, and a deposition all by herself—with time off for a brief jail stay on contempt charges after she resisted the sexual advances of a horny judge. Coming to Kathleen's rescue, in a manner of speaking, was young attorney Will Froman (Tom Everett Scott), who figured that by becoming Kathleen's new partner they'd not only comprise a dynamite defense team, but also allow Will to escape the drudgery of the Public Defender's office. Though Kathleen wasn't exactly enamored of Will, she agreed that a partnership would be mutually beneficial—and besides, bragged Will, "the entertainment factor's free."

Acknowledging that he was not only expected but required to shock his audiences from the first episode onward, Steven Bochco promised potential viewers that *Philly* would not only offer such envelope-pushing nifties as a lady lawyer baring her breasts before a jury and two opposing counsels settling their case by having sex in the conference room, but would also "use a scatological reference that had never before been uttered on an ABC series." To satisfy your morbid curiosity, this occurred in the series opener, when a disgruntled cop, passing Kathleen Maguire in the hallway, called her a bitch—whereupon Kathleen wheeled around and shouted, "Asshole!" Aren't you glad you asked?

Though *Philly* was partially filmed on location in the City of Brotherly Love (resulting in some sticky summer traffic jams on the Ben Franklin Parkway), genuine Philadelphians were not flattered by the compliment. Ellen Gray of the *Philadelphia Daily News* carped, "Just because someone's named a TV show for our town is no reason to think *Philly* is about us," adding that, despite the producers' meticulous recreation of the interior of Philadelphia's City Hall for those scenes shot at California's Culver Studios, "what's happening in those faux-marble hallways is strictly Hollywood." And in the *Philadelphia Inquirer*, it was noted that several local judges despised the show, calling it "insulting" and "unrealistic"; likewise, prominent Philadelphia attorney Jules Epstein found the series "absurd," while an ex-assistant

DA remarked: "I've loved Bochco shows … but this show is unwatchable." This was also the consensus of the general public: when *Philly* was added to ABC's Tuesday-night schedule (in the old *NYPD Blue* timeslot, no less), the evening posted its lowest ratings since 1987.

And what of Bochco's implicit promise that Kim Delaney could return to *NYPD Blue* if *Philly* tanked? Well, the producer had concluded that Delaney was no longer a "right fit" in view of how *Blue* was developing—and with this in mind, he took advantage of a loophole that rivalled anything imagined by *Philly*'s Kathleen Maguire and Will Froman. Reminding Delaney that she would have been reinstated only if *Philly* was cancelled after the first 13 episodes, Bochco pointed out that the new series had somehow limped through 21 installments, making all previous agreements null and void. (Ultimately, Kim Delaney did come back to *NYPD Blue* for a handful of guest appearances before joining the cast of Jerry Bruckheimer's *CSI: Miami*.)

Picket Fences

CBS: Sept. 18, 1992–June 26, 1996. David E. Kelley Productions/Nina Saxon Film Design/20th Century–Fox Television. Executive Producers: David E. Kelley, Alice West, Michael Pressman, Jeff Melvoin. Cast: Kathy Baker (Dr. Jill Brock); Tom Skerritt (Sheriff Jim Brock); Costas Mandylor (Deputy Kenny Lacos); Lauren Holly (Deputy Maxine Stewart); Holly Marie Combs (Kimberly Brock); Adam Wylie (Zach Brock); Justin Shernakow (Matthew Brock); Ray Walston (Judge Henry Bone); Fyvush Finkel (Douglas Wambaugh); Kelly Connell (Dr. Carter Pike); Zelda Rubinstein (Ginny Weedon); Don Cheadle (DA John Littleton); Michael Keenan (Mayor Bill Pugen), Robert Cornthwaite (Mayor Howard Buss); Leigh Taylor-Young (Mayor Rachel Harris); Marlee Matlin (Mayor Laurie Bey); Ann Morgan Guilbert (Miriam Wambaugh, 1992–95); Erica Yohn (Miriam Wambaugh, 1995–96).

Officially, the CBS weekly drama *Picket Fences* was about the family of a small-town sheriff. With the benefit of hindsight, however, the series can be regarded as a dry run for the popular legal shows that would emerge from the David E. Kelley entertainment factory: *Ally McBeal, The Practice,* and *Boston Legal* (all q.v.). *Picket Fences* was in fact the first series to em-

anate from the new production firm headed by former lawyer Kelley after leaving the staff of Steven Bochco's *L.A. Law* (q.v.).

In its "exposé" of the bizarre and sometimes sinister underbelly of Small Town America, *Picket Fences* fell somewhere between the warmhearted eccentricities of Joshua Brand and John Falsey's *Northern Exposure* and the Satanic intrigues of David Lynch's *Twin Peaks*. The setting was the picturesque little community of Rome, Wisconsin—and while there actually are *three* Wisconsin towns named Rome, the series was entirely filmed in California. Tom Skerritt starred as Sheriff Jimmy Brock, with Kathy Baker as his wife, town doctor Jill Brock, and Holly Marie Combs, Justin Shenakow and Adam Wylie as their children Kimberly, Matthew and Zach. No sooner had the series debuted than Rome's thin veneer of cozy normality was stripped away. In one of the most celebrated episodes of the series' first two seasons, Rome's mayor was put on trial after killing a carjacker—but before sentence could be passed, the defendant died of spontaneous combustion. As a matter of fact, the town had a lot of difficulty holding onto its mayors. One was kicked out of office for appearing in a porn movie, another was decapitated and locked in a freezer, and still another was seriously wounded by the Number One Fan of a radio shock jock. But this was kid stuff compared to the fate of the mayor played by Marlee Matlin—a dancing bank robber who was forced to accept the oath of office as part of her 3000 hours' community service.

Other curious happenings in Sheriff Brock's jurisdiction involved the cancellation of a community-theater staging of *The Wizard of Oz* when one of its stars was killed with an injection of pure nicotine, a looney-tune who broke into people's homes just to bathe in their tubs, a woman who claimed she'd flattened her husband with a steamroller because she was undergoing menopause, a cow giving birth to a human baby, and a kid who brought a dismembered hand to school for show-and-tell. All of this weirdness led critics to hosanna the series for its quirky unpredictability, but in one respect it was supremely predictable: the viewer could always count on David E. Kelley to emphasize the ongoing struggle between doing what was right and was was legal, and to affirm such basic human values as family loyalty, mutual trust and respect, and the never-ending pursuit of Justice.

This last aspect was most prevalent in the courtroom scenes presided over by the town's ill-tempered but irrefutably honest Judge Henry Bone, a role which garnered several industry awards for that grand old trouper Ray Walston. It wasn't that Bone was a rigid adherent to the letter of the law, simply that he was guided by a personal and infallible moral compass that always led him to the correct decision no matter how much he irritated and baffled the community. The best scenes found Bone verbally fencing with the town's bow-tied, bloviating defense attorney Douglas Wambaugh (Fyvush Finkel), who neither knew nor cared if his clients were guilty so long as he could clear them with an obscure legal loophole or eleventh-hour grandstand play, and who fed into every "ambulance chaser" stereotype by threatening to sue anyone who disagreed with him or even looked at him cross-eyed (like Walston, Finkel also earned an Emmy for his work). The one-two punch of Bone and Wambaugh resulted in an unusually large turnover of prosecuting attorneys, though DA Jonathan Littleton (Don Cheadle) managed to weather the ordeal and remain on the series for three seasons.

"Off the beam" though it may have been, *Picket Fences* raised some important legal issues regarding such topics as Constitutional rights, freedom of (and freedom *from*) religion, medical ethics, homosexuality, abortion, and Alzheimers. Never more than a few steps away from a soapbox, producer Kelley had a tendency to "speak" through the coutroom summations of Judge Bone, which when they weren't self-congratulatory and longwinded could be quite eloquent and persuasive. In a 1999 Salon.com piece, Joyce Millman wrote of the series' strengths and weaknesses: "Kelley operates within a very narrow comfort zone. The ex-lawyer in him can't resist pushing ethical issues (the right to die, civil liberties vs.

community safety) in viewers' faces in the most sensational way possible…. [T]he crowd-pleasing TV scribe in Kelley knows that voyeurism pays off, often veering into the sort of nasty freak-show comedy that gets viewers talking the next day…. I've come to appreciate *Picket Fences* in cable reruns (9 A.M., weekdays, FX) for its sense of the miraculous in the most mundane of settings, and for its central theme of how communities and parents are torn between doing what's ethically and legally right and what makes them feel safe. But I still cringe, watching *Picket Fences*, when Kelley goes into full-bore message mode, setting up thuddingly obvious 'debates' about, say, religion vs. science or justice vs. revenge."

Never any higher in the ratings than the mid–60s, *Picket Fences* was retooled during its third season when Jeff Melvoin replaced David E. Kelley as executive producer. Much to the dismay of costar Ray Walston, Judge Bone's summations were curtailed as the show downplayed social issues in favor of character-related problems. But Melvoin's changes helped not at all, and as the show entered its fourth and final season, "executive consultant" Kelley reasserted his authority. When the series left its temporary Thursday timeslot and returned to its original Friday berth in December of 1995, there was no question who was back in charge: the episode of December 8 found no less than the Pope standing before Judge Bone, as Douglas Wambaugh tried to impeach His Eminence's testimony by bringing up the Catholic church's anti-gay stance!

It may not have been planned as such, but the 89-episode *Picket Fences* could be described as the first dramatic TV series to depict the administration of justice strictly from the judge's point of view. If one is willing to accept that assertion, then the series can be regarded as the nutty stepfather to the far more successful (but far less strange) *Judging Amy*.

Politics on Trial
ABC: Sept. 4–Oct. 30, 1952.

Telecast in the weeks just prior to the 1952 Presidential election, ABC's *Politics on Trial* was one of several public-affairs series that attempted to enhance viewer interest by staging its debates in a courtroom setting. In each half-hour episode, a prominent representative of one of the two major political parties would articulate his party's platform (in the episode of October 2, 1952, future Secretary of State John Foster Dulles appeared to discuss "The Republican Foreign Policy"). He would then be attacked by an "opposing counsel" from the rival party, with one of his own colleagues acting as defense attorney. An actual judge arbitrated the proceedings and handed down the final decision. Most reference books indicate that *Politics on Trial* was seen on Thursdays at 9 P.M.— or more likely *not* seen, since it was scheduled opposite NBC's *Dragnet*— but other sources indicate that the series first aired locally in New York on Mondays in the summer of 1952, as a lead-in to ABC's coverage of the Republican and Democratic conventions.

Portia Faces Life/The Inner Flame

CBS: Apr. 5, 1954–July 1, 1955. Produced by Beverly Smith. Directed by Hal Cooper. Cast: Frances Reid, Fran Carlon (Portia Blake Manning); Karl Swenson, Donald Woods (Walter Manning); Charles Taylor (Dick Manning); Ginger McManus, Renne Jarrett (Shirley Manning); Patrick O'Neal (Carl); Jean Gillespie (Dorrie Blake); Mary Fickett (Ruth Byfield); Bill Shipley (Announcer); and Eda Heinemann, Richard Kendrick, Teri Keane and Elizabeth York.

Conceived at a time when a female attorney was the exception rather than the rule, *Portia Faces Life* was one of radio's most successful soap operas. Debuting October 7, 1940, on CBS through the courtesy of sponsor General Mills, the series was for many years written by Mona Kent. It was Kent who came up with the idea of naming her protagonist after the heroine of Shakespeare's *The Merchant of Venice*, who while disguised as a male lawyer saved her sweetheart Antonio from surrendering a "pound of flesh" to the usurious Shylock. As noted by radio historian John Dunning, *Portia Faces Life* opened in crisis and ended in turmoil, as beautiful young lawyer Portia Blake waged a never-ending war against corruption in the fictional community of Parkerstown. Widowed in the very first episode, Portia not only fought evil-doers in court and out, but also singlehandedly raised her young son Dickie. The man in Portia's life was "brilliant, handsome" journalist Walter Manning, who like most soap-opera heroes never seemed to be around as the heroine was hit with one crisis after another, ranging from being rendered comatose in a burning house to facing accusations of bearing a child out of wedlock.

Lucille Wall starred as Portia throughout the series' eleven-year run on two different networks (CBS and NBC). The actress became so identified with the role that when she suffered a serious injury at her home and was temporarily replaced on *Portia Faces Life* in the spring of 1948, her hospital room was deluged with flowers, letters and gifts from thousands of anxious fans—prompting Wall to return to the series long before her doctors thought it advisable. Three years later, when *Portia Faces Life* was slated for cancellation, Lucille Wall and the series' producers tried to create another groundswell of listener support by having Portia framed and sent to prison: this time, however, the fans' response was not sufficient to save the series, and on June 29, 1951, *Portia Faces Life* ended its radio run with the heroine's fate still unresolved.

Slightly less than three years later, *Portia Faces Life* was revived as a daytime TV serial, again in a 15-minute timeslot on CBS, still sponsored by General Mills. Described by announcer Bill Shipley as "the story of a not-so-perfect marriage," the TV version of *Portia Faces Life* downplayed the crime-melodrama angle of the radio original, concentrating instead on the shaky union between Portia and her chronically unfaithful husband Walter. Any legal issues raised on the series were invariably linked with the main characters' personal problems: an existing episode shows Walter threatening to bring the full wrath of the courts down on the police department for arresting his current girlfriend. Portia herself continued to perform above and beyond the call of duty, doing her own detective work while pursuing a court case rather than relying on private investigators.

With Lucille Wall committed to other acting assignments, the role of TV's Portia went to Frances Reid, best known to contemporary daytime-drama devotees as Alice Horton on NBC's *Days of Our Lives*. "I was just there!" explained Reid in an Archive of American Television interview when asked how she landed the role. Reid added that her training had not prepared her for the stress of playing a difficult role five times per week, but she came to enjoy it, "in a masochistic kind of way." Costarring as Portia's wayward hubby Walter Manning was Karl Swenson, later seen as Lars Hanson on *Little House on the Prairie*, while the couple's children Dick and Shirley were played by Charles Taylor and Ginger McManus (later replaced by Renne Jarrett). Making trouble for the regulars from time to time was Portia's brother Carl, an early assignment for the versatile Patrick O'Neal.

Six months into the series' run, Frances Reid left *Portia Faces Life*, and not long afterward the title was changed to *The Inner Flame*. Replacing Reid as Portia was Fran Carlon, one of the actresses who had subbed for Lucille Wall while she was recuperating from that 1948 accident; likewise gone was Karl Swenson, with Donald Woods stepping in as Walter. *The Inner Flame* was ultimately extinguished on July 1, 1955, when CBS surrendered the 1–1:15 P.M. timeslot to its local affiliates.

Power of Attorney

Syndicated: Aug. 26, 2000–Jan. 4, 2002. Monet Lane Productions/Twentieth Television. Executive Producers: Jeannine Sullivan, Jill Blackstone. Cast: Andrew P. Napolitano (Himself, judge); Lynn Toler (Herself, judge); Joseph Catalano Jr. (Bailiff); Francine DeGiorgio (Court Reporter).

Like many another syndicated courtroom-reality series making the rounds in 2000, the daily, half-hour *Power of Attorney* featured actual civil and small-claims cases, with real-life litigants airing their disputes before the cameras in anticipation of a quick and binding decision and a cash settlement. The difference here was that the plaintiffs and defendants didn't "tell it to the judge." That responsibility was handed over to such well-known attorneys as Gloria Allred, Christopher Darden, F.

Lee Bailey, Johnnie Cochran and Marcia Clark, who represented the litigants on a pro-bono basis — cross-examining witnesses, delivering summations, and in general doing all the things celebrity attorneys were expected to do. The emphasis in *Power of Attorney* was on its stellar rotating stable of lawyers, who rose to the occasion by giving full-blooded performances of Al Pacino intensity.

The presiding judge during the series' first season was Andrew P. Napolitano, an avowed "pro-life Libertarian" who had been New Jersey's youngest life-tenured judge and who from 1987 until 1995 served on that state's Superior Court. After a few years in private practice, Napolitano joined the Fox News Channel as a legal analyst in January of 1998, holding his own opposite such combustible commentators as John Gibson and Bill O'Reilly. Thus, when *Power of Attorney* rolled around, it was Napolitano's "star power" that sold the series.

In 2001, Andrew Napolitano was promoted to the FNC's Senior Judicial Analyst, forcing him to step down from *Power of Attorney*. His replacement was Lynn Toler, a civil rights attorney and former administrative judge who hailed from Cleveland, Ohio. While Napolitano would undoubtedly be the first to admit that he wasn't as pretty as Lynn Toler, she lacked her predecessor's charisma and stage presence, and before long the series had lost 25 percent of its audience. This, coupled with the expense of keeping as many as ten high-profile attorneys on retainer, compelled Twentieth Television to cease production on *Power of Attorney* three months into its second season. Happily, Lynn Toler went on to greater success when she picked up the gavel on the long-running *Divorce Court* (q.v.) in 2006.

The Practice

ABC: March 4, 1997–May 16, 2004. David E. Kelley Productions/Daydreamers Entertainment/Decode Entertainment/20th Century–Fox Television. Executive Producers: David E. Kelley, Robert Breech, Jeffrey Kramer, John Tinker, Kathy McCormick, Will Ackerman. Cast: Dylan McDermott (Bobby Donnell); Lisa Gay Hamilton (Rebecca Washington); Steve Harris (Eugene Young); Michael Badalucco (Jimmy Berluti); Camryn Manheim (Ellenor Frutt); Kelli Williams (Lindsay Doyle); Lara Flynn Boyle (ADA Helen Gamble); Maria Sokoloff (Lucy Hatcher) Jessica Capshaw (Jamie Stringer); Bill Smitrovich (ADA Kenneth Walsh); Jason Kravits (ADA Richard Bay); Ron Livingston (ADA Alan Lowe); Chyler Leigh (Claire Wyatt); Rhona Mitra (Tara Wilson); James Spader (Alan Shore).

With the exception of Dick Wolf (see *Law & Order*), David E. Kelley is the only American TV producer who has ever managed to sustain two successful prime time legal series simultaneously. From 1998 through 2001, Kelley not only dominated Fox's Monday-night schedule with *Ally McBeal* (q.v.), but also ruled ABC's Sunday-night manifest with *The Practice*. But wait, there's more: going one better than Dick Wolf, David E. Kelley also managed to rescue *The Practice* from the inevitabilty of declining ratings by magically morphing the property into an *entirely different* legal series.

Beginning with a modest six-episode try-out in the spring of 1997, *The Practice* was the polar opposite of Kelley's previous project *L.A. Law* (q.v.), which featured a high-priced law firm trafficking in high-profile clients. The central setting of *The Practice* was the struggling Boston firm of Donnell & Associates, whose clientele was basically limited to the scum of the earth. The lowlifes and subhumans represented by the firm were generally guilty — or as *Doonesbury's* Garry Trudeau might put it, "*GUILTY! GUILTY! GUILTY!*" — frequently forcing the young and essentially decent-minded attorneys to park their ideals and ethics at the courthouse door. The firm's senior partner was Bobby Donnell (Dylan McDermott), who spent as much time wrestling with his conscience as sparring with prosecutors: a nervous breakdown waiting to happen.

Donnell's associates — and later partners when the firm changed its name to Donnell, Young, Dole & Frutt — included Eugene Young (Steve Harris), a former private eye who likewise suffered from bouts of conscience, but who unlike Bobby strictly adhered to the letter of the law, having seen his older brother languish and die in prison on the basis of a coerced confession. Also on the staff was Jimmy Berluti (Michael Badalucco), a pugnacious

blue-collar type who produced his own low-budget TV ads to promote his practice, and who alleviated the tensions of his job by gambling. Then there was Ellenor Frutt, a brilliant, full-figured female attorney with a cute habit of pushing people across the room when they displeased her, and whose closest friends all seemed to be murder suspects; she was played by actress-playwright Camryn Manheim, who two years before *The Practice's* debut had summed up her life philosophy — and Ellenor's — with her off-Broadway show *Wake Up: I'm Fat*. (Chances are that most of the series' fans wouldn't have given her weight a second thought if Manheim hadn't kept obsessing over it.) The firm's other female associate was Harvard-educated Lindsay Dole (Kelli Williams), a mass of insecurities who was something of a magnet for the series' most melodramatic plot contrivances. Finally, serving as the firm's receptionist and paralegal was Rebecca Washington (Lisa Gay Hamilton), who unbeknownst to her coworkers was taking law courses in night school, ultimately passing her bar exams and rejoining the firm as a full-fledged lawyer.

When *The Practice* was renewed for its first full season in the fall of 1997, Lara Flynn Boyle was added to the cast as glamorous district attorney Helen Gamble, who despite her antagonistic courtroom demeanor retained close friendships with her perennial opponents at Donnell, Young, Dole & Frutt (Helen and Lindsay had been law-school classmates); and in the case of handsome Bobby Donnell, Helen's relationship periodically went far beyond the good-buddy stage. Another romance developed between Bobby and Lindsay, culminating in a lavish wedding at Fenway Park; alas, the couple eventually split up due to Bobby's infidelities — over which he characteristically anguished.

In the last three episodes of its first season, *The Practice* tackled the first of its many "vital issue" story arcs as the firm mounted a case against a tobacco company represented by Lindsay's imperious former law professor Anderson Pearson (Edward Herrmann). In typical David E. Kelley fashion, this continuity evolved into a murder story when Lindsay and her colleagues defended Pearson on a charge of killing a man who had threatened his family. Though such storylines might have tempted Kelley to proselytize, during the first few seasons of *The Practice* he admirably avoided the sort of agit-prop speechifying that occasionally slowed his other series *Picket Fences* (q.v.) down to a walk. Lita Scottoline, author of *Rough Justice*, called the show "the single best, all-substance, no-hype lawyer series anywhere," bestowing extra points upon Kelley for his refusal to dumb down the show with long and wearisome explanations of each and every bit of legalese: "It's TV without training wheels." And in a 1999 Salon.com article, Joyce Millman noted: "I don't know how *The Practice* has managed so far to escape Kelley's preachy and sensationalistic tendencies. He writes every episode, and the show *has* pushed its share of hot buttons, with story lines about lawsuits against tobacco companies and gun makers, about nannies accused of murder and Kevorkian–like assisted-suicide crusaders. The show has also featured such sideshow attractions as a severed head in a medical bag, a lawyer who barfs before every big court appearance and an erudite gay serial killer who keeps getting away with murder. But somehow, none of this seems freakish. *The Practice* is comparatively restrained, somber, realistic. Maybe Kelley identifies too strongly with his hard-working, conscience-stricken Boston lawyer characters (Dylan McDermott's Bobby Donnell in particular) to want to screw with them, *[Ally] McBeal*-style."

Which is not to say that Kelley wasn't above a soupçon of old-fashioned ballyhoo to hype ratings. Some of his stunting was sexual in nature, from the occasional flash of nudity to a subplot wherein Lindsay's current boyfriend slipped her the "date-rape" drug. Other times, the grandstanding was stylistic: on two separate occasions, Kelley staged an issue-heavy episode in the manner of a *cinema verite* documentary, with hand-held cameras and in-depth interviews of the principals. And as was Kelley's habit, the series sporadically featured crossover episodes with the producers' other shows, beginning with the now-famous 1998 storyline introduced on *Ally McBeal* and resolved on *The Practice*. This was the first time in TV history a two-parter that began on a comedy series concluded on a dramatic program—and on a different network to boot. This was also the *last* time a gimmick of this nature ever occurred in any of David E. Kelley's legal shows: it was decided never to repeat the *McBeal/Practice* crossover because, according to coproducer Jeffrey Kramer, "It was like planning the Gulf War."

Although *The Practice*'s ratings were disappointing during the 1997–98 season, the series won the first of its many Emmy awards, encouraging ABC to renew for a third season. The timeslot was switched from Monday to the series' familiar Sunday-night berth, where it flourished for the next 5½ years. Added to the cast for Season Three was Maria Sokoloff as jokey, nosy receptionist Lucy Hatcher, replacing the newly-promoted Rebecca Washington; Lucy would later become a part-time counsellor to rape victims.

After two seasons' worth of highlights involving a podiatrist (Michael Monks) suspected of decapitation, a dentist (Henry Winkler) with a grape-jelly and cockroach fetish, an oversexed female judge (Holland Taylor), and Ellenor Frutt's artificially-inseminated pregnancy—with time off for a minor controversy involving a pedophile-priest story arc—*The Practice* reached Valhalla during its fifth season, regarded by fans and critics alike as the series' best year (or at the very least, the busiest!). This was the season in which Jason Kravits joined the regular cast in the previously recurring role of flamboyant, short-statured ADA Richard Bay, soon to depart in spectacular fashion when he was gunned down by a vengeful hitman. This was the season in which Bobby hired a man to scare off a wife-beater, only to stand trial for being accessory to murder when the strategy went too far and the hired goon killed the abusive husband. And this was the season in which *The Practice* emulated the just-released theatrical feature *Erin Brockovich* with a compelling storyline about a pesticide company accused of causing infant deformities.

Season Five also yielded what is arguably the quintessential *Practice* episode, which managed to be issue-driven, story-driven and character-driven all at once. Telecast February 11, 2001, "The Day After" took place in the wake of a near-tragedy when Rebecca Washington opened a letter bomb intended for the pregnant Lindsay Dole. All that could save the comatose Rebecca was a blood transfusion, but her mother (CCH Pounder), a former nurse, refused to grant permission on religious grounds: as a Jehovah's Witness, Rebecca was forbidden to undergo "heroic" medical treatment. Taking Mrs. Hamilton to court, Bobby Donnell reasoned that the only way to save his colleague's life was to assassinate her character. With heated eloquence, he proved that Rebecca had not honored the tenets of her faith because of her tolerant attitude towards divorce, premarital sex and homosexuality—and then, as her Mom looked on in dumbstruck horror, he revealed that Rebecca had had an abortion. Just at the point when it seemed that Rebecca's mother was going to remain a bigoted, close-minded villain, Bobby's wife Lindsay went into premature labor in the middle of the courtroom—whereupon Mrs. Hamilton proved her essential humanity by delivering the baby! All that was missing from the episode, besides the proverbial kitchen sink, was a curtain speech about the right to live vs. the right to die, and this is probably because there wasn't any time: after all, Kelley was also using the episode as a "crossover" for several actors from *another* of the producers' series, *Boston Public*.

It was hard to keep up the pace after this, and *The Practice* began its slow downward slide in Season Six—though not through any fault of the newest regular actor Ron Livingston, cast as the late Richard Bay's successor, ambitious ADA Alan Lowe. As in seasons past, Number Six ended with a cliffhanger, with Lindsay Dole on trial for the murder of a dangerous stalker. Kelley shot two different endings for this finale, one in which Lindsay was found guilty, another in which she was freed; none of the viewers—and none of the actors—would learn Lindsay's fate until the episode was actually telecast. So as not to keep *you* in sus-

pense, the poor girl was found guilty, but the conviction was overturned due to prosecutorial misconduct. Though she was invited to rejoin the firm, Linday opted to start up her own private practice instead.

With ratings flagging, ABC was reluctant to pony up the $6.5 million per episode for what was now the most expensive series on network television. It has been claimed that ABC's decision to move *The Practice* from its comfortable Sunday-night timeslot to a less desirable Monday-night berth opposite CBS' sitcom lineup and WB's *Seventh Heaven*—described by Kelley as "an act of stunning stupidity"—was a master plan to deliberately force the ratings to drop even further so that the network would be in a position to slice *The Practice*'s budget in half. If true, the plan succeeded: the series continued to bleed viewers even after ABC relented and returned it to Sundays. To counter this, Kelley began soft-pedalling "issue" episodes and increased the series' quotient of supermarket-tabloid murder cases. He also added two attractive new cast members: Jessica Capshaw as Jamie Stringer, an idealistic newcomer to the firm whose insecurity over working with her jaded colleagues was offset by an extremely active libido; and Chyler Leigh (late of David E. Kelley's failed legal series *girls club* [q.v]) as Claire Wyatt, Lindsay's associate at her new practice.

For a while, it looked as if *The Practice* would be cancelled at the end of its seventh season, and everyone connected with the show was resigned to this apparent inevitability. At the last minute, ABC reversed itself and renewed the series—on the proviso that Kelley would slash the weekly budget to a little over $3 million per show. Unfortunately, the only way to accomplish this was to stage what has now become infamous as "The Purge." In one fell swoop, Kelley eliminated six regulars from the cast: Dylan McDermott, Kelli Williams, Lisa Gay Hamilton, Lara Flynn Boyle, Maria Sokoloff and Chyler Leigh. "Basically, I sat down and looked at where I thought the creative heart of this series would be in the future," Kelley was quoted as saying in *The Pittsburgh Post-Gazette*. "The decisions were creatively driven, with the exception of Dylan's character. Dylan had such a deal in place that it was pretty much prohibitive to bring him back under our current fee structure." All of which was small comfort to the six dismissed performers, whose mass firing took place in a particularly cold-blooded fashion. Not wishing to leak to the press what was about to happen, ABC invited the actors to a big meeting before the network's advertisers, ostensibly to give them the good news of the series' renewal: only after everyone was seated and the doors were closed did the network drop the bomb.

Even more callously, there was never any explanation as to what had happened to the six missing characters during *The Practice*'s eighth season, except for a throwaway line acknowledging their absence. In the long run, the series lost far more than 60 percent of its cast: worried that NBC's flashy new legal drama *Lyon's Den* (q.v.) posed a serious threat to *The Practice*, Kelley all but abandoned the issue-driven episodes in favor of increasingly lurid and sensationalistic story arcs involving such A-list guest stars as Sharon Stone and Chris O'Donnell. And in a spectacular move to confirm a complete break from the standards set by the fundamentally ethical, conscience-plagued Bobby Donnell, Kelley introduced the first regular character who could truly be described as ethically bankrupt and utterly devoid of conscience: James Spader as the firm's sharkish new partner Alan Shore, whose reputation was so tarnished in legal circles that he literally begged his way into the firm. (He turned out to be another of Ellenor Frutt's unsavory close friends!) Once on board, Shore wasted no time fouling the nest for his colleagues, pushing and prodding them down a dark and winding road they had taken great pains to avoid during the reign of Bobby Donnell; even straight-arrow Eugene Young, now promoted to senior member, was unable to curb Shore's bottom-feeding excesses. The only person whom Shore could call a kindred spirit was newly hired paralegal Tara Wilson (Rhona Mitra). Getting wind that Alan was about to dismissed, Tara helped him steal files that would allow the firm to take such ac-

tion—and in consequence ended up being fired herself.

By the time this continuity was up and running, the whole world knew that David E. Kelley had introduced Alan Shore and Tara Wilson merely to segue into the *Practice* spin-off *Boston Legal* (q.v.), which premiered in the fall of 2004. In this respect, *Boston Legal* can be seen as a continuation of *The Practice*—albeit one festooned with the sort of overwrought histrionics and fatuous speechmaking that *The Practice* had for the most part managed to do without quite nicely.

The Prosecutors: In Pursuit of Justice

Discovery Channel: 2000–2002. New Dominion Pictures. Produced by David O'Donnell, Elizabeth Browde, Jerome Delayen, Mike Sinclair, Jamie Smith.

A presentation of cable's Discovery Channel, the weekly documentary series *The Prosecutors: In Pursuit of Justice* was set amidst what its producers called "grim reality of backlogged courts, the overworked police stations and the hardening streets of America." Each hour-long episode followed an established prosecutor or a talented newcomer during the investigation of a serious crime, combining actual footage with the occasional re-enactment. Representative episode titles included "The Missing Heiress," "Without a Trace," "Infant Mortality," "An Unlikely Suspect," "Halloween Ambush" and "The Fast-Food Killer."

In production for two seasons, the 31-episode *The Prosecutors: In Pursuit of Justice* was later rerun on Discovery Channel's sister digital service, Investigation Discovery.

Public Defender

CBS: March 11, 1954–June 23, 1955. Hal Roach Studios/Official Films. Created by Mort Lewis and Sam Shayon. Produced by Hal Roach Jr. Technical advisor: Edward N. Bliss Jr., LA County Public Defenders Office. Cast: Reed Hadley (Bart Matthews).

It is quite understandable that many casual TV fans tend to confuse the half-hour legal drama *Public Defender* with the more famous crime series *Racket Squad*. Both shows were produced at the old Hal Roach studios; both were introduced by CBS as summer replacements, with *Racket Squad* subbing for the popular *I Love Lucy* and *Public Defender* rushed in as successor to the cancelled *Philip Morris Playhouse*; both were sponsored by Philip Morris cigarettes, with the company's adopted theme song "The Grand Canyon Suite" frequently heard over the closing credits; and both starred the mellifluous Reed Hadley, delivering essentially the same performance as two different characters.

Public Defender boasted a memorable opening signature, with Reed Hadley, smoking his ubiquitous cigarette and standing before an impressionistic courtroom setting dominated by a statue of Lady Justice, explaining, "A public defender is an attorney employed by the community and responsible for giving legal aid without cost to a person who seeks it and is financially unable to employ private council. It is his duty to defend those accused of a crime until the issue is decided in a court of law." We then cut to an animated map of the West Coast, with Hadley continuing: "The first public defender's office in the United States was opened in January, 1913. Over the years, other offices were opened and today that handful has grown to a network ... a network of lawyers cooperating to protect the rights of our clients." And that was usually the last we would see of Hadley for several minutes, as the case at hand took center stage.

Each of the series' 69 episodes focused on an actual or accused lawbreaker in desperate need of assistance from the Public Defender's office: a boxer framed on a drug charge for refusing to throw a fight, a suspected child abuser, a young punk charged with murdering his best friend in an alley rumble, a condemned man facing execution within a few hours, etc. Reed Hadley played public defender Bart Mathews, who did his best to either prove his client wasn't guilty, or at least negotiate a reduced sentence. Though Mathews seldom betrayed his own emotions or indicated whether he was truly convinced of his client's innocence, he was shown expressing frustration over certain defendants who tied the court's hands by refusing to reveal all the facts, forcing our hero to leave his office and conduct his own personal investigation.

Although there were sporadic bursts of violence and melodrama — a handful of episodes were directed by Western specialist Budd Boetticher — *Public Defender* was distinguished by a mile-wide streak of maudlin sentimentality. In one episode, Mathews ended up playing family counselor to help a troubled teen who had turned to a life of crime because he was ashamed of his father, a cheap burlesque comedian. And in another, Mathews defended a washed-up Hollywood director on a charge of assaulting an arrogant young filmmaker who had callously refused to give a deserving starlet a break. (This episode was a special treat for movie buffs, featuring extensive "inside" footage of the fabled Hal Roach lot, once the home of such iconic funmakers as Harold Lloyd and Laurel & Hardy).

Curiously, even after *Public Defender* left CBS and entered local syndication, the series continued invoking the name of its original sponsor: each episode ended with a "Philip Morris salute," as Reed Hadley paid tribute to a real-life public defender.

Public Prosecutor

Syndicated: earliest known telecast, 1948; DuMont (as *Crawford Mystery Theater*): Sept. 6–27, 1951. Jerry Fairbanks Productions/NBC Films. Executive Producer: Jerry Fairbanks. Cast: John Howard (Stephen Allen); Anne Gwynne (Patricia Kelly); Walter Sande (Lt. Evans); Warren Hull (host of DuMont run).

Public Prosecutor is historically significant as television's first filmed, syndicated dramatic series. Produced between 1947 and 1948, it was one of several properties created by Jerry Fairbanks, who'd built his reputation on such theatrical short-subject series as *Popular Science* and *Speaking of Animals*. One of the first moviemakers to recognize the potential of television, Fairbanks churned out a assortment of filmed programs before 1950, among them the pioneering sitcom *Jackson and Jill*, the semi-travelogue *Goin' Places with Uncle George*, the musical variety effort *Club Paradise*, and, most famously, the seminal animated series *Crusader Rabbit*. To save time and money, Fairbanks used his own patented multiple-camera system, shooting every scene in a single take with as many as five cameras simultaneously grinding away from a variety of different angles; it was then up to the editors to break the scene down into long, medium and close shots. (In 1951, Desi Arnaz was praised for his "innovational" three-camera process on *I Love Lucy*, but Fairbanks had beaten him to it by at least three years.)

Starring in *Public Prosecutor* as DA Stephen Allen was John Howard, who'd previously headlined a series of "Bulldog Drummond" B-pictures. Former Universal starlet Anne Gwynne appeared as Allen's spunky secretary Pat Kelly, while utility player Walter Sande was seen as police lieutenant Evans. The supporting actors were recruited from the ranks of Hollywood's $100-per-day players, all familiar if not famous faces. (Though a few did achieve stardom in later years: one episode featured *Star Trek*'s DeForest Kelley as a phony swami—complete with turban.) The series' courtroom scenes were minimal, usually no more than a quick succession of closeups; canny directors like Lew Landers used shadows and sound effects to create the illusion of a crowd of spectators or a full complement of jurors.

Truth to tell, *Public Prosecutor* was less a legal series than a mystery-whodunnit. Each of the 26 episodes began with Stephen Allen sitting behind his desk, speaking directly to the viewer as he recalled the murder case at hand, which was dramatized in flashback. Approximately two-thirds of the way into the episode, Allen returned to recap the clues and list of suspects, whereupon he invited the viewer to help solve the mystery. It was at this point that local stations were supposed to cut away to a "live" panel of mystery experts who would mull over the evidence and attempt to identify the guilty party — and presumably the home audience would do the same. Once the conclusions had been reached in the studio, it was back to the filmed portion of the series and the actual outcome of the mystery, which was often predicated on a hitherto unrevealed clue (talk about

unfair play!). To accommodate the live portion of *Public Prosecutor*, each episode ran a flat 17½ minutes: it was up to the local stations to come up with their own panel of experts, though many station managers opted instead to pad out the excess 12½ minutes with commercials.

Evidently, the series was originally intended to air on the NBC network; in fact, existing prints carry an NBC copyright. But the only documented network exposure of *Public Prosecutor* occurred in September of 1951, when DuMont picked up the show for Prime Time play under the title *Crawford Mystery Theater*— a nod to the sponsor, Crawford Clothes. Actor and game-show host Warren Hull moderated the weekly panel of crime experts (mostly detective-fiction writers) during the live portion of the show. Though *Crawford Mystery Theater* ended its network run after three weeks, the series continued to be seen locally on New York's DuMont flagship station WABD until February 28, 1952.

Queens Supreme

CBS: Jan 10–24, 2003. Shoelace Productions/Spelling Television/Red Om Films/Revolution Studios/Shadowland Productions/CBS Television Distribution. Created by Peter and Daniel Thomas, and Kevin Fox. Executive Producers: Julia Roberts, Kevin Fox, Elaine Goldsmith-Thomas, Deborah Schindler. Cast: Oliver Platt (Judge Jack Moran); Robert Loggia (Judge Thomas O'Neill); Annabella Sciorra (Judge Kim Vicidomini); L. Scott Caldwell (Judge Rose Barnea); James Madio (Mike Powell); Marcy Harriell (Carmen Hui).

Co-executive produced by actress Julia Roberts, the seriocomic CBS legal series *Queens Supreme* took place at the Queens County Courthouse in New York, a locale chosen by the producers because Queens was reputed to be America's most ethnically diverse community. The four jurists presiding over the courthouse were officially designated as Supreme Court Judges, hence the title. Oliver Platt headed the cast as Jack Moran, the motliest of a motley crew of judges who spent as much time working apart as they did working together. Acknowledged as a legal genius and a model of integrity by his peers, Judge Moran was also a relentlessly cynical noncomformist, famous for his eccentric and sometimes un-fathomable behavior in the courtroom. Moran's colleagues included his superior, the pompous but evenhanded Judge Thomas O'Neill (Robert Loggia); Judge Rose Barnea (L. Scott Caldwell) whose blunt outspokenness and brutal honesty sent many defendants and attorneys scurrying from her courtroom with tails between legs; and a new appointee to the staff, idealistic young Judge Kim Vicidomini (Annabella Sciorra), who coasted by on the strength of her political connections. The court was also served by a brace of imperturbable legal clerks, Carmen Hui (Marcy Harriell) and Mike Powell (James Madio).

The concept of *Queens Supreme* was attributed to Peter and Daniel Thomas, the twin lawyer sons of real-life Queens County Supreme Court Justice Charles J. Thomas. In an interview with *Entertainment Weekly*, co-producer Elaine Goldsmith-Roberts insisted, "[the show] does for the judicial system what *M*A*S*H* did for the medical world. As the doctors were operating, they were talking about other things. Here, it won't really be about the cases, it'll be about the characters." "Characters" is right. *Queens Supreme* was predicated on the assumption that the words "wacky," "irreverent" and "obnoxious" are synonymous with "lovable." They aren't.

The series' uneasy mixture of comedy and drama—and its smug showbiz-liberal stance—was established in the debut episode, wherein a chain-smoking jury member, incensed over recent multimillion-dollar damage suits against Big Tobacco, held his fellow jurors at gunpoint and defiantly puffed away on a cigarette just to annoy them—whereupon Judge Moran defused the situation by "trying" the renegade juror's case and persuading him not to go to extremes. Actually, this wasn't *really* the first episode, only the first one shown. At the last minute it was decided to run the series' pilot episode second, prompting legal writer Libby A. White to complain in a *Picturing Justice* article that it was nearly impossible to figure out which characters were judges and which were law clerks. Once the pilot *was* shown, it turned out that the climax involved a rape victim's fa-ther who stole Judge Vicodomini's gun and threatened to kill the rapists—prompting critics to bemoan the fact that they'd been barraged by two overbaked hostage situations in a row.

Of the thirteen hour-long episodes filmed, only three saw the light of day before *Queens Supreme* was cancelled. Some observers have attributed the series' quick demise to the producers' apparent inability to decide whether the show was supposed to be funny or serious. Of course, there's always the possibility that viewers took one look at the series' title and thought it was about a female impersonator doing a Diana Ross imitation.

Raising the Bar

TNT: debuted Sept. 1, 2008. Bochco Productions/ABC Studios. Creators/Executive Producers: Steven Bochco, David Feige. Executive Producers: Jesse Bochco, Dayna Bochco. Cast: Mark-Paul Gosselaar (Jerry Kellerman); Gloria Reuben (Rosalind Whitman); Jane Kaczmarek (Judge Trudy Kessler); Teddy Sears (Richard Patrick Woolsley IV); Melissa Sagemiller (Michelle Ernhardt); Currie Graham (Nick Balco); J. August Richards (Marcus McGrath); Jonathan Scarfe (Charlie Sagansky); Natalia Cigliuti (Roberta "Bobbi" Gilardi).

The weekly, hour-long *Raising the Bar* was in one respect an innovation: it was the first TV drama produced exclusively for a basic-cable network (TNT) by the prolific Steven Bochco (see also *L.A. Law, Civil Wars, Murder One* and *Philly*). We'd give anything to report that the series itself was equally innovational ... but we can't.

Created in collaboration with lawyer David Feige, author of *Indefensible* and a former public defender of twelve years' standing, *Raising the Bar* was something of a family affair for Steven Bochco: the series was co-produced by his wife Dayna, while his son Jesse directed several of the first-season episodes. On the series' website, Bochco explained, "the primary characters are a group of young public defenders and prosecutors who go up against each other during the day, but they're friends and hang out with each other at night," adding, "They argue and negotiate with each other. They are dedicated players within what we feel is a broken criminal justice system." The producer's avowed purpose was to present a balanced view of the system. "*Raising the Bar* is not geared specifically toward the public defenders or the prosecutors," he said. "We try to give equal time to both points of view, with an eye toward revealing the extent to which the system doesn't work very well. It certainly doesn't have all that much to do with justice. It has more to do with keeping the conveyor belt turning, the idea that if every case in the system goes to trial, the system grinds to a halt."

Most of the characters had been friends since law school, and remained so even though they were now on opposite sides of the legal fence. Representing the Defense were the members of the New York City Public Defenders office (which doesn't exist in real life, since most indigent defendants are handled by Legal Aid, but we'll allow Bochco this bit of artistic license). Rosalind Whitman (Gloria Reuben) was the tough-but-compassionate boss, whom the audience instantly recognized as a Liberal because of the "Greenpeace" bumper stickers on her desk; Roz trusted her employees, was as protective of them as a mother hen, and gave them as wide a berth as legally possible. Jerry Kellerman (Mark-Paul Gosselaar) was the hot-tempered chief public defender, who evidently had neither cut nor washed his hair since passing the Bar. In contrast, fellow PD Richard Patrick Woolsey IV (Teddy Sears) was a strait-laced, immaculately groomed Preppie, who'd idealistically given up a posh job at his father's upscale firm to dedicate his energies to the Unfortunate Masses. New to the staff was perkily ambitious Bobbi Gilardi (Natalia Cigliuti), a transfer from the Brooklyn office who boasted a seemingly endless supply of valuable contacts — and, unfortunately for her moonstruck male coworkers, she also had a husband tucked away somewhere.

The Prosecution team was headed by ADA Nick Balco (played by *Boston Legal* alumnus Currie Graham), a cutthroat competitor who cared only about winning cases and who possessed a mile-wide streak of chauvinism, alternating between hitting on the females in his office and accusing them of being "too emo-

tional" to do their job to their utmost ability. Balco's underlings included cool blonde Michelle Ernhardt (Melissa Sagemiller), who was described by her old college chum (and erstwhile lover) Jerry Kellerman as if she were a cup of cappuccino: "Light, sweet, and with just a hint of bitterness." (There was something about Michelle that aroused everyone's appetite: her lascivious boss Graham characterized her as, "two perfect mounds of creamy yogurt, each topped with a pert succulent raisin.") Though she felt a "moral obligation" towards justice, the pressure to win at all costs frequently caused her to compromise her ethics. Also on Balco's team was solemn, scrupulously honest African-American prosecutor Marcus McGrath (J. August Richards), who was intended as the series' "anchor." We were tipped off from the get-go that McGrath was a rock of integrity as we watched him relentlessly prosecuting various black defendants on charges of rape and hate crimes — leaving himself wide open for accusations that he was specifically targeting "his own people" to survive in a system controlled by old white guys.

Then there was the billiously belligerent Judge Trudy Kessler, played by Jane Kaczmarek, who nearly three decades earlier had made her series-TV debut as a teenage law student on *The Paper Chase* (q.v.). A former public defender who "never kidded myself that any of my clients were innocent," Kessler had both eyes on the District Attorney's office, and was determined to carve out a reputation as a "hanging judge": she treated her courtroom as her own private fiefdom, sparing no mercy and cutting no slack no matter what the extenuating circumstances. She also seemed to be carrying on a personal vendetta against the brash, unkempt Jerry Kellerman, at one point citing him for contempt and sentencing him to spend the night in the same jail cell as his client! (Why didn't Kellerman ever ask for a different judge, you ask? Well, then there wouldn't be any show, you silly goose.) For all that, Trudy Kessler was Mother Theresa compared to her oily, conniving law clerk Charlie Sargansky (Jonathan Scarfe), who evidently graduated Sigma Cum Louse from the Iago School of Law. Subtly pitting one side against the other as he pretended to play arbitrator between his quarrelling school buddies, Charlie also strove to advance his career via a clandestine romance with Judge Kessler — all the while spending his nights cruising gay bars, a secret pursuit that literally came back to bite him on several occasions.

If the above-mentioned characters sound like stereotypes, that's because they were. To be sure, the scripts relied upon recent headlines for their story material (one early episode was a deft reworking of the "Jena 6" civil rights demonstrations, which arose from an apparent act of racial terrorism on the campus of a Louisiana high school). And as he'd done so often in the past, Bochco stocked the series with trendy camerawork and eye-popping special effects (for example, characters would "dissolve" onto the screen against a static background to indicate the passage of time). Otherwise, *Raising the Bar* was the Same Old Same Old, far below the usual cutting-edge Steven Bochco standards. The fraternity of professional TV critics, who in the past had invariably rallied in defense of Bochco even for his least successful efforts, uniformly dismissed his newest offering with such phrases as "shockingly ordinary" (*TV Guide*), "irritatingly unorginal" (*The Boston Globe*) and "ancient and worn out" (*USA Today*). The most devastating blow was delivered by Maureen Ryan, who in the August 29, 2008, edition of *The Chicago Tribune* commented, "Bochco has made the most cutting-edge drama — of 1984."

The Public, however, paid no heed to the critics: the debut episode of *Raising the Bar* was seen and enjoyed by the largest opening-night audience in the history of the TNT network. (PS: The ratings quickly plummeted in the weeks that followed, suggesting that those pesky critics may have known what they were talking about.)

Reasonable Doubts

NBC: Sept. 26, 1991–July 31, 1993. Warner Bros. Television. Creator/Executive Producer: Robert Singer. Cast: Marlee Matlin (ADA Tess Kaufman); Mark Harmon (Det. Dicky Cobb); William Converse-Roberts (Arthur Gold); Tim Grimm (Bruce Kaufman); Nancy Everhard (Kay Lockman); Kay Lenz (Maggie Zombro); John Mese (Sean Kelly).

NBC's weekly, 60-minute *Reasonable Doubts* would have been unremarkable but for one important historical distinction: it was the first American TV dramatic series with a deaf star. Cast as Assistant DA Tess Kaufman, attached to the Chicago Police Department's Felony Division, was actress Marlee Matlin, whose brilliant and incisive performance as a deaf-mute in *Children of a Lesser God* (1986) had won her an Academy Award. During filming of the *Reasonable Doubts* pilot episode, a story began circulating that an NBC executive turned to his assistant and exclaimed, "You know, Marlee Matlin is great. Is she going to be deaf for the entire series?"

Strictly a by-the-book type, Tess Kaufman performed her job with admirable efficiency and expertise, though because of her disability she felt uncomfortable verbally conversing with her colleagues and clients, preferring to use the American Sign Language system. Her police partner was rule-bending maverick detective Dicky Cobb (Mark Harmon), who'd been *persona non grata* with the Chicago PD ever since exposing some crooked cops, and had been exiled to Felony. Division chief Arthur Gold (William Converse-Roberts) teamed Dicky with Tess not because they were a "perfect fit" — in fact, they barely got along at all — but because Dicky was fluent in sign language. As a result, while Tess Kaufman's character was not defined by her handicap, Dicky Cobb's *was*, since he spent most of his screen time interpreting his partner's hand gestures for the benefit of others.

Although theirs was an oil-and-water relationship, Tess and Dicky frequently found themselves turning to each other for moral support, unable to rely upon their respective life partners: Tess' lawyer husband Bruce (Tim Grimm) was a serial philanderer, while Dicky's girlfriend Kay (Nancy Everhard) was whiny and manipulative. Kay was killed off in a robbery at the beginning of the series' second season, a move presumably dictated by the producers to strengthen the bond between Tess and Dicky; but the team never truly went beyond the "friendly enemy" stage. Added to the cast during Season Two was Tess' frequent courtroom antagonist, defense attorney Maggie Zombro, a role that earned an Emmy award for the always-underrated Kay Lenz.

Reasonable Doubts never built up much of an audience, partly because NBC was unable to find a suitable timeslot (the show was scrambled all over the Prime Time schedule during its 44-episode run), but mostly because the scripts were banal and hackneyed, making things easy for the audience with such unambivalently hissable villains as Holocaust deniers. Too, Marlee Matlin's performance, though fundamentally realistic and straightforward, grew increasingly preachy and sanctimonious as the show progressed. When it became obvious that the series was doomed, executive producer Bob Singer (*Lois and Clark*) planned a spin-off costarring Mark Harmon with a new character played by actor and film historian Jim Beaver (*Deadwood*), but this never came to fruition. On a more positive note, it was while working on *Reasonable Doubts* that Marlee Matlin met her future husband Kevin Grandalski, an LAPD officer hired as the series' technical adviser.

Rivera Live

CNBC: Feb. 7, 1994–Nov. 16, 2001. Cast: Geraldo Rivera (Himself).

Peripatetic gonzo journalist Geraldo Rivera was still headlining his own syndicated "trash talk" show *Geraldo* when he undertook the additional responsibility of the daily, hour-long CNBC cable chatfest *Rivera Live* in early 1994. During its first few months on the air, *Rivera Live* was a typically wide-ranging, confrontational Rivera effort, with the host and his guests thrashing out current events on a single-topic-per-episode basis. All this changed on that fateful June day when O.J. Simpson climbed into his white Bronco and attempted to flee the clutches of the LAPD, who wanted the ex-athlete to assist them in their inquiries concerning the murders of Simpson's estranged wife Nicole and her friend Ron Goldman. At this point, *Rivera Live* switched to "All O.J., All

the Time," with the host—who happened to be a licensed attorney—bearing down on the legal ramifications of the Simpson case with such obsessiveness that one could envision Geraldo spending his off-hours pasting clippings in a scrapbook and covering his bedroom walls with pictures of "The Juice."

Even after Simpson faded from the headlines, *Rivera Live* continued to be dominated by legal issues. Rivera was provided with a cornucopia of material in the months and years that followed, thanks to the various and sundry peccadilloes of President Bill Clinton and the controversial 2000 election of Clinton's successor George W. Bush. This might have gone on indefinitely had not Rivera left the series to cover the war in Afghanistan in November of 2001, whereupon *Rivera Live* was immediately replaced by *CNBC's America Now*. Since that time, Geraldo Rivera has launched a similar weekly program, *Geraldo at Large*, on the rival Fox News Channel.

Rosetti and Ryan

NBC: Sept. 22–Nov. 10 1977. Heyday Productions/Universal Television. Executive Producer: Leonard B. Stern. Supervising Producers: Don M. Mankiewicz, Gordon Cotler. Produced by Jerry Davis. Cast: Tony Roberts (Joseph Rosetti); Squire Fridell (Frank Ryan); Jane Elliot (Jessica Hornesby); Dick O'Neill (Judge Hardcastle); William Marshall (Judge Black).

Executive producer Leonard Stern (*Get Smart!*) hatched the idea for the lighthearted legal drama *Rosetti and Ryan* back in 1972, when the TV scene was dominated by issue-driven sitcoms featuring fashion-challenged characters: "I was always wondering when the pendulum would swing back to romantic comedy and men in three-piece suits." Evidently, that time came on May 23, 1977, when *Rossetti and Ryan* was test-marketed with the TV-movie pilot *Men Who Love Women*, in which a pair of devilishly handsome young attorneys defended an heiress accused of murdering her husband. Well received by reviewers and whodunit fans alike—the film won the Edgar Allan Poe award, an honor bestowed by the Mystery Writers of America—*Men Who Love Women* matriculated into the weekly, hour-long *Rosetti and Ryan* series the following September.

Described by producer Jerry Davis as "a cerebral *Starsky and Hutch*," the series starred Tony Roberts as Joe Rosetti, charmingly arrogant scion of an upper-middle-class family, and Squire Friddell as his partner Frank Ryan, a brooding ex-cop with a blue-collar background. While Rossetti used his smooth line and winning way to earn the confidence of the witnesses for the prosecution, Ryan, who was still taking law courses at night, did most of the "grunt" research work. Our two heroes' principal antagonist was another carryover from the pilot film, gorgeous Assistant DA Jessica Hornesby (Jane Elliot).

Both Tony Roberts and Squire Friddell had worked with Leonard Stern before: Roberts starring in a failed pilot based on the wartime cartoons of Bill Mauldin, and Friddell guesting on the short-lived comedy/fantasy *Holmes and YoYo*. In particular, Tony Roberts felt an affinity with his screen character, claiming in a *TV Guide* article that he'd been inspired to become an actor after watching his father, radio announcer Ken Roberts, testify on behalf of blacklistee John Henry Faulk in the latter's lawsuit against CBS. Impressed by Faulk's attorney Louis Nizer, Roberts recalled, "He was tremendous, just masterful. He had great poise and a sense of drama. I realized then that performance in the courtroom isn't too different from performance in the theater." (Tony Roberts would ultimately emerge as the more famous of the two *Rosetti and Ryan* stars by virtue of his work in the films of his close friend Woody Allen.)

Outside of this quote from Roberts and the invocation of such series highlights as the two lawyers earning an acquittal for their client by proving the length of time it takes for ice cream to melt, the producers of *Rosetti and Ryan* seemed to go out of their way to downplay the fact that it was a legal series. Describing it as "a cops and robbers show without violence," Jerry Davis added, "The show will just be a good caper. They'll depend on ingenious legal ploys to free their clients, not gunshots. We'll be elegant and sophisticated. And of

course we'll surround them with the prettiest girls we can find." Even Tony Roberts indicated that it was this rather than the series' courtroom trappings that attracted him to the project: "New girls every week! I'm gonna love this show!"

Charlie's Angels to the contrary, pretty girls alone do not a successful series make. *Rosetti and Ryan* ended up as NBC's biggest disappointment of the 1977–78 season, lasting only six episodes before it was plowed under by its CBS competition *Barnaby Jones*.

Sam Benedict

NBC: Sept. 15, 1962–Sept.7, 1963. MGM Television. Created by E. Jack Neuman. Produced by William Froug. Theme music by Nelson Riddle. Cast: Edmond O'Brien (Sam Benedict); Richard Rust (Hank Tabor); Joan Tompkins (Trudy Wagner).

In order to discuss the hour-long NBC legal drama *Sam Benedict*, it is first necessary to discuss the series' technical advisor, celebrated defense attorney Jacob W. "Jake" Ehrlich. Known to his contemporaries as "The Master," Ehrlich built his entire forty-year career on two mottos: "Never plead guilty," and "Things are almost never what they appear to be." He defended 63 clients charged with first-degree murder, of which 59 were acquitted while the remaining four had their charges reduced to manslaughter. He represented actors James Mason and Errol Flynn in their respective divorce cases, helped billionaire Howard Hughes secure a release for his controversial film *The Outlaw*, defended poet Lawrence Ferlinghetti against accusations of pornography and entertainers Gene Krupa and Billie Holliday on drug charges, and might possibly have been able to save convicted rapist Caryl Chessman from the gas chamber had not Chessman decided to act as his own attorney. And for 35 years, he served as head attorney for the San Francisco Police Officer's Association, somehow finding time to write a dozen books and novels. Among Ehrlich's many protégés were F. Lee Bailey and Melvin Belli.

The conventional wisdom is that the TV series *Perry Mason* (q.v.) was partially inspired by Ehrlich's career; indeed, series star Raymond

Burr spent two weeks with Ehrlich just before filming started, studying the attorney's movements and mannerisms in and out of the courtroom. But Ehrlich ultimately had little to do with *Perry Mason*, believing he himself could turn out a better TV series — and that's how *333 Montgomery* came into being. Its title based on the real-life address of Ehrlich's San Francisco offices (actually 300 Montgomery Steet), *333 Montgomery* was written with the attorney's blessing by Gene Roddenberry as the half-hour pilot film for a proposed series based on Ehrlich's 1950 autobiography *Never Plead Guilty*. Starring as dynamic, two-fisted criminal lawyer Jake Brittin was Roddenberry's future *Star Trek* colleague DeForest Kelley. Though *333 Montgomery* didn't sell, the pilot was telecast June 13, 1960, as an episode of the NBC anthology *Alcoa Theater*, and later became a perennial attraction at the various *Star Trek* conventions throughout the land, even though there wasn't a Klingon or Tribble in sight. In 1962, Roddenberry tried again with another Ehrlich-based pilot, this one running an hour and variously titled *Defiance County* and *Ty Cooper*. By this time, however, Ehrlich had hitched his wagon to producers E. Jack Neuman and William Froug, who were readying *Sam Benedict* for its fall 1962 debut.

Largely filmed on location in San Francisco, the series starred Edmond O'Brien as Ehrlichesque trial lawyer Sam Benedict, who never spoke when shouting would do and who tempered his courtroom bombast with a robust sense of humor. While he had an assistant-cum-legman named Hank Tabor (Richard Rust), Sam preferred to do his own detective work, the better to clear his clients or secure a reduced sentence (unlike Perry Mason, many of Sam's clients were unquestionably guilty). And though he backed down from no one in the pursuit of justice, Sam was putty in the hands of his perky, all-wise secretary Trudy Wagner (Joan Tompkins). Beyond his salty delivery of lines and bulldozer courtroom demeanor, Sam Benedict shared with Jake Ehrlich an up-from-poverty background (including a short stint as a prizefighter!), and a sincere empathy for underdog clients, be they wealthy and famous or poor and obscure. Like Ehrlich, Sam expended as much energy helping a client square an overdue rent bill as he did rescuing a widowed heiress from her avaricious brother-in-law.

Among the series' 28 episodes was "So Various, So Beautiful," featuring horror-movie "scream queen" Hazel Court as a pathological liar, which the actress would cite as her favorite role in a 2007 interview for Sierra Nevada College's *Moonshine Ink*. Court also admired Edmond O'Brien, who "had the whole cast in awe" as he recited Shakespeare in the courtroom finale. Another highlight was the episode "17 Gypsies and a Sinner Named Charlie," which not only featured O'Brien in the dual role of Sam Benedict and a soft-hearted con artist, but also represented the acting debut of the star's 12-year-old daughter Maria O'Brien.

The cancellation of *Sam Benedict* after a single season has been blamed on the series' powerhouse CBS competition, *Jackie Gleason's American Scene Magazine*. However, it is also likely that the series simply came on way too strong, alienating or even scaring off some of the more sensitive viewers. The most successful legal shows of the period, *Perry Mason* and *The Defenders* (q.v.), were careful to withhold their flashiest histrionics until the climax of each episode, allowing the audience to savor the slow, steady build-up. But *Sam Benedict* invariably shot its wad a few seconds after the opening credits faded, as confirmed by the review in the October 12, 1962, edition of *Time* magazine: "TV's new professional men include a lawyer and a clutch of newsmen. The tough, quick-thinking, steel-trap lawyer is NBC's Sam Benedict, played by Edmond O'Brien with sheer nervous drive, solving ten cases an hour, picking up phones, barking, slamming them down, dictating letters at 200 words a minute, grabbing punks by the throat, and so on. Statistically, a man like that ought to have a nervous breakdown at least once a week. Not Sam."

Sara

NBC: Jan. 23–May 8, 1985; June 2–15, 1988. MTM Enterprises. Produced by Gary David Goldberg and Linda Nieber. Cast: Geena Davis (Sara McKenna);

Alfre Woodard (Rozalyn Dupree); Bill Maher (Marty Lang); Bronson Pinchot (Dennis Kemper); Mark Hudson (Stuart Webber); Matthew Lawrence (Jesse Webber); Ronnie Claire Edwards (Helen Newcomb).

Advertised as "the first Yuppie TV series" (this is something to be proud of?), the NBC sitcom *Sara* was also touted as, "a *Mary Tyler Moore Show* for the 1980s." But talented and charming though she may be, *Sara* star Geena Davis was no more successful as Mary Tyler Moore than she would be as Hillary Rodham Clinton in her failed 2005 series *Commander in Chief*.

Set in San Francisco, the series cast Davis as Sara McKenna, a recent law school graduate who, taking the first job that came along, ended up working in a storefront legal-assistance office. Sara shared this space with three other lawyers: her hard-drinking best friend Rozalyn Dupree (Alfre Woodard), sleazy skirt-chaser Marty Lang (Bill Maher), and flighty homosexual Dennis Kemper (Bronson Pinchot). Acting as den mother for this motley crew was secretary Helen Newcomb (Ronnie Claire Edwards). At home, our heroine spent much of her time resisting the efforts by her cute little 4-year-old neighbor Jesse Webber (Matthew Lawrence) to play matchmaker for Sara and his divorced dad Stuart (Mark Hudson).

The series is noteworthy today only for its remarkable confluence of young actors on their way up. Both Geena Davis and Alfre Woodard were on the verge of stellar movie careers; Bronson Pinchot was only a year away from his signature role as funny-foreigner Belki in the long-running sitcom *Perfect Strangers*; and Bill Maher would eventually forsake acting to become cable TV's foremost Republican-bashing political satirist. In the light of Maher's vigorous Liberalism, it's ironic that he was cast in *Sara* as the totally non–PC Marty Lang — described by Maher as "the office schmuck" — who relentlessly bullied the wistfully gay Dennis Kemper. (Typical dialogue: Told that he's "slime," Marty shoots back "At least I'm *straight* slime.")

Though in later years Bill Maher would recall, "Everyone was sure it was going to be a huge hit," *Sara* went hugely in the opposite direction. The reviews were tepid at best, damning with faint praise: one critic described the series as "at least better than *The Dukes of Hazzard*," while another opined it was "worth 30 minutes of your time unless you are writing a book or remodeling the bathroom." There is every likelihood that *Sara* would have been totally forgotten after its five-month run if NBC hadn't briefly pulled it out of storage in the summer of 1988 to capitalize on the latter-day celebrity of Geena Davis, Alfre Woodard and Bronson Pinchot.

Second Chances

CBS: Dec. 2, 1993–Jan. 27, 1994; Feb. 10 1994. Lorimar Productions. Executive Producers: Lynn Marie Latham, Bernard Lechowick. Cast: Connie Sellecca (Dianne Benedict); Matt Salinger (Mike Chulack); Megan Follows (Kate Benedict); Jennifer Lopez (Melinda Lopez); Pepe Serna (Salvador Lopez); Daniel Gonzalez (Cesar Lopez); John Chaidez (Johnny Lopez); John Schneider (Pete Dyson); Justin Lazard (Kevin Cook); Frances Lee McCain (Felicity Cook); Ronny Cox (George Cook); Ray Wise (Judge Jim Stinson); Michelle Phillips (Joanna); Erich Anderson (Bruce Christiansen); Sean Fitzgerald (Brian Christiansen); Ramy Zada (Det. Jerry Kuntz); Chuck Bulot (Off. Parker).

Slapped together as a replacement for the failed CBS series *Angel Falls*, the serialized drama *Second Chances* was designed as a vehicle for Connie Sellecca, a proven audience favorite since her costarring turn on Aaron Spelling's moderately successful mid–1980s series *Hotel*. Sellecca was cast as attorney Dianne Benedict, public defender in the small California community of Santa Rita. When the series began, Dianne was the wife of construction executive Bruce Christianson (Erich Anderson), the mother of 8-year-old Brian (Sean Fitzgerald), and a candidate for the office of city judge, running opposite the corrupt incumbent Jim Stinson (Ray Wise). By the end of the first episode, Dianne was a widow, her husband having been murdered in the "Parties Galore" catering service run by her mercurial sister Kate (Megan Follows), who drove a pink Cadillac and pouted a lot. Though she was a prime suspect in her husband's death — especially since she'd had a reputation as a "bad girl" before passing her bar exam — Dianne curtailed her legal and political activities to help police detective Jerry Kuntz (Ramy Zada) investigate the murder. In addition to Dianne, Kate and Judge Stinson, the suspects included Dianne's high-school sweetheart Mike Chulack (Matt Salinger), freshly released from prison after serving ten years of manslaughter; and Bruce's secretary/mistress Joanna (Michelle Phillips).

The mystery came to an abrupt end when Detective Kuntz himself was exposed as the killer (trafficking in illegal drugs, he'd been robbed by the duplicitous Bruce); at the same time, the series' "legal" angle was also dropped, as Dianne joined forces with newcomer Pete Dyson (John Schneider) to save her late husband's failing construction business. Throughout all of this intrigue, a subplot wove its way through the proceedings involving a Latino family named Lopez. In one of her first significant acting assignments, Jennifer Lopez was seen as *Melinda* Lopez, who was working her way through college as a waitress. Melinda had married Kevin Cook (Justin Lazard), a socially well-connected law student, but the marriage had been sabotaged by Kevin's bigoted parents George (Ronny Cox) and Felicity (Frances Lee McCain), forcing Melinda to return to her domineering parole-office father Sal (Pepe Serna).

The cancellation of *Second Chances* after ten episodes was not brought about by audience indifference or bad reviews, but by a much Higher Authority. On January 17, 1994, all of the series' sets were demolished in the Northridge earthquake. By the time the sets could have been rebuilt, both Connie Selleca and Megan Follows would have been in such an advanced state of pregnancy that their condition couldn't possibly have been camouflaged — and since neither character was supposed to be married or sporting, explaining the ladies' condition was a Herculean task that CBS was simply not willing to undertake. Thus, *Second Chances* was axed in late January, with the last remaining episode burned off in February as part of CBS' late-night omnibus *CrimeTime After PrimeTime*. Jennifer Lopez and Pepe Serna would later recreate their *Second Chances* characters in the short-lived spin-off series *Hotel Malibu*.

Science Court [*aka* Squigglevision]

ABC: Sept. 13, 1997–Sept. 2, 2000. Tom Snyder Productions/Burns & Burns Productions. Created by Tom Snyder and Bill Braudis. Executive Producers: Tom Snyder, Bonnie Burns. Voices: Paula Poundstone (Judge Stone); Bill Braudis (Doug Savage); Paula Plum (Alison Krempel); H. Jon Benjamin (Prof. Parsons); Fred Stoller (Stenographer Fred); Jennifer Schulman (Mary Murry).

We would have listed *Science Court* along with *Gary the Rat* and *Harvey Birdman, Attorney at Law* (both q.v.) as one of TV's handful of animated lawyer series, but for two qualifications: (a) It wasn't so much a lawyer series as a courtroom show, and (b) it wasn't really animation, but "Squigglevision." Developed for instructional-film use by Tom Snyder (*not* the late-night talk show host), Squigglevision is a process whereby the illusion of movement was conveyed by running five separate drawings of squiggly black lines, called "flicks," in a continuous loop. Except for the black lines zigzagging around the mouths and bodies of the characters, there is no real movement at all, so the process barely even qualified as "limited" animation. Most famously used on the Comedy Central series *Doctor Katz, Professional Therapist*, Squigglevision enabled Tom Snyder and his staff to turn out an entire season's worth of episodes in a fraction of the time needed to put together a traditional animated series. Adored by the wine-and-cheese crowd and despised by animation purists, Squigglevision was ultimately abandoned by Snyder while he was working on the Cartoon Network property *Home Movies* in 2001; explained the producer, "For every person who though it was cool there were three who said, 'Oh, yeah, the thing where it's kind of blurry and you get a headache.'"

Science Court was the only Squigglevision effort designed for the children's-show market. Introduced as a series of educational CD-ROMs in 1995, the series' concept was developed while Snyder was watching the 24/7 coverage of the O. J. Simpson case. The producer, a former science teacher, reasoned that certain "abstract" basic-science ideas could be clarified for younger viewers within the framework of a mock courtroom setting. Each half-hour *Science Court* featured the voice of comedienne Paula Poundstone as a Squigglevisioned judge named Stone, presiding over "trials" of various scientific concepts that had been dragged into court by a character named Mary Murry. Acting as opposing counsel were brainy prosecutor Alison Krempel (voiced by improv actress Paula Plum), who always took Mary's side, and dimwitted defense attorney Doug Savage (voiced by Tom Snyder's producing partner Bill Braudis). Arguments over such concepts as "evaporation and gravity" and "inertia" were framed in mock legalese which *sounded* complicated, but invariably ended up making cogent educational points in an entertaining and easily graspable manner for the kiddie viewers.

In 1997 *Science Court* was picked up for Saturday-morning TV exposure by the ABC network, whose P.R. staff described the series as being geared for a demographic of "age six to bright adults." However, the network apparently felt the courtroom setting was too limited in scope, thus in 1998 the series was completely reformatted and retitled (what else?) *Squigglevision*. The "Science Court" segment was now merely one of three weekly educational components, along with "See You Later, Estimator" (a weekly math lesson) and "Fizz and Martina's Last Word" (focusing on vocabulary). In this form, *Squigglevision* flourished for two full seasons.

Sex Court

Playboy Channel: debuted Aug. 7, 1998. Created by Jim English. Executive Producer: Tammara Wells. Cast: Julie Strain (Judge Julie); Alexandra Silk (Bailiff); Jon St. James (Private Dick); Linda O'Neil (Female Bodyguard); Erroll Sean Warfield, Emery Guillory (Male Bodyguards); Nakita, Skye Ashton (Stenographers); Brittany Andrews (Sex Expert).

Some may find it unfathomable that there were actually people other than sweaty 10-year-old boys and repressed middle-aged men who were still buying *Playboy* magazine as late as the 1990s (unless they really *did* just buy it for the articles). But the fact remains there was enough of a market out there to support a pay-cable Playboy Channel, which of course trafficked almost exclusively in nude women doing the sort of things men imagine nude women do when no one's looking. To break up the monotony, the Playboy Channel offered a few daily and weekly half-hour series ... trafficking almost exclusively in nude women etc. etc. etc.

Sex Court (aka *Playboy's Sex Court*) was the channel's answer to such mainstream courtroom offerings as *The People's Court* and *Judge Judy*—and no, we don't recall the question. Promising "Ardor in the court," the series featured well-known nude models and porn stars in a ribald lampoon of the entire judge-show genre. Actress Julie Strain, whose previous cinematic triumphs included *Lethal Seduction* and *Heavenly Hooters*, starred as "Judge Julie," presiding over bawdily amusing sex-related cases while wearing a revealing dominatrix outfit. In keeping with this S&M ambience, Judge Julie referred to her bailiff Alexandra Silk (garbed in a policeman's cap and not much else) as her "servant," expressing inordinate interest in Alexandra's arsenal of restraining devices. All the litigants appearing before Judge Julie were prominent (in more ways than one) skin-flick celebrities, and the verdict nearly always required the loser to remove all clothing and/or perform a public sex act while the rest of the courtroom went into full orgy mode and classical music was heard in the background. For international distribution to less permissive TV outlets, the more explicit sequences were edited and digitally blurred.

Sex Court almost didn't make it to cable, due to a legal imbroglio between *Playboy* and the online service sexcourt.com. *Playboy* won this battle, which is either good or bad news depending on one's point of view. For the record, many viewers found the whole experience to be unsettlingly unsexy and downright dull—but there were many others who demanded more of the same, resulting in the R-rated 2001 theatrical release *Sex Court: The Movie*.

Shannon's Deal

NBC: Apr. 13–May 16, 1990; March 23–May 21, 1991. TV–movie pilot episode telecast June 4, 1989. Stan Rogow Productions/NBC Television Productions. Creators/Executive Producers: John Sayles, Stan Rogow. Theme song: Wynton Marsalis. Cast: Jamey Sheridan (Jack Shannon); Elizabeth Peña (Lucy Acosta); Richard Edson (Wilmer Slade); Jenny Lewis (Neala Shannon); Miguel Ferrer (DA Todd Spurrier); Martin Ferrero (Lou Gondolf).

The original TV–movie pilot for the weekly legal series *Shannon's Deal* stirred up a lot of excitement when it first aired in June of 1989. The film was written by John Sayles, whose theatrical-feature credits included such critical favorites as *The Return of the Secaucus Seven* and *Eight Men Out*, and the director was Lewis Teague, best known for helming the *Romancing the Stone* sequel *The Jewel of the Nile*. *Shannon's Deal* was co-created by Sayles and Stan Rogow, the latter a former lawyer who had worked in Boston's Roxbury district in the 1970s. Stan Rogow insisted that the character of protagonist Jack Shannon was semi-autobiographical: if so, Rogow should be commended for his startling candor.

It was quickly established in the pilot that Jack Shannon, played by Jamey Sheridan, had once been a prominent lawyer in a prestigious Philadelphia firm: "I thought I was a big shot. Big money. Big house. Big car. I thought I held all the cards. I thought I could pick the winner every time. I thought I could smell it. But the whole thing was built on garbage." Having mistreated his wife to the point that she walked out on him and took their daughter Neala (Jenny Lewis) with her, Jack began gambling heavily until he couldn't stop himself. "I had turned into the kind of man that I'd grown up hating." Disgusted with himself and what his life had become, Jack quit his lucrative partnership to start from scratch as a solo practitioner in a tiny, dingy walkup practice. He may have been broke and lonely, but now at least he could be ethical with his clients and true to himself. Ken Tucker of *Entertainment Weekly* described Jack Shannon as something out of Raymond Chandler, and indeed the character comported himself more like a seedy private detective than a lawyer. Likewise, the bluesy theme music by the great Wynton Marsalis immediately invoked memories of such *film noir* shamuses as Philip Marlowe and Sam Spade: all that was missing were the neon lights flashing outside the Venetian blinds and Mike Mazurki threatening to rearrange Jack's face. But for all of his "hard-boiled" verbiage, Jack Shannon went out of his way to avoid conflict and was fiercely opposed to soiling his hands in the grimier aspects of the law, enabling *New York Times* critic John J. O'Connor to characterize Jack as "a relatively clean sort of guy."

The *Shannon's Deal* series proper began in April of 1990, right after a rebroadcast of the well-received pilot film. Jamey Sheridan was back as Jack Shannon, as was Jenny Lewis as his estranged daughter Neala (their relationship would grow warmer and more mutually protective as the series progressed). Other cast members included Elizabeth Peña as Lucy Acosta, who worked as Shannon's part-time secretary to pay off a debt owed by her boyfriend; Richard Edson as Wilmer Slade, a philosophy-spouting collector for the loan shark who'd been staking Jack since his gambling days; Martin Ferrero as Shannon's doppelganger Lou Gondolf, an unregenerate shyster; and Miguel Ferrer as antagonistic DA Todd Spurrier.

Amidst a lot of self-righteous palaver about "looking out for the little guy" and "beating the big guys at their own game," many of the earlier episodes dwelled upon Shannon's struggle to overcome his gambling habit. To prepare for his role, star Jamey Sheridan attended meetings of Gamblers Anonymous. Though the other members quickly found out he didn't share their addiction, they allowed Sheridan to stay because he was honest with them.

After a six-week trial run, *Shannon's Deal* was withdrawn for extensive retooling by producers John Sayles and Stan Rogow, who felt the show was not living up to its potential (most viewers and critics agreed with them). Upon the series' return ten months later in March of 1991, hopes were high that the second season would be an improvement on the first, especially since

Sayles had engaged such strong writers as Joan Tewkesbury (*Nashville*) and Kit Carson (*Paris, Texas*). But this second go-round was no more successful than the first, and critic Ken Tucker summed up the problem: "*Shannon's Deal* is great on atmosphere, and in Jamey Sheridan the show has a charmingly raffish hero. But … the show's plots tend to meander annoyingly, and Sheridan's lines aren't as snappy as the way he delivers them."

Shark

CBS: Sept. 21, 2006–May 20, 2008. Imagine Television/20th Century–Fox Television. Creator/Executive Producer: Ian Biederman. Executive Producers: Brian Grazer, David Nevins, Ed Redlich, Rod Holcomb. Produced by Robert Del Valle. Cast: James Woods (Sebastian Shark); Danielle Panabaker (Julie Shark); Jeri Ryan (Jessica Devlin); Sam Page (Casey Woodland); Sophina Brown (Raina Troy); Alexis Cruz (Martin Allende); Sarah Carter (Madeline Poe); Henry Simmons (Isaac Wright); Kevin Alejandro (Danny Reyes); Kevin Pollak (DA Leo Cutter).

Despite the combined credentials of creator Ian Biederman (*Crossing Jordan*) and producer Brian Grazer (*24*), the 60-minute legal drama *Shark* would never have been sold to CBS if James Woods hadn't agreed to star. Viewer and critical response to having so prestigious a stage and film star honoring a humble TV series with his presence was so effusive that Woods might have been forgiven had he gone the route of so many other A-list actors who have issued lofty pompous, self-aggrandizing pronouncements of how they intended to "elevate" the quality of television and bring "Hollywood values" to so disreputable an entertainment medium. Not James Woods, who cheerfully admitted that he'd accepted the series because he loved to work and the money was good.

Heralded with the tagline, "Just when you thought it was safe to go into the courtroom again…," *Shark* was the saga of famously arrogant, notoriously ruthless and wholly unethical Los Angeles defense attorney Sebastian Shark, who specialized in clearing his celebrity clients by whatever means necessary — including the deliberate suppression of damning evidence. Living up to both his name and reputation, Shark was fond of declaring, "I eat prosecutors for breakfast. They're my only source of fiber." But after one of his clients beat his wife to death immediately upon being cleared of a domestic-abuse charge (gee, we wonder who *that* guy was based on), Shark experienced an epiphany, vowing ever afterward to work tirelessly on behalf of the Prosecution. At the suggestion of the city's mayor, the apparently chastened attorney offered his services to his longtime enemy, DA Jessica Devlin (Jeri Ryan) of the High Profile Crime Unit. But though he feigned humility to win Devlin's confidence, Shark hadn't really changed at all: he merely redirected his nasty, underhanded courtroom tactics toward the cause of Good rather than Evil. To his minions in the Unit, Shark preached the four-point doctrine of what he called the Cutthroat Manifesto: "Trial is War, second place is Death"; "Truth is relative. Pick one that works"; "In a jury trial, there are only twelve opinions that matter and yours is not one of them"; and finally, "Your job is to win. Justice is God's problem."

Working side by side with Shark (and helping him undermine the rule-bound Jessica Devlin whenever possible) were the four members of his elite legal team: rich, egotistical Casey Woodland (Sam Page), ultra-ambitious Madeline Poe (Sarah Carter), eloquently streetwise Martin Allende (Alexis Cruz) and tough-talking rookie Raine Troy (Sophina Brown). Also in the cast was Danielle Panabaker as Shark's estranged teenage daughter Julie, whose presence "humanized" the hard-shelled protagonist: since the only person Sebastian Shark truly cared about was not afraid of him, the audience wasn't afraid of him either, and was theoretically more willing to accept the character's frequently venomous behavior.

Referencing a like-vintage medical series dominated by an antiheroic hero, the *Chicago Tribune* reviewer described *Shark* as "*House* with lawyers." Like *House*, *Shark* frequently found the protagonist and his team following the wrong trail of evidence, only to stumble upon the right road near the very end of the episode. And also like *House, Shark* pretended to be an ensemble series when it fact it was basically a one-man show; indeed, the *USA Today* review stated flatly that James Woods was the only reason to tune in. *Shark* also took a page from the long-gone *Perry Mason* (q.v.) by having the hero extract witness-stand confessions from the real culprits, which was one of the many reasons genuine lawyers hesitated to endorse the series.

For his part, star James Woods told Frank Barron of *Entertainment Weekly*: "I've always felt the slight miscarriage of justice. DA's are often political offices throughout the country. And oftentimes cases where it's so clear that a crime has been committed and the perpetrator of that crime is ignored because, as a political move, if the DA thinks he or she can't win the case, they often don't prosecute if there's a chance they could lose. Because they don't want to look bad. So the criminal justice system is kind of the dog being wagged by the tail of political ambition. And that's not a good thing." Once he'd gone over to the Prosecution, one could ever accuse Shark of being "wagged" by anyone — especially his superiors.

The series' pilot episode garnered more press attention than usual because it was directed by Spike Lee (*Do the Right Thing, Malcolm X*), who must have had some fascinating political discussions with his right-of-center star. Production of *Shark* proceeded smoothly throughout its two-season run, except for a brace of unavoidable interruptions: during Season One, the company shut down briefly after the death of James Woods' younger brother; and during Season Two, all of Hollywood shut down, not so briefly, for the 2007–2008 writer's strike.

When Sam Page decided to leave the show at the end of the first season, it was explained that his character Casey had taken a leave of absence to join his senator father's re-election campaign. And having gone on maternity leave just before the strike, Jeri Ryan also had to be removed from the action, with the explanation that Jessica Devlin had returned to Chicago to look after her sick dad; she was temporarily replaced by Paula Marshall as paralegal Jordan Westlake, who joined the show during a Season Two story arc that had been concocted to bolster the series' faltering ratings. Jordan teamed up with Shark as penance for having unearthed evidence indicating that, back in his "bad lawyer" days, Shark may have conspired with a client to cover up a murder — an indiscretion that also resulted in a second epiphany for our hero when, in retaliation for that long-ago murder, Shark's former partner was killed in broad daylight. Disbarred in California, Shark sullenly relocated to Las Vegas, where he once again became a criminal lawyer — only now he didn't have the stomach to make excuses for his mob-connected clients. Just before the series finale, Shark's LA license was reinstated, but his troubles were far from over: now he was being stalked by an escaped killer, providing the lawyer with *still another* crisis of conscience when it appeared he'd have to kill his pursuer in order to save the life of his daughter Julie.

It wasn't that viewers avoided *Shark* so much as they had trouble finding it: CBS shuttled the show all over its Prime Time schedule throughout the second season, at one point running it in a suicide slot opposite the Sunday NFL games. Though ratings actually picked up during its final weeks on the air, the series was cancelled because the CBS executives had determined their network was glutted with crime dramas, and the very expensive *Shark* was the easiest one to eliminate.

Sirota's Court

NBC: Dec. 1, 1976–Apr. 13, 1977. Peter Engel Productions/Universal Television. Produced by Harvey Miller and Peter Engel. Cast: Michael Constantine (Matthew J. Sirota); Cynthia Harris (Maureen O'-Connor); Kathleen Miller (Gail Goodman); Fred Willard (Bud Nugent); Ted Ross (Sawyer Dabney); Owen Bush (Bailiff John Belson).

Regarded by TV and legal-drama historians as the first network sitcom to really show lawyers at work, *Sirota's Court* starred Michael Constantine as Matthew J. Sirota, a slightly eccentric and highly nontraditional night court judge. The tone of the series was set by the opener, in which an offscreen announcer was "kibbitzed" by a likewise unseen man and woman, who described Sirota as "tough but fair" and "a nice man." A self-righteous TV reporter, played by Victor Buono, showed up in Sirota's court while doing an in-depth study on "the country's worst judges." Sirota seemed to be determined to live up (or down) to this designation, even interrupting the courtroom proceedings at one point to watch a football game. But by the end of the debut episode, both the reporter and the home viewers were aware that, for all his shortcomings, Sirota was the right man on the right job.

Like many another sitcom of the late 1970s, Sirota's Court was basically an ensemble effort. Cynthia Harris was seen as caustic court clerk Maureen O'Connor, with whom Sirota had been sporadically carrying on a romance for years; Kathleen Miller played public defender Gail Goodman, an idealistic liberal hampered by a streak of clumsiness; Fred Willard did his Fred Willard schtick as ambitious, womanizing, essentially incompetent DA Bud Nugent; Ted Ross darted in and out of the courtroom as ambulance-chasing lawyer Sawyer Dabney; and Owen Bush rounded out the regulars as bailiff John Belson, who worshipped the ground Judge Sirota walked on.

One of several comedies introduced as midseason replacements during the 1976–77 season, Sirota's Court attracted enough critical adulation to earn a Golden Globe nomination, but unfortunately couldn't draw viewers. After 14 episodes (including the obligatory scenario in which a nutty defendant took everyone in the courtroom hostage), the series was cancelled — but not forgotten. Seven years later, producer Reinhold Weege folded many of the characterizations and story elements seen on Sirota's Court into his own, far more successful legal sitcom Night Court (q.v.)

Slattery's People

CBS: Sept. 21, 1964–Nov. 28, 1965. Pendick Enterprises/Bing Crosby Productions. Created by James Moser. Executive Producer: Matthew Rapf. Cast: Richard Crenna (James Slattery); Edward Asner (Frank Radcliff); Paul Geary (Johnny Ramos); Maxine Stuart (B.J. Clawson); Tol Avery (Speaker Bert Metcalf); Kathie Browne (Liz Andrews); Alejandro Rey (Mike Valera); Francine York (Wendy Wendkowski).

Yes, we know that the Judicial branch of the American government determines the laws of the land, but it is the Legislative branch that writes and enacts those laws. The protagonist of the hour-long CBS dramatic series Slattery's People was the minority leader in an unnamed state legislature (the series was partially location-filmed in Sacramento, California). Further, the character was a trained lawyer. So I say that Slattery's People was a legal series, and it's my book, so there.

Assembled by the same team responsible for the popular medical series Ben Casey, Slattery's People was created by James Moser of Medic fame. The star was Richard Crenna, making a definitive break from such previous TV-sitcom characters as adenoidal high-schooler Walter Denton on Our Miss Brooks and easygoing hayseed Luke McCoy on The Real McCoys. Crenna's James Slattery was a thirtysomething Liberal idealist, firm in his convictions and utterly incorruptible; he was also a fallible human being, capable of such well-intentioned but muddle-headed legislation as trying to ban the sport of professional boxing. Though he had as many setbacks as successes, Slattery could take heart in the words of Winston Churchill, invoked at the beginning of each hour-long episode: "Democracy is a very bad form of government, but I ask you never to forget it, all the others are far worse."

Television had sufficiently matured by 1964 to allow James Slattery to tackle issues that, for their time, represented the ne plus ultra of controversy: race relations, capital punishment, sex education, abortion, medical malpractice, the broken foster-care system, tax rebates and kickbacks for Big Business, questionable campaign funding practices, electronic eavesdropping, right- and left-wing extremism, and the persecution of academicians for "subversive" political leanings. Occasionally the seriousness was leavened by mildly comic episodes centering around eccentric characters: Ed Wynn played a pensioner who bicycled all the way to the state capitol to protest the destruction of a beloved state institution; Forrest Tucker showed up as a Hughes–like billionaire who inexplicably purchased a building in his own neighborhood for the express purpose of tearing it down; and Elsa Lanchester dithered about as an kooky conservationist who tried to block construction of a highway in order to save her private animal sanctuary. There were other interesting casting choices involving younger, newer performers: Sally Kellerman was seen as a crafty lobbyist, Barbara Feldon played the wife of an anti-death-penalty activist, and a pre–*Jeannie* Barbara Eden donned street clothes as the spouse of a Good Samaritan doctor facing malpractice charges.

The regular supporting cast during the first season included Edward Asner as Frank Radcliff, jaded political reporter for the *Times-Chronicle*; Maxine Stuart as Slattery's sarcastic middle-aged secretary B. J. Clawson; Paul Geary as Slattery's eager-beaver aide Johnny Ramos; and Tol Avery as Speaker of the House Bert Metcalf, who though he was a member of the "loyal opposition" was also one of Slattery's few close friends.

Though its ratings were nothing to brag about during its first season (it never rose any higher than the Nielsen "cutoff" mark of 17 points), *Slattery's People* was regarded by CBS as a prestige item to offset such less prestigious offerings as *Gilligan's Island* and *My Living Doll* in the eyes of the critics, and as such earned a last-minute renewal for a second season. At the same time, new producer Irving Elman, acting on orders from CBS programming chief Michael Dann, set about to give the series "mass audience appeal" with a few cast changes. The talented but unprepossessing Paul Geary was replaced by Alejandro Rey as Slattery's roguishly handsome new aide Mike Valera, whom it was hoped would attract young female viewers; and Maxine Stuart was succeeded by Francine York as Slattery's curvaceous new secretary Wendy Wendkowski, who was supposed to lure in a few thousand more male fans. Slattery himself became less priggish and more "humanistic"— his colleagues were now allowed to call him "Jim" instead of "sir" or "representative"— and he was further humanized with the addition of a girlfriend, TV newscaster Liz Andrews (a clear

conflict of interest, but what the hell), played by Kathie Browne.

But there was little chance of *Slattery's People* capturing the audience of its NBC competition *The Man From U.N.C.L.E.*, and as result of dismal overnight ratings for the second-season opener, the series was cancelled. Curiously, its replacement was another CBS legal series that was languishing in the ratings, *The Trials of O'Brien* (q.v.).

Sparks

UPN: Aug. 26, 1996–March 2, 1998. MTM Enterprises. Creators/Executive Producers: Bob Moloney and Bruce Johnson. Executive Producer: Ed. Weinberger. Theme song: Billy Preston. Cast: Miguel Nuñez (Maxey Sparks); Terrence Howard (Greg Sparks); James Avery (Alonzo Sparks); Robin Givens (Wilma Cuthbert); Kym A. Whitley (Darice Mayberry); Arif S. Kinchen (La Mar Hicks).

Created by comedians Bob Moloney and Bruce Johnson, the UPN sitcom *Sparks* was set in Compton, California, where the African-American law firm of Sparks, Sparks & Sparks maintained offices in the Inner City. Patriarch Alonzo Sparks (played by former *Fresh Prince of Bel-Air* costar James Avery) would have liked nothing better than to retire, but he was reluctant to leave the family practice in the hands of his eternally squabbling sons: Maxey (Miguel Nuñez, late of *Tour of Duty*), a devil-may-care womanizer, and Greg (future *Hustle and Flow* star Terrence Howard), a serious-minded prude. Though as different as night and day, both Maxey and Greg fell madly in love with Stanford Law School graduate Wilma Cuthbert (actress-supermodel Robin Givens), whom Greg had hired as an associate after an extensive three-year search. Sparks flew between Sparks and Sparks as they competed both in and out of the courtroom to curry favor with their dad and win the heart of Wilma — who for her part didn't need any more stress in her life thanks to her various loser boyfriends. (In the pilot, Wilma was engaged to a troublesome pro basketball player, a none-too-subtle reference to Robin Givens' real-life tribulations with her ex-husband, prizefighter Mike Tyson.) Also around for laughs — or hoots, hollers, and "awwws," as provided by the series' shrill pre-recorded laughtrack — were the firm's vituperative secretary Darice Mayberry (Kym A. Whitley) and hip-hop assistant LaMar Hicks (Arif S. Kinchen), who'd earned the name "Porky" in law school because he always stuttered in the presence of authority figures.

All vestiges of the series' legal ambience were effectively buried as Wilma and Maxey began a secret romance during the second season; also, Darice somehow found herself in bed with Greg the morning after a party celebrating her engagement to someone else, and Alonzo, aka Porky, impulsively married an old acquaintance (not named Petunia) in Las Vegas. Things got back on track when Wilma broke up with Maxey and joined another firm, whereupon she and her former lover wound up as opposing counsels in a flashy court trial. But the Law was again trumped by Love when Wilma and Maxey "got it on" in mid-trial. (Hoot, Hoot, Hoot! Holler, Holler, Holler! Awwwwwww....)

Lasting two seasons and 40 episodes, *Sparks* never ranked any higher than 125th in the ratings with general audiences. However, it was the eighth most popular show with black viewers, giving one a pretty good idea of UPN's target demographic.

State V.

ABC: June 19–July 17, 2002. A production of the ABC News Division. Executive Producer: Rudy Bednar. Cast: Cynthia McFadden (Narrator).

One of two legal docudramas unveiled on network television in the summer of 2002, ABC's *State V.*, like NBC's *Crime & Punishment* (q.v.), chronicled actual courtroom proceedings, from pre-trial to verdict. Unlike *Crime & Punishment*, in which the raw "reality" footage was re-edited by producer Dick Wolf to conform to the dramatic structure popularized by his *Law & Order* franchise (see separate entry), the ABC series, created by the network's news division, presented its material "straight." Also, whereas Wolf's show was told from the viewpoint of the prosecution, *State V.* afforded a generous amount of camera time to the Defense.

Each of the five hour-long *State V.* episodes chronicled the prosecution of a single homicide case in Arizona's Maricopa County, a jurisdiction that includes the city of Phoenix. The Arizona Supreme Court issued special orders to allow ABC full access to their courtroom system, permitting audiences to see virtually everything: the preparation of arguments by the opposing counsels, private conversations between the Defense and their clients, highlights from the actual trials, etc. Also, for the first time in TV history, cameras were allowed in the jury room. The series was narrated by former Court TV anchor Cynthia McFadden, who fulfilled the same duties for ABC's 2004 documentary special *In the Jury Room*, and later co-anchored the network's *Nightline* and *Primetime*. McFadden was especially skilled at maintaining a level of suspense in each episode, even though the conclusion was foregone — that is, the State never lost.

Comparing *State V.* to *Crime & Punishment*, *Time* magazine's TV columnist James Poniewozik observed: "It's a bit stiffer than [*Crime & Punishment*], with conventional news narration by Cynthia McFadden, but it's as interesting and better balanced. And it will almost certainly do worse in the ratings. Why? It's on Wednesdays at 10 P.M. E.T.—opposite *Law & Order*. And Americans will take a fictional prosecution, even in reruns, over a real defense any day." Unfortunately, Poniewozik was right.

Storefront Lawyers/Men at Law

CBS: Sept. 16–Dec. 9, 1970 (as *Storefront Lawyers*); Feb. 3–Sept. 1, 1971 (as *Men at Law*). Leonard Freeman Enterprises/National General Corporation. Executive Producer: Leonard Freeman. Cast: Robert Foxworth (David Hansen); Sheila Larkin (Deborah Sullivan); David Arkin (Gabriel Kaye); A Martinez (Roberto Alvarez); Pauline Myers (Gloria Byrd); Gerald S. O'Loughlin (Devin McNeil).

1970 was supposed to have been the "Year of Relevance" on network TV, resulting in several series featuring young and "with-it" protagonists dealing compassionately with the crucial issues of the day. One of the results of this line of thinking—hatched by old and "not-with-it" network executives — was the weekly, hour-long CBS series *Storefront Lawyers*, not to be confused with the similar ABC property *The Young Lawyers* (q.v.). In its original form, *Storefront Lawyers* starred Robert Foxworth as David Hansen, a dynamic young attorney who had relinquished a lucrative position at the Century City (California) firm of Horton, Troy, McNeill & Carroll to set up the Neighborhood Law Clinic, a nonprofit organization in a low-rent section of town. Helping Hansen provide pro-bono legal service to the poor and underprivileged were two more under-30 law firm "dropouts," Deborah Sullivan (Sheila Larken) and Gabriel Kaye (David Arkin).

In his *Time* magazine review, critic Richard Burgheim lumped the new CBS legal series together with another of the network's latest bids at "relevance," noting *"The Storefront Lawyers* (CBS) and *The Interns* (CBS) both exploit *Mod Squad's* multihero angle, but neither one is genuinely mod or engrossing." After 13 weeks of diminishing ratings, the network decided to take *Storefront Lawyers* in another direction, consigning its youth-and-relevance angle to the background. Retitled *Men at Law*, the series found the three protagonists rejoining Horton, Troy, McNeill & Carroll, with Gerald S. O'Loughlin added to the cast as the young lawyers' crusty-but-likeable older boss David McNeal. These changes were intended to expand the series' appeal by providing a mixed clientele for the attorneys, no longer confining them to poverty-stricken wretches but also permitting them to occasionally mingle with the Century City upper crust. On both sides of the transition, the series featured several future stars in the guest casts, including Melinda Dillon, Anne Archer, Kurt Russell, Dan Travanti and Jan–Michael Vincent.

Although *Men at Law* was no more successful than *Storefront Lawyers*, the series must have had a profound effect on costar Sheila Larkin, who later put her acting career on temporary hold to earn a Master's degree in social work.

Style Court!

Style Network/E!: debuted August 2003. Style Network Productions. Executive Producer: John Tom-lin. Cast: Henry Roth (Judge); Berglind Icey (Bailiff); Doug Llewelyn (Commentator).

When asked about Judge Joseph Wapner's cable-TV series *Animal Court*, announcer Doug Llewelyn huffily dismissed this latest effort from his former *People's Court* colleague as "just plain stupid." A few years later, Llewelyn himself signed up for another cable series, *Style Court!*— and it is to Wapner's everlasting credit that he didn't phone Llewelyn to say, "Hello, Pot. This is Kettle. Who's black now?"

A production of cable's Style Network, this faux courtroom show was presided over by "Judge" Henry Roth, a real-life Australian lawyer who augmented his income as a dress designer. In each episode, people from every walk of life would haul their friends, family members, neighbors and coworkers before Judge Roth, charging them with "crimes of fashion"— usually bad grooming or tacky clothes or both. Those who were found "guilty" were given a complete makeover, with all expenses absorbed by the series' producers. Featured in the cast were 5'11" Icelandic model Berglind Icey as the sexy court bailiff (complete with rhinestone handcuffs!), and hip DJ "Phat Waxy," who on closer inspection turned out to be none other than Judge Roth.

For his part, Doug Llewelyn claimed that the most interesting aspect of *Style Court!* was: "The payoff! If a husband convinces the judge his wife really needs a makeover and she's found guilty, the makeover team goes into action. And they do magic. It brings tears to people's eyes." That's for sure.

In addition to its Style Network run, *Style Court!* was also seen on the E! network as a respite from that cable service's round-the-clock barrage of celebrity biographies and showbiz gossip.

Sugarfoot

ABC: Sept. 17, 1957–July 3, 1961. Warner Bros. Television. Executive Producer: William T. Orr. Producer, Carroll Case. Cast: Will Hutchins (Tom "Sugarfoot" Brewster).

Introduced as a 1957 episode of the Warner Bros.–produced anthology *Conflict*,

Sugarfoot was the most successful of the five TV series that combined the "western" and "legal" genres (see also *Black Saddle, Dundee and the Culhane, Judge Roy Bean* and *Temple Houston*). Though some observers of the period assumed *Sugarfoot* was inspired by the popular 1939 comedy-western *Destry Rides Again*, it was actually based on *The Boy from Oklahoma*, a 1954 Warner Bros. theatrical feature starring Will Rogers Jr. as Tom Brewster, a young aspiring attorney reluctantly pressed into service as sheriff of a wide-open frontier town. (Another WB western titled *Sugarfoot* bore no relation to the series, and was prudently retitled *Swirl of Glory* in TV syndication to avoid confusion.)

For the TV version, Will Hutchins took over from Rogers as Tom Brewster, a sarsaparilla-drinking Easterner derided by the rough-and-tumble westerners as "Sugarfoot," or one step lower than a tenderfoot. Having taken a correspondence-school law course, Tom hoped to become a practicing attorney, but most everyone wanted to hire him as a peacekeeper instead — a rather embarrassing situation, since in the early episodes Tom wasn't terribly handy with a six-gun. Though he never backed down from a fight, our hero preferred to think his way out of trouble, and generally succeeded; he also possessed an endearing sense of humor and a charming way with the ladies.

Only a handful of *Sugarfoot*'s 69 hour-long episodes had all that much to do with lawyers and courtrooms. Examples include "Deadlock," in which Brewster was tricked into serving as a juror on a case in which his own boss was the plaintiff; "The Gaucho," wherein Tom found himself in the middle of a bitter battle over control of a cattle ranch; "Captive Locomotive," which pitted him against a greedy railroad executive who was gobbling up homesteads by foul and sinister means; "A Noose for Nora," placing Tom between a rock and a hard place when, after witnessing a murder, he was ordered by the judge to defend the murderer; "Vinegaroon," an unattributed reworking of the 1940 film *The Westerner* in which Brewster crossed paths with the infamous Judge Roy Bean (here played by Franklin Ferguson); and the series finale "Trouble of Sand Springs,"

which climaxed in a murder trial where defense attorney Tom discovered that the beautiful female prosecutor was the daughter of the victim. Otherwise, many of the best episodes completely ignored the series' "legal" groundwork: there was nary a judge nor jury in sight in "Apollo with a Gun," a seriocomic yarn directed by no less than Robert Altman in which Tom Brewster was pressed into service as an actor by legendary stage star Adah Isaacs Menken (Mari Blanchard).

Sugarfoot bore the distinctive brand of the Warner Bros. TV western product of the 1950s: the austere, woodcut-style credit titles, the catchy theme song (by Mack David and Jay Livingston), and the utilization of miles of stock footage from earlier Warners films — to say nothing of recycled plotlines, especially during the 1960 Hollywood writers strike, when several episodes of WB's detective shows were reformatted as westerns and vice versa. There was also a near-incestuous relationship with the studio's other western series, with Will Hutchins making crossover appearances as Tom Brewster (or a reasonable facsimile) on such programs as *Bronco* and *Maverick*— while *Bronco*'s Ty Hardin, *Maverick*'s James Garner, *Colt .45*'s Wade Preston and *Lawman*'s Peter Brown likewise showed up in their familiar TV roles on *Sugarfoot*.

Outside of Tom Brewster, there were no other regular characters on the series, though Jack Elam made recurring appearances as inveterate hellraiser Toothy Thompson, while Will Hutchins himself periodically showed up as Brewster's evil look-alike, an outlaw known as The Canary Kid. In the realm of guest appearances, *Sugarfoot* emulated many another western of the era by providing a showcase for handsome young actors on the threshold of bigger things: Dennis Hopper was seen as Billy the Kid in the series opener (an abbreviated remake of *The Boy from Oklahoma*), while others showing up in supporting roles included Charles Bronson, Adam West, Pernell Roberts, Troy Donahue and Richard Long. And in the episode "Man from Mendora," Tom Brewster made the acquaintance of a two-fisted Eastern dude named Teddy Roosevelt, played by Peter

Breck — who had earlier starred in his own "cowboy lawyer" series, *Black Saddle*.

The fading memories of baby-boomers notwithstanding, *Sugarfoot* was never telecast on a weekly basis during its four-season run. From 1957 through 1959, the series was seen on an alternate-week basis with *Cheyenne*. When the star of *that* series, Clint Walker, walked off the show during the 1959–60 season, the *Cheyenne* timeslot was alternately filled by *Sugarfoot* and by WB's new *Cheyenne* clone *Bronco*. And from the fall of 1960 until its cancellation in late 1961, *Sugarfoot* was seen on a rotating basis with both *Cheyenne* and *Bronco* under the blanket title *The Cheyenne Show*.

Superior Court

Syndicated: Sept. 1986–Sept. 1989. Edwards-Billett Productions/Lorimar-Telepictures. Executive Producer: Stu Billett. Cast: William D. Burns (Judge Burns, Season 1); Jill Jakes (Judge Jakes, Season 2); Raymond St. Jacques (Judge Clayton C. Thomas, Season 3).

Hastily assembled in the late summer of 1986 for an early September release, the daily, half-hour *Superior Court* was designed by producer Stu Billett as a companion series to Billett-Edwards' courtroom syndie *The People's Court* (q.v.). But while *The People's Court* featured actual, unscripted small-claims cases presided over by a genuine judge, *Superior Court* was totally scripted, trafficked in heavily fictionalized versions of "drawn from the headlines" courtroom litigation, and featured a man who had never personally tried a case in his life.

Appearing as the judge during the series' first season was William D. Burns Jr., a former police officer and Beverly Hills Municipal Court Manager who noticeably bristled whenever the series' P.R. staff identified him as a real judge. The litigants and lawyers were all professional actors of the "Wave the Arms, Shout the Lines" School of Dramatic Art, while the soapy scripts were penned by daytime-drama veterans Joyce and John William Corrigan, whose joint resume included *One Life to Live* and *General Hospital*. The series' managing script editor was TV legal reporter Harvey

Levin, who also worked as legal adviser on *The People's Court*. Levin did his best to bring up a number of potent legal issues of the 1980s on *Superior Court*, but he was fighting a losing battle against the series' mustache-twirling melodramatics.

In February of 1987 Levin came up with a stunt intended to boost ratings. In honor of the Bicentenary of the U.S. Constitution, each week's quota of episodes featured such real-life guest judges as Mildred Lillie, Arthur Alarcon, Shirley Hufsedler, Bernard Jefferson and Rose Bird, all presiding over cases top-heavy with controversy. For example, ex–California Court of Appeals Justice Jefferson tackled an abortion-rights case, while former California Supreme Court Justice Bird wielded the gavel over a case involving book-banning and freedom of speech. Each episode ended with a short information "bite" about the Constitution, delivered by such genuine jurists as Harry Blackmun of *Roe v. Wade* fame. Alas, simply having these authentic judges on set did not automatically guarantee authenticity: particular, the book-banning episode was rife with such clichéd howlers as showing a snippy prosecution witness breaking down in tears and confessing she'd lied — a particularly egregious moment considering that Rose Bird had demanded the script be rewritten before she agreed to appear. *LA Times* TV columnist Harvey Rosenberg found this entire cycle of episodes to be ridiculous, hackneyed, and unworthy of the celebrity-judge lineup, adding, "The opposing attorneys must fit their closing arguments into a combined 65 seconds — no need to get wordy — and all [the judge] has to do is stay awake, make an occasional comment and deliver a brief, eloquent judgment." Worst of all, this programming stunt did not improve the series' ratings one teeny, tiny bit.

For the second season of *Superior Court*, William Burns was replaced by former Beverly Hills Municipal Court judge Jill Jakes, who in turn was occasionally spelled by guest judges and by actors cast as jurists. By the third and final season, all pretense of reality had been cast to the four winds, as character actor Raymond St. Jacques stepped into the role of fictional judge Clayton C. Thomas. Despite almost unanimous condemnation from TV critics and the legal establishment, *Superior Court* managed to grind out 218 episodes over a three-year period.

Sweet Justice

NBC: Sept. 15, 1994–Apr. 22, 1995. Trotwood Productions/Columbia TriStar Television. Executive Producer: John Romano. Cast: Cicely Tyson (Carrie Grace Battle); Melissa Gilbert (Kate Delacroy); Ronny Cox (James-Lee Delacroy); Cree Summer (Reese Daulkins); Greg Germann (Andy Del Sarto); Jim Antonio (Ross A. Ross); Jason Gedrick (Bailey Connors); Megan Gallivan (Anne Foley); Scott Paetty (Harry Foley); Michael Warren (Michael "T-Dog" Turner).

The hour-long NBC legal drama *Sweet Justice* was set in an Unnamed Southern City, but you can't fool us: the series was extensively location-filmed in New Orleans, including portions of the Garden District and the surrounding Bayou country. Melissa Gilbert was cast as lawyer Kate Delacroy, who after four years with a prominent Wall Street firm returned to her home town in the South to attend the wedding of her belle-of-the-ball sister Anne (Megan Gallivan) and attorney Harry Foley (Scott Paetty). Kate's father James-Lee Delacroy (Ronny Cox), a wealthy contract lawyer and rock-ribbed Conservative, hoped that his daughter would join his firm. But after Kate became embroiled in a bitter child-custody battle involving an impoverished waitress and her evil-millionaire former husband, our heroine opted to cast her lot with the Underdog. Accordingly, she signed with Battle-Ross & Associates, a firm located in an old Victorian mansion and run by legendary Civil Rights activist Carrie Grace Battle (Cicely Tyson), who'd been the best friend and onetime law partner of Kate's late mother. This did not sit well with Kate's father, who'd been at loggerheads with Ms. Battle ever since the Desegregation days of the 1950s and 1960s, and who wistfully recalled how glorious the "Old South" had been before those uppity protestors changed everything. (Hollywood writers will apparently never get past the misapprehension that all Southern White Males over the age of 60 still fly the Confederate flag on their lawns and display a picture of Orval Faubus over the fireplace.)

Also sharing office space with Kate and Carrie were Ms. Battle's partner, the venerable, methodical Ross A. Ross (Jim Antonio); passionate "upstart" attorney Andy Del Sarto (played by *Ally McBeal's* future "Fish," Greg Germann); and the youngest member of the firm, single mom Reese Dawkin (Cree Summer). Adding a dash of romance to the proceedings were crusading journalist Bailey Connors (Jason Gedrick), Kate's former high-school crush; and community activist "T-Dog" Turner (Michael Warren), who was sweet on Grace.

In all fairness, a little bit of attention should have been paid former child actress Melissa Gilbert in her first starring dramatic series. Instead, all eyes were on Cicely Tyson, admittedly the more celebrated of the series' two leading ladies, in her first weekly-series assignment since *East Side—West Side* way back in 1964. In an interview with *Jet* magazine, the multi-award-winning Tyson described her character Carrie Grace Battle as having "gained a reputation as one of the few local attorneys that people are advised not to mess with." She added, "[Carrie] is an exceptional woman because she has an incredible passion for justice. She went through all the sit-ins, the lie-ins. She is determined to buck the system. She is a no-nonsense woman. We need to project strong women, especially strong black women. It's imperative for our young to see women like that." Withal, admitted Tyson, "[Carrie] isn't a perfect person by any stretch of the imagination. Carrie Battle will, hopefully, surprise everyone. She is a voracious kind of human being, who takes no nonsense, but is still extremely vulnerable in many instances." To prepare for her role, Tyson solicited the input of 81-year-old DC lawyer Lovey Roundtree, who had been a key player in the desegregation of the Trailways bus line; others who helped Tyson mold her characterization were former U.S. Congresswomen Barbara Jordan and Yvonne Brathwaite-Burke. With all this extra input, it's hardly surprising that Tyson's costar Melissa Gilbert was somewhat lost in the shuffle. (Years later, Gilbert became president of the Screen

Actors Guild, where she was overshadowed by no one!)

For most of its run, *Sweet Justice* was seen on Saturdays, first scheduled opposite CBS' *Dr. Quinn, Medicine Woman*, then opposite Fox's *America's Most Wanted*—which may explain why the series' first season was also its last.

T. and T.

Syndicated: Jan. 11, 1988–Sept. 1989. Family Channel: Jan. 6–Sept. 1, 1990. Tribune Entertainment. Produced by John Ryan. Cast: Mr. T (T.S. Turner); Alex Amini (Amanda "Amy" Taler); Kristina Nicoll (Terri Taler); David Nerman (Danforth "Dick" Decker); Ken James (Detective Jones); Jackie Richardson (Aunt Martha Robinson).

The syndicated action-adventure series *T. and T.* marginally qualifies as a legal show — just as its star marginally qualifies as an actor. Ex–boxer and ex-bouncer Mr. T (né Laurence Tureaud), he of the Mandinkan haircut and jewelry-bedecked biceps, had recently wrapped up four seasons on *The A-Team* as the euphemistically yclept "B.A." Baracus when he was approached by the producers of *T. and T.* to star as T.S. Turner, a street scrapper framed for a crime he didn't commit. Sprung from durance vile by young crusading lawyer Amy Taler (Alex Amini), T.S. went to work for the lovely Amy as a private detective. It was a copacetic arrangement: while Amy fought for the rights of the underprivileged and oppressed, T.S. kicked the butts of the privileged oppressors. Acting as "T. and T.'s contacts and erstwhile assistants were David Nerman as Dick Decker, owner of the gym where T.S. worked out; Catherine Disher as Amy's secretary Sophie; Ken James as Detective Jones, who didn't entirely trust T.S. but certainly admired his strength and perseverance; Jackie Richardson as T.S.'s Aunt Martha, with whom he lived; and on a recurring basis, Rachael Crawford as Martha's teenage daughter Renee.

Though telecast exclusively outside the network mainstream during its first year on the air, *T. and T.* almost landed a one-shot appearance on ABC, when the series' producers asked the network to showcase the debut episode right after Super Bowl XXII on January 31, 1988. Despite a million-dollar incentive, ABC turned down the request, forcing broken-hearted viewers to abandon all hopes of seeing Mr. T and settle instead for the Redskins and the Broncos.

Each episode of the series' first season opened with a ritualistic sequence wherein T.S. was seen incongruously dressed in a suit and tie, only to swagger into a locker room, shed his professional accoutrements, and don his familiar studded-leather street jacket. This opener was abandoned for Season Two, which downplayed the action sequences while focusing on legal, social and environmental issues. At that time, Joe Casper was occasionally seen as T.S.'s equally pugnacious roommate Joe, Aunt Martha and Renee apparently having moved elsewhere.

After two seasons in syndication, *T. and T.* returned for a third season of half-hour episodes on cable's Family Channel (now ABC Family). Despite its new kid-friendly environs, the series increased its level of violence, all but eliminating the original "activist lawyer" premise. Alex Amini left the show (no explanation was given for her character's absence), and her replacement was Kristina Nicoll as Amy's cousin Terri Taler, enabling the producers to retain the "T. and T." designation. Also new to the cast was David Hemblen, popping in and out as T.S.'s police contact, Detective Hargrove.

Filmed in Toronto, the 65-episode *T. and T.* could never be accused of failing to deliver what its fans wanted. Still, this extremely pedestrian series might have had more lasting interest if, instead of being cast as Amy's legman, Mr. T himself had played a lawyer — thus treating us to climactic courtroom scenes in which the bombastic hero leaned into the jury's faces and screamed, "I *PITY* THE FOOL WHO VOTES GUILTY!"

Temple Houston

NBC: Sept. 19, 1963–September 10, 1964. Rancom Productions/Apollo Productions/Temple Houston Company/Warner Bros. Television. Executive Producer: Jack Webb. Produced by James Lydon. Theme song by Frank Comstock and Ned Washington.

Cast: Jeffrey Hunter (Temple Houston); Jack Elam (George Taggart).

Temple Houston has earned a special niche in the history of network television not because it was one of the rare programs to combine the "western" and "legal" genres (see also *Black Saddle, Dundee and the Culhane, Judge Roy Bean* and *Sugarfoot*), but because it was the only TV western ever produced by Jack Webb.

Appointed executive in charge of television production for Warner Bros. in early 1963, former *Dragnet* star Webb wasted no time pitching new series ideas to the three major networks. Among the proposals was one that had been in development since 1957: a weekly, hour-long drama based on the life based on the life of Temple Houston, the son of fabled Texas revolutionary Sam Houston. Graduating from Baylor University in 1880, the younger Houston traveled the circuit courts of Texas' 35th Judicial District as a licensed attorney. Renowned in legal circles for his grandiose courtroom tactics and his fluency in several languages (most of them Native American), Temple ultimately served in the state's House of Representatives, then set up a practice in Oklahoma Territory. Before his premature death from a cerebral hemorrhage at age 35, Temple achieved nationwide fame for his eloquent defense of prostitute Minnie Stacy, popularly known as "The Soiled Dove Plea." He also gained notoriety as a fast gun, allegedly shooting it out with the likes of Billy the Kid and Bat Masterson.

Cast as Temple Houston was Jeffrey Hunter, who after wrapping up work on the series' 57-minute pilot in March of 1963 was poised to costar in John Ford's theatrical feature *Cheyenne Autumn*. It was assumed at the time that NBC would not begin running the *Temple Houston* series until the 1964–65 season, but at the last minute the network scuttled plans for a private-eye show starring Robert Taylor. Ordered to immediately commence production on *Temple Houston* for a September 1963 release, Jack Webb began shooting with what his biographer Michael Hayde has described as "swift professionalism" on August 7. Unfortunately this required Jeffrey Hunter to drop out of *Cheyenne Autumn*, but the actor accepted this professional setback—and the series' hectic production schedule—in stride. As noted by columnist Terence O'Flaherty while watching production of the first episode "The Twisted Rope," "Hunter is a genial sort of chap and seemed completely unruffled by the fact that the writers are handing out new pages of script. This has changed not only the lines but the condition of one of the principals—a character who is supposed to be carrying a child—being made un-pregnant at the sudden change of a writer's pen." This sort of self-censorship may explain why the "Soiled Dove Plea" was *not* reenacted on the series.

Described by critics as a combination *Perry Mason* (q.v.) and *Have Gun—Will Travel*, *Temple Houston* was characterized by star Hunter as "a different kind of western, a whodunit on horseback. And it has humor in it." Also, as originally conceived, Temple was supposed to have been as erudite and stylishly dressed as his real-life counterpart, not unlike *Have Gun*'s gentleman gunslinger Paladin. But NBC removed these distinctive touches, insisting that the series be treated as a traditional western with mild legal undertones. This proved a major miscalculation, and when the October ratings came out the series was ranked 31st of the 32 new programs. At this point a format change was imposed, restoring the character's sense of humor and opting for a *Maverick* approach to the story material: in this spirit, the love-hate relationship between Houston and his sidekick, burned-out gunfighter George Taggart (Jack Elam), took on Bret-and-Bart *Maverick* dimensions. Much to the dismay of legal-show devotees, the series now also spent less time in the courtroom than in the Wide Open Spaces, though Temple still lent his expertise to a handful of offbeat court cases involving such clients as a circus elephant and a female prizefighter.

Temple Houston lasted only 26 hour-long episodes, its failure attributed to the network's inability to make up its mind what sort of show it was supposed to be. Also as noted by Jeffrey Hunter, viewers were confused by the series' title: indeed, a 1964 *Mad* magazine article satirized this confusion by illustrating *Temple Houston* with a picture of a Texas synagogue attended by dozens of middle-aged Jewish cowboys.

Curiously, the series' pilot episode, which featured James Coburn in what was to have been a recurring role, was not shown during the network run of *Temple Houston*. Instead, it was released theatrically in December 1963 as *The Man from Galveston*, playing in a double bill with the Frank Sinatra-Dean Martin vehicle *Four for Texas*. To obscure the film's TV origins, new dialogue was looped in, changing the name of protagonist Temple Houston to Timothy Higgins!

Texas Justice

Syndicated: Jan. 2001–Sept. 2005. 20th Television. Executive Producer: Karen Melamed. Cast: Judge Larry Joe Doherty (Himself); William Bowers (Bailiff); Randy Schell (Narrator).

One is tempted to observe that Texas Justice is to Justice what Texas Politics is to Politics, but now that we've got that out of our system, let's discuss the daily, syndicated courtroom series titled *Texas Justice*. Its star was Larry Joe Doherty, eminent Houston trial lawyer, singer, poet, and sometimes actor in such cinematic masterworks as *Booty Call*. Using his legally registered persona "Judge Larry Joe Doherty," the star wielded his gavel over a raucous studio courtroom that would not have been out of place on *The Jerry Springer Show*. In keeping with his motto, "Without love there is no justice," Judge Doherty referred to the plaintiffs by their first names, glad-handing the men and flirting with the ladies, all the while cracking jokes for the benefit of the spectators. Whenever things got out of control, resident sheriff-bailiff William Bowers would lean forward and offer the loving advice, "SHUD-DUP!"

For all its farcical aspects, *Texas Justice* didn't kid around when the Judge imposed binding arbitration on the various small-claims cases brought before him, with the series' distributors footing the bill for any awards or damages. And though he liked to play-act the guffawing "good ole boy," Larry Joe Doherty was

an expert on all aspects of Texas jurisprudence, plugging holes in the conversation with such trivia as, "Back in the Republic days, y'all could get yourselves hanged for stealin' a horse."

Initially shipped out on a limited market-by-market basis in early 2001, *Texas Justice* entered nationwide syndication the following year. The series lasted until 2005, when its timeslot was taken over in several markets by the new *Judge Alex* (q.v.)

They Stand Accused

CBS: Jan. 18–May 1949 (as *Cross Question*). DuMont: Sept. 11, 1949–Oct. 5, 1952 (title reverted to *They Stand Accused* in January of 1950); Sept. 9–Dec. 30 1954. Directed by Sheldon C. Cooper and Bruno VeSota, among others. Written and supervised by William C. Wines. Cast: Charles Johnston (Judge); Harry Creighton (Announcer).

Until another candidate emerges from the mists of the forgotten past, *They Stand Accused* qualifies as television's first live dramatic courtroom series. The program originated from the Chicago studios of WGN-TV, where it ran on an exclusively local basis from April 11, 1948, through January 11, 1949. Supervised by William C. Wines, Assistant Attorney General of Illinois, the weekly, half-hour series featured fictional courtroom cases with professional actors as the litigants, but with actual prosecutors and defense lawyers arguing before a judge played by genuine attorney Charles Johnston. Though there was no script, the participants were carefully prepped by William C. Wines in the hours preceding each telecast. Not so the jurors, who were selected from the studio audience — and though announcer Harry Creighton bent over backward to assure the viewers that the jury made the final decision, in several episodes they were instructed to give only an "advisory" verdict, with Judge Johnson himself determining the outcome. Among the production staffers during the series' initial Chicago run was actor Bruno VeSota, later a regular in the films of director Roger Corman.

On January 18, 1949, one week after Chicago and New York were linked by coaxial cable, *They Stand Accused* was beamed out nationally by the CBS network, even though

WGN was affiliated with the rival DuMont hookup. Evidently because of a sponsor conflict, the short-lived CBS version was retitled *Cross Question*, as was the series' subsequent revival on DuMont in the fall of 1949. By January of 1950, however, the title had reverted to *They Stand Accused*, remaining as such for the duration of its DuMont contract.

One of the few surviving kinescopes of *They Stand Accused* comes from the series' final DuMont run in 1954, which commenced after a nearly two-year hiatus. Curiously, no one is "accused" in this entry, which focuses on a complex child-custody case. Technically the production is quite proficient, with the occasional unrehearsed noises — creaky floors, loud paper-rustling, microphone feedback etc. — adding to the series' documentary feel. Similarly, the lawyers perform with authority and authenticity, save for a few minor gaffes: after being lauded by the off-screen announcer for his eloquence, one of the attorneys mixes up the litigants in his summary, declaring emphatically, "Fred realized that he was Tom's wife."

The Tony Randall Show

ABC: Sept. 23, 1976–March 10 1977; CBS: Sept. 24, 1977–March 25, 1978. MTM Enterprises. Executive Producers: Tom Patchett, Jay Tarses, Gary David Goldberg. Cast: Tony Randall (Judge Walter Franklin); Barney Martin (Jack Terwilliger); Allyn Ann McLerie (Janet Reubner); Devon Scott (Roberta "Bobby" Franklin: 1976–77), Penny Peyser (Roberta "Bobby" Franklin: 1977–78); Brad Savage (Oliver Wendell Franklin); Rachel Roberts (Mrs. Bonnie McClellan); Zane Lasky (Mario Lanza); Diana Muldaur (Judge Eleanor Hooper); Hans Conried (Wyatt Franklin).

With four successful seasons of TV's *The Odd Couple* behind him, Tony Randall went solo for his own eponymous starring sitcom. Produced by the same team responsible for *The Mary Tyler Show* and *The Bob Newhart Show, The Tony Randall Show* cast the actor as Walter Franklin, a widowed judge serving in Philadelphia's Court of Common Pleas. A stern and imposing figure on the bench, Judge Franklin nonetheless was fair-minded and compassionate in his rulings, occasionally lightening the solemnity of the proceedings with a wry sense of humor. But though master of his domain in the courtroom, Walter was no match for his wisecracking daughter Bobby (played first by Devon Scott, then by Penny Peyser) and son Oliver (Brad Savage) in the confines of his own home. A typical exchange between Walter and Oliver went thusly:

> WALTER: There was a woman in my court today. I can't get her out of my mind. She was very attractive ... and ... I don't know what to do.
> OLIVER: Well, if she's guilty, you gotta throw her in the slammer!
> WALTER (after a short pause): I guess it's not as complicated as I thought.

This little snatch of repartee reveals the basic comic thrust of the series: Walter's search for an ideal mate after two years of widowhood. The leading candidate was his fellow jurist Eleanor Hooper, played by Diana Muldaur long before she achieved notoriety as the barracuda like Rosalind Shays on *L.A. Law* (q.v). The other significant women in Walter's life (albeit platonically) were his mother-hen British secretary, played by musical-comedy favorite Allyn Ann McLerie (who landed the role because Tony Randall recalled her work in the short-lived 1949 Broadway musical *Miss Liberty*); and a slightly battier Briton, Walter's housekeeper Mrs. McClellan, played by Rachel Roberts (the former Mrs. Rex Harrison). Rounding out the regular cast were Barney Martin (Jerry Seinfeld's future TV dad) as intimidatingly accurate court reporter Jack Terwilliger, and Zane Lasky as arrogant law clerk- turned-DA's assistant Mario Lanza (!). Later on, Hans Conried showed up as the Conservative-minded Walter Franklin's Liberal father Wyatt Franklin — an amusing bit of casting, in that Conried was all of three years older than Randall.

Though the titular star earned a Golden Globe nomination for his performance in *The Tony Randall Show*, the series was dropped by ABC at the end of its first season. CBS picked up the property in the fall of 1977, for another single-season run.

Traffic Court

ABC: June 18, 1958–March 30 1959. Executive Producer: Selig J. Seligman. Cast: Edgar Allan Jones

(The Judge); Frank Chandler McClure (Bailiff); Samuel Whitson (Court Clerk).

One of several locally-produced Los Angeles courtroom series to achieve national exposure, *Traffic Court* originated as an unsponsored public-service weekly, debuting in June of 1957 on LA's ABC flagship KABC. This version aired live, and was presided over by Los Angeles Municipal Court Judge Evelle J. Younger, former cohost of the TV quiz show *Armchair Detective* and later the Attorney General of California. Each episode was a dramatization of actual traffic court cases, featuring professional actors as defendants and witnesses. There was no rehearsal and no script; the actors were required to improvise their dialogue from a printed outline, which had been vetted for accuracy by the meticulous Judge Younger.

Traffic Court ceased to qualify as a public-service entry when it picked up a sponsor, whereupon Younger left the series, feeling that his continued appearance was a "violation of judicial ethics." His replacement was Edgar Allan Jones Jr., not a member of the California Bar but instead the assistant dean of the UCLA Law School. Utilizing the newly perfected videotape process, Jones recorded his appearances on weekends to accommodate his academic schedule.

"My lifetime career is teaching law," Jones told *TV Guide* in a 1960 interview. "I cannot and will not jeopardize that, though television is interesting and remunerative. If I am to be a good teacher—and that's the important thing—I can't afford to become enmeshed on TV." Proof that Jones meant every word he said was provided during one well-publicized taping session, when baseball star Sandy Koufax showed up before the "Judge" after being ticketed for stopping his car in mid-traffic. Koufax explained that he'd done so because his car had been struck by a Little Leaguer's baseball, and brought in the boy's entire team as witnesses. The director had told the kids not to take their hats off in court even when ordered to do so, hoping to elicit a stern ad-libbed admonition from Jones. Instead, after several futile attempts to get the youngsters to doff their caps, Jones

angrily walked off the set—bringing production to a screeching halt!

A year or so after its local LA bow, *Traffic Court* joined the Wednesday-night lineup of the ABC network as a replacement for the cancelled espionage series *O.S.S.* Edgar Allan Jones remained on the bench, joined by Frank Chandler McClure as the bailiff and Samuel Whitson as the court clerk. Each episode of the ABC version featured up to five separate reenacted cases per half-hour. *TV Guide's* synopsis for the broadcast of June 25, 1958, reads as follows: "1. An interpreter comes to court to plead the case of a girl who is a deaf-mute. 2. A man is charged with speeding in a 35-MPH traffic zone. 3. A showgirl is arraigned on reckless driving charges. 4. A man tries to prove his fidelity to his wife. 5. A public defender asks leniency for a man." (Wouldn't you just kill to see a videotape of Case Number Four?)

Used as a convenient stopgap to fill the occasional holes in ABC's nighttime schedule, *Traffic Court* was seen variously on Sundays, Mondays and Thursdays before its cancellation in March of 1959. Though never a huge ratings magnet, *Traffic Court* was popular enough to spawn a cycle of semi-documentary ABC courtroom series, two of which also starred the Honorable Edgar Allan Jones Jr. (see entry for *Day in Court*).

Trial and Error

CBS: March 15–29, 1988. Embassy Television/Columbia Pictures Television. Creators/Executive Producers: Donald L. Seigel, Jerry Perzigian. Executive Producers: Michael Moriarty, Howard Brown, Tommy Chong. Cast: Eddie Velez (John Hernandez); Paul Rodriguez (Tony Rivera); John deLancie (Bob Adams); Debbie Shapiro (Rhonda); Stephen Elliott (Edmund Kittle); Susan Saldivar (Lisa); John deLancie (Bob Adams).

The CBS sitcom *Trial and Error* was a halfhearted effort to stem complaints from the news media that Hispanics were getting shortshifted on Prime Time TV. Former *A-Team* costar Eddie Velez and standup comedian Paul Rodriguez were respectively cast as John Hernandez and Tony Rivera, lifelong *amigos* who had grown up together in the East LA barrio.

Though John and Tony now shared the same apartment, they couldn't have been farther apart temperamentally. John was an uptight young attorney, newly hired by the prestigious firm of Kittle, Barnes, Fletcher & Grey; Tony was a freewheeling funster who made his living as "the hottest T-shirt hustler" on Olvera Street. Knowing full well that he had been hired because the firm needed a token Latino, John nonetheless enjoyed the full support of senior partner Edmund Kittle (Stephen Elliott), though his patronizing fellow associate Bob Adams (John deLancie) constantly undercut John's self-esteem. Meanwhile, Tony and his *mucho loco* buddies were an endless source of embarrassment for his upwardly mobile roommate.

Others in the cast were Debbie Shapiro as John's sympathetic secretary Rhonda and Susan Saldivar as his girlfriend Lisa. The series' executive-producer lineup included actor Michael Moriarty, soon to costar on the legal drama *Law & Order* (q.v.), and counterculture comedian Tommy Chong, who has had plenty of first-hand courtroom experience in his lifetime.

Designed as a midseason replacement in case any of CBS' new 1987–88 entries bombed, *Trial and Error* had taped only a handful of episodes when Hollywood was hit with a writer's strike. Had the series clicked with viewers, the producers would have lensed all subsequent episodes in both English and Spanish–language versions. But critical reaction was uniformly abysmal — "dopey" was the operative word in the *New York Times* review — and thus *Trial and Error* bade the viewers *adios* after a mere three telecasts.

Trial by Jury

Syndicated: 1989–1991. Bob Stewart Productions/ Dick Clark Productions/Viacom. Executive Producers: Dick Clark and Bob Stewart. Produced by Al Schwartz. Cast: Raymond Burr (Judge Gordon Duane).

Debuting nationally in 1989, the strip-syndicated courtroom drama *Trial by Jury* had been in development for at least five years. In 1984, producers Dick Clark and Bob Stewart taped the pilot for a proposed "interactive" network anthology titled *You Are the Jury*. Directed by Burt Brinckerhoff and hosted by Efrem Zimbalist Jr., the pilot featured a dramatization of an actual trial, with flashbacks of the events leading up to the crime, after which the home viewers were invited to phone in their verdicts — not that it would make any difference, since the TV jurors had already made their decision. Featured in this hour-long presentation were Joe Regalbuto, Cindy Fisher, Nita Talbot, Judith Light, M. Emmet Walsh, Richard McKenzie and Thomas Hill. Though this version of *You Are the Jury* never aired, the producers tried again with a second 60-minute pilot in 1987, this one hosted by Peter Graves and starring Kelsey Grammer, John Amos, Audrey Landers, Robert Mandan, Lee Meriwether, Doris Roberts, Max Wright and Adrian Zmed. Once again the folks at home were invited to call in their verdicts, the results of which were to be superimposed over a neutral shot of the courtroom wall (the currently available copy of this pilot features a blank screen). This second incarnation of *You Are the Jury* managed to get a single Prime Time airing on NBC, but no further sales.

Finally, Clark and Stewart dropped the "phone-in" gimmick and opted for off- network syndication. This version, whittled down to half an hour and retitled *Trial by Jury*, retained the dramatized portion of the proceedings, again with flashbacks; the viewer was designated "the twelfth juror," and was beckoned to an empty seat in the jury box. As before, the viewer's verdict was solicited, even though the outcome of the trial was predetermined: This time, however, the "Twelfth Juror" could only talk back to the TV, since the show had been recorded several weeks before the airdate and there was no way to contact the studio.

Serving as host of *Trial by Jury* was Perry Mason himself, Raymond Burr, sporting his late-1980s beard and impersonating fictional judge Gordon Duane (one wonders if, during their first meeting, Burr and producer Dick Clark reminisced over the final episode of the original *Perry Mason* series, in which Clark played the murderer). Existing episodes suggest that *Trial by Jury* was a tad over-rehearsed, with lawyers and plaintiffs launching into their speeches by rote, as if appearing in an Easter pageant: but Raymond Burr was his usual relaxed, authoritative self. Though the supporting casts were largely comprised of unknowns who stayed that way, an episode in which singer Tori Amos appeared as the defendant in a murder trial has been widely circulated on such pastiche programs as VH1's *Before They Were Rock Stars.* (Note: some sources indicate that Joseph Campanella appeared as the prosecutor, Charles Siebert as the defense attorney, and Madlyn Rhue as the Judge; if so, they were not regulars, since none of these actors show up in the episodes I've seen.)

The Trials of O'Brien

CBS: Sept. 18, 1965–May 27, 1966. Mayo Productions/The O'Brien Company/Filmways. Executive Producer: Richard Alan Simmons. Creator/producer: Gene Wang. Cast: Peter Falk (Daniel J. O'Brien); Joanna Barnes (Katie); David Burns (The Great McGonigle); Elaine Stritch (Miss G.); Ilka Chase (Margaret); Dolph Sweet (Lt. Garrison).

Three years before his ascension to National Treasure status in the role of the hopelessly disheveled but brilliant Lieutenant Columbo, Peter Falk landed his first starring TV series, *The Trials of O'Brien.* Falk's portrayal of high-priced New York City defense attorney Daniel J. O'Brien was the polar opposite of Columbo: sharp-witted and fast-talking instead of vague and rambling, impeccably dressed and groomed instead of appearing as if he'd slept in his raincoat, and brash and abrasive instead of humble and obsequious. Virtually the only similarities between the two characters were Falk's distinctive delivery of the phrase, "Oh, that's terrific!," the actor's patented hand-to-forehead gesture, and the fact that the inside of O'Brien's snazzy new sports car was as ungovernably messy as the interior of Columbo's beat-up old Peugeot. Also, it was while starring in the weekly, hour-long *Trials of O'Brien* that Falk first exercised his future *Columbo* prerogative of maintaining complete approval over script content and vetoing any directorial ideas that didn't feel good to him.

"The character is a serious one," Falk explained in a 1965 *TV Guide* interview, "but he can be played for laughs. About three laughs, I guess for every moment of seriousness. No crusading in any of the scripts—nothing but a straightforward offering of entertainment. When they start to get serious on TV, they either get pretentious or they take a simple subject and overcomplicate it with a lot of jazz it doesn't need." He added, "That's not going to happen to our show if I can help it. Each show has a point—but we keep it comic as well as meaningful."

A skilled courtroom combatant, crafty conniver and roguish charmer who could switch moods at the flick of an eye, Daniel J. O'Brien demanded an enormous fee from his clients for two reasons: he invariably secured their release via his encyclopedic knowledge of legal loopholes, and he was eternally in debt. A chronic gambler and "soft touch" for every sob story in Manhattan, O'Brien owed so much money to various bookies and loan sharks that he was generally three rent payments and six alimony payments behind, infuriating his lovely ex-wife Katie (Joanna Barnes)—who, though clearly still quite fond of O'Brien, didn't trust him any farther than she could throw him. And not without reason: on several occasions, O'Brien was seen merrily violating the ethics of his profession, misrepresenting himself to various file clerks and government functionaries in order to secure necessary information. (In one episode, O'Brien gets his hands on a sensitive document by claiming that his client's wife had run off with a Portuguese magician!) This was one of the many reasons real-life lawyers were disdainful of *Trials of O'Brien*, also citing the character's grandstand tactics such as barreling into a courtroom at the very last minute with a sheaf of crumpled papers in his hand.

Working overtime to be "colorful," the series' producers provided O'Brien with a humdinger of a private office, decorated with such props as a vintage dart board, a fireplug, an Indian headdress, a stuffed owl, a human skull, a duck decoy, a telescope, a microscope, a fireman's helmet, a cider press and an antique iron. Series art director Robert Gundlach explained that every prop was related to the main character: referencing the fireman's helmet, Gundlach said, "Peter might put it on the back of his head, to demonstrate how unsedate he is," while he might use the skull as a cue to quote Shakespeare. Equally colorful were the series' supporting characters: David Burns as O'Brien's street informant "The Great McGonigle," a con artist, petty thief, forger and all-around scamp who sported thick glasses and a Ben Turpin moustache; Elaine Stritch as Miss G, O'Brien's secretary and severest critic; Ilka Chase as his former mother-in-law Margaret, who clearly enjoyed O'Brien's company more than did her daughter Kate; and Dolph Sweet as police lieutenant Garrison, something of a dry run for the actor's dyspeptic characterization of Chief Karnansky on the popular sitcom *Gimme a Break!*

Created by *Perry Mason* alumnus Gene Wang and produced by Richard Alan Simmons, who'd written the 1962 episode of *The Dick Powell Theater* that won Peter Falk the first of his many Emmy awards, *The Trials of O'Brien* was filmed in New York, using the same Filmways Studios facilities as the recently cancelled legal series *The Defenders* (q.v.). Being in close proximity with New York's theater district, the series made extensive use of top Broadway talent, including such soon-to-be stars as Faye Dunaway, Martin Sheen, Alan Alda, Tony Roberts and Charles Grodin, as well as established performers like Sheila MacRae, Nehemiah Persoff, Tammy Grimes, Buddy Hackett, Burgess Meredith, Angela Lansbury, Milton Berle and Rita Moreno — the latter two cast as a sad-sack burlesque comic and his faithless stripper spouse.

Reviews for *The Trials of O'Brien* ran the gamut from "The best of all the new offerings" (Anthony La Camera, the *Boston American*), to "The whole show left a bad taste in my mouth" (Terence O'Flaherty, the *San Francisco Chronicle*). Suffering from lukewarm ratings opposite *The Lawrence Welk Show* on Saturdays, the series moved to a Friday-night slot recently vacated by another legal show, *Slattery's People* (q.v.), where it was slaughtered by NBC's *The Man From UNCLE*. A few observers attributed

the show's demise to the supposedly "confusing" fast-cut techniques used in the courtroom scenes, but Peter Falk himself had a more pragmatic explanation: "I talked too much," he insisted in 1976 interview. "We started out at the bottom and stayed there." (Falk didn't even mention the series in his 2007 autobiography).

The penultimate episode of *The Trials of O'Brien*, the two-part "The Greatest Game," was filmed in color but networkcast in black and white. Later on, this episode was syndicated as a 97-minute TV movie, retitled *Too Many Thieves*.

The Trials of Rosie O'Neill

CBS: September 17, 1990–May 30, 1992. The Rosenzweig Company/MTM Productions. Executive Producers: Barney Rosenzweig, Beth Sullivan [Note: the final episode was dedicated to the late actor Ray Danton, who'd been in on the series' development process]. Cast: Sharon Gless (Fiona Rose "Rosie" O'Neill); Dorian Harewood (Hank Mitchell); Ron Rifkin (Ben Meyer); Georgann Johnson (Charlotte O'Neill); Lisa Banes (Doreen Morrison); Lisa Rieffel (Kin Ginty); Bridget Gless (Barbara Navis); Elaine Kagan (Carole Kravitz); Geoffrey Lower (Udell Correy III); Tony Perez (Pete Ramos); Al Pugliese (George Shaughnessy); Edward Asner (Walter Kovatch); Dayna Winston (Valerie Whittaker); Victor Bevine (Mason Pappas); Robert Wagner (Peter Donovan); Barney Rosenzweig (Psychiatrist).

Sharon Gless is nothing if not a survivor. Having entered show business as the very last contract starlet hired by Universal Studios, Gless has matured into a versatile, universally respected character actress, gaining whole new legions of TV fans with each passing year in such incisive roles as Debbie Novotny in *Queer as Folk* and Madeline Western in *Burn Notice*. But when she launched her first solo starring series *The Trials of Rosie O'Neill* back in 1990, Sharon Gless was best known for her lengthy costarring stint opposite Tyne Daly in the feminist cop drama *Cagney and Lacey* — and there were many jaundiced TV reviewers who weren't about to let her forget it.

In *Trials*, the actress was cast as Fiona "Rosie" O'Neill, a 43-year-old Beverly Hills attorney who'd just been divorced by her husband and law partner, leaving her with little more than a Mercedes and a batch of maxed-out credit cards. (Her ex not only got the family practice but also the family dog!) Jump-starting the liberal activism of her youth, Rosie signed on with the Los Angeles Public Defenders Office, where her clients were no longer the rich and famous but the poor and downtrodden (as well as a few unregenerate career criminals). Sharing her cramped new office was ghetto-bred public defender Hank Mitchell (Dorian Harewood), who regarded Rosie as a dilettante and was openly contemptuous of her. Also among Rosie's coworkers were her Orthodox-Jewish supervisor Ben Meyer (Ron Rifkin), receptionist Carole Kravitz (Elaine Kagan), office aide Barbara Navis (played by the star's real-life niece Bridget Gless) and fellow public defenders Udell (Geoffrey Lower), Mason (Victor Bevine) and Valerie (Dayna Winston). On the domestic front, Rosie had to put up with the constant kvetching of her mom Charlotte (Georgann Johnson), given to such highly judgmental and highly inaccurate laments as, "Fifth in your class, and you work for the Sanitation department!" Others in the heroine's private circle were her married sister Doreen (Lisa Banes), her ex-stepdaughter Kim (Lisa Rieffel), and, during the first season, her 25-year-old carpenter boyfriend Steve (Doug Wert).

Many of the initial reviews for *The Trials of Rosie O'Neill* began, "If you liked *Cagney and Lacey...*," harping on the fact that both series were produced by Sharon Gless' companion and later husband Barney Rosenzweig, who also played the psychiatrist in the therapy scenes that opened each hour-long episode. Comparing the two shows, John O'Connor of *The New York Times* wrote: "There's a new line-up of characters. And the setting is Los Angeles instead of New York City. But this is familiar Rosenzweig territory, the liberal terrain of fighting for the underdog and propounding the correct causes, especially feminism." Though O'-Connor liked *Rosie O'Neill*, others tended to downgrade the series at the expense of *Cagney and Lacey*, suggesting that neither Gless nor Rosenzweig were confident enough in the new property to avoid deliberately invoking memories of the earlier show. This was especially true

when Meg Foster began showing up in the recurring role of DA Deb Grant; no one had forgotten that Gless had replaced Foster as Detective Chris Cagney on *Cagney and Lacey* because the latter actress was perceived as being "unfeminine," and this fact was used as a club with which to beat *Rosie O'Neill* to a pulp. Similarly, certain cynical observers felt that Tyne Daly's later guest appearance on *Rosie* merely served to remind viewers how much better the older show had been. (Ironically, no one derided the later *Judging Amy* [q.v.] when *that* series' costar Tyne Daly was reunited in one episode with Sharon Gless.)

Fortunately, there also were plenty of critics who were able to get past *Cagney and Lacey* and enjoy *Rosie O'Neill* on its own terms; certainly, there was a sufficient number of journalists to honor Sharon Gless with a Golden Globe award for her performance. These yea-sayers applauded the series for raising controversial issues in an entertaining and compelling fashion, beginning with the opening episode in which Rosie came to the aid of an unwed mother accused of throwing her newborn baby in a dumpster because her minister father had declared abortion a mortal sin: the story ended with Rosie defending the girl from a charge of involuntary manslaughter by delivering an impassioned summation about our "disposable society."

Even more volatile was the second episode, wherein Rosie was forced to defend an admitted serial rapist because the police had falsified a warrant against him. This particular episode was used as a "test case" by producer Rosenzweig to see how far he could push the profanity envelope. *Rosie O'Neill* had already raised a ruckus with its debut episode, in which the heroine's opening line to her psychiatrist was, "I'm thinking about maybe having my tits done"—the first mainstream-network usage of one of the seven words George Carlin couldn't say on television. Now the CBS standards-and-practices division was on Yellow Alert for any future violations of good taste, and they bore down hard on a scene in the second episode wherein Rosie was taken to task for her initial reluctance to provide her rapist client with

Equal Justice Under Law. "This isn't Beverly Hills," chided coworker Hank Mitchell. "You don't get to pick your clients." At that, Rosie snapped, shouting, "You elitist son of a bitch. You're worse than any snob in my mother's country club." Striking back, Mitchell exclaimed, "I couldn't get into your mother's country club, so kiss my ass!"

Thus in this one scene, two "no-no" words were uttered within the space of ten seconds — and surprise or surprises, Rosenzweig got away with it. When CBS raised objections, the producer replied that the point of the scene was that everyone deserved Due Process, even a scumbag rapist — a brilliant strategic maneuver, in that the network's main issue with episode was not the language, but the fact that the rapist had gone free. Once *this* hurdle was cleared, Rosenzweig was able to score points on the profanity issue, using one of the oldest trade-off ploys in Hollywood: "I will give you every expletive in the script — all the bitches, damns and hells — in exchange for that one scene." As it turned out, CBS' overcautiousness was unwarranted: not one single, solitary viewer complained about the scene in question.

Though it opened to good ratings, *Rosie O'Neill* progressively lost viewers as the first season wore on, barely securing a renewal for a second year. To punch up the numbers, Rosenzweig not only toned down what *Entertainment Weekly*'s Ken Tucker described as "Rosie's post-divorce trauma," but also added Edward Asner to the cast as ex-cop Walter Kovatch, who joined a Public Defenders Office investigative team that already numbered among its members Pete Ramos (Tony Perez) and George Shaughnessy (Al Pugliese). Since Rosie and Hank Mitchell had long since buried the hatchet, Walter served as the heroine's newest adversary, a grouchy Conservative with an inbred dislike of bleeding-heart public defenders. (True to his Liberal convictions — not to mention Ed Asner's — Rosenzweig went out of his way to make Walter as disreputable-looking as possible: in his first appearance, he was arrested after getting drunk and trying to break into his own apartment after being evicted.) The arrival of Asner moved Ken Tucker, otherwise a fan of the series, to fall into the "Quo *Cagney and Lacey*?" trap, describing Walter Kovatch as "Tyne Daly with a paunch."

In an eleventh-hour effort to save the series from cancellation, a tantalizing story arc was introduced late in the second season, as Rosie fell in love with newspaper executive Pete Donovan — a role played by Robert Wagner, formerly the star of the mid–1970s adventure series *Switch*, in which Sharon Gless had gotten her first big break in the continuing role of Wagner's "Gal Friday." The romance between Rosie and Donovan came to an ignominious halt when she discovered that he hadn't divorced his first wife; not long afterward, *The Trials of Rosie O'Neill* was likewise terminated. One imagines Sharon Gless took solace in the fact that *Sister Kate*, the sitcom that she'd turned down in order to star as Rosie O'Neill, didn't even survive its first season.

TrialWatch

NBC: Jan. 28–July 26, 1991. Executive Producers: Donald Kushner, Peter Locke, Adam Shapiro and David R. Sams. Cast: Robb Weller, Lisa Specht (Hosts).

In an effort to diversify its daytime lineup in early 1991, NBC returned to the courtroom-series form that had once served ABC and CBS so well with *Day in Court* and *Verdict is Yours* (both q.v.), and was enjoying a syndication renaissance with *The People's Court* (q.v.). Following in the footsteps of cable's Court TV, the daily, half hour *TrialWatch* featured "Real People, Real Trials" — albeit in short-excerpt form. Covering everything from murder cases to custody battles to job-discrimination complaints, *TrialWatch* also featured interviews with noted legal experts and provided the layman with what the network described as "current legal developments and information." Sharing the hosting duties were *Entertainment Tonight* alumnus Robb Weller and TV journalist Lisa Specht.

With the newest version of *To Tell the Truth* as its lead-in, *TrialWatch* remained on the daily NBC manifest for six months.

Verdict

CBS: June 21–Sept. 5, 1991. Produced by the CBS News Division.

A production of CBS' news division, the weekly *Verdict* showcased videotape footage from actual trials (generally but not always lurid in nature), edited to 22 minutes per episode for public consumption. In addition to the basic trial coverage, the series featured quickie summaries explaining the events leading up to the proceedings and candid glimpses of the strategies used by the attorneys. After the verdict was delivered, the viewer was treated to post-trial comments from the attorneys, taped "on the spot" in the courtroom. Each episode was emceed by a different CBS correspondent, who provided background information on the proceedings and bridged the continuity gaps. Among the contributing correspondents given airtime on the series was Meredith Vieira, soon to be elevated to early-morning stardom on *The CBS Morning News* and NBC's *The Today Show*—not to mention her controversial nine-year stint on the ABC chatfest *The View*.

The Verdict Is Yours

CBS: daytime version Sept. 2, 1957–Sept. 28, 1962; primetime version, July 3–Sept. 25, 1958. Produced by Eugene Burr. Cast: Jim McKay (Himself); Bill Stout (Himself); Jack Whitaker (Himself).

Network television's first daily, half-hour courtroom series, *The Verdict Is Yours* was developed by former State Department attorney Selig J. Silverman, whose other credits included the thematically similar ABC series *Day in Court* (q.v.). Inasmuch as TV cameras were banned from most municipal courtrooms in 1957, Silverman was convinced that viewers would be fascinated by a fictional courtroom show that duplicated the spontaneity and unpredictability of actual trials. Thus, *The Verdict Is Yours*, telecast live throughout its five-year run, was completely unscripted and (except for a single camera run-through) unrehearsed—and not even the series' producer Eugene Burr ever knew what the verdict would be.

Staged on a reasonably authentic courtroom set, each of the series' "trials" unfolded in serial-like fashion, with a new case every three to seven days. After coming up with a "case outline" that often ran as long as 65 pages, Selig Silverman took each cast member aside and privately told him the "truth" about his role in the proceedings—and of course, what was "true" for one participant was "false" for another. The clients would then meet with their lawyers to plan their cases.

Though the litigants were professional actors, chosen by producer Burr for their ad-libbing ability, the lawyers were professional attorneys who were expected to prepare their own cross-examinations, and the judge was similarly the genuine article. The jurors were selected from studio visitors, who were required to sit through each episode from start to finish, finally reaching their verdict by majority vote (hence the series' title). The show's director, who was as much in the dark about the outcome of the case as everyone else, did little more than cue the appropriate camera angles. Whenever the time came for a commercial break, the floor manager held up a sign saying "SUSPEND" just outside camera range, whereupon one of the lawyers launched into a lengthy objection, while the Court Reporter—actually the series' announcer—took advantage of this long-winded tirade to accommodate a word from the sponsor (a similar strategy was used on the syndicated *Divorce Court* [q.v.])

In its heyday *The Verdict Is Yours* was seen as both a daytime and weekly Prime Time series on CBS, and was extensively covered by the mainstream press. Both *Time* magazine and *TV Guide* were especially fascinated by a 1958 story arc wherein actress Betsy Von Furstenberg played a woman on trial for shooting her husband, whom she claimed to have mistaken for a prowler. New York City assemblyman Daniel Kelly played the prosecuting attorney, building up a case that was so damaging that he actually reduced Von Furstenberg to tears. Out of frustration, the actress ad-libbed, "It may look very black for us, but wait until I take the stand!"—leading to a terrifyingly realistic shout fest which threatened to degenerate into physical violence. Lawyer Richard Tilden, cast as the defense attorney, later admitted, "I didn't sleep well nights, worrying about that case." Ultimately, the jury voted "not guilty"—whereupon director Byron Paul lost a 25-cent bet with one of the technicians.

With only one or two kinescopes of *The Verdict Is Yours* available for reappraisal, the series has largely been forgotten, save for the fact that it represented the first nationwide TV exposure for sportscaster Jim McKay, who went on to fame and prestige as host of ABC's *Wide World of Sports* and the network's subsequent coverage of the 1972 Munich Olympics. A pioneer of Baltimore TV broadcasting, McKay had moved to New York City in hopes of becoming a CBS news anchor, but the network consigned him to its entertainment division instead. "I was on the fringes of the news department, and I kept trying to push that way, but it just wasn't happening," he recalled in a 1984 *Sports Illustrated* interview. Being trapped in a "velvet-lined rut" would ultimately lead McKay to a nervous breakdown, but before this happened he was tapped to appear as the *sotto voce* announcer on *The Verdict Is Yours*. Wearing earphones and delivering rapid-fire commentary in an unctuous whisper, McKay seemed to regard every episode as if it were a sports event—which, in a sense, it was.

Jim McKay left the series when the producers moved production from New York to Hollywood, though he stuck around long enough to break in his successor Bill Stout, better known to contemporary viewers for his lengthy run as news anchor on CBS' Los Angeles affiliate. Stout took over the announcing chores in July of 1960, and was himself replaced in June 1962 by Jack Whitaker, who like Jim McKay would achieve his greatest fame as a network sportscaster. Whitaker was on hand for the final broadcast of *The Verdict Is Yours* in September of 1962, at which point the series' timeslot was taken over by reruns of *The Real McCoys*—an ironic turn of events, in that Jim McKay had landed his *The Verdict Is Yours* gig on the strength of his local New York City variety series *The Real McKay*.

We the Jury

Syndicated: Sept. 9, 2002–Sept. 2003. Chambers Productions/Sand in My Pants Inc./Telco Productions/TVi Media. Executive Producers: Bill Nagy, Alex Paen, Bill Grundfest. Cast: Chris Kitchel (Judge); James Flagg (Court Officer).

This daily, half-hour syndicated series promised a "new spin" on the traditional courtroom drama. Instead of concentrating on the judge or the litigants, *We the Jury* lived up to its title by focusing on the jurors. In fact, it was one of the first TV series to go inside the jury deliberation room, where eight Good and True Men and Women argued over the finer points of the case at hand. But though the jurors were "real people," the court cases were scripted and the litigants played by actors. The jury members were allowed to question the litigants directly, while the folks at home were given the opportunity to vote for their favorite juror online, with daily results posted on the series' website.

Acknowledging the public's interest in news stories about "corporate fraud, religious scandals, and celebrity trials," the court cases were calculatedly sensational in nature. A perusal of sample episode titles should give the reader an idea of what *We the Jury* had to offer: "Topless Mayor," "Gender Bender," "Chimps are People Too" and "Suing God."

We the Jury was produced at the studios of Chambers Communications in Eugene, Oregon. Acting as judge was Portland trial lawyer Chris Kitchel, who was billed as "kinder" and "gentler" than your average TV jurist (does the name "Judge Judy" ring a bell?). Oddly enough, the series has to this date never been mentioned in Kitchel's online resume, but *We the Jury*'s production team—including former *Mad About You* scriptwriter Bill Grundfest—have not evinced a similar modesty in *their* Internet bios.

Willy

CBS: Sept. 18, 1954–July 7, 1955. Desilu Productions. Produced by William Spier. Cast: June Havoc (Willa "Willy" Dodger); Danny Richards Jr. (Franklin Sanders); Mary Treen (Emily Dodger); Lloyd Corrigan (Papa Dodger); Whitfield Connor (Charlie

Bush); Hal Peary (Perry Bannister); Sterling Holloway (Harvey Evelyn).

When a few well-meaning reviewers labeled the 1975's *Kate McShane* (q.v.) as the first Prime Time series to feature a female lawyer as protagonist, they were only half-right. While *Kate McShane* was the first *dramatic* series to do this, it was preceded by the 1954 CBS sitcom *Willy*.

Produced by William Spier, whose radio credits included the classic anthology *Suspense*, *Willy* was a vehicle for Spier's then wife June Havoc, who up to that point was best known for her non-comic roles in such films as *Gentleman's Agreement* and *The Iron Curtain*. The actress was cast as Willa "Willy" Dodger, a recent law school graduate who opened her first practice in her home town of Renfrew, New Hampshire. Lloyd Corrigan costarred as Willy's supportive father, while Mary Treen (aka "Cousin Tilly" in the imperishable *It's a Wonderful Life*) played the heroine's aunt Emily. Whitfield Connor was seen as Willy's veterinarian boyfriend Charlie Bush, and making occasional appearances as town dogcatcher Homer was an up-and-coming young actor named Aaron Spelling (wonder whatever became of him?).

Though the presence of a lady attorney in a 1950s comedy series might suggest that *Willy* struck an early blow for feminism, this wasn't entirely the case. The original *TV Guide* listings dripped with chauvinistic condescension, labeling the heroine as "New Hampshire's adorable female lawyer" and "feminine legal eagle"; and hardly an episode went by without Charlie Bush urging Willy to give up her practice and become his wife, or showing a particularly lascivious male client ogling the poor girl's legs. Also, the point was made in the first episode that the Renfrew locals weren't entirely comfortable with a beauteous barrister in their midst: after months and months of inactivity, Willy finally landed her first case when a little kid showed up in her office begging her to spring his dog from the pound. A few other episodes at least acknowledged the possibility that Willy knew her business and was as capable as any man in her profession, if not more

so: she proved that a kleptomaniac client committed a crime without criminal intent, purged the local statute books of such obsolete laws as forcing motorists to be preceded by a person waving a red flag, battled a bean-counting building inspector, and defended herself against a charge of assault and battery (though this last case arose from a "typically" female fracas in a French hat shop).

Halfway through the series' run, the folks at Desilu Productions decided that the "small-town lawyer" premise was going nowhere, so Willy moved from Renfrew to New York City, there to take a job as counsel to a theatrical organization run by Perry Bannister, played by Hal Peary of *The Great Gildersleeve* fame. Also added to the cast at this juncture was the ubiquitous Sterling Holloway as Harvey Evelyn, manager of a rundown stock company. Being in close proximity with showbiz types enabled Willy to occasionally kick up her heels and perform a snappy song-and-dance number; to explain the heroine's hitherto untapped reserve of talent, it was revealed that Willy had been a child performer in vaudeville — hardly a stretch for June Havoc, who, as anyone who has ever seen the musical *Gypsy* knows full well, was once a headliner in a traveling kiddie troupe featuring her equally famous sister Gypsy Rose Lee.

Though *Willy* lasted but a single season on CBS, the series was widely syndicated on local stations well into the early 1960s. Even those baby boomers who can't remember a single plotline will readily recall the series' bouncy theme music, adapted from Hugo Aflen's "Swedish Rhapsody No. 1, Opus 19" (aka "Midsommarvaka").

The Witness

CBS: Sept. 29, 1960–Jan. 26, 1961. Talent Associates. Executive Producer: David Susskind. Creator/Producer: Irve Tunick. Cast: Verne Collett (Himself, court reporter); William Griffis (Court Clerk).

If sports fans can have their Fantasy Baseball League, it is only fair that legal-show aficionados be allowed to have a Fantasy Courtroom Series. That's the best way to describe the ambitious CBS dramatic anthology *The*

Witness, which though it featured actual historical figures in a setting resembling an authentic Congressional investigation, had about as much relation to reality as *Alice in Wonderland*.

Produced on videotape by no less than David Susskind, *The Witness* was careful to specify at the beginning of each hour-long episode that what we were about to see was "not a trial," but a simulated hearing before a select committee "representing the morality and conscience of today, with the power to call before it the famous, the infamous, the evil and controversial figures of yesterday." The drama that followed was staged exactly like a televised hearing, replicating the "organized chaos" of the recent Kefauver and McClellan Committees with a full complement of reporters and crowd extras, random flash bulbs, rapping gavels, objections, points of order etc. Ensconced in a soundproof booth, court reporter Verne Collett would give a meticulously researched thumbnail biography of the Defendant of the Week at the beginning of each episode, occasionally pop up in mid-proceedings to break for a commercial, and reveal the ultimate fate of the Defendant at episode's end. In some episodes, the Committee members were played by actual members of the New York Bar association; in others, they were portrayed by such character actors as Paul McGrath, William Smithers, Charles Haydon and Frank Milan, using their real names.

Unlike the similar "on-the-spot" historical anthology *You Are There*, in which the actors were garbed in the costumes of the era wherein the story took place, no attempt was made to create a period feel on *The Witness*. Though defendants and witnesses were drawn from the ranks of the "celebrity offenders" of the years 1910 through 1935, all of them were dressed in the style of the early 1960s. To avoid possible defamation-of-character lawsuits, most of the defendants were safely dead by the time they stood before Mr. Susskind's faux Committee. In the now-famous debut episode, Telly Savalas achieved overnight stardom in the role of vice lord Charles "Lucky" Luciano, playing his part with such icy conviction that for years afterward, Savalas would be stopped in the street by former mobsters who wanted to congratulate him for a job well done! Later episodes featured Joan Blondell as Ma Barker, Warren Stevens as John Dillinger, and Peter Falk as Abe "Kid Twist" Reles, the same role he essayed in the 1961 theatrical feature *Murder Inc.* Also making ersatz guest appearances were the likes of Arnold Rothstein, Al Capone, Bugsy Siegel, Roger Touhy, Dutch Schultz and Mayor Jimmy Walker, most of whom had been summoned before the committee not so much to explain and defend their activities as to justify their very existence. (Even back then, David Susskind had a predilection for playing God.)

For all its efforts to sustain a feeling of spontaneous verisimiltude in the courtroom — right down to the casual deployment of "Damn" and "Hell," epithets rarely heard on 1960s network television — *The Witness* could not help but come off as corny and contrived. The defendants were prone to grandstand speeches and operatic overacting ("I'm a box-office attraction!" boasted John Dillinger with a movie-star grin), the witnesses for the prosecution were mercilessly badgered and insulted by the Committee members (Legs Diamond had a particularly rough time of it in the Arnold Rothstein episode), all action would stop in its tracks for heated confrontations between witnesses and defendants in which numerous old scores were settled (Dillinger vs. "The Lady in Red" was a riot!), and subtle-as-a-meat-cleaver dramatic irony was heaped upon the proceedings at every possible opportunity, usually just before the commercial break.

In retrospect, *The Witness'* most controversial episode was the series finale, in which Biff McGuire portrayed disgraced baseball player "Shoeless Joe" Jackson. This episode was an outgrowth of a planned *DuPont Show of the Month* dramatization of the 1919 "Black Sox" scandal, in which several members of Chicago White Sox accepted bribes to throw the World Series. Scriptwriter Eliot Asinof had spent two weeks laboring over the proposed 1960 special when producer David Susskind was ordered to kill the project by dictatorial baseball commissioner Ford Frick, who felt that dredging up the scandal was not in the best interests of the

Game. Though Asinof would ultimately parlay his fascination with the Black Sox into his 1963 bestseller *Eight Men Out,* Susskind used the existing material as the basis for a *Witness* episode, which though credited to Asinof bore little relation to the author's original concept — and even less relation to the known facts. The conventional wisdom is that Shoeless Joe was an innocent dupe who refused to accept the bribe offered him and even begged to be benched during the Series so there would be no doubt of his honesty; it is also the general consensus that the illiterate Jackson was coerced into signing a confession in exchange for immunity from prosecution. But if we were to believe *The Witness,* Jackson was an active participant and perhaps even one of the masterminds of the conspiracy, this "evidence" provided by a fictional hotel chambermaid who claimed to have been present at the first meeting between the ballplayers and the gamblers who arranged the fix. (Since many key players in the scandal were still alive, Susskind was limited to characterizations of the long-dead Jackson, White Sox manager Kid Gleason, team owner Charles Comiskey and gambler Billy Maharg; other characters were either composites or conjured from thin air.) In the absurd climax, Shoeless Joe not only made a full confession, but did so twice — the second time to a sniveling kid who somehow managed to appear in the courtroom long enough to deliver the obligatory "Say it ain't so, Joe."

This particular episode was mercilessly panned by virtually every major sports publication of the era, bringing *The Witness* to a most ignominious close. One week later, the high-budgeted but low-rated series was replaced by a filmed western, *Gunslinger.* (In a postscript that would have made a fascinating *Witness* episode in itself, Eliot Asinof turned down David Susskind's request to produce a filmed version of *Eight Men Out*— whereupon Susskind sued for $1.2 million, claiming that the basic story material still belonged to him!)

Work with Me

CBS: Sept. 29–Oct. 20, 1999. Calm Down Productions/Nat's Eye Productions/Stephen Engel Productions/Universal Television. Executive Producer: Stephen Engel. Co-executive Producers, Nancy Travis, Kevin Pollak, Lucy Webb. Cast: Nancy Travis (Julie Better); Kevin Pollak (Jordan Better); Ethan Embry (Sebastian); Emily Rutherfurd (Stacy); Bray Poor (A.J.)

Describing the half-hour CBS legal sitcom *Work With Me* in the pages of *TV Guide,* executive producer Stephen Engel said, "I don't view this as a show about lawyers, but a show about married couples who work together. They could easily be butchers." Sure. We can envision Kevin Pollak in a bloody apron and Nancy Travis wielding a carving knife. Come to think of it, that would have been a heck of a lot funnier than *Work With Me.*

Pollak was cast as attorney Jordan Better, who after being denied a promotion at a high-profile Wall Street firm (seems he had the temerity to beat his boss at golf) went into partnership with his lawyer wife Julie (Travis), who maintained an "alternative" legal office specializing in underdog cases. Julie's headquarters were so tiny that Jordan's desk was in the lunchroom, sharing space with several unopened boxes of junk food and a malfunctioning microwave. The couple's "voice mail" consisted of each lawyer yelling at the other through closed doors (a harkback, perhaps, to the "Buzz me, Miss Blue" running gag on radio's *Amos 'N' Andy*)—yet somehow they were able to afford state-of-the-art computers, allowing the inclusion of another regular, Jordan's techno-geek assistant Sebastian (Ethan Embry). Also in the cast was Julie's simpering assistant Stacy (Emily Rutherfurd), who for idiotic reasons of her own didn't want anyone to know that she was dating Sebastian—who in turn was forced to endure humiliation after humiliation to keep Stacy's secret. The building where the Jordans worked was adjacent to the office of a masseuse (aka "relaxation therapist") named A.J. (Bray Poor), who periodically climbed through the couple's window just to say hello (yet another harkback to a better comedy property, TV's *He and She,* in which Paula Prentiss and Dick Benjamin's apartment was similarly "invaded" by their fireman neighbor Kenneth Mars).

272   The Wright Verdicts

Although the family partnership was Jordan's idea, he couldn't resign himself to the fact that most of Julie's clients were men, some of them quite handsome. Conversely, Jordan gave Julie plenty to worry about when he went gaga over such female clients as guest star Lynda Carter, whose appearance triggered a lengthy discussion about the couple's sexual fantasies. Matter of fact, virtually all Jordan and Julie ever talked about was sex, making one wonder how they managed to get any work done, or even how they managed to pass their bar exams. In an interview with CNN Entertainment News, Kevin Pollak acknowledged the series' preoccupation with what Preston Sturges used to call "Topic A," explaining that this was a counteroffensive against *Work With Me*'s racy Fox Network competition *Beverly Hills 90210*. "The first thought was to have a lot of nudity and sex," noted Pollak, admirably articulate despite the tongue in his cheek, "and then we realized after I'd been cast, no one would watch. So we did an about-face, as it were, and it'll just be Nancy who's nude and talking about sex."

Despite the superb comic timing of Kevin Pollak and Nancy Travis, there was no chemistry at all between the stars — one of the series' many demerits, which included arbitrary gay jokes, tiresome contradictory flashbacks (a character says that one thing happened, then we see that the exact opposite occurred) and situations that were ridiculous for their own sake (on the night they were engaged, Julie accidentally swallowed the wedding ring and Jordan ended up proposing in the ER, whereupon the couple's picture was taken by a man with a lawn dart in his head). *Work With Me* was off the CBS Wednesday-night schedule before most people knew it had even been on, leaving unresolved a most baffling mystery: how could A.J. the masseuse be located in an office directly across from the Better's lunchroom window, when the exterior establishing shots clearly showed that the couple's office faced an open intersection, with no other building in sight?

The Wright Verdicts

CBS: March 31–June 11, 1995. Wolf Films/Universal TV Productions. Executive Producer: Dick Wolf.

Cast: Tom Conti (Charles Wright); Margaret Colin (Sandy Hamor); Aida Turturro (Lydia).

Produced by *Law & Order*'s Dick Wolf and filmed in New York, the hour-long drama *The Wright Verdicts* was a vehicle for Tony Award–winning Scottish actor Tom Conti, cast as British barrister Charles Wright. Having relocated to the Big Apple in 1980 or thereabouts, Wright was firmly established as a top-of-the-ladder criminal attorney — and also, rather surprisingly, as an expert special prosecutor for the DA's office. From all outward appearances, Wright seemed vague, stumble-tongued and a bit muddle-headed, but woe betide the opposing counsel who mistook his absentmindedness as a sign of weakness or ineptitude. And despite his habit of garbling and swallowing his words, Wright's histrionic cross-examinations and Shakespearian summations suggested that, at the very least, he had passed his bar exams at the Royal Academy of Dramatic Art. Featured in the cast were Margaret Colin (formerly the star of her own legal series, *Foley Square* [q.v.]) as Wright's investigator Sandy Hamor, an ex-cop and chronic gambler; and Aida Turturro (cousin of John and Nicholas Turturro) as the protagonist's levelheaded assistant Lydia.

Describing *The Wright Verdicts*, Dick Wolf insisted, "It's not as serious as *Law & Order*, but not as light as *Matlock*." Sadly, it was also not as good as either of those shows. Clichés abounded, right down to a surly judge warning Wright that he was on "a very short leash" and the last-minute confessional breakdown in the witness box. And whenever the series tackled a headline-driven storyline (the O.J. Simpson case showed up in thinly disguised form), it only served to remind us how much better Wolf pulled off this sort of thing on *Law & Order*.

In his review of *The Wright Verdicts* in *Entertainment Weekly*, Ken Tucker wrote: "Executive producer Dick Wolf has cannily combined two genres — *Murder, She Wrote*'s warm coziness and his own *Law & Order*'s cold, complex cases — and come up with a lukewarm show that's nonetheless pretty irresistible." Knowing as we now do that the series barely lasted two and a half months, we can safely observe that Ken Tucker was only half-wright — er, half-right.

The Young Lawyers

ABC: Sept. 21, 1970–May 5, 1971. Crane Productions/Paramount Television. Executive Producers: Herbert Hirschman and Matthew Rapf. Cast: Lee J. Cobb (David Barrett); Zalman King (Aaron Silverman); Judy Pace (Pat Walters); Philip Clark (Chris Blake).

Like CBS' *The Storefront Lawyers* (q.v.), which also debuted in 1970, ABC's *The Young Lawyers* was a desperate bid for "relevance" and a balm to the Youth Culture that predominated the entertainment scene of the period — and typically, it was concocted by Hollywood insiders who were not only well past the age of consent, but were already halfway down the road to dotage. According to *Time* magazine, the main distinction between the two new legal series was that, unlike the characters on *Storefront Lawyers*, the protagonists of *The Young Lawyers* "are at least half interesting."

The ABC series was set at the Neighborhood Law Office in Boston, a nonprofit legal-aid firm headed by crusty old private practitioner David Barrett, played by Lee J. Cobb as if he was worried that the microphones weren't picking up his every word. It was explained by the series' promotional packet that in the Commonwealth of Massachusetts, law students were allowed to go right into court and "take a case all the way, win or lose." Accordingly, Barrett's youthful staff consisted of three Bercol University law students of mixed backgrounds and ethnic origins. The resident idealist was Aaron Silverman (Zalman King), Barrett's star pupil, who never combed his hair and seemed forever poised to lead the masses to the barricades; the obligatory African American — also female, so as not to miss any of the accepted "with-it" stereotypes — was Pat Walters (Judy Pace), who for all her streetwise patois possessed a brilliantly analytical legal mind; and the standard well-meaning WASP, who joined the series halfway through its run, was Chris Blake (Philip Clark), whose wealthy upbringing was held in amused disdain by his oh-so-hip

coworkers. (In the series' 90-minute pilot episode, telecast October 28, 1969, Zalman King was already in place as Aaron Silverman, but Jason Evers was seen as the senior partner/father figure character, while white actress Anjanette Comer played Aaron's female coworker.)

Never did the Young Lawyers accept a case that didn't provide the opportunity to attack and debate a vital issue of the day, beginning with the debut episode in which they defended a long-haired holistic doctor accused of malpractice by the hidebound Medical Establishment. Later episodes dealt with a pot-smoking teenager turned in by his unforgiving father, an unwed couple suing for the right to keep their child, and a disturbed Vietnam veteran (was there any other kind on TV in the early 1970s?) accused of committing murder during a blackout. In keeping with the series' strenuous solicitation of the Under-30 set, the guest casts included such twentysomethings as Richard Dreyfuss, Peter Strauss and Scott Glenn. The producers also truckled to an even younger demographic group by licensing a *Young Lawyers* comic book — making this the second occasion since *The Virginian* that Lee J. Cobb appeared in cartoon form, unless there's a *Death of a Salesman* comic book that we don't know about.

One of the handful of journalistic pundits who had anything truly positive to say about the short-lived *The Young Lawyers* was Harlan Ellison, who saw "greatness" in series star Zalman King. As it turned out, King went on to scale the showbiz heights, not as an actor but as the producer of such critically acclaimed soft-core productions as the theatrical feature *9½ Weeks* and the adult cable-TV anthology *Red Shoe Diaries*.

Your Witness

ABC: Sept. 19, 1949–September 26, 1950.

One of the first network series to be nominated for an Emmy award, *Your Witness* originated as a local program from the Chicago studios of WENR. The weekly, half-hour series featured reenactments of actual trials and hearings, with professional actors impersonating the litigants, jurors and judges. Some sources have listed veteran Hollywood leading man Edmund Lowe as the host, thought it is unlikely that he would have relocated to Chicago for this purpose; Lowe was probably just one of the series' several guest stars, which included such familiar character players as Vivi Janiss, A. Cameron Grant and Robert Boon.

When it was picked up for network play on the fledgling ABC hookup, *Your Witness* was telecast opposite another series of courtroom re-enactments, DuMont's *Famous Jury Trials*— meaning that few viewers ever saw either program, since ABC and DuMont had been effectively crowded out of the marketplace by the twin titans NBC and CBS.

Bibliography

Books

Bergman, Paul, and Michael Asimow. *Reel Justice: The Courtroom Goes to the Movies.* Rev. ed. Kansas City, MO: Andrews McMeel, 2006.

Brooks, Tim, and Earl Marsh. *The Complete Directory to Prime Time Network and Cable TV Shows, 1946–Present.* 8th ed. New York: Ballantine, 2003.

Bruskin, David N. *The White Brothers: Jack, Jules & Sam White.* Metuchen, NJ: Scarecrow Press, 1990.

Cotter, Bill. *The Wonderful World of Disney Television: A Complete History.* New York: Hyperion, 1997.

Crook, Tim. *International Radio Journalism: History, Theory and Practice.* London, New York: Routledge, 1998.

Dooley, Roger. *From Scarface to Scarlett: American Films in the 1930s.* New York: Harcourt Brace Jovanovich, 1981.

Dunning, John. *On the Air: The Encyclopedia of Old-Time Radio.* New York: Oxford University Press, 1998.

Ehrlich, Jacob W. *A Life in My Hands: An Autobiography.* New York: Putnam, 1965.

Erickson, Hal. *Syndicated Television: The First Forty Years.* Jefferson, NC: McFarland, 1989.

_____. *Television Cartoon Shows: An Illustrated Encyclopedia, 1949 through 2003.* 2d ed. Jefferson, NC: McFarland, 2005.

Fireman, Judy. *TV Book: The Ultimate Television Book.* New York: Workman Publishing, 1977.

Gazzara, Ben. *In the Moment: My Life As an Actor.* New York: Carroll & Graft, 2004.

Goldberg, Lee. *Unsold Television Pilots: 1955 through 1989.* Jefferson, NC: McFarland, 1990.

Harmon, Jim. *The Great Radio Heroes.* New York: Ace Books, 1967.

Hatchett, Glenda. *Say What You Mean and Mean What You Say! 7 Simple Strategies to Help Our Children Along the Path to Purpose and Possibility.* New York: Morrow, 2003.

Hayde, Michael J. *My Name's Friday: The Unauthorized but True Story of Dragnet and the Films of Jack Webb.* Nashville: Cumberland House, 2001.

Kelleher, Brian, and Diana Merrill. *The Perry Mason TV Show Book: The Complete Story of America's Favorite Television Lawyer.* New York: St. Martin's Press, 1987.

Kisseloff, Jack. *The Box: An Oral History of Television, 1920–1961.* New York: Viking, 1995.

Lane, Mills, with Jedwin Smith. *Let's Get It On: Tough Talk from Boxing's Top Ref and Nevada's Most Outspoken Judge.* New York: Crown Publishers, 1998.

Larka, Robert. *Television's Private Eye: an Examination of Twenty Years Programming of a Particular Genre: 1949 to 1969.* New York: Arno, 1979.

Levin, Harvey; from cases adjudicated by Joseph A. Wapner. *The People's Court: How to Tell It to the Judge.* New York: Quill, 1985.

Levine, Josh. *David E. Kelley: The Man Behind Ally McBeal.* Toronto: ECW Press, 1999.

Mathis, Greg, with Blair S. Walker. *Inner City Miracle.* New York: Ballantine, 2002.

McKay, Jim. *The Real McKay: My Wide World of Sports.* New York: Dutton, 1998.

McNeil, Alex. *Total Television.* 4th ed. New York: Penguin, 1995.

Metz, Robert. *CBS: Reflections in a Bloodshot Eye.* Chicago: Playboy Press, 1975.

Miller, Arthur R. *Miller's Court.* Boston: Houghton Mifflin, 1982.

Mills, John. *Up in the Clouds, Gentlemen Please.* New Haven, CT: Tickner and Fields, 1981.

Museum of Television History. *Worlds Without End: The Art and History of the Soap Opera.* New York: Harry N. Abrams Inc., 1998.

O'Brien, Pat. *The Wind at My Back.* New York: Doubleday, 1964.

Osborn, John Jay, Jr. *The Paper Chase.* New York: Popular Library, 1970.

Rapping, Elayne. *Law and Justice as Seen on TV.* New York: New York University Press, 2003.

Repperto, Thomas A. *American Mafia: A History of Its Rise to Power.* New York: H. Holt, 2004.

Sheindlin, Judy, with Josh Getlin. *Don't Pee on My Leg and Tell Me It's Raining: America's Toughest Family Court Judge Speaks Out.* New York: HarperCollins, 1996.

Shirley, Glenn. *Temple Houston: Lawyer with a Gun.* Norman: University of Oklahoma Press, 1980.

Shulman, Arthur, and Roger Youman. *How Sweet It Was!* New York: Bonanza, 1966.

Thomey, Ted. *The Glorious Decade.* New York: Ace, 1970.

Wapner, Joseph A. *A View from the Bench.* New York: Simon and Schuster, 1987.

Wilson, Steve, and Joe Florenski. *Center Square: The Paul Lynde Story.* New York: Advocate Books, 2005.

Wolper, David L., with David Fisher. *Producer: A Memoir.* New York: Scribner, 2003.

Articles (Print and Online)

Abernathy, Michael. "An Inch Every Ten Years." *Pop Matters,* Fall 2000. http://www.popmatters.com/tv/reviews/j/judging-amy.html.

Berman, Marc. "The Scoop: Lucky Rob Lowe." *MediaWeekl,* July 25, 2003. http://www.mediaweek.com/mw/esearch/article_display.jsp?vnu_content_id=1942490.

Carney, Two Finger. "The Eliot Asinof Factor." *Notes from the Shadows of Cooperstown: Observations from Outside the Lines* website, no. 276, November 22, 2002. http://www.baseball1.com/carney/index.php?storyid=190.

"Cicely Tyson and Melissa Gilbert Star in New TV Drama Sweet Justice." *Jet,* September 12, 1994.

Flood, Mary. "Cop-and-lawyer shows stretch real-life truth." *Houston Chronicle,* June 18, 2007.

Frolich, Robert. "Lake Tahoe's Scream Siren." *Moonshine Ink* website, October 11, 2007. http://moonshineink.com/archives.php/40/474.

Gunter, Mark. "The Little Judge Who Kicked Oprah's Butt: Daytime Television's Hottest Property." *Fortune,* May 10, 1999.

Haltom, Bill. "I'm Not a Judge, but I'll Play One on TV!" *Tennessee Law Journal,* December 1999.

Hayes, Cassandra. "The Verdict: Jones & Jury (Television Personality Star Jones)." *Black Enterprise,* December 1, 1994.

Horwitz, Simi. "Interview with Michael Rispoli." *BackStage,* February 20, 2004.

Joseph, Paul R. "Saying Goodbye to Ally McBeal." *University of Arkansas at Little Rock Law Review* 25, no. 3, 2003.

"Jurors Take Center Stage on *We The Jury.*" *Market Wire* website, December 2002.

Kramer, Irwin. "Kramer vs. Judging Judy." *The Daily Record,* November 10, 2003.

Kredens, Julie. "TV Station 'Teases' Suicide." *FineLine: The Newsletter on Journalism Ethics* 2, no. 3, June 1990.

Lee, Patrick. "*Century City* Takes a Great Leap Forward Into the Future of Jurisprudence." *Science Fiction Weekly* website, no. 360, March 2004.

Marino, Jacqueline. "*Ally McBeal* (review)." *Memphis Flyer,* May 18, 1998.

McFarlane, Melanie. "It Could Be Worse, You Could Be on TV." *Seattle Times,* February 12, 1999.

Mikva, Abner. "*LA Law*— Is It Law or Is It Just L.A.?" *Chicago Daily Law Bulletin,* March 15, 1987.

Mosely, Glenn A. "*Temple Houston*: The Story Behind a Forgotten Television Western." *Wildest Westerns,* Collectors Issue #2, 2000.

Owen, Rob. "Analysis: ABC Series Boston Legal Defies Conventions: TV Show Ignores Young Audience, is Unafraid to Flaunt its Politics." *Pittsburgh Post-Gazette,* February 21, 2006.

Popenoe, David. "Remembering My Father, Paul Popenoe." *Popenoe/Popnoe/Poppino & Allied Families* website. 1992.

Smith, Janet. "Interview with Elfego Baca." Library of Congress. http://lcweb2.loc.gov/ammem/wpa/20040209.html.

Taaffe, William. "You Can't Keep Him Down On the Farm." *Sports Illustrated,* July 18, 1984.

Thompson, Kevin D. "Daring *Day* Tackles Racial Issue." *Palm Beach (FL) Post,* March 18, 2001.

Walter, Tom. "Racism Gives Edge to Drama of Friendship." *Memphis Commercial- Appeal,* Aug. 16, 1998.

Watson, Bruce. "Hang 'Em First, Try 'Em Later." *Smithsonian,* June 1998.

Watson, Debra. "*InJustice* Dramatizes Reality of US Criminal Justice System." *World Socialist Web Site,* February 22. 2006.

Wohl, Alexander. "And the Verdict Is..." *The American Prospect,* November 30, 2002.

TV Program Transcripts

The Larry King Show. CNN, 2000–2001.

"Monkey Trial." *The American Experience,* PBS, 2002.

"TV JURY: Have Cameras in the Courtroom Undermined the US Justice System?" *PBS Newshour,* January 20, 1998.

Periodicals and Online Magazines

The following sources are referenced so often in the text that to list the articles singularly would be redundant:

Broadcasting & Cable magazine
The Chicago *Sun-Times*
The Chicago *Tribune*
Entertain Your Brain
Entertainment Weekly
The Harvard Law Bulletin

The Harvard Law Review
The Hollywood Reporter
The Independent (London, U.K.)
The Los Angeles *Times*
Media Life
Metropolitan News-Enterprise (Los Angeles)
The Milwaukee *Journal-Sentinel*
The New York Times
Newsweek
Newsday (Long Island, NY)
Picturing Justice: The On-Line Journal of Law and Popular Culture.
Reality TV Magazine.
The San Francisco *Chronicle*
Time magazine
TV Guide
USA Today
Variety
The Wall Street Journal
The Washington *Post*
The Weekly Alibi

Selected Internet Websites

In addition to the sites listed below, the major TV networks and cable services have their own official websites, as do many individual TV programs and production companies.

All-Movie Guide. http://www.allmovie.com
Archive of American Television. http://tvinterviewsarchive.blogspot.com
Behind the Scenes. http://employees.oxy.edu/jerry/bts/dc.htm
The Black Saddle. http://blacksaddlewestern.com
CNN.com. http://www.cnn.com
CTVA: The Classic TV Archive. http://aa.1asphost.com/CTVA/index.htm
The Edge of Night Homepage. http://lavender.fortunecity.com/casino/403
FindLaw. http://www.findlaw.com
Internet Movie Database. http://www.imdb.com
Judge Roy Bean's History Page. http://www.judgeroybean.com/jrbhistory.html
Museum of Broadcast Communications Archive. http://archives.museum.tv
"Roaring Rockets." http://www.slick-net.com/space
Rotten.com. http://www.rotten.com
Salon.com. http://www.salon.com
Tarlton Law Library: University of Texas School of Law. http://tarlton.law.utexas.edu
Thrilling Detective Web Site. http://www.thrillingdetective.com
TV Acres. http://www.tvacres.com
TV.com. http://www.tvcom.com
TV4U. http://www.tv4u.com
The Ultimate Judge Show Page. http://www.tvjudgeshows.com

Index

A & E *see* Arts & Entertainment channel 26
A. U.S.A. 15, 28–29
Aalda, Mariann 94
Aames, Willie 94
Abbott, Bruce 75, 76
ABC *see* American Broadcasting Company
ABC Album 155
ABC Films 217
ABC Mystery Movie see *Christine Cromwell*
ABC News Division 252
Abele, Robert 105
Abernathy, Donzaleigh 38, 39
Abernathy, Ralph 39
Abrams, Dan 17
Abramson, Leslie 17
Abruzzo, Ray 198
Abugov, Kelly 109
Acapulco 27
Accused 77, 78; see also *Day in Court*
The Accused (1988 film) 16
Acevedo, Kirk 168, 176
Acker, Sharon 217, 224
Ackerman, Harry 210
Ackerman, Will 231
Actman, Jane 210–211
Acuff, Eddie 219
Adams, Julie 223
Adam's Rib (film and TV series) 13, 29–30, 58
The Addams Family 26
Addy, Wesley 93
Adler, Luther 182
Adonis, Joe 8
Adrian, Iris 79
Adult Swim 119
Adventures of Brisco County Jr. 27
Adway, Dwayne 59
Affirmative Action 143, 208
Aflen, Hugo 269
Against the Law 30–31
Agee, James 117
Alarcorn, Arthur 255
Alaskey, Joe 118
Albee, Denny 93
Albice, Chrissy 55
Alcoa Theater 243
Alda, Alan 264
Alejandro, Kevin 248

Alexander, A.L. 5
Alexander, Ben 77
Alexander, Denise 44
Alexander, Joan 219
Alexander, Millette 93
Ali, Muhammad 115
Alice, Mary 123, 124
All in the Family 211–212
Allen, Elizabeth 210–211
Allen, Krista 120
Allen, Thomas 118
Allen, Woody 242
Allenby, Peggy 93
Allred, Gloria 33, 193, 231
Ally 35; see also *Ally McBeal*
Ally McBeal 13, 19, 31–35, 51, 97, 107, 111, 158, 162, 177, 204, 228, 232, 233
Allyson, June 211
Almost Perfect 26
Alpine Medien Productions 187
Altman, Robert 254
Amador, Mark 71
The Amazing Mr. Malone: radio series 6, 8; 36; TV series 8, 35–36, 57, 192
The Amazing Mr. Tutt 6
Ambuehl, Cindy 127, 130
Ameche, Jim 6
America West airlines 136
American Academy of Pediatrics 98
American Bar Association 5
American Board of Trial Advocates 85
American Broadcasting Company (ABC) 6, 7, 8, 10, 17, 18, 25, 29, 35, 36, 37, 38, 40, 43, 44, 45, 46, 48, 57, 58, 60–64, 66, 68–69, 73, 74, 77–79, 86, 92, 93, 96, 97–98, 103, 110, 115–116, 126, 133, 134, 149, 155, 166, 167, 170, 178, 182, 184, 185, 186, 188, 189, 190, 191, 193, 195–196, 200, 201–202, 206, 210–212, 217, 227–228, 229–230, 231, 232, 234–235, 245–246, 252, 253, 257, 260–261, 266, 267, 273–274
American Civil Liberties Union (ACLU) 3
American Courtroom Network *see* Court TV

American Institute of Marital Relations 89, 90
American Justice 24, 36
American Lawyer magazine 16
American Movie Classics 203
American Polygraph Association 178
The American Prospect 21
The American Woman's Jury 7
Ames, Teal 93, 95
Amini, Alex 257
Amory, Cleveland 42, 224–225
Amos, John 262
Amos, Tori 263
Anastopoulo, Judge "Extreme" Akim 99–100
Anatomy of a Murder 120
Anderson, Anthony 168, 171
Anderson, Cmdr. Bob 129
Anderson, Erich 244
Anderson, Harry 15, 197–199
Anderson, Melody 131, 132
Anderson, Rhonda 216
Anderson, Richard 217, 223
Anderson, Robert 64
Andes, Keith 77
Andrew Solt Productions 36
Andrews, Brittany 246
Andrews, David 37, 127, 130
Andrews, Edward 53
Andrews, Patricia 94
Andrews, Ralph 89, 90
Andrews-Wolper-Spears Productions 89
The Andy Griffith Show 183–186
Angel 27
Angel, Heather 79
Animal Court (aka *Judge Wapner's Animal Court*) 20, 36–37, 85, 253
Animal Planet 36–37
Anna-Lisa 45
Anonymous Content 70
Anspach, Susan 135
The Antagonists 37
Antonio, Jim 256
Any Day Now 13, 38–40
AOL Video Service 96
Apostle Productions 54
Appel, Rich 28
The Apprentice 177
Arbus, Loreen 55

Archer, Anne 253
Archive of American Television 169
Arena Productions 119
Argosy magazine 64, 65
Arizona Supreme Court 252
Arkin, Alan 203, 205
Arkin, David 252–253
Armenante, Jillian 146, 149
Arms, Russell 77
Armstrong, Kerry 94
Army-McCarthy Hearings 8–9
Arnaz, Desi 237
Arndt, Dennis 159
Arnold, Victor 94
Arrest and Trial (1963) 40–42, 169
Arrest and Trial (2000) 42–43, 71
Arthur, Jean 132–133
Arts & Entertainment channel (A&E) 24, 36, 203
Aryan Nation 17
As the World Turns 94, 96, 121
Asher, William 210–212
Ashmont Productions 210
Ashton, Skye 246
Asimow, Michael 1, 19, 106, 173, 176
Asinof, Eliot 270–271
Asner, Edward 108, 250–251, 264, 266
The Associates 15, 43–44
Astaire, Fred 116
Astin, John 197
Astin, Mackenzie 106, 107
Astor, Mary 219
Atkins, Dr. Stuart 46
Atlas Worldwide Syndication 99
Attanasio, Paul 56
Attorney at Law 6
Atwater, Edith 157
Auberjonois, Rene 48, 49, 52
Aubrey, Jim 80–81
Aubuchon, Remi 180, 181
Austin, Karen 197, 199
Avery, James 251
Avery, Tex 19, 51
Avery, Tol 250–251
Ayres, Leah 94
Ayres, Lew 120

Baa Baa Black Sheep 127
"Baby Cha Cha" 32
Baby Owl Productions 180
Baca, Elfego 200–201
Bach, Jillian 67, 68
Bacharach, Burt 38
Bachelor Father 26, 211
BackStage magazine 113
Badalucco, Michael 231, 232
Baer, Neal 168
Baggetta, Vincent 92
Bailey, F. Lee 42, 115, 134, 163, 177–178, 221, 225, 226, 231, 242
Bailey, Julian 109
Baird, Jeanne 44
Baker, Dylan 103–104, 193–195
Baker, Kathy 228
Baker, Simon 113, 114
Bakula, Scott 96–97

Balcer, Rene 168
Balfour, Eric 168, 176
Ball, Lucille 133
Banes, Lisa 111, 264–265
Banks, Bo 138, 143
Banks, Gene 77, 79
Banner Films 196, 197
Barba, Norberto 168
Barbara Hall-Joseph Stern Productions 146
Barbee, Carol 146
Barker, Ma 270
Barnaby Jones 27
Barnes, Joanna 263
Barney Miller 198, 199, 209–210
Barrett, Ben 44
Barrett, Laurinda 94
Barron, Frank 249
Barron, James 141
Bartlett, Martine 44
Bartold, Norman 29, 30
Baruchel, Jay 154
Basche, David Alan 189
Bashoff, Blake 146, 148
Baskin, John 69
Bateman, Jason 185
Bates vs. State Bar of Arizona 15
Bathe, Ryan Michelle 48, 50
Battles, Wendy 168
Baumgarten-Prophet Entertainment 187
Baxter, Meredith 206
Baylor University 258
BBC2 106, 107
Beach, Adam 168
Bean, "Judge" Roy 145, 254
Beaudine, Deka 208
Beaumont, Charles 77
Beaver, Jim 241
Beck, Glenn 52
Becker, Fred 210
Bednar, Rudy 252
Beetem, Chris 127, 131
Before They Were Rock Stars 263
Begley, Ed, Jr. 206
Begley, Ed, Sr. 53, 155
Belisarius Productions 105, 126
Bell, Catherine 126, 128, 130, 131
Bell, Lake 48, 49, 50, 189
Bell, Michael 118
Bell, Warren 178, 179
Bellamy, Ralph 51, 57, 80, 81
Belli, Melvin 115, 163, 242
Bellisario, Donald P. 21, 105, 106, 126–129, 131
Bellows, Gil 31
Belzer, Richard 168, 174
Ben Casey 250
Ben Jerrod 8, 44
Bender, Jim 103
Benjamin, H. Jon 245
Bennett, John Aaron 123
The Bennetts (aka *The Bennett Story*) 8, 44
Benny, Jack 184
Bensinger, Tyler 155
Benson, Shaun 153
Benton, Douglas 206
Bentzen, Jayne 93

Benzali, Daniel 193–195
Beradino, John 77
Berendt, John 184
Bergen, Candice 48, 49, 52
Berger, Warren 84
Bergman, Andrew 6
Bergman, Paul 1
Berlanti, Greg 97
Berlanti Television 97
Berle, Milton 84, 264
Berman, Marc 181
Bernardi, Herschel 53
Berner, Fred 168
Bernsen, Corbin 159, 161, 185
Berwick, Vi 44
Besch, Bibi 93
Beswick, Martine 69
Bettger, Lyle 64–66, 111–112
Betz, Carl 11, 134–135, 224
Beutel, Jack 145
Beverly Hills Municipal Court 255
Bevine, Victor 264–265
Bewitched 211
Bianco, Robert 74
Biederman, Ian 248
Bierbauer, Charles 105, 106
Bierko, Craig 48, 50, 52, 62, 64
Big Ticket Television 139, 140
Bill, Tony 108
Billett, Stu 193, 212–214, 255
Bing Crosby Productions 166, 250
Birch, Paul 64
Bird, Rose 255
Birdman and the Galaxy Trio 118
Black, Karen 135
Black, Lewis 118
Black, Roy 177
Black Entertainment Network (BET) 142
Black Mask magazine 217
The Black Robe (aka *Police Night Court*) 7, 44–45
Black Saddle 12, 45–46, 255
Blackmon, Gus L. 143
Blackmun, Harry 255
Blackstone, Jill 85, 231
Blair, Linda 121, 225
Blair Television 85, 87
Blake, Al 89
Blake, Amanda 94
Blake, Robert 53, 173
Blake, Timothy 30
Blakely, Susan 93
The Blame Game 46
Blanchard, Mari 254
Blatt, Daniel H. 30
Bliss, Edward N., Jr. 235
BlitzDS 110
Blockbuster Video 195
Blomquist, Tom 209
Blondell, Joan 270
Blood, Colonel 202
Bloom, Lisa 17, 22, 72
Bloomberg, Stu 126
Blue Man Group 150
Bob Stewart Productions 262
Bobbitt, John and Lorena 17
Bochco, Dayna 238

Bochco, Jesse 238
Bochco, Steven 16, 19, 57–58,
    159–164, 193–196, 227–228,
    238–240
Body of Evidence 23
Boetticher, Budd 236
Bogart, Humphrey 6, 191
Boggie, Clarence 64, 65
Bogosian, Eric 168
The Bold Ones [includes *The
    Lawyers* and *The Protectors*] 13,
    46–48
Bon Jovi, Jon 31, 35
Bonds, Barry 53
Bonin, William ("The Freeway
    Killer") 87
Bonneville, Hugh 67, 68
Boon, Robert 274
Boreanz, David 27
Borg, Richard 94
Bork, Robert 16
Bosco, Philip 108
Bosley, Tom 83
Bosnak, Karen 72
Bosson, Barbara 193, 194
Boston American 264
Boston Globe 240
Boston Law School 142
Boston Legal 19, 23, 48–53, 80,
    154, 162, 176, 228, 235, 238
Boston Public 234
Botkin, Perry, Jr. 30
Bowen, Julie 48, 50, 52
Bowers, William 259
Bowie, John Ross 28
Bowman, Blaine 71
The Boy from Oklahoma (aka *Swirl
    of Glory*) 254; see also *Sugarfoot*
Boyle, Lara Flynn 232, 234
Brackelmans, Walter E. 88
Bracken, Kathleen 93
Brad Grey Television 180
Brady, Ben 220
Brady, Derrex 38
Brady, Orla 101, 102
Brand, Joshua 123, 228
Brando, Christian 17
Brathwaite-Burke, Yvonne 256
Bratt, Benjamin 168, 171
Braudis, Bill 245–246
Bravo Cable Channel 177
Brawley, Tawana 170
Braxton, Stephanie 94
Breck, Peter 45, 254–255
Breech, Robert 231
Breen, Patrick 158
Brennan, Peter 71, 72, 140
Brennan, Walter 145
Brenneman, Amy 146–147
Brenneman, Judge Frederica
    Shoenfield 146–147, 148, 150
Brewster, Paget 118
Brian, David 190, 192
Bride, DeWitt 103
Bridges, Beau 84, 186
Bridges, James 206, 208
Bridges, Jordan 168, 176
Bridges, Todd 152
Briggs, Donald 219

Bright Horizon 6
Brill, Steve 16, 17
Brillstein, Bernie 180
Brinckerhoff, Burt 262
Briscoe, Brent 186
Briskin, Mort 111–112
Brockovich, Erin 104–105, 153
Broderick, Betty 17
Brodkin, Herbert 79–82, 108–109,
    154
Brokenshire, Norman 110
Bronco 254, 255
Bronson, Betty 79
Bronson, Charles 254
Bronx Zoo 27
Brooklyn Civil-Small Claims
    Court 137
Brooks, Albert 43
Brooks, David 94
Brooks, James L. 43
Brooks, Richard 168, 172
Broun, Heywood Hale 83
Browde, Elizabeth 235
Brown, Alonzo 143
Brown, Blair 103–104
Brown, Georg Stanford 135
Brown, Himan 121–122
Brown, Howard 261
Brown, Judge Joe 139–140, 141
Brown, Kathryne Dora 146, 149
Brown, Roger Aaron 123
Brown, Russ 167
Brown, Russell 193
Brown, Shane 111
Brown, Sophina 248
Brown, Tom 79
Browne, Angela 68, 69
Browne, Kathie 250
Bruckheimer, Jerry 59, 60, 154,
    155–156, 228
Bruno, Catherine 94
Bruskin, David N. 218
Bryan, William Jennings 3
Buccieri, Paul 46
Buccieri and Weiss Entertainment
    46
Buchanan, Edgar 145
Buchman, Harold 226
Buena Vista Television 126, 137
Bugliosi, Vincent 36, 106
Bula, Caroline 55
Bulluck, Vicangelo 143
Bulot, Chuck 244
Buono, Victor 223, 250
Burck, Robert [The Naked Cow-
    boy] 72
Burden of Proof 53, 136
Burgess, Hedy 105, 106
Burgheim, Richard 253
Burgi, Richard 147
Burke, Billy 63, 150
Burke, David J. 168
Burke, Michael Reilly 205
Burke, Paul 77
Burke, Walter 45
Burns, Allan 96, 97
Burns, Bonnie 245
Burns, Brooke 31
Burns, David 263–264

Burns, William D., Jr. 255
Burns & Burns Productions 245
Burr, Eugene 267
Burr, Raymond 66, 94, 131, 217,
    220–222, 224, 225, 242–243,
    262–263
Burrell, Rusty M. 36, 37, 85, 87,
    212–213, 215
Burrows, Saffron 48, 52
Busch, Adam 150
Bush, George W. 52, 231
Bush, Owen 249
Bushman, Francis X. 85
Butters, Tara 168
Byers, Ralph 94
Byrd, Dan 38
Byrne, Rose 74, 75
Byron, Edward A. 190–191

C-SPAN 16
"Cable in the Classroom" 159
Cable News Network (CNN) 16,
    17, 18, 26, 53, 56, 106, 179
Cablevision 16
Cadorette, Mary 197
Cagney and Lacey 264–266
Cahn, Barry 135
Cain's Hundred 53–54
Caldwell, L. Scott 237–238
Calfa, Don 209–210
California Commission on the Fair
    Administration of Justice 126
California Judges' Association 87
California State Bar 72, 217
Call, Anthony 93
Callaway, Bill 118
Callinan, Dick 94
Calloway, Vanessa Bell 205
Callum, Myles 58
Calm Down Productions 271
Calvin, John 210–211
Cameron, D.G. 34
Campanella, Joseph 46, 47
Campbell, Alan 131, 132
Campbell, Kay 93
Campbell, Ken 94
Campos, Bruno 74
Canalos, Asha 159
Cannavale, Bobby 31, 35. 203,
    204
Cannell, Stephen J. 115
Cannon, Dyan 31, 32, 180
Canon 35 (*aka* Canon 3A[7]) 5, 7,
    9, 12, 13
Canterbury Productions 54
Canterbury's Law 22, 23, 54–55
Cantor, Charles 117
Capone, Al 270
Capshaw, Hank 42
Capshaw, Jessica 232, 234
Carbonell, Nestor 56
Cardea, Frank 69, 70, 112
Cardea Productions 112
Cardille, Lori 94
Carey, Harry 180
Carey, Macdonald 179, 180
Carey, Olive 179, 180
Carleton, Claire 79
Carlin, George 265

Carlon, Fran 230, 231
Carlson, Amy 168, 176
Carlson, Linda 157
Carmichael, Hoagy 202
Carrington, Chuck 127
Carroll, Bob 212
Carroll, Diahann 62, 63
Carroll, Helena 94
Carroll, Ron 94
Carroll, Steve 71
Carson, Kit 248
Carson, Lisa Nicole 31, 32
Carter, Conlan 166
Carter, Dixie 93, 101
Carter, Hurricane 36
Carter, Jade 130
Carter, June 93
Carter, Lynda 272
Carter, Sarah 248
Carter, Thomas 98, 99
Cartoon Network 118, 119
Caruso, David 187–188
Caruso, Paul 85
Case, Carroll 253
Case Closed 55
The Case of the Black Cat 219
The Case of the Curious Bride 218–219
The Case of the Dangerous Dowager 219
The Case of the Howling Dog: film 6, 218–219; novel 6
The Case of the Lucky Legs 218–219
The Case of the Stuttering Bishop 219
The Case of the Velvet Claws: film 218–219; novel 218
Catalano, Sgt. Joseph, Jr. 85, 231
Cavanagh, Tom 97
CBS *see* Columbia Broadcasting System
CBS News Division 267
CBS Paramount Television 139, 140
CBS Productions 146, 209
CBS: Reflections in a Bloodshot Eye 80
CBS Television Distribution 237
Celebrity Deathmatch 144
Celebrity Jury 138, 152
Celebrity Justice 55–56
Center for the Study of Popular Television (Syracuse University) 52
Century City 23, 56–57
Cezon, Connie 224
Chais, Bill 120–121
Chalfont, Kathleen 113, 114
Chambers, Nanci 127, 130
Chambers Productions 268
Chandler, Kyle 180, 181
Chandler, Raymond 247
Chandler et al. vs. Florida 14
Chang, Christina 111
Chaplin, Curt 212, 215
Chapman, Lonny 108
Charde, Matthew 118
Charles, Keith 93
Chase, Ilka 263–264

Cheadle, Don 228–229
Chemerinsky, Erwin 20, 141
Chernuchin, Michael S. 103, 168, 187
Cherry, Jake 120
Chessman, Caryl 16, 48, 242
Cheyenne 255
Cheyenne Autumn 258
Chicago Daily Law Bulletin 159
Chicago Daily News 167
Chicago Tribune 191, 240, 248
Children of a Lesser God 240
Childress, Alvin 197
Chong, Tommy 261
Christine Cromwell 57, 169
Churchill, Winston 250
Cigliuti, Natalie 238–239
Cilento, Diane 69
Cincinnati Enquirer 39
Civil Wars 57–58, 165, 238
Claire, Ludi 121
Clare, Diane 68, 69
Clark, Dane 154, 155, 217, 224
Clark, Dick 262
Clark, Marcia 106, 231
Clark, Philip 273
Clarke, Richard 93
Clark, Justice Tom 12
Clarke, William Kendall 189
Clarkson, Patricia 193–195
Clay, Juanin 93
Clemenson, Christian 48, 51
Clemons, Inny 146
Clennon, David 209–210
The Client: film and TV series 13, 58–59; novel 59
Clinton, Bill 22, 173, 241
Clinton, Hillary 138
Clooney, Nick 202–203
Close, Glenn 74–75
Close to Home 59–60
The Closer 54
Closing Arguments 22, 72
Club Paradise 236
Cluess, Chris 197
CNBC Cable Channel 136, 241
CNBC's America Now 241
CNN *see* Cable News Network
CNN.com 49, 174
CNN Entertainment News 272
Coates, Alan 94
Cobb, Julie 73
Cobb, Lee J. 73, 273
Cobb, Mel 94
Cobbs, Bill 123, 124
Coburn, James 53, 83
Cochran, Johnnie 18, 36, 106, 231
Code, Kathy 93
Cohan, George M. 116
Cohen, Eric 31, 33
Cohen, Scott 104, 168, 176
Cohen, Steven 31, 33
Colbert, Stephen 118
Colby, Jamie 179
Cole, Gary 118
Coleman, Dabney 67, 68, 97, 113, 114
Coleman, Gary 89
Colin, Margaret 92, 93, 107, 272

Colin, Michaela 74
Collett, Verne 269–270
Collier, Marian 62
Collins, Ray 217, 220, 223
Collins, Robert 196
Colt .45 254
Columbia Broadcasting System (CBS) 4, 6, 7, 8, 10, 16, 18, 25, 37, 44, 56, 57, 59, 60, 66, 69, 75, 76, 77, 79–82, 84, 86, 87, 90, 91, 92, 93–97, 101, 102–104, 105, 107–109, 113–114, 119–120, 121, 126, 128, 131, 132, 140, 146, 150, 156–157, 170, 179, 183, 187–188, 190, 202, 205, 206–209, 210, 211, 212, 217, 219–223, 228, 230–231, 232–232, 235–236, 237–238, 242, 244–245, 248–249, 250–251, 252, 258–260, 261–262, 263, 264–266, 267–268, 269, 271–273, 274
Columbia Pictures Television 66, 69, 92, 131, 261
Columbia TriStar Television 101, 187, 256
Columbo 92, 120, 134, 175, 185, 263
Combs, Holly Marie 228
Comer, Anjanette 273
Comiskey, Charles 271
The Commish 27
Common Law 15, 60–61
Compton, Forrest 93, 95
Compton, John 76, 77
Compton, O'Neal 205
Comstock, Frank 257
Conference of State Chief Justices 13
Confession 61–62
Confessions 17, 22
Conflict 253
Congressional Investigator 9, 62, 196
Conn, Eileen 67
Connell, Kelly 228
Connelly, Marc 83
Connolly, Kevin 106
Connor, Whitfield 268–269
Connors, Chuck 40–42, 202
Connors, Mike 77, 222
Conrad, David 189
Conrad, Robert 73, 180
Conrad, William 131–132
Conried, Hans 260
Constantine, Michael 249
Conte, Richard 132, 133
Conti, Tom 272
Converse-Roberts, William 240
Conviction 168, 176–177; see also *Law & Order*
Conway, Gerry 168
Conwell, Pat 93
Cook, Daniel 184
Cook, Linda 93
Cook, Rachael Leigh 121
Cooper, Chuck 203
Cooper, Hal 230
Cooper, Jackie 155
Cooper, Sheldon C. 259

Coopersmith, Jerome 122
Corallo, Jesse 196
Corcos, Christine Alice 29, 105, 179
Cornthwaite, Robert 77, 228
Coronet Blue 81
Corrigan, John William 255
Corrigan, Joyce 255
Corrigan, Lloyd 268–269
Corsaut, Aneta 185
Corte de Familia, La 72
Corte de Pueblo 71
Cortez, Ricardo 219
Cosgrove, Daniel 126
Cosgrove, John 104
Cosgrove/Meurer Productions 104
Cossack, Roger 53
Costello, Frank 8
Costello, Mark 105
Costello, Ward 93
Cotler, Gordon 241
Cotten, Joseph 202
Cotton, Nancy 158
The Court 23, 62–64, 105
Court, Hazel 243
Court of Current Issues (aka *Court of Public Opinion*) 7, 64
The Court of Human Relations 5
The Court of Last Resort 9, 64–66
Court TV 16–18, 19, 20, 22–23, 24, 26, 59, 66, 72, 134, 179, 189, 252, 266
Court TV—Inside America's Courts 66
Court TV—Justice This Week 66
Courthouse 66–67
Courting Alex 15, 67–68
Court-Martial 21, 26, 68–69
Courtroom U.S.A. see *Night Court* (1958)
Cox, Brian 63
Cox, Ronny 244–245, 256
Craig, Tony 94
Crais, Robert 190
Cramoy, Michael 122
Crane Productions 273
Craven, Matt 180, 181
Crawford, Joan 202
Crawford, Rachael 257
Crawford Mystery Theater see *Public Prosecutor*
Crazy Like a Fox 69–70
Crehan, Joseph 219
Creighton, Harry 259
Crenna, Richard 146, 149, 250
Crier, Catherine 17, 22
Crime & Punishment (1993) 70
Crime & Punishment (2002) 70–71
Crime Takes a Holiday 6
Crimetime After Primetime 76, 245, 252
Criscuolo, Lou 94
Cristina's Court 20, 71–72, 142
Crofford, Keith 118
Crosby, Bing 166
Cross, Alison 227
Cross, Rebecca 57
Cross, Roger C. 153

Cross Question see *They Stand Accused*
Crothers, Joel 94
Crowley, Matt 219
"Crown Heights" Incident 133
Crusader Rabbit 7, 237
Cruz, Alexis 248
Cryder, Missy 193, 195
CSI: Miami 188
Cubitt, David 187, 188
Cukor, George 29
Cullen, Brett 205
Cullen, Mark 110
Cullen, Rob 110
Cullum, John 83
Culp, Robert 53, 225
Culp, Steven 127, 130
Cummings, Bob 155
Cumpsty, Michael 159, 165
Cumulus Productions 109
Cunningham, T. Scott 186
Curtis, the Honorable James E. 72–73
Curtis Court 20, 72–73
Cutler, Bruce 151–152
CW Network 25, 138, 139, 159

The D.A. (1971) 73, 74
The D.A. (2004) 73–74
The D.A.: Conspiracy to Kill 73
The D.A.: Murder One 73
Dade County State's Attorney's Office 216
Dahmer, Jeffrey 17
Dailey, Irene 93
Daily Record 214, 215
Dallas 205
Daly, Joseph 121
Daly, Tyne 135, 146, 148, 149, 264, 265, 266
Damages 23, 74
Dana, Viola 79
Dandridge, Dorothy 53
Dane, Frank 219
Danforth, Harold 76
Daniels, Samantha 189
Daniels, Stan 43
Daniels, William 134
Dann, Michael 80, 91, 109, 251
Danner, Blythe 29
Danner, Hillary 187, 188
Danson, Ted 74, 75, 111
Danton, Ray 264
Danza, Tony 101, 102
Darcy, Georgine 116, 117
Darden, Christopher 231
Dark Justice 27, 75–76
Darling, Joan 206
Darren Star Productions 189
Darrow, Clarence 3
The D.A.'s Man 9, 76–77
da Silva, Howard 83, 108
Davenport, Jack 107
Davey, Bruce 158
David, Mack 254
David E. Kelley Productions 31, 48, 111, 228, 231
David Hollander Productions 113
David Salzman Entertainment 75

Davidovich, Lolita 114
Davies, Brandon 105
Davis, Bette 222
Davis, Geena 243–244
Davis, George T. 16
Davis, Herb 94
Davis, Jerry 241–242
Davis, Madelyn 212
Davis, Ossie 58, 59
Davis, Terry 94
Davis, Viola 56
Davy Crockett 200
Day in Court 10, 77–79, 261, 266
Daydreamers Entertainment 231
Deadlock 47; see also *The Bold Ones*
Dean, Amy 151
Dean Hargrove Productions 131, 182
Death and the Maiden see *Hawkins on Murder*
DeCarlo, Jill 71
Decision House 89
DeCorsia, Ted 79
Deeb, Gary 9
Deering, Mary Lee 121
"The Defender" 51, 80, 81, 108
The Defenders 11, 51, 79–85, 108, 264
The Defenders: Choice of Evil 84
The Defenders: Payback 84
The Defenders: Taking the First 85
Defense Attorney 6
Defiance County (aka *Ty Cooper*) 243; see also *Sam Benedict*
DeGeneres, Vance 110
DeGeorgio, Francine 231
DeGore, Janet 166
DeHuff, Nicole 62, 63
DeKoven, Roger 103
de la Fuente, Christian 101
de la Fuente, Joel 203
de la Garza, Alana 168, 173
deLancie, John 261–262
Delaney, Kim 227–228
DeLay, Tom 52
Delayen, Jerome 235
Delisle, Grey 118
Delta Airlines 137, 195
Del Valle, Robert 248
Dennehy, Brian 42
Dennis, Sandy 42
DeNoon, Dawn 168
Denton, Jamie 227
de Pablo, Cote 150
de Rossi, Portia 31, 34
Dershowitz, Alan 36, 141, 163, 199
DeSalvo, Albert 178
DeSando, Anthony 159
De Santis, Stanley 206
Desilu Productions 111, 112, 116, 268–269
Detmer, Amanda 28
Detroit Free Press 181
Detroit Neighborhood City Hall 143
Devault, Calvin 38
DeVicq, Paula 203, 204
Devine, Loretta 97

DeVries, Hilary 33
DeWalt, Kevin 153
Dewey, Thomas E. 6, 191
Dey, Susan 159, 161–162
Dharma and Greg 26, 67, 68
Dhiri, Amita 107
Diagnosis: Murder 132, 186
Diamond, Legs 270
Diamond, Reed 146, 149, 198
Diamond, Selma 197
Diary of a Perfect Murder 183; see also *Matlock*
Diaz, Carlos 55
DiCenzo, George 98, 103–104
Dick Clark Productions 48, 262
Dickinson, Angie 180
Dietrich, Dena 29, 30, 227
Diggs, Taye 31, 35, 158
Dillahunt, Garrett 186
Diller, Barry 170
Diller, Phyllis 152
Dillinger, John 270
Dillman, Bradford 68
Dillon, Melinda 253
DiMaggio, John 60, 61
Dinner, Michael 62
Discovery Channel 235
Disney, Walt 200
Disney Channel 64
Disney Corporation 137
Disneyland 200
Divorce Court: (1958–1969) 10, 37, 78, 85–87, 89, 213, 267; (1985) 10–11, 15, 85, 87–88, 159; (1999) 1, 20, 85, 88–89, 142, 231
Divorce Hearing 10, 89–90
Dixon, Ivan 53, 83
Doctor Katz, Professional Therapist 245
The Doctors and the Nurses 108
Dodd, Alice 146
Dodd, Clare 218
Dodson, Jack 185
Doherty, Larry Joe 259
Dolling, Elliott and Jordan 101
Dominick Dunne's Power, Privilege and Justice 22
Donahue, Troy 254
Donlevy, Brian 36
D'Onofrio, Vincent 168, 175
Donohoe, Amanda 159, 164
Donovan, Tate 74
Don't Pee on My Leg and Tell Me It's Raining 20, 140
Doocy, Steve 214
Doran, Ann 77
Dorfman, David 101
Dothit, Randy 140
Doucette, John 179, 180
Douglas, Burt 93
Douglas, Diana 206
Douglas, Melvyn 207
Douglas, Robert 68
Douglas, Suzzanne 30, 31
Downey, Robert, Jr. 31, 35, 92, 93
Doyle, Len 190–192
Drachkovitch, Rasha 100
Drachkovitch, Stephanie 100
Dragnet 18, 73, 76, 169, 224

Drake, Ellen 159
Drake, Larry 159, 161
Drew, Barbara 117
Dreyfuss, Richard 135, 273
Driver, John 93
Drummond, Alice 199, 209–210
Drury, James 180
Dubin, Conor 59, 60
Dubois, Marta 187
Du Brow, Rick 79
Duff, James 73–74
Duffy, Mike 181
Dufour, Val 93
Dukan, Sy 122
Dukoff, Eddie 133
Dulles, John Foster 230
Dummar, Melvin 178
DuMont Network 7, 64, 103, 236–237, 258–259, 274
Dunaway, Don Carlos 157
Dunaway, Faye 264
Dunbar, Rockmond 120
Dundee and the Culhane 12, 90–92
Dunn, James 201
Dunn, Michael 42
Dunn, Trieste 54, 55
Dunning, John T. 191, 219, 230
Dupont Show of the Month 270
Durkin, Betsy 121
Durning, Charles 105, 106, 205
Dutton, Tim 31
Duvall, Robert 42, 53
Dvorak, Ann 219
Dylan, Bob 114
Dylan, Jakob 114
Dymon, Linda 151, 152
Dymon, Vincent 151, 152
Dysart, Richard A. 159, 161, 164
Dzundza, George 168, 170, 171

E! Entertainment Network 26, 72, 253
East Side—West Side 256
Eastern Michigan University 143
Eaton, Meredith 101, 102
Ed 27
The Eddie Capra Mysteries 12, 92
Eddie Dodd 58, 92–93; see also *True Believer*
Edelstein, Lisa 31
Eden, Barbara 53, 251
Edgar Allan Poe Award 241
The Edge of Night 8, 44, 93–96, 121, 190, 220
Edge of Night Homepage 95
Edgerly, Chris 118
Edson, Richard 247
Edwards, Anthony 59
Edwards, Ralph 212–213
Edwards, Ronnie Clare 244
Edwards-Billett Productions 193, 212–214, 255
Efron, Edith 108
Eglee, Charles H. 193
Ehrlich, Jacob W. "Jake" 242–243
Eid, Rick 168
Eight Faces at Three 35, 36
Eight Men Out 271

Eikenberry, Jill 159, 161
Eisenhower & Lutz 15, 96–97
Elam, Jack 254, 258
Elder, Larry 193
Elfman, Jenna 67–68
Eli Stone 22, 23, 97–98
Elise, Kimberly 59
Elizondo, Hector 56, 57, 107
Ellery Queen 92
Elliot, Jane 241
Elliott, David James 59, 60, 126, 130, 131
Elliott, Stephen 261–262
Ellis, Antony 45
Ellis, Herb 76
Ellis, Patricia 219
Elman, Irving 251
Ely, Rick 47
Embassy Television 261
Embry, Ethan 271
Emerson, Karrie 94
Emory University School of Law 137
Engel, Peter 249
Engel, Stephen 271
English, Diane 107–108
English, Jim 246
Entertainment Weekly 34, 37, 46, 58, 93, 101, 104, 109, 111, 113, 125, 128, 147, 179, 181, 203, 238, 247, 249, 266, 272
Ephraim, Mablean 85, 88–89
Epstein, Jacob 161
Epstein, Jules 227
Equal Employment Opportunity Commission 139
Equal Justice 98–99, 104, 209
Erbe, Kathryn 168, 175
Erickson, Dan 54
Erin Brockovich 105, 233
Esposito, Giancarlo 111
Esposito, Jennifer 146, 148
Estes, Billie Sol 12
Estes v. Texas, 381 U.S. 532 12
Evans, David 217
Evans, Derek 94
Evans, Estelle 123
Evans, "Miss Holly" 139
Evans, Orrin B. 217
Everhard, Nancy 240
Evers, Jason 273
Ewell, Tom 29, 30
Exclusive Story 6
Exiled 175; see also *Law & Order*
Eye for an Eye 99–100
Eyherabide, Eugenia 71

Faded Denim Productions 137
Fairbanks, Jerry 7, 236
Faison, Sandy 94
Falk, Peter 134, 263–264, 270
Falk, Rick 91
Fallon, William J. 6
Falls, Kevin 180
Falsey, John 123, 125, 228
Falwell, Jerry 33
Family 27
Family Channel (*aka* ABC Family) 257

Family Court with Judge Penny 18, 100–101
Family Law 13, 101–103, 188
Famous Jury Trials: radio series 5, 103; TV series 7, 103, 274
Fancher, Hampton 45
Fancy, Richard 205
Farentino, Debra 98, 99
Farentino, James 46, 47
Farina, Dennis 168, 171
Farley, Elizabeth 93
Farrell, Tim 77
Fatal Vision 183
Father of the Bride 211
Faulk, John Henry 108, 242
Faulkner, Ed 196
Faulkner, Mark 95, 121
Fawcett, Allen 94
Fawcett, Farrah 114, 206
Fay, Meghan 122, 123
Fay, Patrick 61
Fazekas, Michael 168
FCC *see* Federal Communications Commission
Federal Communications Commission (FCC) 11, 80, 208
Federal Writers Project 200
Feds 103–104
Feige, David 238
Feinstein, Alan 93
Feldon, Barbara 251
The Felony Squad 135
Felton, Norman 119
Fender, Raoul Lionel 163
Ferdin, Pamelyn 118, 210–211
Ferguson, Franklin 254
Ferguson, Jay R. 146
Ferlinghetti, Lawrence 242
Ferrell, Conchata 159, 165
Ferrell, Will 114
Ferrer, Alex E. (aka "Judge Alex") 135, 141
Ferrer, Jose 131
Ferrer, Miguel 247
Ferrero, Martin 247
Ferris, Bruce 179
Fickett, Mary 230
Field, Sally 62–64
Figgis, Mike 54
Filmways Studios 83, 264
Filmways Television Productions 90, 263
Fimerman, Michael 58
Final Appeal: From the Files of Unsolved Mysteries 104
Final Justice with Erin Brockovich 24, 104–105
Find Me Guilty 150, 204
Finkel, Fyvush 228–229
Finkelstein, William M. 57, 58, 159, 165, 168, 193
Finley, Eileen 94
Finnegan, Bill 38
Finnegan-Pinchuk Company 38
Finneran, Mike 197
Finnigan, Jennifer 59, 60
Fino, James 110
First Monday 23, 64, 105–106
First Years 15, 106–107

Fischer, Mary 174
Fischer, Peter S. 92
Fisher, Bob 212
Fisher, Buddy 151
Fisher, Charles 93, 121
Fisher, Cindy 262
Fisher, Frances 93, 180, 181
Fisher, Terry Louise 159–160, 163
Fitzsimmons, Tom 206, 208–209
Fix, Paul 221
Flanigan, Joe 105
Fleck, John 193, 194
Flint, Carol 62, 63
Flockhart, Calista 31, 32, 34, 35
Flohe, Charles 94
Flood, Ann 93
Flood, Mary 173
Florek, Dann 159, 162, 168, 170, 172, 174, 175
Florida's 11th Judiciary, Miami-Dade County 135
Floyd, Jami 66
Flynn, Colleen 205
Flynn, Errol 115, 219, 242
Flynn, Steven 94
Flynt, Larry 33
Foley, Ellen 197, 198
Foley, Scott 28–29
Foley Square 15, 107–108
Foley Square Courthouse (New York City) 8, 79
Follows, Megan 244–245
Fonda, Bridget 31
Fonda, Henry 207
Fonda, Peter 83
Fontaine, Joan 202
Fontana, Tom 150–151, 156
For the People (1965) 11, 81, 108–109
For the People (2002) 109–110
Ford, Harrison 226
Ford, Jack 18
Ford, John 117, 258
Fordes, William N. 167
Fordyce Enterprises 202
Foreman, Percy 134
Forensic Files 23
Forney, Arthur W. 103, 167
Forsythe, Henderson 96
Fortunato, Ron 204
Fortune magazine 141
44 Blue Productions 100
Foster, Diana 151
Foster, Meg 265
Foster-Tailwind Entertainment 151
Four for Texas 259
Four Point Entertainment 55
The Four Seasons 27
Four Seasons Hotel (Westlake Village CA) 152
Four Square Court 7, 110, 210
Four Star Productions 45, 166
Fowkes, Conrad 93, 121
Fox, Kevin 237
Fox Network 28, 30, 31, 35, 54, 60, 107, 111, 112–113, 120, 136, 150, 151, 155–156, 170, 194, 233
Fox News Channel 24, 52, 53, 100, 138, 179, 231, 241

Fox Reality Channel 43
Foxworth, Robert 252–253
Foxx, Redd 212
Frank, Carl 93
Frank, Joanna 159
Frank, Sandy 135, 178
Frankel, Ernie 217, 224
Franklin, Maurice 192
Franks, Fred 145
Fraser, Woody 202
Frawley, James 146
Fred Silverman Company 182
Freeman, Al, Jr. 108
Freeman, Leonard 252
Freeman, William 202
Frei, Cindy 36
Freil, Anna 150
Freilich, Jeff 75
Fresco, Michael 193
Fresh Prince of Bel-Air 27
Frick, Ford 270
Friddell, Squire 241–242
Friedman, James T. 87
Friedman, Olivia 38
Friedman, Steve 141
Friends 106
Froman, David 94, 182
Frost, Warren 184
Froug, William 242–243
Fuller, Penny 93
Fulton County (Georgia) Juvenile Court 137
Funicello, Annette 201
Furie, Sidney J. 226
Futterman, Dan 146, 148
FX Network 26, 54, 74, 229

Gabel, Martin 53
Gabet, Sharon 93
Gabor, Eva 155
Gabor, Zsa Zsa 16
Gainey, M.C. 30
Galahad Productions 121
Galeota, Jimmy 187, 188
Gallivan, Megan 256
Gallo, Christopher 142
Galloway, Don 40, 41
Gam, Rita 121
Gamblers Anonymous 247
Ganaway, Gary 135
Gang Bullets 6
Gangbusters 191
Gannett Company Inc. vs. DePasqualé 14
Gansa, Alex 186
Garber, Victor 97, 155
Garcetti, Gil 74
Garde, Betty 93
Gardell, Billy 110
Gardenia, Vincent 159
Gardner, Erle Stanley 6, 7, 9, 64–66, 94, 217–221, 223–224
Garfield, Alan 30
Garfield, Gil 30
Garner, James 105, 254
Garret, Spencer 110
Garrison, Sean 90–91
Gary the Rat 23, 110–111, 245
Gast, Harold 134, 135

Gauthier, Dan 66, 67
Gazzara, Ben 40–42
Geary, Paul 250–251
Gedrick, Jason 193–195, 256
Gelb, Ed 177–178
Gemignani, Rhoda 112
General Electric Theater 202
General Hospital 44, 79
General Mills 230
General Service Studios 91
Genesis-Colbert Productions 115, 135
Georgia State Court 137
Geraldo 241
Geraldo at Large 241
Germann, Greg 31, 32, 256
Gerringer, Bob 94
Getachew-Smith, David 137
Getter, Leo 96–97
Ghalayini, Shukri 55
Giamati, Marcus 146, 148
Giannini, Justice Anthony 14
Gibson, Cal 209–210
Gibson, Don 44
Gibson, John (actor) 93
Gibson, John (cable host) 231
Gibson, Mel 173
Gielgud, John 207
Gifford, Alan 93
Gilbert, Herschel Burke 45
Gilbert, Lauren 93
Gilbert, Melissa 256–257
Gilford, Jack 83
Gill, Katherine 189
Gillespie, Jean 230
Gilmore, Violet 77
Gilyard, Clarence, Jr. 182, 184
Ginsburg, Ruth Bader 21, 63
Ginty, Robert 206, 208
Giraldo, Greg 60–61
Girard, Michael 32
Girlfriends 26
girls club 111, 234
Gish, Annabeth 66
Gish, Lillian 83
Givens, Robin 66, 67, 251
Glass, Ned 108
Gleason, Kid 271
Gleason, Regina 44
Glenn, Scott 273
Gless, Bridget 264–265
Gless, Sharon 264–266
Glickman, David R. 78, 85, 86
Godart, Lee 94
Goddard, Trevor 130
Goetz, Bernard 170
Goin' Places with Uncle George 7, 236
Goldberg, Adam 120
Goldberg, Gary David 243, 260
Goldberg, Lee 132
Goldblum, Jeff 168, 175
Goldman, Carole 206, 208
Goldman, Richard S. (Rich) 136, 137, 142
Goldman, Ronald 18, 241
Goldsmith, Jerry 53
Goldsmith-Thomas, Elaine 237–238

Goldstein, Dan 71
Goldstick, Oliver 62
Goldstone, Jules C. 64
Goldwater, Barry 9
Golonka, Arlene 185
Gomberg, Sy 166–167
Gonzago, Julie 97
Gonzalez, Daniel 244
Good Company 178
Good Morning America 18, 189
Goodwill Court 5
Goodwin, Judge Alfred 14
Goodwin, Bob 92
Goodwin, Maya Elise 38, 39
Gordon, Chris 66
Gordon, Gerald 44
Gordon, Herbert 180
Gordon, Jill 106
Gordon, Lester 68
Gordon, Ruth 29
Gordon, William 68
Gorman, Robert 122, 123
Gorshin, Frank 82
Gosselaar, Mark-Paul 238–239
Gossett, Louis, Jr. 206
Gough, Lloyd 108
Gould, Harry 209–210
Gould, Lewis 168
Goulet, Robert 111
Govich, Milena 168, 171, 176
Grace, Nancy 17, 22, 100
Grace, Roger M. 2, 86
Graham, Currie 238–239
Graham, Fred 16, 22
Grammer, Kelsey 110, 111, 262
Gramnet Productions 110
Gran Via 113
Grand Jury 9, 62, 111–112
Grandalski, Kevin 241
Granny Get Your Gun 219, 221
Grant, A. Cameron 274
Grant, Albert 93
Grant, Beth 186
Grant, Lee 83, 108, 155
Grant, Mickey 93, 115
Graphia, Toni 205
Grasso, June 17, 66
Grate, Rachel 105
Graves, Peter 68, 262
Gray, Bruce 94
Gray, Charles H. 46
Gray, Coleen 77
Gray, David Barry 58, 59
Gray, Ellen 227
Gray, Sam 44
Grazer, Brian 189, 248
The Great Defender 12, 112–113
Greaza, Walter 93
Green, Amanda 168
Green, John 44, 45
Green, Walon 54, 168
Greene, Jonathan 168
Greene, Michele 159, 161–162, 164
Gregg, Virginia 202
Gregorisch-Dempsey, Lisa 55
Gregory, Alex 65
Gregory, James 211
Gregory, Simon 93
Grey, Brad 180

Grice, Lara 205
Griffis, William 269
Griffith, Andy 120, 131, 182–186
Grillo, Frank 109
Grimes, Tammy 264
Grimm, Tim 240
Grisham, John 59, 111
Grizzard, George 155
Groch, Michael 71
Grodin, Charles 264
Groom, Sam 93
Group W 133
Groves, Robin 94
Gruber, Frank 182
Gruffudd, Ioan 56
Grundfest, Bill 268
The Guardian 22, 113–115
Guardino, Harry 217, 224
Guggenheim, Marc 97
Guilbert, Ann Morgan 228
Guild Films 85, 86, 217
Guilfoyle, Kimberly 179
Guiliano, Peter 168
Guillory, Emory 246
Guilt by Association 22
Guilty as Sin 203
Guilty or Innocent 115
Gulf War (1990) 17
Gumbel, Andrew 34
Gundlach, Robert 263–264
Gunn, Janet 75, 76
Gunsmoke 133, 168–169
Gunther, Max 14
Gunton, Bob 66, 67
Gurwitch, Annabelle 92, 93
Guttentag, Bill 71
Guzman, Vivian 193
Gwinn, William 77–78
Gwynne, Anne 236–237
Gypsy 269

Haas, Joanne 76
Hack, Olivia 38, 39
Hack, Shelley 198
Hackett, Buddy 264
Hackett, Joan 79, 83
Hadley, Reed 235–236
Haenhle, Garry 71
Hagen, Jean 29, 30
Hagen, Molly 178, 179
Haggis, Paul 101, 102, 187, 188
Hagman, Larry 93, 205
Haid, Charles 62, 63, 156
Hajeno Productions 190
Hal Roach Studios 235–236
Halahan, Charles 206, 208
Haldeman, Meghan 122, 123
Hale, Alan, Jr. 201
Hale, Barbara 217, 220, 222, 224, 225–226
Hall, Albert 31
Hall, Barbara 146, 147
Hall, George 93
Hall, Karen 146
Hall, Monty 212
Hall, Regina 31
Hall, Shashawnee 178
Halley, Rudolph 8
Hallmark Channel 187

Hallmark Entertainment 187
Halmac 45
Halmi, Robert, Jr. 187
Halop, Florence 197, 198
Haltom, Bill 20–21, 216
Hamill, Mark 206, 226
Hamilton, Dan 94
Hamilton, Lisa Gay 231, 232, 234
Hamilton, Murray 68
Hamilton, Neil 79
Hamlin, Harry 159, 161, 163, 164
Hancock, John 115, 116
Handel, Bill 137
Hanley, Robert 69, 70
Hanna-Barbera Studios 118, 119
Hanning, Rob 67
Hansen, Peter 44
Hanson, Hart 146
Harbert, Ted 195
Harbinson, Patrick 168
Hardcastle and McCormick 15, 115–116
Hardin, Jerry 205
Hardin, Melora 205
Hardin, Ty 254
Harewood, Dorian 264–265
Harford, Betty 206, 208–209
Hargitay, Mariska 168, 174
Hargrove, Dean 131, 132, 182, 183, 185, 186, 187, 225
Harlan, Judge John 14
Harlow, Shalom 150
Harmon, Angie 168, 173
Harmon, Mark 130, 240
Harper, Hill 62, 63
Harper, Robert 227
Harper, Ron 132, 133
Harper, Valerie 225
Harrell, Jack 212–213
Harriell, Marcy 237–238
Harrigan and Son 116–118
Harris, Cynthia 159, 249–250
Harris, Julie 120
Harris, Robert H. 64
Harris, Steve 231, 232
Harris County (Texas) District Attorney's Office 98
Harrison, Gregory 146, 149
Harrold, Kathryn 123–125
Hartley, Mariette 225
The Harvard Lampoon 28
Harvard Law Review 207
Harvard Law School 28, 60, 89, 121, 141, 146, 188, 206–207
Harvey, Jerry 44
Harvey, Joan 93
Harvey Birdman, Attorney at Law 23, 118–119, 245
Harvey Levin Productions 55
Haryman, Jim 39
Hasburgh, Patrick 115, 116
Hasselhoff, David 225
Hastings, Don 93
Hatchett, Judge Glenda 100, 137–138, 141
Hatos, Stephen 212
Hauptmann, Bruno Richard 4, 18
Hauptmann Trial (*aka* Lindbergh Kidnapping Trial) 4–5, 12, 14

Have Gun, Will Travel 91, 258
Having Wonderful Crime 36
Havoc, June 268–269
Hawkins 9, 119–120
Hawkins-Byrd, Petri 140
Hawkins on Murder 120
Hayde, Michael 73, 77, 258
Hayden, Mary 93
Hayden, Michael 193, 194
Hayden, Russell 145–146
Hayden Ranch 146
Haydon, Charles 270
Hayes, Jeffrey L. 167
Hazard, Geoffrey 21
Hazel 26
Head Cases 120–121
Headlines on Trial 121
Heard, John 58, 59
Hearst, Patty 157, 178
Heatter, Gabriel 4
Heche, Anne 31, 35
Hecht, Ben 182
Heckart, Eileen 83
Hedwall, Deborah 123
Heel & Toe Films 56
Heinemann, Eda 230
Heinsohn, Elisa 75, 76
Hemingway, Mariel 57–58
Hempkins, Andre 55
Hemus, Percy 5
Hendricks, Christina 62, 64, 158
Hendrix, Leslie 168, 172
Henley, Barry Shabaka 59, 60
Hennessy, Jill 168, 172
Henson, Taraji P. 48, 52
Henstridge, Natasha 97
Hepburn, Katharine 29
Hephner, Jeff 150
Her Honor, Nancy James 7
Herbert, Holly 55
Herrmann, Edward 232
Herzberg, Jack 179, 190
Hetrick, Jennifer 159
Heyday Productions 241
Hickman, Gail Morgan 153
Hidden Faces 8, 121
Higgins, Joe 40, 41
Higgins, John Michael 49, 118
Higher and Higher 223
Highway to Heaven 27
Hilden, Julie 49
Hildreth, Mark 153
Hill, Arthur 13, 206, 224
Hill, Bill 68
Hill, Jackson 85, 86
Hill, Jenny 206
Hill, Steven 168, 170, 171, 182
Hill, Thomas 262
Hill Street Blues 15, 16, 160, 161
Hilliard, Bob 38
Hingle, Pat 62, 63
Hirschman, Herbert 273
Hirson, Alice 93
His Honor, Homer Bell 15, 121–122
His Honor, the Barber 7
Hodson, Jim 77
Hoffa, Jimmy 9
Hoffman, Cecil 159, 164
Hoffman, David 197

Hoffman, Dustin 83, 216, 223
Hoffman, Isabella 127, 130
Hoffman, Jane 93
Hoffman, Rick 227
Hogan, Jack 131
Holbrook, Hal 226
Holcomb, Rod 248
Holder, Christopher 94
Hollander, David 113, 114
Holliday, Billie 242
Holliday, Judy 29, 30
Holliday, Kene 182, 184
Holliday, Polly 58, 59
Hollitt, Ray 127
Holloway, Natalie 179
Holloway, Sterling 269
Holly, Lauren 37, 228
Hollywood & Crime 17
Hollywood Foreign Press Association 120
The Hollywood Reporter 40, 152
The Hollywood Squares 21, 211
Holm, Celeste 57
Holmes, Ed 93
Holt, Jack 6
Holyfield, Evander 144
The Home Court 15, 122–123
Homeier, Skip 155
Homicide: Life on the Street 22, 150, 174
Honey I Shrunk the Kids 26
Hooks, Kevin 227
Hope, Bob 55
Hopkins, Josh 31
Hopper, Dennis 254
Hopper, Hal 145
Hopper, William 217, 220, 224
Horan, James B. 76
Horn, Michelle 101
Horowitz, Simi 113
Horton, Michael 92
Hotel Malibu 245
House 248–249
House UnAmerican Activities Committee (HUAC) 83
Houseman, John 13, 43, 206–209
Houston, Paula 44
Houston, Sam 258
Houston, Temple 258
Houston Chronicle 173
Houts, Marshall 65
Howard, Cy 116
Howard, D.D. 199
Howard, John 236–237
Howard, Ken 29
Howard, Lisa 93
Howard, Sandy 62, 196
Howard, Susan 208, 226
Howard, Terrence 251
Howie see *The Paul Lynde Show*
Hubbard, Liz 93
Huddleston, David 226
Hudson, Hal 45
Hudson, Mark 244
Huffman, Felicity 74
Hufsedler, Shirley 255
Huggins, Roy 46, 68
Hugh-Kelly, Daniel 15, 115–116
Hughes, Howard 178, 242

Hughes, Jason 107
Hugo, Larry 93
Hull, Warren 236–237
hulu.com 196
Humphries, Barry [Dame Edna Everage] 31, 35
Hunt, Peter H. 29
Hunter, Henry 6
Hunter, Jeffrey 258
Hunter, Kim 94
Hurricane Entertainment Corporation 144
Huston, Carol 182, 184
Huston, Patricia 159
Hutchins, Will 253–254
Hyde-White, Wilfred 43
Hyser, Joyce 159

I Am the Law 6, 191
I Detective 17
I Married Joan 27
Ice-T 168, 174
Icey, Berglind 253
Icon Productions 158
Ihnat, Steve 47
I'll Fly Away 104, 123–126, 172
I'll Fly Away: Then and Now 125–126, 147
Imagine Entertainment 189, 248
Imperioli, Michael 171
In-Court Television *see* Court TV
In the Jury Room 252
Indefensible 238
The Independent 34
Indictment 7
Ingram, Rex 46
Inherit the Wind 3
Injustice 23, 126
Inner City Miracle 144
The Inner Flame see Portia Faces Life
Inside Edition 18
Instant Justice 17
International Brotherhood of Teamsters 9
The Interrogation of Michael Crowe 22
Interstate Television 89
Investigation Discovery 235
ION Network 193; *see also* PAX Network
Iran-Contra 16
Ironside 27
ITC 68
Itkin, Paul Henry 93
Ito, Judge Lance 19, 23
ITV 216
Itzen, Gregory 193, 194
Ivanek, Zeljko 74, 75
Ives, Burl 46, 47, 48
Ivey, Leli 94

Jackson, Chuck 38
Jackson, Cornwell 217, 220, 224
Jackson, Gail Patrick 217, 220, 224
Jackson, John M. 127
Jackson, Kate 33
Jackson, Mary 115, 116

Jackson, Michael 22, 72, 173
Jackson, "Shoeless Joe" 270–271
Jackson and Jill 236
Jacobs, Katie 56
Jacobson, Peter 28
Jacott, Carlos 60, 61
Jaeger, Sam 97, 111
Jaffe, Sam 53
Jaffee/Braunstein Films 203
JAG 21, 69, 106, 114, 115, 126–131
Jake and the Fatman 15, 131–132, 170, 185
Jakes, Jill 255
James, Ken 257
Janes, Loren 161
Janiss, Vivi 274
Jankowski, Peter 42, 168
Janssen, David 53
Jarrett, Chris 94
Jarrett, Gregg 17, 18, 66
Jarrett, Renne 93, 230, 231
Jarvis, Jeff 128, 164, 165
Jaskulski, Jerry 137
The Jean Arthur Show 15, 132–133
Jeannette-Myers, Kristin 66
Jefferson, Bernard 255
"Jena 6" Demonstrations 240
Jenkins, Allen 219
Jenner, Bruce 152
Jennings, Brent 37
Jennings, Peter 18
Jensen, Sanford 107
Jerry Bruckheimer Television 59, 154, 155
Jerry Fairbanks Productions 236
Jet magazine 256
Jeter, Ida 81
Jewell, Isabel 79
John Charles Walters Productions 43
John Grisham's The Client see The Client
John Rust Productions 154
John Wells Productions 62, 63
Johns, Glynis 83
Johnson, Brad 66
Johnson, Bruce 251
Johnson, Charles 259
Johnson, Don 154
Johnson, Georgann 264–265
Johnson, Lyndon B. 12
Johnson, Mark 113
Johnson, Penny 206, 209
Johnson, Raymond Edward 191
Johnson, Russell 45
Johnson, Taneka 38
Jones, Chuck 19
Jones, Davey 212, 216
Jones, Edgar Allan, Jr. 10, 77–79, 260–261
Jones, James Earl 83
Jones, Khaki 118
Jones, Renée 159
Jones, Richard T. 146, 149
Jones, Star 18, 133–134
Jones, Tommy Lee 59
Jones and Jury 20, 133–134
Jones-Moreland, Betsy 77
Jordan, Barbara 256

Jordan, S. Marc 197, 199
Joseph, Paul R. 33
The Joseph Cotten Show and *Joseph Cotten Show: The On Trial* see *On Trial* (1956)
Josephson, Barry 186
Jostyn, Jay 190–192, 196–197
Joyce, Stephen 121
Juarbe, Israel 107
Judd, Wynona 104
Judd for the Defense 9, 11, 134–135, 224
The Judge 135, 178
Judge Advocate General (JAG) Division, American Armed Forces 21, 68–69, 127
Judge Alex 20, 135–136, 142, 259
Judge and Jury 136
Judge David Young 20, 136–137, 142
Judge for Yourself 137
Judge Hatchett 20, 137–138, 143
Judge Jeanine Pirro 20, 138–139
Judge Joe Brown 20, 139–140
Judge Judy 11, 20, 72, 139, 140–142, 214–215, 246
Judge Karen 20, 142
Judge Maria Lopez 20, 142–143
Judge Mathis (aka *Judge Greg Mathis*) 20, 143–144
Judge Mills Lane 20, 144–145
Judge Roy Bean 12, 122, 145–146
Judging Amy 13, 114, 146–150, 229
Judith Paige Mitchell Productions 58
Juilliard School of Fine Arts 207
Jury Duty 20, 138, 151–152
The Jury 150–151, 156
Just Cause 153
Just Legal 22, 154, 155
Justice (1954) 9, 154–155, 182
Justice (2006) 23, 60, 155–156

KABC-TV (Los Angeles) 9, 10, 77, 261
Kaczmarek, Jane 98, 206, 209, 238–239
Kaelin, Kato 99, 100
Kafka, Franz 110
Kagan, Daryn 56
Kagan, Elaine 264–265
Kahn, Madeline 30
Kallis, Stanley 190
Kane, Christian 59, 60
Kanin, Garson 29
Karen, James 105
Karlen, John 121
Kate McShane 13, 156–157, 269
Katt, William 225
Katz, Burton 136
Katz, Joel 108
Katzman, Leonard 226
Kaufman, Judge Charles 143
Kaye, Virginia 93
Kaz 12, 157–158
Kazan, Lainie 178, 179, 206, 209
KCOP-TV (Los Angeles) 10, 217
Keach, Stacy 55
Keane, James 51, 206, 208–209

Keane, Teri 230
Keaton, Michael 111
Kedzie Productions 66
Keefer, Mel 220
Keen, Noah 40
Keenan, Michael 228
Keene, William B. 85, 87–88
Kefauver, Sen. Estes 8
Kefauver Committee 8, 270
Keith, Brian 15, 115–116, 201, 225
Kellerman, Sally 223, 251
Kelley, David E. 18, 19, 23, 31, 32, 34, 35, 48, 50–52, 111, 159, 160, 162, 164, 177, 181, 228–229, 231–235
Kelley, DeForest 237, 243
Kelley, Sheila 159
Kelly, Daniel 267
Kelly, Paula 197, 198
Kelly, Rae'Ven Larrymore 123, 186
Kemmer, Ed 93
Kendrick, Richard 230
Kennedy, John F. 9
Kennedy, Laurie 93
Kennedy, Robert F. 9
Kent, Mona 230
Kerr, John 40, 41
Kesler, Henry S. 179
Kessler, Glenn 74
Kessler, Jacque 139
Kessler, Todd A. 74
Kevin Hill 158–159
Kibbee, Lois 93
Kid's Court 159
Kieswetter, John 39
Kiger, Robby 69, 70
Kiley, Richard 83, 112, 155
Kim, Jacqueline 66
Kincannon, Kit 75
Kinchen, Arif S. 251–252
Kind, Richard 119–120
Kinetix Character Studio 32
King, Albert 43
King, Larry 18, 58, 88, 143
King, Martin Luther, Jr. 134, 139
King, Michelle 126
King, Robert 126
King, Rodney 17, 165
King, Zalman 273
King of the Hill 28
King World Productions 72
Kings County (NY) District Attorney's Office 133
Kinoy, Ernest 83
Kirchner, Jaime Lee 154
Kirkconnell, Claire 206, 209
Kiser, Terry 197
Kisseloff, Jeff 80, 155
Kitchel, Chris 268
Kleeb, Helen 116, 117
Klugman, Jack 83, 155
Knievel, Evel 178
Knopf, Christopher 98
Knotts, Don 185
Knudsen, Albert T. 110
Koberg, Mark 85
Koch, Ed 20, 141, 212, 215–216
"Kojak Defense" 15
Konte, Sandra Hansen 16

Kopell, Bernie 152
Kotcheff, Ted 168
Kraft Suspense Theatre 68
Krakowski, Jane 31, 32
Kramer, Irwin 214, 215
Kramer, Jeffrey 32, 231, 233
Kramer, Mandell 93, 219
Kramer, Richard 146
Krause, Peter 112, 113
Kravits, Jason 232, 233
Kreisman, Stu 197
KRIV-TV (Houston TX) 72
Krumholtz, David 180, 181
Krupa, Gene 242
Kruschen, Jack 77
KTLA-TV (Los Angeles) 10, 196, 212
KTTV (Los Angeles) 10, 85–86
Kulik, Buzz 68
Kumagel, Denice 197
Kummer, Eloise 44
Kunstler, William 36
Kurland, Seth 67
Kurtis, Bill 36
Kushner, Donald 85, 87, 266
KZK Productions 74

L.A. Law 16, 19, 30, 31, 58, 98, 99, 101, 114, 159–166, 170, 172, 188, 194, 195, 228, 232, 238, 260
L.A. Law: The Movie 165–166
Labyorteaux, Patrick 126–127
La Camera, Anthony 264
Lackey, Lisa [Elizabeth] 153
Ladies Home Journal 89
Lafferty, Perry 223
LaGioia, John 93
Laird, Jack 46
Lalama, Pat 55
LaMarche, Maurice 118
LaMarr, Phil 118
Lambie, Joe 94
Lanchester, Elsa 251
Landau, Martin 83
Landers, Audrey 262
Landesberg, Steve 118
Landgraf, John 74
Landis, James M. 207
Lane, Judge Mills 144–145
Langtry, Lily 145
Lansbury, Angela 176, 264
Lansbury, Bruce 109
LaPaglia, Anthony 193, 195
Larch, John 40
Larka, Robert 182
Larkin, John 93, 219
Larkin, Sheila 252–253
Larkin, Victoria 93
Larroquette, John 15, 48, 52, 84, 187, 197–198
Larry Levinson Productions 187
Lasker, Lawrence 92
Lasky, Zane 260
Latessa, Dick 94
Latham, Lynn Marie 244
LaTourette, Frank 76
Lau, Wesley 217, 223
Launer, S. John 64

Lavardera, Jeana 146
Law and Justice as Seen on TV 1
The Law and Mr. Jones 12, 116, 166–167
Law & Order 18, 19, 34, 38, 40, 41, 42, 43, 54, 57, 58, 70, 73, 84, 104, 123, 126, 132, 147, 150, 167–177, 181, 190, 232, 252, 272
Law & Order: Criminal Intent 19, 168, 174, 175, 176, 177, 181
Law & Order: Special Victims Unit 19, 150, 168, 174–175, 176, 177
Law & Order: Trial by Jury 168, 174, 175–176
The Law Firm 23, 177
Lawrence, Matthew 244
Lawrence, Scott 127, 130
The Lawyer 226; see also Petrocelli
Lawyer Q 7
Lawyer Tucker 6
The Lawyers see The Bold Ones
Layne, Mary 94
Lazard, Justin 244–245
LaZebnik, Rob 60–61
Lazzo, Michael 118
Leach, W. Barton 207
Leachman, Cloris 53, 83, 182
Leary, Denis 54
Leavenworth, Scott 227
Lechowick, Bernard 244
Ledeyo, Khaira 153
Lee, Chol Sol 92
Lee, Gypsy Rose 269
Lee, Harper 123
Lee, Irving 94
Lee, Peggy 206
Lee, Spike 249
Leerhsen, Erica 113, 114
Legal Café 17
LeGros, James 31, 35
Lehman, Kristin 56, 146
Lei, Lydia 69, 70
Leibman, Ron 157–158
Leibowitz, Samuel J. 4
Leigh, Chyler 111, 232, 234
Leight, Warren 168
Leister, Johanna 93
Leitch, Donovan 111
Lembeck, Harvey 155
Lembeck, Michael 107
Lennix, Harry J. 38
Lenz, Kay 240–241
Leonard, Elmore 186
Leonard, Jim 59
Leonard, Lu 131, 132
Leonard, Sheldon 185
Leonard Freeman Enterprises 252
LePard, Catherine 109
Lerner, Michael 66, 67
LeRoy, Gloria 157
Lester, Jack 44
Lethin, Lori 184
Let's Get It On 144
Levering, Kate 158
Leversee, Loretta 135
Levin, Harvey 55–56, 212, 214, 215, 255
LeVine, Deborah Joy 38, 66
Levine, Gary S. 60

Levinson, Barry 150–151, 156
Levinson, Chris 168
Levinson, Larry 187
Levitan, David 201
Levitsch, Ashlee 123
Lewis, Elliott 64
Lewis, Jennifer 66
Lewis, Jenny 247
Lewis, Lightfield 178, 179
Lewis, Mort 235
Liberty Media 16
Lie Detector 115, 135, 177–178
A Life in Your Hands 7
Life's Work 178–179
Lifetime cable channel 24, 38, 104, 105, 109–110
Light, Judith 262
Lighthearted Entertainment 133
Lillie, Mildred 255
Lincoln see *The Law and Mr. Jones*
Lincoln, Andrew 107
Lindberg, Chris 71
Linden, Hal 210
Linder, Cec 93
Lindfors, Viveca 83
The Line-Up 24, 179
Linkletter, Art 90
Lipman, Jonathan 59
Lipsky, Eleazer 7
Littlefield, Warren 124, 172
Littman, Jonathan 154, 155
Liu, Lucy 31, 34
Livingston, Jay 254
Livingston, Ron 232, 234
Lizer, Kari 182, 184
Llewelyn, Doug 37, 212–213, 215, 253
Lloyd, Kathleen 98, 99
LoBianco, Tony 121
Lock & Key 17
Locke, Peter 85, 87, 266
Lockhart, Gene 121–122
Lockhart, June 155, 202
Lock-Up 9, 179–180
Lodge, Stephen 91
Lofgren, Charles 220
Loggia, Robert 200, 237–238
Loken, Kristianna 227
London, Jason 126
London, Jeremy 123, 126
London Daily Mail 91
London Daily Mirror 106
Long, Jodi 187, 188, 189
Long, Nia 146, 149
Long, Richard 254
Long Beach Independent 167
Longfield, Michael 94
Longobardi, Josephine Ann 212, 216
Lopez, Jennifer 244–245
Lopez, Judge Maria 141, 142–143
Lord, Phillips H. 44, 45, 190–191
Lorimar Productions (*aka* Lorimar Television) 75, 123, 157, 244
Lorimar-Telepictures 212, 255
Los Angeles County Bar Association 86
Los Angeles District Attorney's Office 73, 74

Los Angeles Municipal Court 10, 261
Los Angeles Superior Court 136, 213
Los Angeles Times 140, 255
Los Lobos 60
Lost 23
Lott, Trent 52
Loughery, Jackie 145, 190, 192
Loughlin, Lori 94
Lovejoy, Frank 36
Lovitz, Jon 107, 108
Lowe, Edmund 274
Lowe, Rob 180–182
Lowell, Carey 168, 172
Lower, Geoffrey 264–265
Lowery, Andrew 178, 179
Lown, David 110
Loy, Myrna 218
Luciano, Charles "Lucky" 6, 191, 270
The Lucky Stiff 36
Luddy, Barbara 79
Lugones, Omar 55
Lumet, Sidney 150–151, 203–205
Lutz, Jolene 197, 199
Lutz, Matt 187
Lydon, James 257
Lyman, Dorothy 93
Lynch, David 228
Lynch, Jerome 216
Lynch, John Carroll 59, 60
Lynde, Paul 210–212
Lynn, Betty 185
Lynn, Diana 202
Lyon's Den 180–182, 235
Lyttle, Larry 20, 140

MacDonald, Philip 182
MacDonnell, Ray 93
MacKenzie, Gisele 155
MacKenzie, J.C. 193, 194
MacKenzie, Robert 190
MacLachlan, Kyle 126
MacLeod, Gavin 180
MacNeille, Tress 118
MacNicol, Peter 31, 118
MacRae, Sheila 264
Macready, George 77, 182
Mad magazine 84, 258–259
Mader, Rebecca 155, 156, 168
Madio, James 237–238
Madison, Sarah Danielle 146, 148
Magnum P.I. 127, 132
Magnum Productions 75
Maharg, Billy 271
Maher, Bill 18, 244
Mahoney, John 111
Mahoney, Maggie 77
Majors, Lee 206
Makkena, Wendy 146
Makris, Constantine 168
Male, Colin 85, 87
Malloy, Larkin 94
The Man from Galveston see *Temple Houston*
The Man from U.N.C.L.E. 91
Mancuso 27
Mandan, Robert 262

Mandylor, Costas 228
Manheim, Camryn 231, 232
Mankiewicz, Don M. 241
Manning, Jack 206
Manson, Charles 87, 136
Mantegna, Joe 105
Manza, Ralph 76
March, Stephanie 168, 175, 176
Marciano, David 57
Marcus Welby, M.D. 206
Margulies, Julianna 54
Marin, Cheech 146, 149
Marine Corps Times 129
Marino, Dennis 94
Marino, Jacqueline 33
Marino, Ken 106, 107
Maris, Herbert L. 179–180
Mark VII Productions 73, 76
Marked Woman 6, 191
Markey, Christian 213
Markham 27
Markham, Monte 217, 224
Markinson, Brian 111
Markle, Stephen 105
Marsalis, Wynton 247
Marsden, James 31
Marshall, E.G. 79, 81, 84–85, 155
Marshall, E. Pierce 23
Marshall, Paula 249
Marshall, William 241
Martin, Barney 260
Martin, Bruce 93
Martin, Carol 212, 215–216
Martin, George 104
Martin, Jesse L. 31, 34, 168, 171
Martin, Julie 168
Martin, Kiel 94
Martin, Rosemarie 101
Martin, Strother 119, 120, 166
Martindale, Margo 203
Martinez, A 109, 159, 165, 252
Martinsville U.S.A. see *Miss Susan*
Marvin, Lee 68
Marx, Arthur 212
*M*A*S*H* 238
Masius, John 159, 165
The Mask 9, 155, 182
Mason, James 207, 242
Massachusetts Supreme Court 143
Massimei, Gerald G. 42
Masters, William 62
Masterson, John 212
Masterson, Mary Stuart 111, 168
Masty & Vititoe (law firm) 104
Matchett, Christine 206
Matheson, Richard 77
Mathis, Judge Greg 138, 141, 143–144
Mathis, Samantha 106, 107
Matlin, Marlee 228, 240–241
Matlock 9, 120, 131, 182–186, 272
Matlock Company 182
Matt Houston 27
Matteson, Ruth 93
Matthews, DeLane 96–97
Mauldin, Bill 242
Maverick 254, 258
Maximum Bob 186–187
Maxwell, John 64

May, Donald 93
"Mayflower Madam" 170
Mayo Productions 263
Mazar, Debi 57, 159, 165
Mazur, Michelle 135
McAdams, James 92
McBride 52, 84, 187
McCain, Frances Lee 244–245
McCambridge, Mercedes 83
McCarthy, Sen. Joseph 8–9
McCarthy, Kevin 83
McClellan, Sen. John 9
McClellan Committee 9, 62, 270
McClendon, Riley 154
McClintock, Eddie 28
McClory, Sean 156
McClure, Frank Chandler 261
McCommas, Merrilee 101
McCormack, Mary 193, 194
McCormick, Carolyn 172
McCormick, Kathy 231
McCreary, Judy 168
McDermott, Dylan 231–233, 234–235
McDonald, Christopher 101
McDonald, Seven Ann 92
McDowell, Roddy 41
McEachin, James 105, 106
McFadden, Cynthia 17, 18, 66, 252
McGavin, Darren 69
McGrath, Paul 270
McGraw, Dr. Phil 100
McGreevey, John 45
McGuire, Biff 270
McGuire, Maeve 93
McHugh, Frank 117
McIntosh, Douglas 212, 216
McKay, Bruce 85, 87
McKay, Jim 267–268
McKean, Michael 118
McKenzie, Richard 262
McKeon, Doug 94
McLellan, Zoe 127, 130
McLerie, Allyn Ann 260
McLoughlin, Hilary Esty 138
McMartin, John 223
McMaster, Niles 94
McNally, David 155
McNamara, Kara 46
McNeely, Jerry 206
McQueen, Steve 51, 80
McVey, Patrick 108
Meara, Anne 156–157
Media Life 114, 204
mediatrip.com 111
MediaWeek1 181
Medina, Ernest 178
Melamed, Karen 135, 259
Meloni, Christopher 168, 174
Melrose Place 31
Melvoin, Jeff 126, 228, 229
Memphis Flyer 33
Men at Law see *Storefront Lawyers*
Men Who Love Women 241; see also *Rosetti and Ryan*
Menendez, Lyle and Erik 17
Mentell, Justin 48, 50
Mercurio, Micole 59

Meredith, Burgess 264
Meredith, Charles 64
Meredith, Madge 79
MERIDIA(r) Audience Response 46
Meriwether, Lee 152, 262
Merkerson, S. Epatha 168, 172
Merrill, Gary 154, 155, 182
Merson, Marc 157
Mese, John 66, 240
Meskill, Katherine 93
Metcalf, Mark 127
Metromedia 188, 189
Metropolis Studios 137, 142
Metropolitan News-Enterprise 2
Metz, Robert 80
Meyer, Breckin 122
Meyer, Marlene 168
MGM (Metro-Goldwyn-Mayer) 183
MGM Television 29, 53, 68, 119, 242
Michael, George 97
Michael, Gertrude 79
Michael Fimerman Productions 58
Michael Hayes 101, 187–188
Michalka, Amanda 113, 114
Michel, Werner 93
Michele, Michael 158
Middleton, Mae 38, 39
Mikva, Abner 159–160, 163, 164
Milan, Frank 270
Milchan, Arnon 58
Miles, Joanna 93
Miles, Vera 68
Milian, Marilyn 212, 216
Milkus, Edward J. 226
Miller, Arthur R. 121, 188–189
Miller, Barry 98, 99
Miller, Harvey 249
Miller, Jonny Lee 97
Miller, Kathleen 249–250
Miller, Larry 49
Miller, Lee 217
Miller, Linda G. 190
Miller, Nancy 38, 40
Miller, Ron 144
Miller, Thomas L. 226
Miller-Milkis Productions 226
Miller's Court 121, 188–189
Millian, Andra 206, 209
Millman, Joyce 229, 233
Mills, Alley 43
Mills, Frank 197
Mills, Hayley 91
Mills, John 90–91
Mills, Juliet 91
Mills-Francis, Karen (aka "Judge Karen) 142
Milman, Joyce 188
Milwaukee Journal-Sentinel 97
Mineo, Sal 69
Miner, Jan 219
Minow, Newton R. 11, 80
Misiano, Chris 97
Miss Match 189
Miss Susan 8, 189–190
The Mississippi 9, 190
Mister District Attorney: radio se-

ries 6, 7, 8, 191–192, 196; TV series 8, 36, 179, 190–192, 196
Mr. T 257
Mitchell, Cameron 69, 120
Mitchell, Elizabeth 180, 181
Mitchell, Judith Paige 58
Mitchelson, Marvin 163
Mitra, Rhona 48, 49, 232, 235
Miya, Jo Ann 40
Mod Squad 253
Moessinger, David 131, 132
Mohr, Gerald 77
Mol, Gretchen 111
Moll, Richard 197–198
Moloney, Bob 251
Monash, Paul 53, 134
Monet Lane Productions 85, 231
Moniz, Wendy 113, 114
Monks, Michael 233
Montalban, Ricardo 53
Montejano, Sonia 139
Monti, Mary Elaine 209–210
Moonshine Ink (Sierra Nevada College) 243
Moonves, Les 124, 128
Moore, Julianne 94
Moore, Mary Alice 93
Moore, Mary Tyler 180
Moore, Michael 18
MOPO Entertainment 42
Moral Court 20, 193
Moran, Terry 17
Moreno, Rita 114, 264
Moretti, Willie 8
Morgan, Christopher 190
Morgan, Debbi 109–110
Morgan, Derek 109–110
Morgan, Harry 73
Morgan, Wendell & McNicholas (law firm) 87
Moriarty, Michael 168, 172, 261–262
Mornell, Sara 146
Morning Court 10, 77, 78
Morris, Chester 6
Morrison, Shelley 66
Morton, Joe 98
Moser, James E. 77, 250
Moses, Kim 109
Moss, Carrie-Ann 75, 76
Moss, Denise 122
Mothers Against Drunk Driving (M.A.D.D.) 136
The Mothers-in-Law 26
Mount, Anson 168, 176
The Mouthpiece 5
Moutrie, Jane 121–122
Mrs. O'Malley and Mr. Malone 36
MSNBC 26, 71, 136
MTM Productions (*aka* MTM Enterprises) 96, 243, 251, 260, 264
MTV see Music Television
Muldaur, Diana 159, 162–163, 226, 260
Mulkey, Chris 38, 39
Murder One 193–196, 238
Murder She Wrote 272
Murphy Brown 107

Museum of Broadcast Communications (Chicago, IL) 2, 18, 160, 163
Music for Television Library 192
Music Television (MTV) 26, 46, 144
Mutual Broadcasting System (MBS) 4, 103
My Mother the Car 26
Myers, Pauline 252
Myles, Meg 94
MyNetwork TV 89

Nagy, Bill 268
Nakashini, Nobi 159
Nakita 246
Napier, John 44
Napolitano, Andrew P. 231
Nardini, Daniela 107
Nassau County (N.Y.) Board of Social Services 201
Nathan, Robert 168
Nathan, Stephen 101
National Association for the Advancement of Colored People (NAACP) 38, 125
National Association of Broadcasters (NAB) 15
National Broadcasting Company (NBC) 4, 5, 6, 7, 8, 16, 18, 21, 25, 28–29, 36, 44, 45, 46, 47, 53, 54, 58, 64, 66, 70, 73, 76, 77, 86, 92, 103, 104, 106, 121, 122–124, 126–128, 134, 147, 150, 154–155, 159, 160, 165, 166, 167–168, 170–172, 174, 177, 180, 182–185, 189, 197–198, 202, 203, 212, 220, 225, 226, 230, 235, 237, 240–243, 247, 249, 252, 257–258, 262, 266
National General Corporation 252
National Italian American Foundation 106
The National Lampoon 99
National Legal Aid Society 154
National Telefilm Associates (NTA) 85, 86, 111
Nat's Eye Productions 271
Navedo, Andrea 168
Navlen, Lee 72
Navy NCIs 130
Navy Times 129
Naxan Productions 166
NBC *see* National Broadcasting Company
NBC Blue Network *see* American Broadcasting Company
NBC Films 121
NBC Studios Productions (*aka* NBC Productions *and* NBC Television Productions) 106, 123, 247
NBC Sunday Showcase 81
Neal, Diane 168, 175
Needham, Tracey 126, 128
Needle, Karen 94
Neigher, Geoffrey 168
Nelson, John Allen 94
Nelson, Peter 206, 209

Nelson, Rick 206
Nerman, David 257
Neuman, E. Jack 156, 226, 242–243
Neuwirth, Bebe 168, 176
Never Plead Guilty 243
Neves, Steve 15
Nevins, David 189, 248
New Dominion Pictures 235
New Line Television 66
The New Perry Mason 217, 224–225; see also *Perry Mason*
New Regency Pictures 58, 187
New York City District Attorney's Office 76
New York County Courthouse 84
New York County Lawyers' Association 5
New York Daily News 8
New York Post 72, 170
New York Supreme Court 5
New York Times 11, 28, 30, 56, 66, 67, 84, 99, 106, 125, 141, 202, 210, 247, 262, 265
Newbern, George 66
Newman, Barry 93, 226
Newman, Laraine 118
Newman, Paul 145, 154, 160, 182
Newman, Phyllis 203, 204
Newsday 164
Newsweek 10, 34, 171, 176
Nichols, Marisol 126
Nicholson, Erwin 93, 95, 96
Nicholson, Julianne 31, 35, 168, 176
Nickelodeon Network 159
Nicoll, Kristina 257
Nieber, Linda 243
Nielsen, Leslie 46, 47, 48
Night Court (1958) 10, 62, 192, 196–197
Night Court (1984) 15, 52, 84, 197–200, 210, 250
Night Court Live 17
Night Court U.S.A. see *Night Court* (1958)
Night Games 226; see also *Petrocelli*
Nimoy, Leonard 180
Nina Saxon Film Design 228
9/11 (September 11, 2001) 21–22, 128–129, 130, 171, 204
The Nine Lives of Elfego Baca 200–201
Nip/Tuck 74
Nixon, Agnes 94
Nizer, Louis 11–12, 191, 242
Nolan, Lloyd 68
North, Sheree 120
North Carolina School of the Arts 133
Northern Exposure 123, 228
Noseworthy, Jack 148
Nostalgia Channel 167, 197
Noth, Christopher 168, 170, 171, 174, 175
Nourse, Allen 93
Nowak, Chris 204
NTA *see* National Telefilm Associates

NTA Film Network 86
Nuñez, Miguel 251
NYPD Blue 66, 147, 187, 188, 227–228

O'Brian, Hugh 222
O'Brien, Edmond 242–243
O'Brien, Liam 190
O'Brien, Maria 243
O'Brien, Pat 36, 116–118
The O'Brien Company 263
O'Connor, Carroll 84
O'Connor, John J. 30, 66, 97, 99, 202, 247, 265
O'Connor, Sandra Day 21
O'Donnell, Chris 120
O'Donnell, David 235
O'Dwyer, William 8
Official Films 235
Offsay, Jerry 84
O'Flaherty, Terence 258, 264
O'Hair, Madalyn Murray 36
O'Halloran, Bill 107
O'Hanlon, George, Jr. 131
O'Hara, Jenny 178, 179
O'Keefe, Michael 30, 178, 179
O'Loughlin, Gerald S. 252–253
O'Malley, J. Pat 45
O'Mara, Jason 126
On the Air 191
On Trial (1948) 7, 201–202
On Trial (1956) 202
On Trial (1988) 202–203
100 Centre Street 151, 203–205
O'Neal, Patrice 150
O'Neal, Patrick 157, 225, 230, 231
O'Neal, Ryan 189
O'Neil, Linda 246
O'Neill, Dick 75, 76, 157, 241
Onorati, Peter 57
"Operation Daybreak" 77
Orbach, Chris 174
Orbach, Jerry 168, 170, 171, 174, 176, 225
Orbis Communications 121
O'Reilly, Bill 231
Orenstein, Bernie 107
Orion Television 98
O'Riordan, Tom 137
Orleans 205
Orr, William T. 253
Ortega, Santos 219
Ortiz, Ana 28
Osborn, John Jay, Jr. 43, 206, 208
Osmundson, Timothy 146, 148
Ostroff, Dawn 159
Oteri, Cheri 141
Outerbridge, Peter 187, 188
Ouwleen, Michael 118
Overmeyer, Eric 168
Owen, Garry 219
Owen, Rob 50, 52
Owen Marshall: Counselor at Law 13, 206, 224
Oz 150, 151

Pace, Judy 273
Pacific Gas and Electric 105

Pacino, Al 204
Paen, Alex 268
Paetty, Scott 256
Page, Sam 248–249
Paid Our Dues Productions 38
Paisano Productions 64, 217, 220, 223, 224
Paley, William 80
Palm, Robert 168
Palm Beach Post 40
Palmer, Stuart 36
Palmer, Tom 77, 79
Panabaker, Danielle 248
Panettiere, Hayden 31, 35
Pantoliano, Joe 111
Paolantonio, Bill 104
The Paper Chase: film 13, 43, 206–207; novel 43, 206; TV series 13, 43, 125, 160, 206–209, 239
Paramount Network Television Productions 67, 122, 126, 205
Paramount Pictures 196
Paramount Television 43, 144, 156, 227, 273
Parisse, Annie 168, 173
Park Place 15, 197, 209–210
Parker, Andrea 127
Parker, Corey 92, 93
Parker, Dennis 94
Parker, Norman 94
Parker, Sarah Jessica 98, 99
Parkes, Walter 92
Parole 10, 210
Parole Productions 210
Pasdar, Adrian 103–104, 146, 149
Pasquesi, David 60, 61
Pasquin, Anthony 72
Patchett, Tom 260
Patrick, Butch 72
Patrick, Gail *see* Jackson, Gail Patrick
Patriot Act 184
Patterson, Kelly 94
Paul, Alexandra 225
Paul, Byron 268
Paul Haggis Productions 101
The Paul Lynde Show 15, 27, 210–212
Paulsen, Rob 110
Paulson, Sarah 73
PAX Network 25, 153, 178
PBS *see* Public Broadcasting System
The PBS Newshour 19, 20
Pearce, Richard 168
Peary, Hal 269
Peck, Gregory 123, 154
Peck, Jim 85, 87
Peeples, Nia 66, 67
Peña, Elizabeth 247
Pendick Entertainment 250
Pendleton, Wyman 94
Penn, Arthur 168
Penn, Matthew 168, 170
Penny, Joe 131–132
People vs. Withers and Malone 36
The People's Court: (1981) 11, 15, 16, 36, 37, 55, 85, 87, 140, 159, 193, 212–217, 253, 255, 266; (1997)

20, 71, 72, 88, 141, 212, 215–216, 246
The People's Court of Small Claims 10, 217
Perez, Judge Cristina 71–72
Perez, Manny 203, 204
Perez, Tony 264, 266
Perkins, Voltaire 85, 87
Perlman, Rhea 185
Perry, Mark B. 106
Perry, Roger 40, 116–117
Perry, Sherri Dyon 38, 39
Perry Mason (character) 6, 9, 82, 94–95, 182, 218–219, 222
Perry Mason: made-for-TV movies 131, 159, 183, 225–226; radio series 7, 8, 94, 219–220; TV series 9, 11, 24, 53, 57, 64, 66, 79, 80, 81, 84, 94, 117, 217–225, 242–243, 249, 258, 262, 264; see also *The New Perry Mason*
Persky, Lisa Jane 31
Persoff, Nehemiah 264
Persons Unknown Productions 28
Perzigian, Jerry 261
Peter Engel Productions 249
Peters, Susan 8, 189–190
Petersen, Lisa Marie 168
Peterson, Edgar 35
Petrie, Dan 155
Petrie, George 36
Petrocelli 12, 226–227
Petrovna, Sonia 94
Petty, Lori 38
Peyser, Marc 176
Peyser, Penny 69, 70, 260
Philadelphia Daily News 227
Philadelphia Inquirer 227
Philbin, Regis 225
Philip Morris Cigarettes 236
Philips, Gina 31, 34
Phillips, Clyde 92
Phillips, Grace (Gracie) 103, 193–195
Phillips, Irna 94
Phillips, Michelle 244–245
Phillips, Sugar Ray 99
Phillips, Wendy 92
Philly 149, 227–228, 238
Phipps, Joey Alan 94
Phipps, William 44
Phyllis 27
Picardo, Robert 180, 181
Pickens, James, Jr. 180, 181
Picket, Bronson 38
Picket Fences 19, 51, 228–229, 233
Picturing Justice website 29, 106, 173, 176, 179, 238
Pidgeon, Walter 222
Pierce, David Hyde 111
Pileggi, Nicholas 187
Pinchot, Bronson 244
Pinchuk, Sheldon 38, 109
Pinewood Studios 68
Pioneertown (California) 146
Pirro, Judge Jeanine 138–139, 141, 152
Pittsburgh Post-Gazette 50, 52, 234
Platt, Oliver 237

Plautus Productions 79, 108
Playboy Channel 246
Playboy magazine 23, 58, 246
Plaza, Begonya 75, 76
Plimpton, Martha 84
Plum, Paula 245–246
Poe 111
Poitier, Sydney Tamiia 106, 107
Polen, Nat 121
Police Night Court see *The Black Robe*
Politanoff, Peter 32
Politics on Trial 7, 229–230
Pollak, Kevin 248, 271–272
Poniewozik, James 71, 129, 204, 252
Poor, Bray 271
Popenoe, Dr. Paul 89–90
Port, Robert David 42
Porter, Cole 30
Portia Faces Life: radio series 6, 8, 230–231; TV series 8, 230–231
Portillo, Rose 96–97
Post, Markie 197–199
Post, Mike 160, 167, 170
Potter, Monica 48, 49, 50
Potts, Annie 38, 39, 40
Pounder, CCH 234
Poundstone, Paula 152, 245–246
Powell, Dick 166, 167
Powell, William 218
Power of Attorney 89, 231
Powers, Alexandra 159, 165
Powers, Stefanie 226
Poznick, Barry 46
The Practice 19, 30, 31, 48, 121, 154, 162, 176, 177, 181, 228, 231–235
Prager, Dennis 193
Prager, Emily 93
Pratt, James 200
Press, Gwyn 94
Pressman, Michael 113, 228
Preston, Billy 251
Preston, Wade 254
Price, Megyn 60, 61
Price, Sharon 89
Prime Time Justice 17
Prince, Clayton 75
Prince, William 93, 154, 155, 182
Prince of the City 203
Princeton University 19
Pringle, Edward T. 14
Pritchard, Dr. Edward 202
Procter & Gamble 19, 93–96, 102, 167, 190, 219–220
Program Partners 100, 101
The Prosecutors: In Pursuit of Justice 235
Prosper, Sandra 105
The Protectors see *The Bold Ones*
Provenza, Paul 159
Providence 147
Pryor, Nick 93
Psychic Detectives 17
Public Broadcasting System (PBS) 25, 26, 123, 125, 208
Public Defender 9, 235–236
Public Prosecutor 7, 236–237

Pugliese, Al 264, 266
Purl, Linda 105, 182, 184

Quayle, Anthony 69
Queens County (N.Y.) Supreme
    Court 238
Queens Supreme 237–238
Quilling, Cyd 94
Quine, Richard 132, 133
Quinlan, Kathleen 101, 206
Quinn, Aidan 54
Quintet Productions 145

Rachins, Alan 159, 161
Racket Busters 6
Racket Squad 236
Radar Entertainment 151, 152
Radio Univision 72
Radnor, Josh 62, 63
Raising the Bar 12, 238–240
Rake, Jeff 189
Ralph Waite Productions 190
Ramsey, JonBenet 136, 178
Rancom Productions 257
Randall, Gary A. 38
Randall, Josh 67
Randall, Marilyn 94
Randall, Tony 260
Randolph, John 83
Rapf, Matthew 250, 273
Rappaport, David 162
Rapping, Elaine 1
Rashomon 23, 150
Rawlins, Lester 93
Ray, Helen 189
Ray, James Earl 134, 139
Reagan, Ronald 178
Reasonable Doubts 13, 240–241
Rebeck, Therese 168
Red Om Films 237
Redford, Robert 83
Redlich, Ed 248
Reed, Chris 46
Reed, Donna 134
Reed, Pamela 122–123
Reed, Robert 79, 81, 82–84
Reel Justice 1
Reeves Entertainment 202
Regalbuto, Joe 43, 262
Regler, Timothy 140
Rehnquist, Justice William 16,
    105
Reid, Frances 93, 230, 231
Reid, Tim 225
Reilly, Charles Nelson 88, 212
Reilly, Hugh 93
Reles, Abe "Kid Twist" 270
Renfro, Brad 59
Rennie, Michael 222
Reno, Janet 104, 136, 172
Republic Pictures 202
Resnik, Bert 167
Reuben, Gloria 238–239
Revolution Studios 237
Revue Television Productions 40,
    202
Rey, Alejandro 250
Reyes, Jorge 158
Reynolds, Debbie 152, 225

Reynolds, Penny Brown (aka
    "Judge Penny") 100–101
Rham, Kevin 146, 148
Rhode Island Superior Court 14
Rhodes, Hari 46, 47, 48
Ricci, Christina 31, 35
Rice, Craig [Georgiana Ann Ran-
    dolph] 35–36
Rich, Lee 157
Richards, Addison 44
Richards, Danny, Jr. 268
Richards, J. August 168, 176, 238–
    239
Richards, Matthew 109
Richardson, Bill 200
Richardson, Jackie 257
Richardson, LaTanya 203
Richardson, Patricia 96–97
Richardson, Sally 101
Richman, [Peter] Mark 53
Richmond, Ray 152, 179
*Richmond Newspapers Inc. vs. Vir-
    ginia* 14
Richter, Erik 118
Rieffel, Lisa 264–265
Rifkin, Ron 29, 30, 264–265
Rigg, Rebecca 187, 188
Rispoli, Michael 112–113
Ritchey, Lee 115
Ritter, Bill 18
Ritter, John 227
Riva, Diana-Maria 60, 61, 227
Rivera, Geraldo 18, 173, 241
Rivera Live 136, 241
Rizzi, Trefoni 44
RKO Radio Pictures 112
Roach, Hal, Jr. 235
Roache, Linus 54, 168, 172
Robards, Sam 186
Roberts, Doris 83, 262
Roberts, Julia 105, 237
Roberts, Ken 242
Roberts, Mark 189
Roberts, Pernell 208, 254
Roberts, Rachel 260
Roberts, Stephen 62
Roberts, Tony 93, 241–242, 264
Robertson, Kathleen 111
Robertson, Keith 54, 55
Robin, Michael M. 73
Robin's Hoods 27
Robinson, Bartlett 219
Robinson, Bumper 197
Robinson, Charlie 197, 199
Robinson, Edward G. 6, 191, 207
Robson, May 219
Rocket, Charles 122, 123
Rockford Files 27
Roday, James 106, 107, 189
Roddenberry, Gene 192, 243
Rodriguez, Nina 168
Rodriguez, Paul 261
Rodriguez, Pete 142, 143
Roe vs. Wade 82, 255
Roebuck, Daniel 182, 184
Roger Kilgore: Public Defender 6
Rogers, Brooks 94
Rogers, Shorty 210
Rogers, Will, Jr. 254

Rogow, Stan 84
Röhm, Elisabeth 168, 173
Rohner, Clayton 193
Roland, Gilbert 201
Rolfe, Sam 90–91, 157
Romano, John 187, 256
Romero, Carlos 73
Romero, Ned 73
Roncom 68
Rooney, Mickey 41
Rorke, Hayden 117
Rose, Reginald 79–81
Rose, Si 122
Rosecrans Productions 113
Rosen, Rob 55
Rosenberg, Alan 57, 113, 114, 159,
    165
Rosenberg, Chuck 160–161
Rosenberg, Ethel 36
Rosenberg, Frank P. 40
Rosenberg, Harvey 255
Rosenberg, Julius 36
Rosenman, Leonard 84
Rosenzweig, Barney 18, 264–266
The Rosenzweig Company 264
Rosetti and Ryan 9, 241–242
Ross, Katherine 42
Ross, Ted 249–250
Rossi, Blake 101
Rosten, Peter 92
Roth, Henry 253
Roth, Matt 37
Rothman, John 103–104
Rothstein, Arnold 270
Rough Justice 233
Roundtree, Lovey 256
Rourke, Michael 136, 142
Roush, Matt 34, 175
Rowles, Polly 79, 83
Rubenzer, Eugene 88
Rubin, Benny 79
Rubin, Ellis 15
Rubinstein, John 69–70
Rubinstein, Zelda 228
Rumpole of the Bailey 26
Run for Your Life 27
Runyon, Michael 71
Ruscio, Elizabeth 30, 31
Ruskin, Jeannie 93
Russell, Charles 53, 134
Russell, Dick 157
Russell, Kurt 253
Russell, Nipsey 212
Russell, Olga 131
Rust, Richard 242–243
Rutherford, Kelly 112
Rutherfurd, Emily 271
Ruttan, Susan 159, 161
Ruysdael, Basil 45
Ryan, Amy 124, 203
Ryan, Jeri 248–249
Ryan, John 257
Ryan, Maureen 240
Ryan, Michael 44
Ryan, Quinn 3–4
Rysher Entertainment 144

Sabinson, Allen 203
"The Sacco-Vanzetti Story" 81

Sackheim, Daniel 180
Sackheim, William 37
Sagemiller, Melissa 238–239
St. Jacques, Raymond 255
St. James, John 246
Saito, James 97
Salamon, Julie 106
Saldivar, Susan 261–262
Salem, Jessica 206
Salinger, Matt 244–245
Salmans, Susan 11
Salmi, Albert 226
Salon.com 188, 229, 233
Saltzman, Philip 131
Salzman, David 75
Sam Benedict 9
Samantha's Table 189
Samoset Productions 205
Sams, David R. 266
Sams, Jeffrey D. 66
San Diego District Attorney's
   Office 70–71
San Francisco Chronicle 68, 264
San Francisco Police Officer's Asso-
   ciation 242
Sanchez, Jaime 108
Sand in My Pants Inc. 268
Sande, Walter 236–237
Sanders, Ian 109
Sanderson, William 186
Sandoval, Miguel 62, 63
Sandy Frank Productions 177
Sandy Howard Productions 62,
   196
Santiago-Hudson, Ruben 187, 188
Santoni, Reni 206
Santos, Joe 115, 116
Sanz, Carlos 112
Sara 15, 243–244
Sarabande Productions 30
Sarandon, Chris 62, 63, 146
Sarandon, Susan 59
Sarris, Andrew 68
Saturday Night Live 141–142
Saturday Night Solution 17
Savage, Brad 260
Savage, Susan 110
Savalas, Telly 53, 270
Saviola, Camille 105, 106
Sayles, John 247–248
Sbarge, Rafael 113, 114, 154
Scalia, Anthnoy 63
Scarborough, Chuck 162
Scarfe, Jonathan 238–239
Schaeffer, Eric 56, 106
Schaffel, Hal 82
Schaffer, Kent 24
Scheck, Barry 106
Schell, Randy 259
Schenck, George 69, 112
Schiavo, Terry 173
Schindel, Barry 168
Schlieff, Henry 22
Schlitz Playhouse 202
Schneider, John 244–245
Schombing, Jason 153
Schordinger's Cat Productions 71
Schotz, Eric 104
Schulman, Jennifer 245

Schultz, Dutch 191, 270
Schultz, Howard 133
Schwartz, Al 262
Science Fiction Weekly 56
Sciorra, Annabella 168, 237–238
Scofield, Paul 207
Scopes, John 3
Scopes "Monkey Trial" 3–4
Scott, Devon 260
Scott, Ken 44
Scott, Tom Everett 227
Scottoline, Lita 233
Screen Actors Guild 256–257
Screen Gems Television 210
Screencraft Pictures 145
Scrubs 29
Sears, Teddy 238–239
Second Chances 27, 244–245
Secor, Kyle 227
Seda, Jon 158
Seehorn, Rhee 68
Segal, George 42
Segal, Jonathan 206, 208
Sehorn, Jason 173
Seid, Art 217, 224
Seidlin, Judge Larry 23
Seigel, Donald L. 261
Seinfeld, Jerry 132
Self, William 224
Seligman, Selig J. 10, 77, 260,
   267
Sellecca, Connie 244–245
Selleck, Tom 206
Senger, Frank 104
Sengupta, Stephanie 168
Sergeant Ryker 68; see also *Court-
   Martial*
Serna, Pepe 244–245
Serpico 204
Serpico, Jim 54
Serra, J. Tony 92
Serra, Ray 94
Seven, Johnny 77
1776 (film) 29
Seward, William 202
Sex and the City 26
Sex Court 20, 246
Sex Court: The Movie 246
sexcourt.com 246
Sexton, Bruce 146, 149
Shadowland Productions 237
Shafer, Harry 214
Shaffer, Louise 94, 121
Shales, Tom 61, 111
Shannon's Deal 247–248
Shape Pictures 70
Shapiro, Adam 266
Shapiro, Debbie 261–262
Shapiro, Jonathan 154, 155
Shark 22, 23, 248–249
Sharon, Fran 93
Shatner, William 41, 48–49, 51,
   80, 81, 108–109
Shaw, Ellen Torgerson 214
Shaw, Stan 190
Shayon, Sam 235
Shearin, John 115
Sheen, Martin 83, 108, 264
Sheindlin, Jerry 141, 212, 216

Sheindlin, Judy (aka "Judge Judy")
   20, 21, 72, 88, 122, 139, 140–
   142, 144, 215–216
Shelby County State Criminal
   Courts, Division Nine 139
Sheldon, Lee 93, 96
Shelton, Samantha 146, 148
Shenakow, Justin 228
Shenkman, Ben 54, 55
Shepard, Jan 77, 79
Shephard, Greer 73
Shephard/Robin Company 73
Shepherd, Vonda 31, 32
Sheppard, Dr. Sam 178, 226
Sheridan, Jamey 168, 247–248
Sherin, Edwin 167
Sherman, Ransom 79
Shield, Bob 135
Shields, Brooke 111
Shimerman, Armin 111
Shin, Eddie 111
Shipley, Bill 230
Shoberg, Dick 93
Shoelace Productions 237
Shoemaker, Pamela 94
Shore, David 101
Short, Columbus 146
Short, Martin 43
Showtime Cable 84–85, 206,
   208–209
Shukovsky English Entertainment
   107
Shulman, Roger 69
Siegel, Bugsy 270
Siegel, Sam 44
Sierra, Gregory 60, 61
Sign of the Ram 189
Sikes, Cynthia 127
Sikes, Gini 17
Silberling, Brad 147
Silk, Alexandra 246
Silkwood vs. Kerr-McGee 163
Silverman, Fred 131, 132, 182, 183,
   185, 186, 225
Silvers, Cathy 107
Silverstone, Alicia 189
Silvestre, Armando 73
Simensky, Linda 118
Simmons, Henry 248
Simmons, J.K. 74
Simmons, Jean 225
Simmons, Peter 123, 124
Simmons, Richard Alan 263–264
Simms, Hal 94
Simon, Dan 55
Simons, David A. 98
Simpson, Nicole Brown 18, 241
Simpson, O.J. 17–18, 19, 20, 22,
   23, 36, 53, 66, 82, 84, 100, 106,
   137, 140, 173, 178, 194, 195, 241,
   245, 272
The Simpsons 28
Sims, Jocko 54, 55
Sinclair, Mike 235
Singer, Robert 240, 241
Sirhan Sirhan 115
Sirota's Court 15, 197, 210
Sisters 27
Sisto, Jeremy 168, 171

Six Gun Law see *The Nine Lives of Elfego Baca*
60 Minutes 140
Skakel, Michael 176
Skerritt, Tom 228
Slater, Helen 187
Slattery, John 103–104, 147
Slattery's People 11, 250–251, 264
Slesar, Henry 93, 95
Slezak, Walter 53
Sloan, Lisa 93
Smashing the Rackets 6
Smith, Alexis 202
Smith, Anna Nicole 23
Smith, Beverly 230
Smith, Brady 68
Smith, Cotter 66, 98, 205
Smith, Dennis 183
Smith, Harry 18
Smith, Jaclyn 57
Smith, Jamie 235
Smith, Janet 200
Smith, Kerr 155–156
Smith, Lionel 209–210
Smith, Scott Alan 227
Smith, Shelley 43
Smith, Truman 103
Smith, William Kennedy 17, 134
Smithers, William 270
Smitrovich, Bill 232
Smits, Jimmy 159, 162, 164, 165
Smoking Gun TV 17
Sneed, Johnny 113, 114
Snyder, LeMoyne 65
Snyder, Tom 245–246
Sobocinski-Puchert, Susan 142
"Soiled Dove Plea" 258
Sokoloff, Maria 232, 233, 234
Somers, Brett 217, 224
Sommars, Julie 182, 185
Sonnenfeld, Barry 186
Sonnenfeld Josephson Worldwide Entertainment 186
Sony Pictures Television 54, 74, 113, 136, 137
Sorbo, Kevin 152
Sorvino, Paul 168, 171
Soul, David 206
Space Ghost Coast-to-Coast 118
Spader, James 48–50, 232, 235
Sparks 15, 251–252
Spears, Harry 89, 90
Specht, Lisa 266
Spelling, Aaron 269
Spelling Productions 38, 237
Spence, Gerry 36, 106, 163
Spencer, John 159, 164
Spier, William 268–269
Spike TV 26, 110–111, 145; *see also* The National Network (TNN)
Spillman, Sanford 196
Spiner, Brent 199
Sports Illustrated 268
Squigglevision (animation process) 245
Squigglevision (TV series) see *Science Court*
Stack, Robert 104, 225
Stafford, Nancy 182, 184

Stan Rogow Productions 247
Stanberg, Josh 67, 68
Stanley, Florence 210
Stanley, Pat 94
Stanton, Harry Dean 77
Stapleton, Jean 84
Star, Darren 189
Star Trek 115, 237, 243
Stark, Michael 94
Starr, Deborah 125
Starry Night Productions 197
Starsky and Hutch 242
State V. 252
Steeger, Harry 65
Steiger, Joel 131, 182
Steiger, Rod 155
Stein, Joel 20, 141, 214, 215, 216
Steiner, Fred 222
Stephen Engel Productions 271
Stephen J. Cannell Productions 115
Stephens, James 206, 208–209
Stern, Joseph 146, 168
Stern, Leonard B. 241–242
Steven Bochco Productions (*aka* Bochco Productions) 57, 159, 191, 227, 238
Stevens, Justice John Paul 14
Stevens, Robert 182
Stevens, Shadoe 152
Stevens, Warren 270
Stewart, Bob 262
Stewart, James 119–120, 207
Stewart, Nicola 159
Stewart, Justice Potter 13
Still Crazy Like a Fox 70
Stiller, Ben 157
Stiller, Jerry 157
Stinette, Dorothy 94
Stoler, Shirley 94
Stoller, Fred 245
Stone, Harold J. 112
Stone, Leonard 132, 133
Stone, Peter 82
Storefront Lawyers (aka *Men at Law*) 13, 112, 252–253, 273
Storer Television 85, 87
Stout, Bill 267–268
Strain, Julie 246
Strathmore Productions 131
Stratton, Albert 217, 224
Stratton, W.K. 127
Strauss, Peter 273
Street, Park, Jr. 65
Strickland, Gail 105, 198
Stritch, Elaine 263–264
Strock, Herbert L. 190
Stroka, Michael 94
Stroll, Edson 62
Stroud, Robert 65
Stuart, Maxine 250–251
Studio One 51, 79–80, 108
Studios USA Television 106
Style Court! 20, 253
Style Network 253
Style Network Productions 253
Suarez, Cecilia 109
Sugarfoot 12, 27, 253–255
Sulds, Irvin Paul 64
Sullivan, Barry 222

Sullivan, Beth 264
Sullivan, Brad 123, 124
Sullivan, Jeannine 231
Sullivan, Tim 19, 20
Summer, Cree 256
Summers, Tara 48
Sunset Studios 196
Superior Court 255–256
Supreme Court of the United States 12, 14, 15, 16, 23, 62, 105, 146
Surratt, Mary 202
Susskind, David 154, 269–271
Sussman, Sharon 135
Sutherland, Donald 69
Sweet, Dolph 263–264
Sweet Justice 13, 256–257
Swenson, Karl 230, 231
Swerdloff-Ross, Helaine 55
Sweren, Richard 167
Switzer, Mary Kay 12
Swofford, Ken 92
Sylvester, William 226
Syndicated Productions 143
Syntax 168

T. and T. 257
Tacker, Francie 206, 208
Taggart, Millie 93
Talbot, Nita 262
Talbott, Gloria 77
Talent Associates 154
Tales of Texas John Slaughter 201
Talman, William 217, 220, 222, 224
Tammi, Tom 94
Tarlton Law Library (University of Texas School of Law) 2, 106
Tarses, Jay 260
Tartikoff, Brandon 160, 170, 171, 183
Taub, Alex 146, 158
Tavel, Connie 146
Taylor, Holland 94, 233
Taylor, Jennifer 94
Taylor, Joseph Lyle 203, 204
Taylor, Regina 103–104, 123
Taylor-Young, Leigh 228
TCF Television Productions 217
Teague, Lewis 247
Teitz, Andrew 14
Telco Productions 268
Telephone Time 200
Telepictures 138, 143, 212
Telestar Films 210
Televisio de Cataluna 75
Temple Houston 12, 257–259
Temple Houston Company 257
Tennessee Law Journal 21, 216
Tenney, Jon 98
Terenzio, John 139
Tew, Alan Stanley 213
Tewkesbury, Joan 248
Texas Justice 20, 259
Texas SWAT 17
Thayer, Brynn 182, 184
They Stand Accused 7, 259–260
Thin Man (movie series) 218
thinkfilm 180

Thinnes, Roy 170
This Life 106–107; see also *First Years*
Thomas, Judge Charles J. 238
Thomas, Daniel 237–238
Thomas, Frank, Sr. 44, 45
Thomas, Frankie 103
Thomas, Peter 237–238
Thomas, Richard 153
Thomas, Tony 60
Thomerson, Tim 43
Thompson, Andrea 127, 128
Thompson, Fred Dalton 168, 171, 176
Thompson, Lea 109–110
Thompson, Robert 52
Thompson, Robert C. 206
Thomson, Patricia Ayame 69, 70
Thorne-Smith, Courtney 31, 34
Thorsell, Karen 93
333 Montgomery 243; see also *Sam Benedict*
Thrilling Detective website 132
Tiffany 152
Tighe, Kevin 193–195
Til Death Do Us Part 17
Tilden, Richard 267
Tilton, Charlene 152
Time magazine 20, 33, 37, 61, 71, 129, 141, 204, 214, 215, 216, 243, 252, 253, 267, 273
Time Telepictures Television 55
Time-Warner 16
Tinker, Grant 160
Tinker, John 159, 165, 231
Tinling, Ted 93
TMZ 56
TMZ on TV 56
TNN (The National Network) 26, 110–111, 145
TNT *see* Turner Network Television
To Kill a Mockingbird 123, 154
Tobey, Charles 8
Tobin, Dan 217
Tobin, Genevieve 218
The Today Show 134
Toler, Lynn 85, 89, 231
Tom, Lauren 68
Tom Snyder Productions 245
Tomita, Tamlyn 127
Tomlin, John 144, 253
Tomorrow Entertainment 224
Tompkins, Joan 242–243
Tonight Show with Jay Leno 111
The Tony Randall Show 15, 260
Toobin, Jeffrey 19, 20
Topolsky, Ken 106
Torme, Mel 197, 199
Torn, Rip 181
Torres, Liz 105
Toth, Alex 118
Touchstone Television 62, 67, 97, 126, 137, 158, 178
Touhy, Roger 270
Toussaint, Lorraine 38, 39, 40
Towers, Jonathan 36
Towers Productions 36
Towey, John 121

Townsend, Ernie 94
Tracy, Lee 35, 36
Tracy, Spencer 29
Traffic Court 9–10, 77–78, 260–261
Trailways Bus Line 256
Trammell, Sam 186
Travanti, Daniel J. 83, 253
Travis, June 219
Travis, Nancy 271–272
Treen, Mary 268–269
Trenholme, Helen 218
Trial and Error 15, 261–262
Trial by Jury 262–263
Trial Heat 22
The Trials of O'Brien 12, 251, 263–264
The Trials of Rosie O'Neill 13, 18, 264–266
TrialWatch 266
Tribune Entertainment 257
Trotwood Productions 187, 256
Troupe, Tom 135
True Believer 13, 92; see also *Eddie Dodd*
Trump, Donald 36, 177
Trump, Ivana 36
TruTV 23; *see also* Court TV
Tryon, Tom 182, 201
Tucci, Michael 206, 209
Tucci, Stanley 193–195
Tuck, Jessica 146, 148
Tucker, Forrest 250
Tucker, Ken 34, 37, 93, 101, 111, 113, 115, 125, 147, 203, 247, 248, 266, 272–273
Tucker, Michael 159, 161
Tulane University 136
Tully, Phil 196
Tunick, Irve 269
Tunie, Tamara 168
Turner, Karri 127, 129
Turner, Lana 90, 115, 213
Turner, Terry 167
Turner Network Television (TNT) 26, 58, 150, 238, 240
Turner Productions 118
Turrene, Louis 93
Turtletaub, Saul 107
Turturro, Aida 272
TV Book 83
TV Guide 10, 14, 16, 17, 18, 19, 32, 33, 34, 42, 58, 70, 71, 79, 82, 86, 87, 88, 91, 102, 105, 107, 108, 117, 122, 125, 128, 129, 133, 136, 147, 157, 160, 164, 165, 169, 175, 180, 184, 188, 190, 191, 199, 206, 214, 222, 225, 240, 242, 261, 263, 267
TVi Media 268
Twelve Angry Men 79, 81, 150, 203
20th Century–Fox (film studio) 220
20th Century–Fox Television 28, 31, 48, 57, 111, 120, 134, 135, 146, 150, 159, 180, 189, 193, 206, 208–209, 217, 224, 228, 231, 248

Twentieth [20th] Television 71, 72, 85, 88, 89, 135, 231, 259
24 (TV series) 23, 194
21 Beacon Street 27
Twin Peaks 228
Tyler, Harry 45
Tyson, Cicely 256
Tyson, Mike 144, 251

UCLA (University of California–Los Angeles) 10, 71, 78, 79, 139, 173, 261
Ullman, Tracey 31
Under Fire: Deadliest Police Shootouts 17
Underwood, Blair 159, 161
United Paramount Network (UPN) 25, 141, 158–159, 251–252
United Press International (UPI) 10, 79
United States Court of Appeals 14
Universal-HD channel 56, 57
Universal Television 37, 42, 46, 47, 56, 57, 68, 70, 73, 92, 132–133, 167, 170, 174, 206, 224, 241, 249, 271
University of Arkansas at Little Rock Law Review 33
University of British Columbia 206
University of Detroit Law School 143
University of Houston Law Center 133
University of Miami College of Law 136
University of Pennsylvania Law School 21, 89
University of Southern California (USC) 20, 141, 217
University of Wisconsin 206
Unsolved Mysteries 104
UPN *see* United Paramount Network
Urich, Robert 206
U.S. Comedy Arts Festival 111
USA Cable Network 26, 55, 96, 168, 177, 214
USA Productions 42
USA Today 63, 74, 98, 240, 249
Usher, Guy 219
Utay, William 197

Vaccaro, Brenda 83
Valentine, Karen 206
Valley, Mark 48, 49, 52
Vance, Courtney B. 168
Vance, Pruitt Taylor 194, 195–196
Vandas Productions 53
Van Dyke, Dick 132
Van Dyke, Phillip Glenn 122, 123
Van Fleet, Jo 182
Van Patten, Dick 152
Van Steeden, Peter 192
van Susteren, Greta 53
Van Voorhis, Westbrook 154, 155
Variety 53, 59, 67, 77, 110, 112, 121, 179

Vasquez, Randy 105, 106, 127, 130
Vassar College 133
Vassey, Liz 186
Vaughn, Heidi 93
Vaughn, Robert 53
Vawter, Nancy 159
Vega, Alexa 178, 179
Velez, Eddie 261
Velez-Mitchell, Jane 55
Vendig, Irving 93–95, 121, 219, 220
Vengimiglia, John 104
Verdict 267
The Verdict 150, 154, 160, 203
The Verdict Is Yours 10, 77, 87, 266, 267–268
Verica, Tom 159, 165
VeSota, Bruno 259
VH1 263
Viacom 22, 182, 223, 225, 262
Vick, Michael 53, 173
Victor, David 90–91, 206
Vieira, Meredith 267
The View 133, 134, 267
Viewers for Quality Television 125
Vincent, Jan-Michael 253
Vintage Court U.S.A. see *Night Court* (1958)
The Virginian 27, 273
Viro, Barney 196
Vission, Richard "Humpty" 46
Vitagraph Studios 122
Vives, Vivian 75, 76
Vogt, Paul 186
Vogt, Peter Allen 186
Voight, Jon 83
Vola, Vicki 190–192
Voland, Herb 211
Von Bulow, Claus 14, 16
Von Furstenberg, Betsy 267

WABD-TV (New York City) 237
Waddington, Lawrence 116
Wagner, Lindsay 206, 208
Wagner, Robert 264, 266
Waite, Ralph 190, 193, 195
Wake Up: I'm Fat 232
Walden, Lynette 205
Walden Productions 64
Walker, Clint 255
Walker, Eamonn 155, 156
Walker, Jimmy 270
Walker, Judge John 28
Walker, Texas Ranger 27
Wall, Lucille 230–231
Wall Street Journal 102
Wallace, Don 93
Wallace, George D. 94
Wallace, Jane 203
Wallace, Rick 159, 227
Wallenstein, Andrew 204
The Wallflowers 113
Walsh, M. Emmet 262
Walsh, Sydney 92, 93
Walston, Ray 228–229
Walt Disney Presents 200–211
Walter, Jessica 108
Walter, Lisa Ann 178–179
Walters, Barbara 132, 134

Walther, Gretchen 121
Wanamaker, Sam 83
Wang, Eugene (Gene) 220, 263–264
Wapner, Judge Joseph A. 15, 16, 36–37, 85, 141, 212–217, 253
Wapner, Joseph Max 85
Ward, Lyman 105
Ward, Mary B. 187, 188
Ward, Susan 154
Warden, Jack 69–70, 116
Warfield, Erroll Sean 246
Warfield, Marsha 197, 199
Warner, Gertrude 219
Warner, Julie 101, 102, 154
Warner, Malcolm-Jamal 185
Warner Bros. (film studio, *aka* Warner Bros.-First National) 6, 116, 218–219, 221, 254
Warner Bros. Domestic Television 138, 212
Warner Bros. Television 58, 59, 62, 63, 73, 112, 123, 124, 143, 152, 154, 155, 190, 197, 240, 253–255, 257–258
Warner Bros. Television Network (WB) 25, 154, 159
Warren, Chief Justice Earl 12
Warren, Edward Henry "Bull" 207
Warren, Karle 146, 147
Warren, Kenneth J. 68, 69
Warren, Kiersten 186
Warren, Lesley Ann 108
Warren, Michael 256
Washington, Ned 257
Washington College of Law (American University) 21
Washington Post 46, 111
Washington Watch 17
Watergate Hearings 13, 171
Waters, Ed 131, 190
Waterston, Sam 123, 125, 168, 172
Watson, Debra 126
Watson-Johnson, Vernee 107
Watts, Rolonda 178
Wayne, David 29
WB Network *see* Warner Bros. Television Network
WBNS-TV (Columbus OH) 135
WCBS-TV (New York City) 141
WCVB-TV (Boston) 188, 189
We the Jury 268
Weary, A.C. 94
Weatherfield, Chris 94
Weaver, Fritz 53, 83
Webb, Jack 73, 74, 76–77, 145, 169, 257–258
Webb, Lucy 271
Weber, Steven 74, 181
Webster, Chuck 219
Weege, Reinhold 197–199, 209–210, 250
Weekly Alibi 34
Weibel, Cameron 178, 179
Weibel, Luca 178, 179
Weinberger, Ed. 43, 251
Weinman, Roz 169
Weinreb, Lisa 71
Weintraub, Joanne 97

Weiss, Joanne 110
Weiss, Robert 46
Weitzman, Howard 194
Welch, Joseph N. 9
Welch, Ronnie 93
Welker, Frank 118
Weller, Robb 266
Welles, Halsted 182
Welles, Orson 83, 207
Wells, John (actor) 115
Wells, John (writer) 102
Wells, Mary K. 93
Wells, Tammara 246
Welsh, Bill 85, 87
WENR-TV (Chicago) 273
We're in the Money 6
Wert, Doug 265
West, Adam 254
West, Alice 228
West, Billy 118
The West Wing 180, 181, 182
The Westerners see *Black Saddle*
Westfall, Kay 44
Wettig, Patricia 66, 67
WFAA-TV (Dallas-Ft. Worth) 61–62
WGN (Chicago radio station) 3, 4
WGN-TV (Chicago) 258–259
Wheeler, Burt 135
Whitaker, Jack 267–268
White, Betty 48, 50, 111, 185
White, Byron 105
White, Libby A. 238
White, Michole 203, 204
White, Sam 218, 220
Whitfield, Charles Malik 113, 114
Whitfield, Lynn 98, 99
Whitley, Kym A. 251
Whitmore, James 36, 166–167
Whitson, Samuel 261
The Whole World Is Watching 47; see also *The Bold Ones*
Widdoes, James 209–210
Wiehl, Christopher 105
Wiest, Dianne 168, 171
Wilcox, Dan 96
Wilcox, David 168
Wilder, James 98, 99
Wilding, Michael 202
Wilhoite, Kathleen 159
Wilke, Mrs. Faye 135
Will & Grace 26, 211
Willard, Fred 249–250
William, Warren 5, 6, 218–219
Williams, Ann 94
Williams, Anson 206
Williams, Emlyn 83
Williams, Gary Anthony 48, 51
Williams, JoBeth 58, 59
Williams, John 3
Williams, Kelli 232, 234
Williams, Treat 92
Williams, Vanessa 193–195, 225
Williams Street Productions 118
Williamson, Kathy 104
Willy 9, 15, 268–269
Wilson, Charles C. 219
Wilson, Crawford 146
Wilson, Theodore 69

Winant, Bruce 106
Winer, Jason 46
Wines, William C. 259
Winfrey, Oprah 141
Winkler, Henry 233
Winsor, Roy 44
Winston, Dayna 264–265
Winston, Susan 151
Winter, Edward 29, 30
Winters, Dean 168, 174
Wise, Ray 244
Withers, Mark 157
The Witness 9, 154, 269–271
Witt, Alicia 63, 168
Witt, Paul Junger 60
Witt-Thomas Productions 60
WKRC-TV (Cincinnati OH) 95
WMCA (New York City radio station) 5
WNBC-TV (New York City) 141, 162
WNEW (New York radio station) 4
Wohl, Alexander 21
Wolf, Dick 18–19, 40, 42–43, 57, 70–71, 73, 103, 132, 167–177, 181, 191, 232, 252, 272
Wolf Films 42, 57, 70, 103, 167, 169, 272
Wolfson, Fred 55
Wolk, Andy 84
Wolper, David L. 10, 89, 90
Wong, B.D. 168, 174
Woodard, Alfre 244
Woods, Donald 219, 230, 231
Woods, James 92, 248–249
Woodside, David Bryan 193, 195
Woodward, Edward 83

Woolrich, Cornell 182
Wooly Mammoth Productions 69
Wootton, Nicholas 168
Work with Me 15, 271–272
World Socialist Web Site 126
World's Wildest Police Videos 17
Worldvision 139, 140
Worthy, Rick 193, 195
WPIX-TV (New York City) 8, 144
WPLG-TV (Miami FL) 13
WPTZ-TV [KYW-TV] (Philadelphia) 189
WRC-TV (Washington DC) 121
Wright Judge Bruce 203
Wright, Max 262
Wright, Teresa 120, 225
The Wright Verdict 272–273
Writers Guild of America 102
WWOR-TV (New York City/Secaucus NJ) 141
Wyatt, Jack 61
Wylie, Adam 228
Wyner, George 15, 157
Wynn, Ed 250
Wynn, Keenan 116, 202

Yahee 94
Yates, Andrea 173
Yohn, Erica 228
York, Elizabeth 230
York, Francine 250
Yoshimura, James 150
You Are the Jury see *Trial by Jury*
Young, Bob 144
Young, Carleton 64, 65
Young, Mayor Coleman A. 143
Young, Judge David 136–137, 141, 142

Young, John Sacret 205
Young, Stephen 134, 135
Young, Tawya 136, 137
Young Adults Asserting Themselves 143
The Young Lawyers 13, 85, 253, 273
Younger, Beverly 44
Younger, Judge Evelle J. 10, 78, 261
Your Hit Parade 155
Your Witness 7, 103, 273–274
Youthful Companion magazine 218
Yusen, Susan 94

Zacks, Jerry 94
Zada, Ramy 75, 76, 244–245
Zambrano, Jacqueline 153
Zamora, Ronnie 15
Zane Grey Productions 45
Zane Grey Theater 45
Zane, Lisa 159
Zayas, David 168
Zelman, Daniel 74
Zimbalist, Efrem, Jr. 220, 262
Zimbler, Jason 94
Zimmer, Constance 48, 50, 52, 126
Zirnkilton, Steve 42, 43, 169
Ziskin, Ron 55
Ziv, Frederick 192
Ziv Productions 179–180, 190, 192
Zmed, Adrian 262
Zola, Emile 202
Zoo Productions 46
Zuckerman, Ed 56, 168
Zuvich, Adam 46